Tylavsky

ANALYSIS AND DESIGN OF DIGITAL SYSTEMS WITH VHDL

Allen M. Dewey

IBM Corporation

PWS Publishing Company

International Thomson Publishing

*Boston • Albany • Bonn • Cincinnati • Detroit • Madrid • Melbourne • Mexico City •
New York • Pacific Grove • Paris • San Francisco • Singapore • Tokyo • Toronto • Washington*

PWS PUBLISHING COMPANY
20 Park Plaza, Boston, MA 02116-4324

I(T)P™
International Thomson Publishing
The trademark ITP is used under license

For more information, contact:

PWS Publishing Company
20 Park Plaza
Boston, MA 02116

International Thomson Publishing Europe
Berkshire House I68-I73
High Holborn
London WC1V 7AA
England

Thomas Nelson Australia
102 Dodds Street
South Melbourne, 3205
Victoria, Australia

Nelson Canada
1120 Birchmount Road
Scarborough, Ontario
Canada M1K 5G4

International Thomson Editores
Campos Eliseos 385, Piso 7
Col. Polanco
11560 Mexico D.F., Mexico

International Thomson Publishing GmbH
Königswinterer Strasse 418
53227 Bonn, Germany

International Thomson Publishing Asia
221 Henderson Road
#05-10 Henderson Building
Singapore 0315

International Thomson Publishing Japan
Hirakawacho Kyowa Building, 31
2-2-1 Hirakawacho
Chiyoda-ku, Tokyo 102
Japan

Sponsoring Editor: Bill Barter
Assistant Editor: Ange Mlinko
Editorial Assistant: Monica Block
Production Editor and Interior Design: Pamela Rockwell
Cover Designer: Julia Gecha
Manufacturing Coordinator: Wendy Kilborn
Marketing Manager: Nathan Wilbur
Cover Printer: John Pow Company
Text Printer and Binder: Quebecor/Hawkins

Printed and bound in the United States of America.

96 97 98 99 00 – 10 9 8 7 6 5 4 3 2 1

Library of Congress Cataloging-in-Publication Data
Dewey, Allen M. (Allen Mark)
    Analysis and design of digital systems with VHDL/Allen M. Dewey.
        p.   cm.
    Includes bibliographical references and index.
    ISBN 0–534–95410–3
    1. Digital electronics–Computer simulation. 2. VHDL (Computer hardware description language) 3. Computer aided design.
I. Title.
TK7868. D5D47   1997
621.39′2– –dc20   96–24465
                                                             CIP

This book is dedicated to my family and the love that binds us, the spirit that uplifts us, and the light that guides us.

CONTENTS

Chapter 19 VHDL: A Last Look   620

Appendix A Powers of Two   663

Appendix B VHDL Reserved Keywords   664

Appendix C Introduction to Semiconductor Physics   665

Index   671

PREFACE

This text is based on several premises concerning the future of digital engineering education. First, it is important to not only present information concerning digital engineering but to also present an associated context in which the information can be assimilated and understood. Knowing a set of facts is like knowing how to operate a set of carpenter tools. Knowing how to use individual tools does not imply a mastery of the more valuable skill of constructing a building; similarly, knowing a set of facts about switching algebra and state machines does not imply a mastery of the more valuable skill of constructing digital systems. Thus, this text pays considerable attention to *context,* explaining in a top-down fashion the general process of digital engineering from concept to realization and how various elements and procedures contribute to this process. This context serves to continually orient readers so that they can more easily navigate and understand the material of the text. The digital engineering process context also helps define a more readily extensible and agile skill set, providing a flexible framework for accommodating new information and discarding old information.

Another main premise of this text is that hardware description languages will play an increasingly important part in digital systems engineering. Just as it is now a requirement to learn a software programming language to complete certain computer science courses, it will likely become a requirement to learn a hardware description language to complete certain digital engineering courses. Learning a hardware description language aligns an engineer's skill set with present industry practices and enables an engineer to readily communicate and contribute to design projects. Learning a hardware description language also enables an engineer to specify a digital system in a format that can be processed by computer programs, which introduces the substantial power of design automation technology. Design automation technology allows one to conceive of and explore challenging and innovative digital system designs that would be difficult, if not impossible, to define using conventional pen-and-paper and hardware prototyping techniques.

Thus, this text seeks to combine the synergistic strengths of digital engineering and hardware description languages in a coordinated and unified manner. We use the forum of teaching the analysis and design of digital systems to learn the hardware description language VHDL, and, in a complementary fashion, we use the forum of teaching the concepts and modeling practices of VHDL to learn digital systems design.

ORGANIZATION OF TEXT

Table 1 shows how the general organization of the text supports the goals of presenting digital engineering in a top-down context coupled with a hardware description language. Looking first at the left half of Table 1, digital engineering is divided into three topics: combinational systems, manufacturing technology, and sequential systems. Combinational and

Table I

Integrating Digital Engineering and VHDL

Digital Engineering			VHDL
Sequential Systems	Synthesis & Implementation	Algorithmic Modeling	Sequential Systems
	Definition & Analysis	Synchronous Data Flow Modeling	
Manufacturing Technology	Integrated Circuits	Technology Modeling	Manufacturing Technology
	Logic Families		
Combinational Systems	Synthesis & Implementation	Asynchronous Data Flow Modeling	Combinational Systems
	Definition & Analysis	Introduction & Stuctural Modeling	

sequential systems are, in turn, divided into definition/analysis and synthesis/implementation. Definition/analysis studies existing sample digital systems to understand basic concepts, terminology, and behavior. Then, synthesis/implementation addresses the more challenging task of designing new digital systems. Synthesis translates an initial, general specification into a more detailed description, which is cast in the mathematics of digital systems—switching algebra. Then, implementation maps the detailed description onto hardware. The analysis-synthesis-implementation paradigm helps readers understand the complexities of digital system engineering. The analysis-synthesis-implementation paradigm also emphasizes that there are typically many possible solutions to a given problem and that engineers must compare and contrast possible solutions and make reasonable trade-offs in determining the best solution. Sandwiched between combinational and sequential systems is a discussion of manufacturing technology. Manufacturing technology expands implementation issues, introducing circuit and physical level concepts. The impact of the properties and limitations of physical devices and manufacturing processes on digital systems is examined.

Looking next at the right half of Table 1, we build an understanding of VHDL as we build an understanding of digital engineering; the dance is a simple two-step. Unfolding VHDL from such a design perspective helps readers learn more than just the mechanics of correctly writing VHDL; readers have a deeper contextual understanding of how to use VHDL. As the digital engineering material builds toward more complex systems, the corresponding VHDL material builds toward more abstract modeling techniques. Basic combinational logic operations introduce the general "look and feel" of VHDL models. Then, networks of combinational logic operations introduce structural modeling techniques involving networks of VHDL models. Next, combinational logic expressions introduce data flow modeling techniques involving concurrent VHDL statements. Guarded concurrent VHDL statements are introduced to describe the synchronous behavior of sequential networks. Finally, state diagrams introduce algorithmic modeling techniques involving sequential statements.

Table 2 presents a more detailed topical breakdown on how the text couples and aligns digital engineering and VHDL. Use Table 2 as a road map to keep your bearings in touring the textbook on our journey exploring the world of digital systems

Finally, in addition to digital engineering and VHDL, engineers also need problem-solving skills. Never forget that facts and formulas are just tools that, by themselves, do not solve problems. Addressing the challenges of designing a better widget requires an under-

Table 2

Digital Engineering and VHDL Subject Alignment

Digital Engineering	VHDL
• switching algebra • logic operations • truth tables	• model organization • signal transformations • design units
• logic expressions • active state notation • combinational analysis	• component declaration • component instantiation • packages
• algebraic minimization • Karnaugh maps • Quine-McCluskey tables	• conditional concurrent signal statement • selected concurrent signal statement • bit vectors
• multi-level networks • multiplexers • decoders • memory • programmable logic devices	• port associations • generate statement • generics • arithmetic operators • numeric literals
• MOS logic • bipolar logic • wired logic • speed • noise margin	• physical type • propagation delay • functions • resolution functions • signals, variables, and constants
• diodes • MOS transistor • bipolar transistor • microelectronics • ASIC	• enumeration type • arrays • multi-valued logic • attributes
• latches • master-slave flip-flops • edge-sensitive flip-flops • setup and hold timing • synchronous machines	• blocks • guards • guarded statements • assertions
• state diagrams • state reduction • state assignment • excitation tables	• processes • sequential statements • integer type • floating point type • files
• registers • shift registers • counters • sequential programmable logic devices	• configurations • records • access type

standing of facts and the ability to use this knowledge to understand problems and formulate viable solutions. Although facts are important, they are usually of only temporary significance; as technology advances, old facts are replaced by new ones.

TO THE INSTRUCTOR

The digital engineering and VHDL material is organized into separate chapters to provide flexibility in supporting different course curricula. The text can be used for an introductory digital systems course, a VHDL course, or a two-course sequence of an introductory digital systems course followed by a VHDL course. If the emphasis is on digital engineering, Chapters 2–10 serve as primary reference and Chapters 11–19 serve as secondary reference. If the emphasis is VHDL, Chapters 11–19 serve as primary reference and Chapters 2–10 serve as secondary reference. Between these two ends of the spectrum, the digital engineering and VHDL chapters are given somewhat equal weighting.

ACKNOWLEDGMENTS

Bringing the vision of this text to reality has been an arduous task that would not have been possible without the help of several individuals. I would like to thank the many reviewers who critiqued the manuscript and suggested ways to improve content and presentation:

Don Bouldin
University of Tennessee
Michael Driscoll
Portland State University
Huber Graham
University of Missouri
Martin Kaliski
California Polytechnic State University

Wayne McMorran
California Polytechnic State University
John Pavlat
Iowa State University
Mani Soma
University of Washington

I would also like to thank Mr. Stanley Wagner, U.S. Air Force, Mr. Bob Hunter, Model Technology, and Mr. Rick Herrick, Topdown Design Solutions, for supporting this "grand experiment."

Allen M. Dewey

1
INTRODUCTION

Let us undertake a journey to explore the world of digital systems. We will start by looking at the "big picture." First, we will discuss the general characteristics of digital systems. Next, we will examine the reasons digital systems are so pervasive in many of today's products ranging from wristwatches to satellites. Then, we will investigate the process of designing digital systems and the impact of design automation. Also, we will introduce the hardware description language VHDL as an important international standard for communicating digital design information.

1.1 ELECTRONIC SYSTEMS

To understand *electronic systems,* it is useful to begin by viewing them as "black boxes" that perform a desired input/output transform, as shown in Figure 1.1(a). The inputs and outputs are electrical signals; an *electrical signal* is a current and/or voltage as a function of time. Information enters the electronic system encoded as input electrical signals, is processed or transformed within the electronic system, and leaves the electronic system encoded as output electrical signals. The input information could be two numbers and the output could be their product. Alternatively, the input information could be a radio broadcast signal and the output could be music.

If the magnitude of the current and/or voltage varies in a *continuous* fashion with respect to time, an electronic system is classified as an *analog* electronic system. Figure 1.1(b) shows an example of an analog system called an *amplifier*. The input electrical signal is a voltage (vertical axis) that varies in a continuous, sinusoidal fashion with respect to time (horizontal axis). The output electrical signal is an amplified or enlarged version of the input signal. The world around us is mostly analog, as most events in nature change in a continuous fashion with respect to time. For instance, the ambient temperature of our homes, pitch and volume of our voices, and speed of our automobiles all change continuously with time.

If the magnitude of the current and/or voltage varies in a *discrete* fashion with respect to time and has only a finite set of possible values, the electronic system is classified as a *digital* electronic system. Figure 1.1(c) shows an example of a digital system called a *counter*. The input electrical signal is a voltage (vertical axis) that changes in an abrupt, discrete fashion with respect to time, taking on one of two possible values—"low" or "high." Each transition from low ⇒ high ⇒ low is called a pulse ⌐⌐, and the output electrical signal contains a pulse for every ten input pulses.

1.2 DIGITAL SYSTEMS

The digital system shown in Figure 1.1(c) is incomplete because, as previously mentioned, most aspects of nature that we may be interested in monitoring or controlling change in a continuous fashion and thus do not directly generate digital signals. To control the speed of

Figure 1.1

Electronic systems

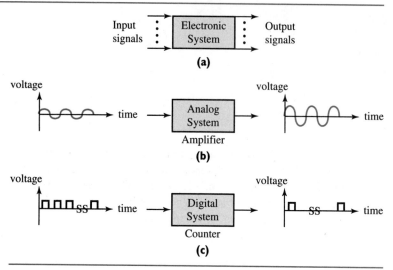

Figure (a) illustrates a *general* electronic system; Figure (b) illustrates an *analog* electronic system; and Figure (c) illustrates a *digital* electronic system.

an automobile (cruise control), we need to encode the rotational speed of the driveshaft as a digital signal. Figure 1.2 expands Figure 1.1(c), using a voice-processing application to illustrate a typical situation of embedding a digital system within an analog environment.

The first step in generating a digital signal involves a *transducer*. A transducer transforms variations in physical properties to variations in electrical analog signals, or vice versa. In Figure 1.2, the microphone is a transducer that converts sound waves into an electrical analog signal. The second step converts the analog signal into a digital signal via a device appropriately called an *analog-to-digital converter* (ADC). An analog-to-digital converter samples the continuous magnitude of the analog signal at regular intervals of time and converts the samples into discrete digital values. Having transformed sound waves into a digital signal, the digital system then performs various voice-processing functions, such as amplifying or noise filtering, and the resulting output digital signal is transformed back into sound waves via a *digital-to-analog converter* (DAC) and another transducer. The digital-to-analog converter transforms a digital signal into an analog signal, and the speaker serves as a transducer to convert an analog signal into sound.

Notice that the analog-to-digital converter and digital-to-analog converter combine to create a temporary, somewhat "artificial," digital world to process information. An alternative approach eliminates the converters and directly processes the electrical signals with an analog system. The decision to use an analog or digital system is not always an easy one because advances are continually being made in both arenas, as evidenced by recent developments in analog neural networks and digital reduced-instruction set processors. Presently, digital technology is generally used more often than analog technology, increasingly so because digital systems offer the following advantages:

* greater system complexities,
* higher reliability, and
* easier programmability.

Problem-Solving Skill

Transforming a problem into a simpler domain or a more developed and better understood domain can often make it easier to find a solution.

Digital system design and manufacturing technology is presently more advanced than analog system design and manufacturing technology, enabling the production of digital sys-

Figure 1.2

Expanded view of a digital system

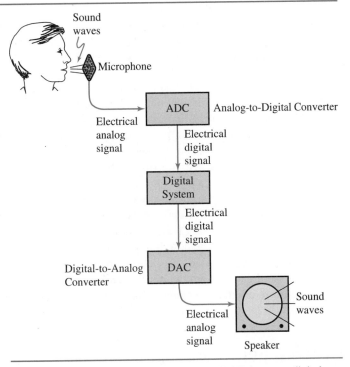

ADC denotes an analog-to-digital converter. DAC denotes a digital-to-analog converter.

tems that are significantly more complex (number of components) than analog systems. Digital systems are often easier to design than analog systems because digital systems are readily partitionable and can be described using a variety of models, ranging from very high level and abstract to very low level and detailed. These properties allow digital systems to be designed in a hierarchical, stepwise fashion. In other words, we typically do not design a digital system containing 1 million processing elements by spreading the processing elements out on a gymnasium floor and hooking them together one-by-one. Instead, we combine groups of processing elements into basic operations, then combine the basic operations into functions, and so on. To further simplify digital system design, many portions of the hierarchical, stepwise design process are being increasingly automated by an extensive set of computer programs. Digital systems are often easier to manufacture than analog systems because digital processing elements operate on discrete and finite-valued electrical signals. Digital devices that operate with a small set of values, typically *two*, are generally easier to manufacture than analog devices that operate with a continuum of possible values.

Digital systems are often more reliable than analog systems because digital systems can tolerate stronger fluctuations in operating conditions. Reliability means that a system continues to function correctly under unexpected or unwanted changes to normal operating conditions. Such unwanted changes may be variations in (1) ambient temperature due to excessive power dissipation, (2) device characteristics due to aging, or (3) signal voltages due to spurious noise or interference. Analog systems are often more sensitive to changes in normal operating conditions and thus potentially less reliable. Analog systems respond to continuous signal changes, and unwanted changes are difficult to filter out. A small increase in an input signal voltage may be a transient power supply surge that should be ignored or a legitimate

Figure 1.3

Programming a digital system via software

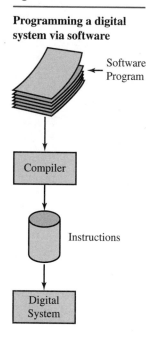

input excitation that should be processed. Digital systems, on the other hand, can be more reliable because ranges or tolerances are typically specified for possible signal values. For example, a popular convention allows any voltage in the range $2V–5V$ to be considered a "high" value and any voltage in the range $0V–1V$ to be considered a "low" value. Signal voltages can vary within these ranges in response to changes in normal operating conditions without adversely affecting system function.

Finally, digital systems are typically more *programmable* than analog systems. The term *programming* refers to modifying system function to perform different tasks. There are various levels or degrees of programming. Both analog and digital systems accommodate relatively simple hardware-oriented programming techniques, such as modifying function by setting switches or changing components. Digital systems, however, can also be designed to accommodate more extensive function modifications via *software*-oriented programming techniques.

Figure 1.3 shows how software-oriented programming works. The digital system is designed to execute a set of instructions. The desired function is described in a programming language, such as Pascal or C; the resulting description is called a *software program* or sometimes just *software*. The description is then translated by a computer program called an *assembler* or a *compiler* into instructions that the digital system executes. Changing the function of the digital system, within the hardware constraints of the system, requires only that the software program be changed and recompiled, generating a new list of instructions.

1.3 DIGITAL SYSTEM DESIGN PROCESS

Having explained the general characteristics and advantages of digital systems, let us next examine how digital systems are designed. A large digital system, such as a general-purpose computer used for personnel accounting or a special-purpose processor used for satellite communications, typically contains millions of components and takes several years to develop. Designing such a system at a detailed level is difficult, if not impossible, because the complexity is overwhelming. As mentioned in the previous section, a series of hierarchically related *levels of abstraction* have been developed to make it easier to conceive and correctly implement large, complex digital systems. A level of abstraction is a formalism for viewing or modeling a digital system. Each organization and/or design team tends to use a unique set of abstractions tailored to their products and personal preferences.

Figure 1.4 shows a typical set of digital system abstractions. As the design progresses from the *stochastic* level of abstraction to the *fabrication* level of abstraction, each succeeding level of abstraction adds more detail to the design. The full complexity of a design is gradually and systematically developed as the stepwise refinement of a series of levels of abstraction.

The *stochastic* or *performance* level of abstraction identifies the major functional units and defines how the units interact to transform the input data into the desired output data. The *algorithmic* level of abstraction decomposes a functional unit into a series of processing steps or procedures and identifies whether these actions are realized in hardware or software. The *data flow* level of abstraction decomposes an algorithm into a series of data movements and transformations and identifies the hardware components that control, operate, and store data. The *logic* level of abstraction decomposes the data flow hardware components into smaller devices, such as *gates* and *flip-flops*. The logic devices are, in turn, decomposed at the *circuit* level of abstraction into electrical elements, such as *resistors* and *transistors*. Finally, the *layout* and *fabrication* levels of abstraction detail how the circuit elements are manufactured on an *integrated circuit*.

Figure 1.4

Hierarchical digital system design process

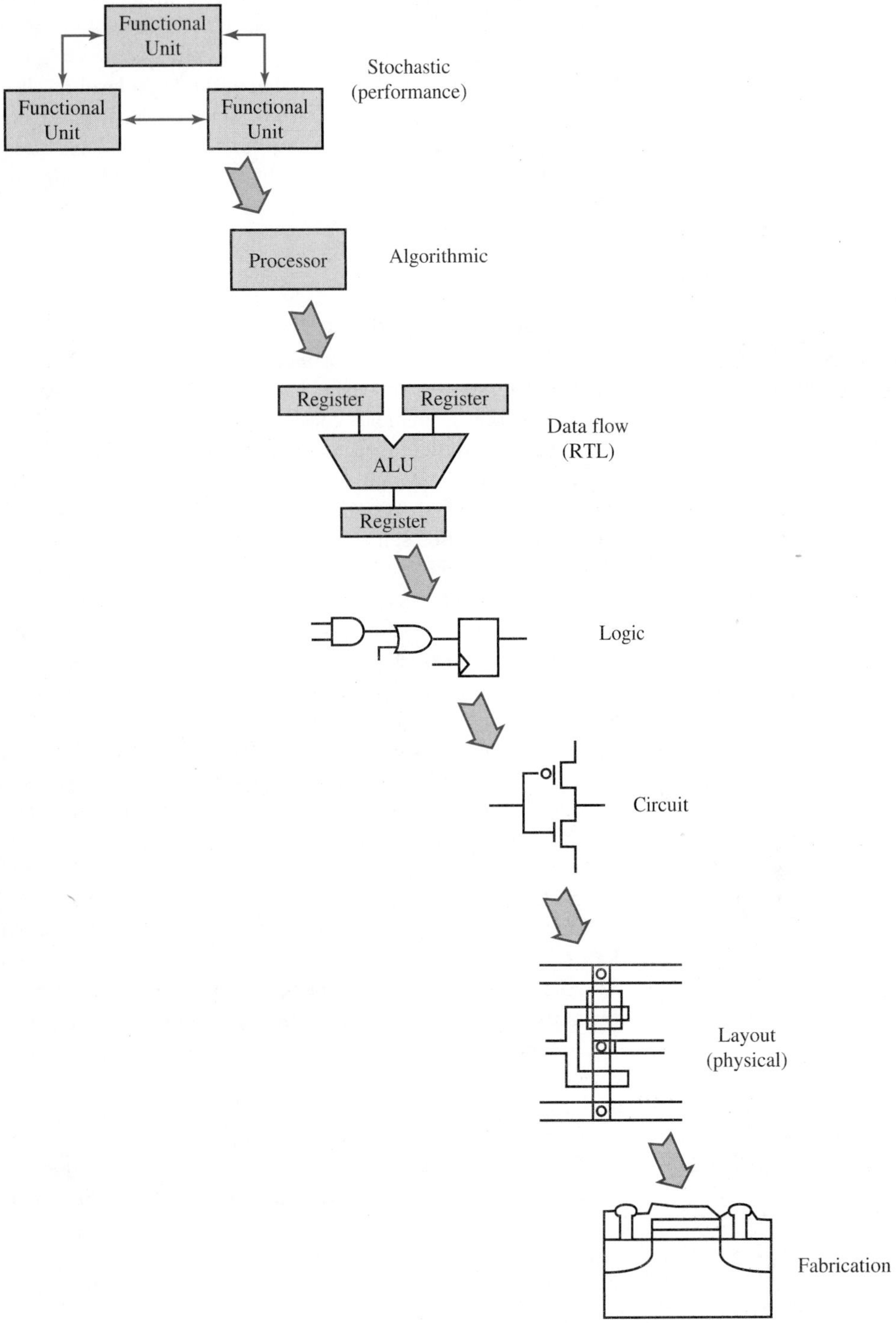
Functional Unit
Functional Unit
Functional Unit
Stochastic (performance)
Processor
Algorithmic
Register
Register
ALU
Register
Data flow (RTL)
Logic
Circuit
Layout (physical)
Fabrication

If the levels of abstraction are unfamiliar, do not be dismayed. At this point, the intent is to present a perspective to orient you on how the material we will study fits into the general design process. No one person is expected to be an expert in all areas. Most engineers have a general understanding of the entire design process but specialize in only one or two areas. This text will primarily discuss design at the logic level of abstraction, but aspects of other levels of abstraction will be addressed. Since the steps in the digital design process build on each other in a hierarchical fashion, understanding logic design facilitates understanding data flow design, which, in turn, facilitates understanding algorithm design, and so on.

1.4 DESIGN AUTOMATION

Problem-Solving Skill
Try to identify aspects of a solution that can be automated to exploit the capabilities of computers.

Design automation (DA) or *computer-aided design* (CAD) is playing an increasingly important role in digital system design and, as such, should be a part of an engineer's skill set. Design automation technology consists of various computer programs that execute the more labor-intensive, routine aspects of digital system design.

Starting from academic research projects in the early 1960s, design automation technology has matured into a billion-dollar, international industry. Figure 1.5 presents a list of classes or types of design automation programs, commonly called *tools*, and the general order in which they are used during the digital design process.

Design entry tools serve as the interface between the design automation/computer system and its users. In other words, design entry helps designers "converse" with design automation tools. There are many types of design entry tools, ranging from simple command-line and menu entry to sophisticated voice-activation and solid (3-dimensional) modeling entry.

Synthesis tools assist in creating a design. Synthesis is a challenging task to automate because it is difficult to capture in an algorithm the creative and innovative decision-making process of design. Consequently, synthesis technology is relatively new to design automation; the first commercial logic synthesis tools were introduced in the late 1980s. As with design entry, there are many types of synthesis tools, ranging from low-level logic optimization to high-level algorithmic partitioning, scheduling, and allocation.

Verification tools assist in checking the correctness of a design, in other words, whether a design implements the desired function and satisfies the desired performance. Timing analysis tools and simulators are common examples of verification tools. A timing analysis tool identifies circuit paths that violate timing requirements, such as paths that have too much delay. A simulator exercises a design over a period of time, generating a set of outputs in response to a set of inputs. There are simulators that address only certain levels of abstraction, such as circuit simulators, switch-level simulators, and logic simulators. There are also mixed-mode or hierarchical simulators that address a range of abstractions.

Having generated a logic design and verified that the design realizes the desired function and performance, *physical design* tools assist in translating the logic schematic into physical devices. Physical design tools generally address the tasks of partitioning, placement, and wiring. Partitioning tools decide how the design function will be apportioned across several physical devices. Placement and wiring tools decide how to arrange and interconnect the physical devices.

Finally, *fabrication* tools assist in manufacturing the design, and *test* tools assist in verifying that the manufactured part—the final product—meets acceptance criteria. Fabrication tools help characterize and define the chemical and physical properties of the manufacturing processes. Test tools help generate test vectors (sample inputs and outputs) to identify faulty or bad parts.

Figure 1.5

Design automation taxonomy

Design Entry

Synthesis

Verification

Physical Design

Fabrication

Test

1.5 VHDL

The digital design process discussed in the previous section requires the basic need to communicate design information. This communication may occur between companies, designers, design automation tools, or some combination of these agents. Starting in the early 1960s, several languages were developed for conveying design information, such as Computer Design Language (CDL), Digital Systems Design Language (DDL), and A Hardware Programming Language (AHPL). These early hardware description languages only partially solved digital system description problems, and their proliferation exacerbated digital system communication problems. Organizations using different languages could not easily communicate design information. Thus, the need to develop a standard language to address the rapidly growing complexity and sophistication of digital system design became the impetus for VHDL.

VHDL is a hierarchical acronym denoting VHSIC Hardware Description Language; VHSIC, in turn, denotes Very High Speed Integrated Circuits. This somewhat strange naming is based on the historical origin of VHDL. The United States Department of Defense Very High Speed Integrated Circuits (VHSIC) Program initiated the design of VHDL to support the development of a new generation of digital system technology. The design of VHDL formally began in 1983 and, after a series of several iterations and versions, culminated in 1987 with the acceptance of VHDL as an IEEE (Institute of Electrical and Electronic Engineers) standard. VHDL is rapidly gaining acceptance and is influencing advancements in design methodologies, design automation technology, and the very nature of the global electronics industry.

An understanding of VHDL is a valuable part of an engineer's skill set because digital system engineering requires the ability to communicate ideas and designs. Since the essence of VHDL is to describe basic digital design concepts and practices, it is advantageous to align learning VHDL with learning digital design. A digital system design perspective makes it easier to understand VHDL, while a VHDL perspective makes it easier to understand digital system design.

In addition to reinforcing digital system engineering education, VHDL also serves, as mentioned earlier, as a linguistic "gateway" to design automation technology. Design automation tools enable engineers to explore truly challenging and creative digital systems that would be difficult, if not impossible, to investigate with the limited and expensive capabilities of traditional hardware-oriented digital design laboratories.

1.6 SUMMARY

Figure 1.6 illustrates that this textbook is organized to bring together three aspects of digital system technology: (1) digital theory and practice, (2) problem-solving skills, and (3) VHDL.

Electronic systems can be classified as analog or digital. Analog systems process electrical signals that vary continuously with respect to time. Digital systems process electrical signals that vary discretely with respect to time.

Digital systems are generally produced using a hierarchical design methodology. A series of models ranging from high-level, abstract, stochastic descriptions to low-level, detailed, circuit descriptions are used to translate a desired specification into a physical implementation. To reduce production time and expense, many aspects of hierarchical design methodologies from initial design entry to final part testing are being automated by a variety of computer programs.

VHDL is a computer language for describing digital systems. VHDL is coupled with dig-

Figure 1.6

Key aspects of digital system engineering

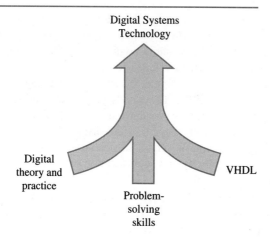

ital engineering to emphasize and exploit their synergistic strengths. This instructional approach is only one of many possible approaches to learning VHDL. Hardware modeling styles, as with software programming styles, are often largely a matter of personal preference. As you build expertise and experience with VHDL and digital systems, you are encouraged to experiment with alternative modeling styles. The world is always looking for a "better mousetrap."

Digital technology is advancing rapidly and is becoming an increasingly pervasive part of the world we live in. Understanding digital system engineering is a challenging task, but its potential raises exciting prospects for the future. Let us begin our journey.

REFERENCES AND READINGS

[1] Barbe, D., ed. *Very Large Scale Integration (VLSI)—Fundamentals and Applications*. New York: Springer-Verlag, 1980.
[2] Gajski, D., ed. *Silicon Compilation*. Reading, MA: Addison-Wesley, 1988.
[3] Pecht, M., ed. *Handbook of Electrical Package Design*. New York: Marcel Dekker, 1991.
[4] Abramovici, M., M. Breuer, and A. Friedman. *Digital Systems Testing & Testable Design*. Piscataway, NJ: IEEE Press, 1994.
[5] Barbe, D. "VHSIC Systems and Technology," *IEEE Computer*, vol. 14, no. 2, pp. 13–22, February 1981.
[6] Coelho, D. *The VHDL Handbook*. Boston, MA: Kluwer Academic Publishers, 1989.
[7] Dewey, A. "VHSIC Hardware Description Language Development Program," *Proc. IEEE/ACM Design Automation Conference*, pp. 625–628, June 1983.
[8] Dewey, A. "Hardware Description Languages: Move from Academic Projects to Industrial Production Tools," *Proc. IEEE 15th Southeastern Symposium on System Theory*, pp. 144–147, March 1983.
[9] Dewey, A. and A. Gadient. "VHDL Motivation," *IEEE Design & Test of Computers*, vol. 3, no. 2, pp. 12–16, April 1986.
[10] Dillinger, T. *VLSI Engineering*. Englewood Cliffs, NJ: Prentice-Hall, 1988.
[11] Hines, J. "Where VHDL Fits Within the CAD Environment," *Proc. IEEE/ACM 24th Design Automation Conference*, pp. 491–494, June 1987.
[12] Lipsett, R., C. Schaefer, and C. Ussery. *VHDL: Hardware Description and Design*. Boston, MA: Kluwer Academic Publishers, 1989.
[13] Perry, D. *VHDL*. New York: McGraw-Hill, 1991.

[14] Preston, G. Report of the IDA Summer Study on Hardware Description Languages, Institute for Defense Analyses, IDA Paper P-1595, October 1981.

[15] Price, T. *Introduction to VLSI Technology*. Englewood Cliffs, NJ: Prentice-Hall, 1994.

[16] Pucknell, D. and K. Eshraghian. *Basic VLSI Design*. Englewood Cliffs, NJ: Prentice-Hall, 1994.

[17] Sapiro, S. *Handbook of Design Automation*. Englewood Cliffs, NJ: Prentice-Hall, 1986.

[18] Sumney, L. "VLSI with a Vengeance," *IEEE Spectrum*, vol. 17, no. 4, pp. 24–27, April 1980.

[19] Tunick, D. "The Emergence of Two CAE Standards," *Engineering Tools*, vol. 1, no. 2, pp. 63–66, April 1988.

2
REPRESENTING INFORMATION

Before we can begin learning how to design digital systems, we must devise techniques for representing the information that such systems process. The information could be text for document publishing, numbers for mathematical calculations, or events for system monitoring.

We will start by looking at various methods for representing numbers within a digital system. The decimal number system is not used because it is difficult to build physical devices that have 10 different states or operating conditions that can represent the 10 numerals (0, 1, . . . , 8, 9). Even the most powerful supercomputers use physical devices that have only *two* states—on and off. A two-state device can represent only two numerals, which is all that is required in the *base-2* number system, also called the *binary number system*.[1] Although it takes time and effort to learn a new number system, it is easier to change number systems than to change digital device technology. Therefore, the following sections will explain the binary number system by first reviewing the decimal number system conventions and operations and then introducing analogous binary number system concepts. We will present several examples of writing integer/fraction and positive/negative binary numbers. We will also discuss the basic arithmetic operations of binary addition and subtraction.

After gaining an understanding of the binary number system, we will then discuss alternative techniques for representing numbers using the *octal number system*, *hexadecimal number system*, and various *binary codes*.

Finally, we will conclude this chapter with a discussion of representing character information using *alphanumeric codes*, including the *American Standard Code for Information Interchange (ASCII)* and *Extended Binary-Coded-Decimal Interchange Code (EBCDIC)*.

2.1 REPRESENTING POSITIVE NUMBERS IN BINARY

We will begin by looking at positive numbers in both the decimal and binary number systems (negative numbers will be studied in Section 2.4). Figure 2.1 shows an odometer example of counting from zero to ten in both number systems. Decimal and binary are both *juxtapositional* number systems, meaning that numbers are constructed by a series of digits placed side by side. The decimal system uses the conventional ten numerals (0, 1, . . . , 8, 9), whereas the binary system uses only two numerals (0, 1).

---

[1] **binary**—"Characterized by or composed of two different parts." *The American Heritage Dictionary*, Houghton Mifflin Company.

Figure 2.1

Counting in the decimal and binary number systems

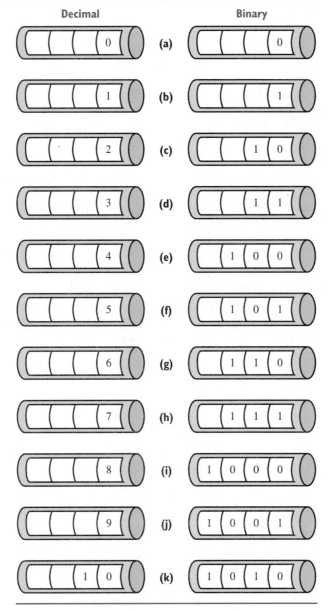

The left-hand sequence of odometer readings counts in the decimal number system. The right-hand sequence of odometer readings counts in the binary number system.

Recall that the decimal number system is a *weighted* number system, with each digit representing units of a power of 10. For example, consider the decimal number 307_{10}. The subscript 10 denotes that 307 is a decimal or base-10 number; we will use subscripts to avoid

$$10^2 \ 10^1 \ 10^0$$
$$\downarrow \quad \downarrow \quad \downarrow$$
$$3 \quad 0 \quad 7$$

confusion over which number system is implied when discussing multiple number systems. The rightmost digit, 7, denotes the number of ones (10^0); the next digit to the left, 0, denotes the number of tens (10^1); and so on. The value of 307_{10} is given by

$$
\begin{aligned}
&= (3 \times 10^2) + (0 \times 10^1) + (7 \times 10^0) \\
&= (3 \times 100) + (0 \times 10) + (7 \times 1) \\
&= 300 + 0 + 7 \\
&= 307_{10}
\end{aligned}
$$

The value of a decimal number

$$
d_{N-1} d_{N-2} \ldots d_1 d_0 . d_{-1} d_{-2} \ldots d_{-(M-1)} d_{-M}
$$

having an N-digit integer part and an M-digit fraction part is given by

$$
\begin{aligned}
&= (d_{N-1} \times 10^{N-1}) + (d_{N-2} \times 10^{N-2}) + \cdots + (d_1 \times 10^1) + (d_0 \times 10^0) \\
&\quad + (d_{-1} \times 10^{-1}) + (d_{-2} \times 10^{-2}) + \cdots + (d_{-(M-1)} \times 10^{-(M-1)}) + (d_{-M} \times 10^{-M}) \\
&= \sum_{i=-M}^{N-1} d_i \times 10^i
\end{aligned}
\tag{2.1}
$$

The binary number system is also a weighted number system, but each digit represents units of a power of 2. Also, digits can have only two possible values: 0 or 1. For example, consider the binary number given in Figure 2.1(k), 1010_2, where the subscript 2 denotes a binary or base-2 number. Reading the binary number from right to left, the rightmost digit,

$$
\begin{array}{cccc}
2^3 \searrow & 2^2 \downarrow & 2^1 \downarrow & 2^0 \swarrow \\
1 & 0 & 1 & 0_2
\end{array}
$$

0, denotes the number of ones (2^0); the next digit to the left, 1, denotes the number of twos (2^1); and so on. The value of 1010_2 is given by

$$
\begin{aligned}
&= (1 \times 2^3) + (0 \times 2^2) + (1 \times 2^1) + (0 \times 2^0) \\
&= (1 \times 8) + (0 \times 4) + (1 \times 2) + (0 \times 1) \\
&= 8 + 0 + 2 + 0 \\
&= 10_{10}
\end{aligned}
$$

Note that 1010_2 does *not* represent the decimal number one thousand and ten. Return now to Figure 2.1 and study how the decimal numbers 0–10 are represented in binary. Remember that binary uses only 0's and 1's. For instance, the binary representation for 2_{10} is 10_2, denoting one 2 (2^1) and zero 1's (2^0). The binary representation for 3_{10} is 11_2, denoting one 2 (2^1) and one 1 (2^0).

The value (base 10) of a binary number

$$
(b_{N-1} b_{N-2} \ldots b_1 b_0 . b_{-1} b_{-2} \ldots b_{-(M-1)} b_{-M})
$$

having an N-bit integer part and an M-bit fraction part is given by

$$
\begin{aligned}
&= (b_{N-1} \times 2^{N-1}) + (b_{N-2} \times 2^{N-2}) + \cdots + (b_1 \times 2^1) + (b_0 \times 2^0) \\
&\quad + (b_{-1} \times 2^{-1}) + (b_{-2} \times 2^{-2}) + \cdots + (b_{-(M-1)} \times 2^{-(M-1)}) + (b_{-M} \times 2^{-M}) \\
&= \sum_{i=-M}^{N-1} b_i \times 2^i
\end{aligned}
\tag{2.2}
$$

As an example, the value of 1010.101_2 is given by

$$
\begin{aligned}
&= (1 \times 2^3) + (0 \times 2^2) + (1 \times 2^1) + (0 \times 2^0) + (1 \times 2^{-1}) + (0 \times 2^{-2}) + (1 \times 2^{-3}) \\
&= (1 \times 8) + (0 \times 4) + (1 \times 2) + (0 \times 1) + (1 \times 0.5) + (0 \times 0.25) + (1 \times 0.125) \\
&= 8 + 0 + 2 + 0 + 0.5 + 0 + 0.125 \\
&= 10.625_{10}
\end{aligned}
$$

To help you become proficient with the binary number system, positive and negative powers of two are compiled in Appendix A. In working with powers of 2, it is customary to refer to $2^{10} = 1024$ as simply "1K," $2^{11} = 2048$ as simply "2K," and so on. Similarly, $2^{20} = 1,048,576$ is referred to as simply "1M," $2^{21} = 2,097,152$ is referred to as "2M," and so on.

Figure 2.2 illustrates several terms that are commonly used to describe strings of binary digits, in general, and binary numbers, in particular.

Term	Definition
bit	A single binary digit; an acronym for **bi**nary dig**it**.
LSB	**L**east **S**ignificant **B**it; the rightmost bit.
MSB	**M**ost **S**ignificant **B**it; the leftmost bit.
nibble	A group of four bits.
byte	A group of eight bits.

A *bit* is often abbreviated using the lowercase letter b. A *byte* is often abbreviated using the uppercase letter B. For example, a company might sell a computer with "4MB" of memory, meaning that the computer can store 4 million (actually 4,194,304) bytes of information. As the name implies, a *nibble* is smaller than a byte (most engineers are frustrated comedians). The term *word* is also used to describe groups of bits. Common word lengths include 8, 16, 32, and 64 bits.

2.2 CONVERTING BETWEEN BINARY AND DECIMAL

Since humans prefer decimal and digital systems prefer binary, we obviously need to convert between the two number systems if we are to converse. In the course of explaining how the weighted number system convention is used to construct binary numbers, we have already learned binary-to-decimal conversion via Equation 2.2.

Procedure to Convert Binary to Decimal
Multiply each bit of the binary number by the associated weight or power of two and sum the resulting products.

As an example, let us find the decimal equivalent of 110.11_2. Setting $N = 3$ and $M = 2$ in Equation 2.2, the decimal value of 110.11_2 is given by

$$
\begin{aligned}
&= (1 \times 2^2) + (1 \times 2^1) + (0 \times 2^0) + (1 \times 2^{-1}) + (1 \times 2^{-2}) \\
&= (1 \times 4) + (1 \times 2) + (0 \times 1) + (1 \times 0.5) + (1 \times 0.25) \\
&= 4 + 2 + 0 + 0.5 + 0.25 \\
&= 6.75_{10}
\end{aligned}
$$

Figure 2.2

Binary naming conventions

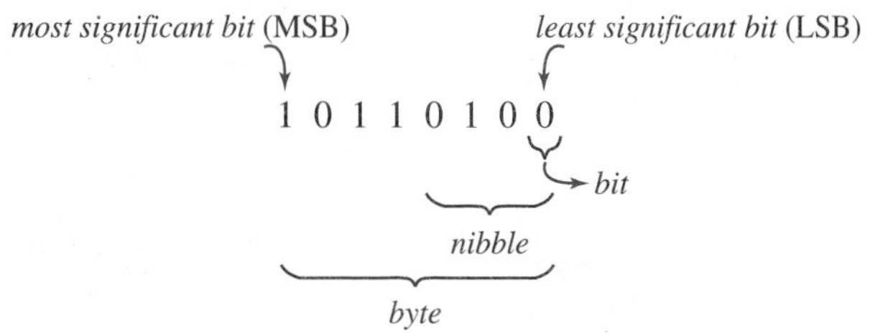

most significant bit (MSB)   *least significant bit* (LSB)

1 0 1 1 0 1 0 0

bit

nibble

byte

How do we convert from decimal to binary? Instead of staring at equations, let us try a more intuitive approach. First, we might correctly suspect that we should divide the task into two separate problems: (1) converting decimal integers to binary integers and (2) converting decimal fractions to binary fractions.

Consider first the task of converting a decimal integer (D_{10}) to a binary integer (B_2), formulated as the following mixed-base algebra problem.

$$D_{10} = B_2$$
$$(d_{M-1}d_{M-2}\ldots d_1 d_0)_{10} = (b_{N-1}b_{N-2}\ldots b_1 b_0)_2$$

Based on our understanding of the binary number system, we know that b_0 denotes whether the number is even or odd: $b_0 = 0$ if the number is even and $b_0 = 1$ if the number is odd. Hence, we now know how to find b_0: Divide D_{10} by two and the remainder equals b_0.

As an example, let $D_{10} = 13$. The decimal-to-binary conversion problem is formulated as

$$13_{10} = (b_3 b_2 b_1 b_0)_2 \tag{2.3}$$

Dividing both sides of Equation 2.3 by two yields

$$\frac{13}{2} = \frac{b_3 b_2 b_1 b_0}{2}$$
$$6.5\,5 = b_3 b_2 b_1.b_0 \tag{2.4}$$

Notice that dividing a binary number by two simply shifts all the bits to the right by one place, just like dividing a decimal number by ten shifts all the digits to the right by one place. Comparing fraction and integer parts of the decimal (left-hand side) and binary (right-hand side) numbers in Equation 2.4 reveals

$$0.5_{10} = (0.b_0)_2$$
$$6_{10} = (b_3 b_2 b_1)_2$$
$$\therefore b_0 = 1 \tag{2.5}$$

Recall that $0.1_2 = 1 \times 2^{-1} = 0.5_{10}$. This result is consistent with the fact that 13_{10} is indeed an odd number.

Having determined b_0, we finish solving Equation 2.5 by finding the binary representation of 6_{10}. Things are looking promising. The value of b_1 can be determined by applying the same procedure of testing whether 6 is even or odd. Dividing both sides of the second line of Equation 2.5 by two yields

$$\frac{6}{2} = \frac{b_3 b_2 b_1}{2}$$
$$3.0 = b_3 b_2 b_1 \tag{2.6}$$

Again, comparing the fraction and integer parts of the decimal and binary numbers of Equation 2.6 reveals

$$0.0_{10} = (0.b_1)_2$$
$$3_{10} = (b_3 b_2)_2$$
$$\therefore b_1 = 0 \tag{2.7}$$

Repeating this divide-by-two operation two more times yields the values for b_2

$$\frac{3}{2} = \frac{b_3 b_2}{2}$$
$$1.5 = b_3 b_2$$
$$\therefore b_2 = 1$$

and b_3.

$$\frac{1}{2} = \frac{b_3}{2}$$
$$0.5 = 0.b_3$$
$$\therefore b_3 = 1$$

**Procedure to Convert
Decimal Integers to
Binary Integers**
*Conduct a series of divide-
by-two operations. The dividend for the first division is
the decimal number. The
dividend for each succeeding division is the quotient
from the previous division.
Stop dividing when the quotient equals zero. The remainders from the series of
divide-by-two operations
form the binary number,
starting with the least significant bit (LSB) and ending with the most significant
bit (MSB).*

Success! Collecting the results of the divisions shows that $13_{10} = 1101_2$.

The procedure for converting decimal integers to binary integers is summarized to the left. Examples of converting 71_{10} and 34_{10} to binary integers are given in Figure 2.3.

Armed with our success at converting a decimal integer to a binary integer, let us next tackle the task of converting a decimal fraction (D_{10}) to a binary fraction (B_2), formulated as the following mixed-base algebra problem.

$$D_{10} = B_2$$
$$(0.d_{-1}d_{-2} \ldots d_{-(M-1)}d_{-M})_{10} = (0.b_{-1}b_{-2} \ldots b_{-(N-1)}b_{-N})_2$$

Correctly guessing that there is symmetry between the integer/fraction decimal-to-binary conversion problems reveals that the same basic approach used for converting integers can be used for converting fractions provided the *divide*-by-two operations are replaced with *multiply*-by-two operations. The procedure for converting decimal fractions to binary frac-

Figure 2.3

**Integer decimal-to-binary
conversion examples**

	Remainders
2⌊71	$1 = b_0 = $ LSB
2⌊35	$1 = b_1$
2⌊17	$1 = b_2$
2⌊8	$0 = b_3$
2⌊4	$0 = b_4$
2⌊2	$0 = b_5$
2⌊1	$1 = b_6 = $ MSB
0	

$$\therefore 71_{10} = 1000111_2$$

	Remainders
2⌊34	$0 = b_0 = $ LSB
2⌊17	$1 = b_1$
2⌊8	$0 = b_2$
2⌊4	$0 = b_3$
2⌊2	$0 = b_4$
2⌊1	$1 = b_5 = $ MSB
0	

$$\therefore 34_{10} = 100010_2$$

Figure 2.4

Fraction decimal-to-binary conversion examples

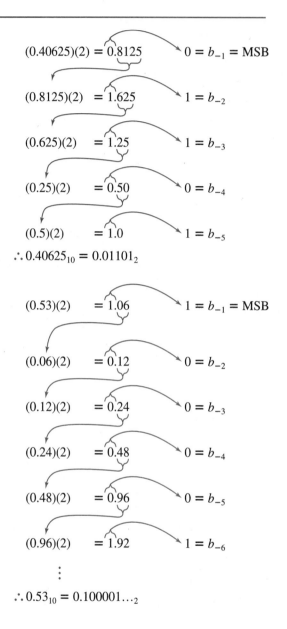

$$(0.40625)(2) = 0.8125 \qquad 0 = b_{-1} = \text{MSB}$$

$$(0.8125)(2) = 1.625 \qquad 1 = b_{-2}$$

$$(0.625)(2) = 1.25 \qquad 1 = b_{-3}$$

$$(0.25)(2) = 0.50 \qquad 0 = b_{-4}$$

$$(0.5)(2) = 1.0 \qquad 1 = b_{-5}$$

$$\therefore 0.40625_{10} = 0.01101_{2}$$

$$(0.53)(2) = 1.06 \qquad 1 = b_{-1} = \text{MSB}$$

$$(0.06)(2) = 0.12 \qquad 0 = b_{-2}$$

$$(0.12)(2) = 0.24 \qquad 0 = b_{-3}$$

$$(0.24)(2) = 0.48 \qquad 0 = b_{-4}$$

$$(0.48)(2) = 0.96 \qquad 0 = b_{-5}$$

$$(0.96)(2) = 1.92 \qquad 1 = b_{-6}$$

$$\vdots$$

$$\therefore 0.53_{10} = 0.100001\ldots_{2}$$

tions is summarized on the next page. Examples of converting 0.40625_{10} and 0.53_{10} to binary fractions are given in Figure 2.4. Notice that 0.53_{10} does not have an exact binary representation. A decimal fraction that is not the sum of a series of negative powers of 2 does *not* have an exact binary representation. Simple decimal fractions, such as $0.2_{10} = 0.\overline{0011}_{2}$ and $0.7_{10} = 0.1\overline{0110}_{2}$, do not have exact binary representations.

Figure 2.5 combines integer and fraction decimal-to-binary conversion to find the binary representation for 23.1875_{10}. The decimal-to-binary conversion is conducted in two steps. The first step converts the integer part 23_{10} to 10111_{2} by a series of divide-by-two operations. The second step converts the fraction part 0.1875_{10} to 0.0011_{2} by a series of multiply-by-two operations.

Figure 2.5

Integer and fraction decimal-to-binary conversion example

Convert integer part	$2\lfloor 23$	$1 = b_0$
	$2\lfloor 11$	$1 = b_1$
	$2\lfloor 5$	$1 = b_2$
	$2\lfloor 2$	$0 = b_3$
	$2\lfloor 1$	$1 = b_4$
	0	

Convert fraction part	$(0.1875)(2) = 0.375$	$0 = b_{-1}$
	$(0.375)(2) = 0.75$	$0 = b_{-2}$
	$(0.75)(2) = 1.5$	$1 = b_{-3}$
	$(0.5)(2) = 1.0$	$1 = b_{-4}$

$$\therefore 23.1875_{10} = 10111.0011_2$$

Procedure to Convert Decimal Fractions to Binary Fractions
Conduct a series of multiply-by-two operations. The multiplicand for the first multiplication is the decimal fraction. The multiplicand for each succeeding multiplication is the fractional part of the product from the previous multiplication. Stop multiplying when the fractional part of the product equals zero. The integer parts of the products from the series of multiply-by-two operations form the binary fraction, starting with the most significant bit (MSB) and ending with the least significant bit (LSB).

As a final note on decimal/binary conversion, a decimal-to-binary or binary-to-decimal conversion can be verified by performing the reverse conversion and checking that the original number is regenerated. In other words, to verify a decimal-to-binary conversion, transform the resulting binary number back to decimal and check that the result equals the original decimal number. Similarly, to verify a binary-to-decimal conversion, transform the resulting decimal number back to binary and check that the result equals the original binary number.

2.3 BINARY ARITHMETIC

To show that the binary number system is indeed a viable number system and to set the stage for discussing negative binary numbers, we will examine the arithmetic operations of binary addition and subtraction.

2.3.1 Binary Addition

The rules for binary addition are similar to the rules for decimal addition. Recall that two decimal numbers are added column by column, moving right to left, starting with the least significant digits. Anytime the sum of the digits in a column equals or exceeds the decimal base of 10, 10 is subtracted from the column sum and a 1 is carried forward to the next column.

$$
\begin{array}{r}
1 \\
6 \quad 6_{10} \quad augend \\
+1 \quad 7_{10} \quad addend \\
\hline
13 \\
-10 \\
\hline
8 \quad 3_{10} \quad sum
\end{array}
$$

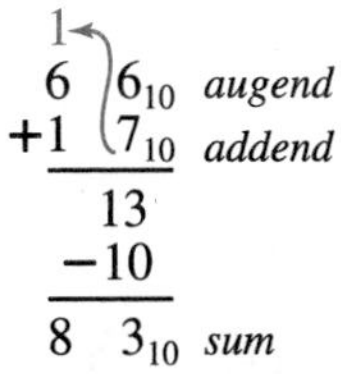

Figure 2.6

Binary addition

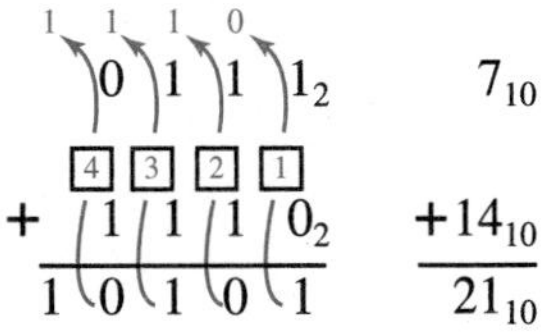

$$
\begin{array}{r}
\overset{1}{}\overset{1}{}\overset{1}{}\overset{0}{} \\
0 \ 1 \ 1 \ 1_2 \qquad 7_{10} \\
\boxed{4}\ \boxed{3}\ \boxed{2}\ \boxed{1} \\
+ \ 1\ 1\ 1\ 0_2 \qquad +14_{10} \\
\hline
1\ 0\ 1\ 0\ 1 \qquad 21_{10}
\end{array}
$$

The arrows indicate carries.

Now think binary. Two binary numbers are added column by column, moving right to left, starting with the least significant bits. Anytime the sum of the bits in a column equals or exceeds the binary base of 2, 2 is subtracted from the column sum and a 1 is carried forward to the next column. Figure 2.6 shows an example of how $(7_{10} + 14_{10})$ is performed in binary.

Let us go through this binary addition example column by column. In the first column, $1 + 0 = 1$ with no carry, denoted by the arrow labeled ☐. In the second column, $1 + 1 = 2_{10} = 10_2$. This column sum equals the binary base 2, so 2_{10} is subtracted from the second column sum (leaving a column sum of 0) and a carry is forwarded to the third column. This carry is denoted by the arrow labeled ②. In the third column, the augend, addend, and carry-in bits sum to $1 + 1 + 1 = 3_{10} = 11_2$. This column sum exceeds the binary base 2, so 2_{10} is again subtracted from the third column sum (leaving a column sum of 1) and a carry is forwarded to the fourth column. This carry is denoted by the arrow labeled ③. In the fourth column, the augend, addend, and carry-in bits sum to $1 + 0 + 1 = 2_{10} = 10_2$. Subtracting 2_{10} from this column sum leaves a column sum of 0. Also, a carry is forwarded to the last column, denoted by the arrow labeled ④. In the last column, the augend, addend, and carry-in bits sum to $0 + 0 + 1 = 1$, which completes the addition. You are encouraged to use the binary-to-decimal conversion method to check that 10101_2 equals 21_{10}.

Based on the example in Figure 2.6, binary addition can be defined by a set of rules that specify the value of the column sum and carry-out based on the *augend*, *addend*, and *carry-in* bits. Such a set of rules is given in Table 2.1.

Several binary addition examples are given in Figure 2.7. Use Table 2.1 to understand each addition example.

2.3.2 Binary Subtraction

As was the case with binary addition, binary subtraction is best explained by first reviewing decimal subtraction. Recall that two decimal numbers are subtracted column by column,

Table 2.1

Rules for adding two binary numbers. These rules describe how to generate the sum and carry out bits on a column-by-column basis when adding two binary numbers.	Carry-in bit from previous column	Addend bit	Augend bit	Sum bit	Carry-out bit to next column
	0	0	0	0	0
	0	0	1	1	0
	0	1	0	1	0
	0	1	1	0	1
	1	0	0	1	0
	1	0	1	0	1
	1	1	0	0	1
	1	1	1	1	1

Figure 2.7

Binary addition examples

$$
\begin{array}{cc}
5_{10} & 101_2 \\
+\ 6_{10} & +\ 110_2 \\
\hline
11_{10} & 1011_2
\end{array}
\qquad
\begin{array}{cc}
10_{10} & 1010_2 \\
+\ 4_{10} & +\ 100_2 \\
\hline
14_{10} & 1110_2
\end{array}
$$

(a) (b)

$$
\begin{array}{cc}
7_{10} & 111_2 \\
+\ 7_{10} & +\ 111_2 \\
\hline
14_{10} & 1110_2
\end{array}
\qquad
\begin{array}{cc}
5_{10} & 101_2 \\
+\ 12_{10} & +\ 1100_2 \\
\hline
17_{10} & 10001_2
\end{array}
$$

(c) (d)

moving from right to left, starting with the least significant digits. Anytime the subtrahend digit is more than the minuend digit, a borrow is made from the next column to the left. The minuend digit contributing the borrow is decremented by 1 and the minuend digit receiving the borrow is incremented by the decimal base of 10.

$$
\begin{array}{r}
\overset{8}{\cancel{9}} \overset{\nearrow 10}{\cancel{0}} \,_{10} \quad \textit{minuend} \\
-\quad 7_{10} \quad \textit{subtrahend} \\
\hline
8 \quad 3_{10} \quad \textit{difference}
\end{array}
$$

Now, think binary again. Two binary numbers are subtracted in a similar fashion, except a borrow is accomplished by incrementing the minuend bit by the binary base of 2_{10} or 10_2. Figure 2.8 shows an example of how $(12_{10} - 7_{10} = 5_{10})$ is performed in binary.

Let us go through this subtraction example column by column. A borrow is required for the first column because the subtrahend bit, 1, is greater than the minuend bit, 0. This borrow is generated in two steps. First, the minuend bit in the *third* column is decremented by 1 and a borrow of 2 (binary base) is added to the minuend bit in the second column, denoted by the arrow labeled ☐. Then, the new minuend bit in the second column is decremented by 1 and a borrow of 2 is added to the minuend bit in the first column, denoted by the arrow labeled ②. Now, the subtractions for the first column, $2 - 1 = 1$, and the second column, $1 - 1 = 0$, are performed. A borrow is again required for the third column because the subtrahend bit, 1, is greater than the minuend bit, 0. The minuend bit in the fourth column is decremented by 1 and a borrow of 2 is added to the minuend bit in the third column, denoted by the arrow labeled ③. Finally, the bits in the third column are subtracted $(2 - 1 = 1)$, which completes the subtraction.

Several binary subtraction examples are given in Figure 2.9.

Figure 2.8

Binary subtraction

The arrows indicate borrows.

Figure 2.9

Binary subtraction examples

$$
\begin{array}{cc}
10_{10} & 1010_2 \\
-\ 4_{10} & -\ 100_2 \\
\hline
6_{10} & 110_2
\end{array}
\qquad
\begin{array}{cc}
7_{10} & 111_2 \\
-\ 7_{10} & -\ 111_2 \\
\hline
0_{10} & 000_2
\end{array}
$$

$$
\text{(a)} \qquad\qquad \text{(b)}
$$

$$
\begin{array}{cc}
8_{10} & 1000_2 \\
-\ 1_{10} & -\quad 1_2 \\
\hline
7_{10} & 111_2
\end{array}
\qquad
\begin{array}{cc}
12_{10} & 1100_2 \\
-\ 5_{10} & -\ 101_2 \\
\hline
7_{10} & 111_2
\end{array}
$$

$$
\text{(c)} \qquad\qquad \text{(d)}
$$

2.4 REPRESENTING NEGATIVE NUMBERS IN BINARY

What if we try to subtract $101_2 - 111_2$ ($5_{10} - 7_{10}$)? The binary subtraction process explained in the previous section does not work because we have not yet discussed how to represent negative numbers in binary. Of the several schemes for representing negative binary numbers, we will discuss the following:

- signed-magnitude representation,
- two's complement representation (radix complement), and
- one's complement representation (reduced radix complement).

Signed-magnitude representation will be discussed first because it is similar to how we represent negative decimal numbers and is conceptually the simplest scheme. Although conceptually simple, signed-magnitude representation is generally not the most popular binary number scheme because it is not convenient for arithmetic operations. Arithmetic operations using signed-magnitude representations typically require more digital hardware than similar arithmetic operations using complement representations. More digital hardware translates into longer design times, bigger space requirements, higher power consumption, and higher product costs.

As a lead into discussing these representation schemes, Table 2.2 presents a summary comparison of the two's complement, one's complement, and signed-magnitude 4-bit encodings for the decimal numbers -8_{10} to $+7_{10}$. Notice that the binary representation schemes differ only for zero and negative numbers.

2.4.1 Signed-Magnitude Representation

In the decimal system, negative numbers are represented by appending a minus sign to the number, for example, $5_{10} - 7_{10} = -2_{10}$. Binary signed-magnitude representation uses a similar approach by designating a bit to represent the sign. The sign bit is typically the most

Table 2.2

Methods for representing positive/negative binary numbers. Codeword length is set to 4 bits. Signed-magnitude and one's complement representations do not have unique encodings for the number 0.

Decimal number	Binary representations		
	Two's complement	One's complement	Signed-magnitude
+7	0111	0111	0111
+6	0110	0110	0110
+5	0101	0101	0101
+4	0100	0100	0100
+3	0011	0011	0011
+2	0010	0010	0010
+1	0001	0001	0001
0	0000	0000	0000
		1111	1000
−1	1111	1110	1001
−2	1110	1101	1010
−3	1101	1100	1011
−4	1100	1011	1100
−5	1011	1010	1101
−6	1010	1001	1110
−7	1001	1000	1111
−8	1000	X	X

Note: An X denotes that the representation scheme does not have a 4-bit encoding for the respective decimal number.

significant bit (MSB), as illustrated by the signed-magnitude column in Table 2.2. The high-order bit encodes the sign and the low-order three bits encode the magnitude, hence the name "signed-magnitude." By convention, the sign bit is 0 if the number is positive and 1 if the number is negative. For example, the signed-magnitude 4-bit representation for -5_{10} is

$$-5_{10} = \underbrace{1}_{} \; \underbrace{1 \;\; 0 \;\; 1}_{}{}_2$$

Sign bit $\longleftarrow \underbrace{}$ $\underbrace{}\longrightarrow$ Magnitude

" $-$ "   5

Unfortunately, there are two possible encodings for zero: one for $+0$ (0000_2) and one for -0 (1000_2). This "feature" is cumbersome and complicates arithmetic operations.

Signed-magnitude arithmetic is typically performed by removing the sign bits from the operands, manipulating the magnitudes, and then appending the appropriate sign bit to the result. For example, ($5_{10} - 7_{10}$) or $0101_2 - 0111_2$ is accomplished by removing the sign bits, switching the minuend and subtrahend, and subtracting: $111_2 - 101_2 = 010_2$. Then, a sign bit is appended to the result and set to 1 to denote the negative difference 1010_2 or -2_{10}. The actions of checking magnitudes, switching operands, and manipulating sign bits are difficult to implement in digital hardware. Thus, signed-magnitude is a simple "book-keeping" scheme for representing negative binary numbers but is typically not used for binary arithmetic.

2.4.2 Two's Complement Representation

Another way to denote negative binary numbers is by the two's complement representation, more generally called a *radix complement* representation. Generating the two's complement representation is more complex than generating the signed-magnitude representation; however, two's complement arithmetic is simpler than signed-magnitude arithmetic. We will introduce two's complement by first illustrating the radix complement for the decimal system, called the *ten's complement*. Although ten's complement is typically not used to represent negative decimal numbers, it is instructive to study the scheme to understand the general concept and properties of a radix complement.

As the name implies, ten's complement is a negation or invert operation. The representation of a negative decimal number is obtained by forming the ten's complement of the associated positive decimal number. For instance, the representation for -10 is obtained by forming the ten's complement of 10 and vice versa. The ten's complement of an N-digit decimal integer D is

$$\text{ten's complement} = \begin{cases} 10^N - D, & D \neq 0 \\ 0, & D = 0 \end{cases} \qquad \textbf{(2.8)}$$

Rearranging Equation 2.8

$$\text{ten's complement} = \begin{cases} (10^N - 1) - D + 1, & D \neq 0 \\ 0, & D = 0 \end{cases}$$

and noting that ($10^N - 1$) is a number consisting of N 9's shows that the ten's complement of a decimal integer D can be formed by subtracting each digit of D from 9 and then adding 1 to the total N-digit resulting difference.

Examples of 2-digit ten's complement operations are shown in Figure 2.10. Figure 2.10(a) shows that -4_{10} is represented by 96_{10}, obtained by forming the 2-digit ten's complement of 4_{10}. Similarly, Figure 2.10(b) shows that -7_{10} is represented by 93_{10}, obtained by forming the 2-digit ten's complement of 7_{10}. Figure 2.10(b) and Figure 2.10(c) illustrate that the radix complement, in general, and the ten's complement, in particular, obey the identity property, that is, $-(-7_{10}) = 7_{10}$.

An N-digit ten's complement number represents decimal integers in the range -10^{N-1} to $10^{N-1} - 1$. The ten's complement representations for -10_{10} to 9_{10} ($N = 2$) are given in Table 2.3.

Although thinking of 96_{10} as representing -4_{10} is admittedly a little strange, ten's complement is a legitimate representation for negative decimal numbers that supports conventional arithmetic operations. For example, Figure 2.11 shows that decimal subtraction can be performed by taking the ten's complement of the subtrahend and then adding this number to the minuend, that is, $X - Y = X + (-Y)$.

The resulting sum of 97_{10} is the correct answer, but the answer is encoded in ten's complement because it is a negative number. That is, 97_{10} is the ten's complement representation for -3_{10}. This example illustrates that ten's complement arithmetic offers no advantage over standard practices and is consequently not used to represent negative decimal numbers. However, as we will discover, the radix complement turns out to be an efficient, desirable representation for denoting negative numbers and performing arithmetic operations in binary.

Like ten's complement, the radix complement (or two's complement) for the binary system is a negation or invert operation. The representation of a negative binary number is obtained by forming the two's complement of the associated positive binary number. Referring to Table 2.2, we see that the binary representation for -5_{10} is obtained by forming the 4-bit two's complement of 0101_2, that is, $+5_{10}$. The two's complement of an N-bit binary integer B is

$$\text{two's complement} = \begin{cases} 2^N - B, & B \neq 0 \\ 0, & B = 0 \end{cases} \qquad \textbf{(2.9)}$$

Figure 2.10

Examples of forming the ten's complement

$$\begin{array}{ll} 99_{10} & \\ -\ 04_{10} & \text{step \#1} \\ \hline 95_{10} & \\ +\ \ 1_{10} & \text{step \#2} \\ \hline 96_{10} & \end{array} \qquad \begin{array}{ll} 99_{10} & \\ -\ 07_{10} & \text{step \#1} \\ \hline 92_{10} & \\ +\ \ 1_{10} & \text{step \#2} \\ \hline 93_{10} & \end{array}$$

$$\textbf{(a)} \qquad\qquad\qquad \textbf{(b)}$$

$$\begin{array}{ll} 99_{10} & \\ -\ 93_{10} & \text{step \#1} \\ \hline 6_{10} & \\ +\ \ 1_{10} & \text{step \#2} \\ \hline 7_{10} & \end{array}$$

$$\textbf{(c)}$$

Two-digit ten's complement representations of -4_{10} and -7_{10}.

Table 2.3

Two-digit ten's complement representation

Decimal number	Ten's complement representation
+ 9	09
+ 8	08
+ 7	07
+ 6	06
+ 5	05
+ 4	04
+ 3	03
+ 2	02
+ 1	01
0	00
− 1	99
− 2	98
− 3	97
− 4	96
− 5	95
− 6	94
− 7	93
− 8	92
− 9	91
− 10	90

Two's Complement Operation

For an N-bit binary integer B, invert all bits and then add 1. Inverting all bits means that 0's change to 1's and 1's change to 0's.

or

For an N-bit binary integer B, scan the binary integer from the least significant bit to the most significant bit and invert all the bits following (but not including) the first 1.

Rearranging Equation 2.9

$$\text{two's complement} = \begin{cases} (2^N - 1) - B + 1, & B \neq 0 \\ 0, & B = 0 \end{cases}$$

and noting that $(2^N - 1)$ is a binary number consisting of N 1's shows that the two's complement of a binary integer B can be formed by subtracting each bit of B from 1 and then adding 1 to the total N-bit resulting difference. Moreover, since $1 - 0 = 1$ and $1 - 1 = 0$, subtracting each bit from 1 is equivalent to simply flipping or inverting each bit. The two's complement operation is summarized to the left.

For example, Figure 2.12 (p. 24) illustrates how to form the two's complement binary representation for -5_{10}. Figure 2.12(a) forms the two's complement by inverting each bit to form 1010_2 and then adding 1 to form 1011_2. Figure 2.12(b) demonstrates that the same re-

Figure 2.11

Example of ten's complement subtraction

Standard subtraction	$\begin{array}{r} 4_{10} \\ -\ 7_{10} \\ \hline -\ 3_{10} \end{array}$		
Ten's complement subtraction	$\begin{array}{r} 4_{10} \\ -\ 7_{10} \\ \hline \end{array}$ $\rightarrow$	$\begin{array}{r} 4_{10} \\ +\ 93_{10} \\ \hline 97_{10} \end{array}$	*10's complement of 7 = –7* *10's complement of 3 = –3*

Figure 2.12

Examples of forming the two's complement for -5_{10}

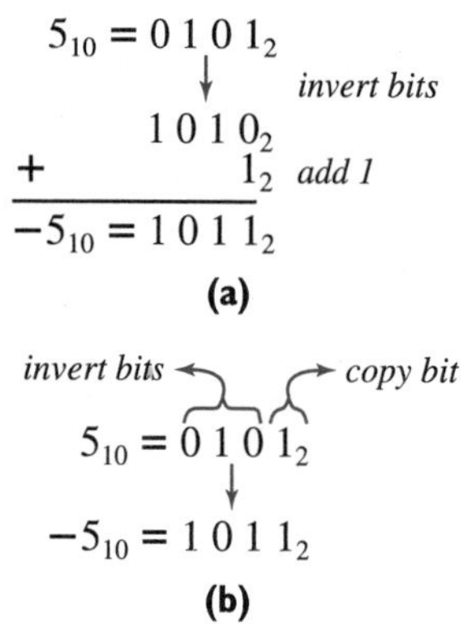

$$5_{10} = 0\ 1\ 0\ 1_2$$
$$\downarrow \quad \textit{invert bits}$$
$$1\ 0\ 1\ 0_2$$
$$+ \qquad\qquad 1_2 \ \textit{add 1}$$
$$\overline{}$$
$$-5_{10} = 1\ 0\ 1\ 1_2$$

(a)

$$\textit{invert bits} \longleftarrow \qquad \longrightarrow \textit{copy bit}$$
$$5_{10} = 0\ 1\ 0\ 1_2$$
$$\downarrow$$
$$-5_{10} = 1\ 0\ 1\ 1_2$$

(b)

sult can be obtained by scanning 0101_2 from right to left and inverting all bits following the first 1 (the first bit is copied and the remaining bits are inverted).

It is important to recognize the significance of the number of bits N in the two's complement operations given above. For example, generating the two's complement of 7_{10} by flipping all the bits in 111_2 and adding 1 yields the incorrect result of 001_2 because the number of bits N in the binary representation of 7_{10} was not properly set. Recall from the definition of two's complement in Equation 2.9 that a fixed number of bits N is *always* used. Fixing the number of bits fixes the range of decimal integers that can be represented. An N-bit two's complement binary number represents decimal integers in the range -2^{N-1} to $2^{N-1} - 1$. Thus, the two's complement representation for -7_{10} requires at least four bits because three bits covers only the range -2^2 to $2^2 - 1$ or -4 to $+3$. Setting $N = 4$ and flipping all the bits in 0111_2 and adding 1 yields the correct result of 1001_2.

To verify a two's complement operation, an N-bit binary integer $b_{N-1}b_{N-2} \dots b_1 b_0$ in two's complement notation can be converted to decimal by

$$= (b_{N-1} \times -2^{N-1}) + (b_{N-2} \times 2^{N-2}) + \cdots + (b_1 \times 2^1) + (b_0 \times 2^0)$$
$$= b_{N-1} \times -2^{N-1} + \sum_{i=0}^{N-2} b_i \times 2^i \qquad\qquad \textbf{(2.10)}$$

We can verify that 1011_2 is the two's complement of -5_{10} by

$$= (1 \times -2^3) + (0 \times 2^2) + (1 \times 2^1) + (1 \times 2^0)$$
$$= (1 \times -8) + (0 \times 4) + (1 \times 2) + (1 \times 1)$$
$$= -8_{10} + 0 + 2_{10} + 1_{10}$$
$$= -5_{10}$$

The most significant bit has a weight of -2^{N-1} instead of 2^{N-1}.

You are encouraged to examine and verify the two's complement representations given in Table 2.2 to gain familiarity and proficiency with this numbering scheme. Table 2.2 shows that a 4-bit, two's complement binary number represents decimal integers in the range -2^3 to $2^3 - 1$ or -8_{10} to $+7_{10}$. Observe that the most significant bit indicates the sign of the number: 0 indicates positive and 1 indicates negative.

2.4.3 Two's Complement Addition and Subtraction

We stated earlier that the principal advantage or utility of the two's complement representation is its convenience for binary arithmetic. Let us check this claim by examining two's complement addition and subtraction.

Two's complement addition uses the rules discussed in Section 2.3.1 for standard binary addition, as shown below. Notice that the addition process correctly computes the sign bit,

$$
\begin{array}{r} 4_{10} \\ + \ 2_{10} \\ \hline 6_{10} \end{array}
\quad\longleftrightarrow\quad
\begin{array}{r} 0100_2 \\ + \ 0010_2 \\ \hline 0110_2 \end{array}
\qquad
\begin{array}{r} 3_{10} \\ + \ (-6)_{10} \\ \hline -3_{10} \end{array}
\quad\longleftrightarrow\quad
\begin{array}{r} 0011_2 \\ + \ 1010_2 \\ \hline 1101_2 \end{array}
$$

(a)   **(b)**

which is more efficient than the separate logical operations required by signed-magnitude arithmetic.

There are, however, two situations where two's complement addition differs from standard binary addition. The first exception is that any carry-outs of the most significant bit are ignored. For instance, ignoring the carry-out of the MSB in adding $-3_{10} + -4_{10}$ yields the

$$
\begin{array}{r} -3_{10} \\ + \ (-4)_{10} \\ \hline -7_{10} \end{array}
\quad\longleftrightarrow\quad
\begin{array}{r} 1101_2 \\ + \ 1100_2 \\ \hline \cancel{1}\ 1001_2 \end{array}
$$

correct answer of -7_{10} or 1001_2 in two's complement notation. Ignoring the carry-out of the MSB makes the mathematics of two's complement work out.

$$
\begin{aligned}
(-3_{10}) + (-4_{10}) &= (-0011_2) + (-0100_2) \\
&= (2^4 - 0011 - 2^4) + (2^4 - 0100 - 2^4) \\
&= (1101_2 - 2^4) + (1100 - 2^4) \\
&= (11001_2 - 2^4 - 2^4) \\
&= (1001_2 - 2^4)
\end{aligned}
$$

Ignoring the carry-out of the MSB is equivalent to subtracting 2^4, which accounts for the final term.

The second exception concerns *overflow* and *underflow* conditions. Recall that an N-bit two's complement binary representation encodes decimal numbers in the range -2^{N-1} to $2^{N-1} - 1$. *Overflow* occurs when an arithmetic operation yields a result that is greater than the range's positive limit of $2^{N-1} - 1$. *Underflow* occurs when an arithmetic operation yields a result that is less than the range's negative limit of -2^{N-1}. The example shown below demonstrates an overflow error condition. Two's complement addition yields the incor-

$$
\begin{array}{r} 5_{10} \\ + \ 6_{10} \\ \hline 11_{10} \end{array}
\quad\longleftrightarrow\quad
\begin{array}{r} 0101_2 \\ + \ 0110_2 \\ \hline 1011_2 \end{array}
= -5_{10}
$$

rect answer of 1011_2 or -5_{10} because the correct sum of 11_{10} exceeds the positive 4-bit range limit of $+7_{10}$. The next example demonstrates an underflow error condition. Two's

$$\begin{array}{r} -5_{10} \\ + \ (-7)_{10} \\ \hline -12_{10} \end{array} \quad \longleftrightarrow \quad \begin{array}{r} 1011_2 \\ + \ 1001_2 \\ \hline \cancel{\ 0100_2\ } = 4_{10} \end{array}$$

complement addition here yields the incorrect answer of 0100_2 or 4_{10} because the correct sum of -12_{10} exceeds the negative 4-bit range limit of -8_{10}.

Overflow can only occur when adding two positive numbers and underflow can only occur when adding two negative numbers. Overflow and underflow error conditions can be easily detected by comparing the most significant bits of the operands and sum: When the most significant bits of the addend and augend are equal and different from the most significant bit of the sum, then an overflow or underflow error has occurred and the sum is invalid. If an overflow or underflow error occurs, extend the range of the two's complement system by using more than N bits and redo the addition. It is advantageous to address overflow/underflow conditions *early* in the digital system design process because extending ranges may be difficult or impossible if word sizes and associated hardware resources have already been committed. Two's complement addition is summarized to the left.

Similar to ten's complement subtraction discussed earlier, two's complement subtraction is performed by inverting (two's complementing) the subtrahend and adding the new subtrahend to the minuend. Converting subtractions into additions saves hardware; only one hardware component is needed to perform addition, whereas two are needed to perform addition and subtraction. Examples of two's complement subtraction are shown below.

Two's Complement Addition

Add two N-bit two's complement numbers by using standard binary addition rules and ignoring any carry-outs of the most significant bit. The sum is invalid if an overflow or underflow error occurs.

$$\begin{array}{r} 4_{10} \\ - \ 7_{10} \\ \hline \end{array} \quad \longrightarrow \quad \begin{array}{r} 4_{10} \\ + \ (-7)_{10} \\ \hline -3_{10} \end{array} \quad \longleftrightarrow \quad \begin{array}{r} 0100_2 \\ + \ 1001_2 \\ \hline 1101_2 \end{array}$$

(a)

$$\begin{array}{r} -6_{10} \\ - \ (-7)_{10} \\ \hline \end{array} \quad \longrightarrow \quad \begin{array}{r} -6_{10} \\ + \ 7_{10} \\ \hline 1_{10} \end{array} \quad \longleftrightarrow \quad \begin{array}{r} 1 \leftarrow \text{Carry-in} \\ 1010_2 \\ + \ 0110_2 \\ \hline \cancel{\ 0001_2\ } \end{array}$$

(b)

In the example (a), the subtraction $4_{10} - 7_{10}$ is converted to an addition $4_{10} + (-7_{10})$ by obtaining the 4-bit two's complement representation for -7_{10}, which is 1001_2. Then, 0100_2 is added to 1001_2. This two's complement subtraction example requires two addition operations: (1) one addition to form the two's complement of the subtrahend and (2) one addition to sum the minuend and complemented subtrahend. These two addition operations can be combined into one addition, as illustrated in the example (b), $-6_{10} - (-7_{10})$. First, all the bits of the subtrahend $1001_2 = -7_{10}$ are inverted to form 0110_2, which is the first step in obtaining the 4-bit two's complement of the subtrahend. Then, the minuend, new subtrahend, and a carry-in in the rightmost (least significant) column are added; the carry-in completes the 4-bit two's complement of the subtrahend. Two's complement subtraction is summarized to the left.

Two's Complement Subtraction

Subtract two N-bit two's complement numbers by forming the two's complement of the subtrahend and then summing the numbers using two's complement addition.

2.4.4 One's Complement Representation

An alternative approach to the radix complement for representing negative numbers is the *reduced-radix complement*. For the binary system, the reduced-radix complement is called the *one's complement*. Generating the one's complement representation is simpler than generating the two's complement representation. However, one's complement arithmetic is more complex than two's complement arithmetic. Thus, like the signed-magnitude representation, one's complement is not as widely used as two's complement. Consequently, we will take only a cursory look at one's complement.

The one's complement of an N-digit binary integer B is

$$\text{one's complement} = (2^N - 1) - B \tag{2.11}$$

Comparing Equation 2.11 and Equation 2.9 reveals that the one's complement is just the first step of the two's complement. That is, the one's complement of a binary integer B can be formed by subtracting each bit in B from 1 or, equivalently, inverting each bit in B. Recall that $1 - 0 = 1$ and $1 - 1 = 0$. For example, the 4-bit one's complement representation for $-5_{10} = 1010_2$ is obtained by inverting each bit of the binary representation of $5_{10} = 0101_2$. Look at the 4-bit one's complement representations for the decimal numbers -7_{10} to 7 listed in Table 2.2. Unlike the two's complement representation, the one's complement representation has two encodings for zero: one for $+0$ (0000_2) and one for -0 (1111_2). Also, an N-bit one's complement binary representation encodes decimal numbers in the symmetric range $-2^{N-1} + 1$ to $2^{N-1} - 1$.

Examples illustrating one's complement addition and subtraction are given below. Exam-

$$
\begin{array}{rcl}
-3_{10} & & 1100_2 \\
+\,-3_{10} & \longleftrightarrow & +\ 1100_2 \\
\hline
-6_{10} & & 1\ \overline{1000_2} \\
& & +\ \raisebox{-0.5ex}{$\llcorner\!\!\!\to$}\ 1_2 \\
\cline{3-3}
& & 1001_2
\end{array}
$$

(a)

$$
\begin{array}{rcl}
4_{10} & 4_{10} & 0100_2 \\
-6_{10} \longrightarrow & +\,(-6)_{10} \longleftrightarrow & +\ 1001_2 \\
\hline
-2_{10} & & 1101_2
\end{array}
$$

(b)

$$
\begin{array}{rcl}
6_{10} & & 0110_2 \\
+\ 5_{10} & \longleftrightarrow & +\ 0101_2 \\
\hline
11_{10} & & 1011_2 = -4_{10}
\end{array}
$$

(c)

$$
\begin{array}{rcl}
-3_{10} & & 1100_2 \\
+\,(-5)_{10} & \longleftrightarrow & +\ 1010_2 \\
\hline
-8_{10} & & 1\ \overline{0110_2} \\
& & +\ \raisebox{-0.5ex}{$\llcorner\!\!\!\to$}\ 1 \\
\cline{3-3}
& & 0111_2 = 7_{10}
\end{array}
$$

(d)

ple (a) illustrates the principal difference between two's complement arithmetic and one's complement arithmetic: A carry-out of the most significant bit is ignored in two's complement addition, whereas a carry-out of the most significant bit in one's complement addition is added back into the sum. This rule is often called the *end-around carry* rule. Example (b) shows that one's complement subtraction is performed by inverting (one's complementing) the subtrahend and adding. Examples (c) and (d) illustrate overflow and underflow error conditions, respectively.

2.5 OCTAL AND HEXADECIMAL NUMBER SYSTEMS

One disadvantage of the binary system that may not have been clearly demonstrated in the simple examples presented thus far is that the representation can readily yield long bit strings. For example,

$$617.171875_{10} = 1001001001.001011_2$$

Wait! That is not correct. Let's try again:

$$617.171875_{10} = 1001101001.001011_2$$

As you can see, it is very easy to make a mistake with long bit strings. To make it easier to read and write large binary numbers, designers often use the *octal* or *hexadecimal* number systems. The octal or hexadecimal representation of a given decimal number typically has many fewer digits than the associated binary representation. For example, the octal and hexadecimal representations of 1001101001.001011_2 are 1151.13_8 and $269.2C_{16}$, respectively. The octal system is base or radix 8 and the hexadecimal system is base or radix 16. These number systems support standard arithmetic operations, but are primarily used by designers as a notational convenience; computers use the binary system.

Octal and hexadecimal are weighted number systems. The octal system uses eight numerals (0–7) to construct digits, and each digit represents units of a power of 8. The hexadecimal system, also called hex, uses sixteen numerals (0–9 and $A–F$) to construct digits, and each digit represents units of a power of 16. The numerals $A_{16}–F_{16}$ respectively represent the decimal values $10_{10}–15_{10}$.

An octal number having an N-bit integer part and an M-bit fraction part

$$(o_{N-1}o_{N-2} \cdots o_1 o_0 \cdot o_{-1}o_{-2} \cdots o_{-(M-1)}o_{-M})$$

is converted to decimal by

$$
\begin{aligned}
&= (o_{N-1} \times 8^{N-1}) + (o_{N-2} \times 8^{N-2}) + \cdots + (o_1 \times 8^1) + (o_0 \times 8^0) \\
&\quad + (o_{-1} \times 8^{-1}) + (o_{-2} \times 8^{-2}) + \cdots + (o_{-(M-1)} \times 8^{-(M-1)}) + (o_{-M} \times 8^{-M}) \\
&= \sum_{i=-M}^{N-1} o_i \times 8^i
\end{aligned}
$$

Similarly, a hexadecimal number having an N-bit integer part and an M-bit fraction part

$$(h_{N-1}h_{N-2} \cdots h_1 h_0 \cdot h_{-1}h_{-2} \cdots h_{-(M-1)}h_{-M})$$

is converted to decimal by

$$
\begin{aligned}
&= (h_{N-1} \times 16^{N-1}) + (h_{N-2} \times 16^{N-2}) + \cdots + (h_1 \times 16^1) + (h_0 \times 16^0) \\
&\quad + (h_{-1} \times 16^{-1}) + (h_{-2} \times 16^{-2}) + \cdots + (h_{-(M-1)} \times 16^{-(M-1)}) + (h_{-M} \times 16^{-M}) \\
&= \sum_{i=-M}^{N-1} h_i \times 16^i
\end{aligned}
$$

For example, the decimal value of the octal number 1151.13_8 is given by

$$
\begin{aligned}
&= (1 \times 8^3) + (1 \times 8^2) + (5 \times 8^1) + (1 \times 8^0) + (1 \times 8^{-1}) + (3 \times 8^{-2}) \\
&= (1 \times 512) + (1 \times 64) + (5 \times 8) + (1 \times 1) + (1 \times 0.125) + (3 \times 0.015625) \\
&= 512 + 64 + 40 + 1 + 0.125 + 0.046875 \\
&= 617.171875_{10}
\end{aligned}
$$

and the decimal value of hexadecimal number $269.2C_{16}$ is given by

$$
\begin{aligned}
&= (2 \times 16^2) + (6 \times 16^1) + (9 \times 16^0) + (2 \times 16^{-1}) + (12 \times 16^{-2}) \\
&= (2 \times 256) + (6 \times 16) + (9 \times 1) + (2 \times 0.0625) + (12 \times 0.00390625) \\
&= 512 + 96 + 9 + 0.125 + 0.046875 \\
&= 617.171875_{10}
\end{aligned}
$$

Notice that the hexadecimal numeral C_{16} is equal to the decimal value 12_{10}.

It is easy to convert between octal or hexadecimal and binary because their bases are all powers of 2. Since $2^3 = 8$, one octal digit represents three binary bits. Thus, to convert from binary to octal, we first group the binary bits into sets of three, as shown in Figure 2.13. We

Figure 2.13

Converting binary to the
octal shorthand notation

$$\underbrace{001}_{1}\ \underbrace{001}_{1}\ \underbrace{101}_{5}\ \underbrace{001}_{1}\ .\ \underbrace{001}_{1}\ \underbrace{011}_{3}{}_2$$
$$_8$$

group the integer part of the binary number into sets of three bits, starting at the binary point and moving toward the most significant bit. Similarly, we group the fraction part of the binary number into sets of three bits starting at the binary point and moving toward the least significant bit. We then pad with zeros when necessary to generate complete groups of three bits. Then, we convert each set of three bits to an octal digit according to Table 2.4.

Since $2^4 = 16$, one hexadecimal digit represents four binary bits. Thus, to convert from binary to hexadecimal, we first group the binary bits into sets of four, as shown in Figure 2.14. Again, we start at the binary point and work outward toward the most significant bit and toward the least significant bit. We pad with zeros when necessary to obtain complete groups of four bits. Then, we convert each set of four bits to a hexadecimal digit according to Table 2.5.

Converting from decimal to octal or hexadecimal is performed similarly to the procedure discussed in Section 2.2. A decimal integer is converted to an octal or hexadecimal integer by a series of divide-by-eight or divide-by-sixteen operations, respectively. The remainders from the divisions form the octal or hexadecimal number, starting with the least significant digit and ending with the most significant digit. A decimal fraction is converted to an octal or hexadecimal fraction by a series of multiply-by-eight or multiply-by-sixteen operations, respectively. The integer parts of the products form the octal or hexadecimal fraction, starting with the most significant digit and ending with the least significant digit. As an example, the decimal number 685.78125_{10} is converted to hexadecimal below.

$$685.78125_{10}$$

Convert integer

Remainder

$16\ \lfloor 685 \quad 13_{10} = D_{16} = h_0$

$16\ \lfloor 42 \quad 10_{10} = A_{16} = h_1$

$16\ \lfloor 2 \quad\ \ 2_{10} = 2_{16} = h_2$

$\quad\quad 0$

Convert fraction   **Integer**

$(0.78125)(16) = 12.5 \quad\quad 12_{10} = C_{16} = h_{-1}$

$(0.5)(16) = 8.0 \quad\quad\quad\ 8_{10} = 8_{16} = h_{-2}$

$$685.78125_{10} = 2AD.C8_{16}$$

Figure 2.14

Converting binary to the
hexadecimal shorthand
notation

$$\underbrace{0010}_{2}\ \underbrace{0110}_{6}\ \underbrace{1001}_{9}\ .\ \underbrace{0010}_{2}\ \underbrace{1100}_{C}{}_2$$
$$_{16}$$

Table 2.4

Converting between bi-nary and octal	Binary 3-bit group	Octal digit
	000	0
	001	1
	010	2
	011	3
	100	4
	101	5
	110	6
	111	7

In summary, octal and hexadecimal are used as shorthand notations for binary. The hexa-decimal system tends to be more popular than the octal system because hexadecimal gener-ally uses fewer digits to represent a binary number. In other words, hexadecimal is more compact than octal. Also, computers use word sizes that are typically 4, 8, 16, 32, . . . , which are conveniently represented in hexadecimal. Do not, however, dismiss the octal num-ber system; it is used in engineering. For example, the popular UNIX[2] operating system uses an octal number to specify file system protection and access privileges (see Problem 14).

Table 2.5

Converting between bi-nary and hexadecimal	Binary 4-bit group	Hexadecimal digit
	0000	0
	0001	1
	0010	2
	0011	3
	0100	4
	0101	5
	0110	6
	0111	7
	1000	8
	1001	9
	1010	A
	1011	B
	1100	C
	1101	D
	1110	E
	1111	F

2.6 BINARY CODES

Thus far we have focused on representing decimal numbers in binary; this is an elementary form of *coding,* often called *straight binary coding*. There are many other types of *binary codes* that have been developed for special-purpose applications, such as error detection and

---

[2] UNIX® is a trademark of Bell Laboratories.

correction, serial communication, and transducer encoding. We will look at the following two examples of other binary codes:

- binary coded decimal (BCD), and
- Gray code.

2.6.1 Binary Coded Decimal

First we will explain how the binary coded decimal (BCD) representation for a decimal number is generated. Then we will look at some of the distinguishing characteristics and applications of BCD. To introduce BCD, we will compare the BCD and straight binary representations for 247_{10} given below. The straight binary representation shown on the right is

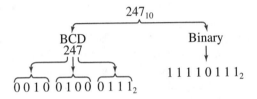

generated by a series of divide-by-two operations, as described in Section 2.2. In contrast, the BCD representation shown on the left is generated simply by converting each digit of the decimal number into its 4-bit binary representation. The digit 7_{10} is represented by the 4-bit binary number 0111_2, the digit 4_{10} is represented by the 4-bit binary number 0100_2, and the digit 2_{10} is represented by the 4-bit binary number 0010_2.

For BCD, each decimal digit is encoded into a string of four bits because four bits is the minimum number required to uniquely encode the ten decimal numerals (0–9). In general, a string of N bits supports a total of 2^N possible unique encodings or bit combinations. BCD is also called the *8-4-2-1* code; the numbers 8, 4, 2, and 1 denote the weightings of each bit in a 4-bit group. Table 2.6 summarizes the BCD code, listing the mapping between BCD codes and corresponding decimal digits.

Table 2.6

Binary coded decimal (BCD)	BCD 4-bit code	Decimal digit
	0000	0
	0001	1
	0010	2
	0011	3
	0100	4
	0101	5
	0110	6
	0111	7
	1000	8
	1001	9
	1010	
	1011	
	1100	not used
	1101	
	1110	
	1111	

The example of generating the BCD and straight binary representations for 247_{10} illustrates that converting between decimal and BCD is easier than converting between decimal and straight binary. Thus, one application of BCD is computer input/output. Since people prefer the decimal system and digital systems prefer the binary system, computational tasks typically execute the following three general actions:

1. *Input*: Input decimal numbers and convert the decimal numbers to binary numbers.
2. *Compute*: Perform the desired computations in binary.
3. *Output*: Convert the binary results back to decimal numbers and output the decimal results.

For tasks that involve a lot of input/output with relatively simple computations, more time can be spent performing the decimal/binary conversions (actions 1 and 3) than executing the desired computations (action 2). Thus, tasks that are input/output-intensive rather than compute-intensive may execute faster using BCD encoding rather than straight binary encoding.

A disadvantage of BCD is that it typically requires more bits than straight binary coding, which, in turn, implies more hardware to store the bits. Also, BCD arithmetic is more complex than straight binary arithmetic. BCD addition, for example, is performed like straight binary addition but has an extra step. If a column sum exceeds 9_{10}, then 6_{10} is added to the sum. For instance, consider the BCD addition shown below. Adding the BCD representa-

Decimal	BCD
7_{10}	0111_2
$+\ 5_{10}$	$+\ 0101_2$
$\overline{12_{10}}$	$\overline{1100_2}$
	$+\ 0110_2$
	$0001\ \overline{0010_2}$

tions for the digits 7_{10} and 5_{10} ($0111_2 + 0101_2$) using straight binary addition yields 1100_2, which is an incorrect answer and an illegal BCD code. However, since the sum exceeds 9_{10} (1001_2), adding 6_{10} (0110_2) yields the correct BCD encoded answer of $0001\ 0010_2$. Adding 6_{10} makes the mathematics of BCD work out by "pushing" column sums past the six unused binary codes of 1010_2 to 1111_2.

2.6.2 Gray Code

The *Gray code* in another example of a binary code. First we will explain how the Gray code representation for a decimal number is generated. Then we will look at some of the distinguishing characteristics of the Gray code.

To represent D decimal values, a Gray code of N bits is required, where $N \geq \log_2 D$.[3] For

---

[3] Since most calculators and logarithm tables support only common (base 10) and natural (base e) logarithms,

$$N \geq \log_2 D$$

can be reformulated as

$$N = \text{ceil}(\log_2 10 \times \log_{10} D)$$

or

$$N = \text{ceil}(3.322 \times \log_{10} D)$$

Where ceil(x) yields the smallest integer not less than x.

example, Table 2.7 illustrates the 4-bit Gray codes for the $2^4 = 16$ decimal numbers 0–15.

One easy way to generate Gray codes is to recognize the column-by-column bit patterns. Starting with the least significant column and moving toward the most significant column, the pattern of 1's and 0's in each column is as follows:

Column 1	01	10	01	. . .
Column 2	0011	1100	0011	. . .
Column 3	00001111	11110000	00001111	. . .

Note that the first column involves the alternate mirroring of the pattern one 0 and one 1. The second column involves the alternate mirroring of the pattern two 0's and two 1's, and so on.

The Gray code is also called a *minimum change code* because only one bit changes value between any two consecutive Gray codes. For example, the difference between the Gray code for 9_{10} $(1101)_2$ and 10_{10} $(1111)_2$ is only one bit; the second bit from the right changes from 0 to 1. This minimum change characteristic is useful in addressing digital system design problems associated with *intermediate states*.

To illustrate intermediate states and their potential problems, consider the hypothetical application of a digital system controlling the ambient temperature of a building. The digital system accepts the ambient temperature as a binary number and generates signals to control heating, ventilation, and cooling. Let us represent the input temperature as a 4-bit straight binary code and as a 4-bit Gray code and compare what might happen when the ambient temperature changes from 30°C to 40°C. Using a 4-bit straight binary code, we let 0000_2 represent 0°C, 0001_2 represent 10°C, and so on. When the ambient temperature changes from 30°C to 40°C, the straight-binary-encoded input changes from 0011_2 to 0100_2. However, the individual bits might not change simultaneously, so the digital system may actually receive a series of intermediary inputs. For example, 0011_2 (30°C) could change to 0010_2 (20°C), 0010_2 could then change to 0000_2 (0°C), and 0000_2 could then change to the final reading of 0100_2 (40°C). During the transition from 0011_2 to 0100_2, two false intermediate inputs are

Table 2.7

Four-bit Gray codes for the decimal numbers 0–15

Decimal value	Gray code
0	0000
1	0001
2	0011
3	0010
4	0110
5	0111
6	0101
7	0100
8	1100
9	1101
10	1111
11	1110
12	1010
13	1011
14	1001
15	1000

generated that may be wrongfully interpreted by the digital system as actual ambient temperatures.

Now consider representing the input temperature as a Gray code. Using the 4-bit Gray codes in Table 2.7, we let 0000_2 represent 0°C, 0001_2 represent 10°C, and so on. When the ambient temperature changes from 30°C to 40°C, the Gray-encoded input changes from 0010_2 to 0110_2. Only the third bit from the right changes, resulting in no false intermediate states.

Another benefit of the Gray code minimum change characteristic is that it helps to minimize the number of transitions or changes on physically adjacent wires. A 4-bit digital signal is typically realized in hardware as four physically adjacent wires. Simultaneously switching the values of physically adjacent wires can result in electrical problems, such as crosstalk, voltage spikes, and current surges. Changing only one bit between consecutive Gray codes reduces simultaneous electrical switching and the associated problems.

2.7 REPRESENTING CHARACTERS IN BINARY

Thus far, we have explored techniques for representing numbers using a variety of binary codes, including straight binary, BCD, and Gray. However, *numbers* are not the only type of information conveyed to digital systems. Look at a computer terminal keyboard. In addition to numbers, there are letters, punctuation marks, and command characters such as "backspace" and "enter." This keyboard information is represented by *alphanumeric* binary codes. There are two prominent alphanumeric codes: American Standard Code for Information Interchange (ASCII) and Extended Binary-Coded-Decimal Interchange Code (EBCDIC). A sampling of ASCII and EBCDIC codes is presented in Table 2.8. The

Table 2.8

Comparing ASCII and EBCDIC alphanumeric codes. A sample of the full set of ASCII and EBCDIC alphanumeric codes is presented in hexadecimal.	Alphanumeric character	ASCII 7-bit code	EBCDIC 8-bit code
	!	21	5A
	"	22	7F
	#	23	7B
	$	24	5B
	0	30	F0
	1	31	F1
	2	32	F2
	3	33	F3
	4	34	F4
	5	35	F5
	6	36	F6
	A	41	C1
	B	42	C2
	C	43	C3
	D	44	C4
	E	45	C5
	F	46	C6
	G	47	C7

ASCII (pronounced "*ask-ee*") code is seven bits, whereas the EBCDIC (pronounced "*ebb-see-dic*") code is eight bits. The EBCDIC code is probably most noted for being used with IBM computers.

2.8 SUMMARY

Because a digital system processes discrete information using electrical signals that have only two possible values, we have explored ways to represent information in the *binary number system*. The binary number system is a weighted, base-2 number system using only two values—0 and 1. We have learned how to represent integers and fractions, as well as positive and negative numbers in binary. To gain familiarity and proficiency with the binary number system, we also discussed the rules for addition and subtraction.

After examining the basic binary number system, we then considered the octal and hexadecimal number systems. These number systems serve as convenient shorthand notations for large binary numbers. Octal is a weighted, base-8 number system. Hexadecimal is a weighted, base-16 number system.

We also studied binary coded decimal (BCD) and Gray code representations for binary numbers. BCD represents each decimal digit as a 4-bit binary number and is used to minimize the amount of processing required to convert from decimal to binary and vice versa. The minimum change property of the Gray code is used to eliminate false intermediate states.

Finally, we discussed how alphanumeric information is represented in a digital system using ASCII and EBCDIC codes. Alphanumeric information includes numbers, letters, punctuation, and command characters.

2.9 PROBLEMS

1. Verify the decimal/binary numbers listed in Figure 2.1 by writing each binary number as a polynomial of the form shown in Equation 2.2 and finding the decimal value.

2. Expand Figure 2.1 by listing the binary representations for the decimal integers $0_{10}-15_{10}$.
   a. Determine the minimum number of bits required for the binary numbers.
   b. Build a table listing the decimal/binary numbers.
   c. Devise a shortcut for generating the binary numbers by noting the column-by-column bit patterns.

3. A digital system is required to amplify a binary-encoded audio signal. The user should be able to control the signal amplitude from minimum to maximum in 100 increments. Find the minimum number of bits required to encode, in straight binary, the user-specified amplitude.

4. Convert the following decimal numbers to binary numbers using the divide-by-two and/or multiply-by-two methods. Check the results by converting the binary numbers to decimal numbers using the weighted-sum-of products method.
   a. 5.625_{10} b. 0.40625_{10} c. 0.6_{10} d. 507_{10} e. 255_{10} f. 63.375_{10}

5. For the decimal numbers in Problem 4,
   a. Convert the decimal numbers to octal and hexadecimal numbers using the divide-by-eight and/or multiply-by-eight and the divide-by-sixteen and/or multiply-by-sixteen methods, respectively.
   b. Check the results in part (a) by converting the corresponding binary numbers generated in Problem 4 to octal and hexadecimal numbers.

6. Perform the following unsigned binary additions and subtractions; show all carries and borrows.

   a. 10110_2 **b.** 110.110_2 **c.** 0.10101_2
      $+\ \underline{10011_2}$ $+\ \underline{11.011_2}$ $+\ \underline{0.1001_2}$

   d. 1101_2 **e.** 100.00_2 **f.** 0.10101_2
      $-\ \underline{\ \ 11_2}$ $-\ \underline{\ 1.01_2}$ $-\ \underline{0.1001_2}$

7. Planet Ork uses the binary number system for everything. On vacation there, you see a souvenir priced at $\$110110.10_2$, with a sale discount of $\$1010.01_2$. How much does the souvenir cost in binary and decimal?

8. Binary multiplication is performed similarly to decimal multiplication, as shown below for the example $4_{10} \times 5_{10}$.

$$
\begin{array}{r}
100_2 \\
\times\ \underline{101_2} \\
100_2 \\
000_2 \\
+\ \underline{100_2\ \ } \\
10100_2
\end{array}
$$

Perform the following binary multiplications.

   a. 110_2 **b.** 0110_2 **c.** 0.1_2 **d.** 101_2
     $\times\ \underline{10_2}$ $\times\ \underline{1010_2}$ $\times\ \underline{0.1_2}$ $\times\ \underline{101_2}$

9. Expand Table 2.2 by listing the 5-bit two's complement, one's complement, and signed-magnitude representations for the decimal integers -16_{10} to $+15_{10}$.

10. Perform the following additions and subtractions using the 5-bit two's complement representation; note any overflows or underflows.

   a. 11_{10} **b.** 10_{10} **c.** -8_{10}
     $+\ \underline{-7_{10}}$ $+\ \underline{10_{10}}$ $+\ \underline{-5_{10}}$

   d. 5_{10} **e.** -9_{10} **f.** -3_{10}
     $-\ \underline{5_{10}}$ $-\ \underline{9_{10}}$ $-\ \underline{-11_{10}}$

11. The two's complement for a binary fraction (B) is

$$
\text{two's complement} = \begin{cases} 1 - B, & B \neq 0 \\ 0, & B = 0 \end{cases}
$$

Find the two's complement representation for the following negative decimal fractions.

   a. -0.125_{10} **b.** -0.25_{10} **c.** -0.5_{10} **d.** -0.75_{10}

12. Perform the following conversions by condensing groups of bits into octal and hexadecimal digits or expanding octal and hexadecimal digits into groups of bits.

   a. $1011.11_2 = $ _____ $_{16}$ **b.** $35.2_8 = $ _____ $_2$
   c. $11111111.1_2 = $ _____ $_{16}$ **d.** $11111111.1_2 = $ _____ $_8$
   e. $1BC_{16} = $ _____ $_2$ **f.** $C.4_{16} = $ _____ $_2$

13. Hexadecimal numbers are often used as shorthand notations for memory addresses. For example, a 4K memory has $2^{12} = 4096$ storage locations, and each location can be uniquely identified by a 3-digit hexadecimal address in the range 0_{10} to 4095_{10}. Which storage location (base 10) is denoted by the memory address BBB_{16}?

14. Basic UNIX file protection can be denoted by three octal digits. The three octal digits represent three sets of users referred to as "owner," "group," and "public." The three bits per octal digit represent three file access privileges referred to as "read" (r), "write" (w), and "execute" (x). Thus, a UNIX file permission given by the octal number $744_8 = 111\ 100\ 100_2$ is interpreted as

"rwx r-- r--," which specifies that the "owner" set of users have read, write, and execute privileges (rwx) and the "group" and "public" sets of users have only a read privilege (r--). Based on this notation, interpret the following popular UNIX file protections:

a. 644_8 **b.** 444_8 **c.** 711_8 **d.** 755_8

15. Octal and hexadecimal arithmetic are similar to binary and decimal arithmetic, except carries and borrows are based on the radices 8 and 16, respectively. Perform the following octal and hexadecimal additions and subtractions; show all carries and borrows.

a. 72_8 $+ 41_8$ **b.** 72_8 $- 41_8$ **c.** 100_8 $- 6_8$

d. $7E_{16}$ $+ 4C_{16}$ **e.** $7E_{16}$ $- 4C_{16}$ **f.** 100_{16} $- 6_{16}$

16. Define a base-18 number system for positive integers and fractions.
    a. Define the digit symbol set.
    b. Define base-18 to base-10 conversion.
    c. Define base-10 to base-18 conversion.
    d. Define base-18 addition and subtraction.

17. Consider a digital system that counts in 8-bit binary coded decimal (BCD).
    a. What is the decimal range of this counter?
    b. Show the output of this counter for the decimal numbers 0_{10} to 20_{10}.
    c. List all illegal outputs.

18. The *excess-3 code* is a variation of the binary coded decimal (BCD) code. In excess-3, each decimal digit is represented by a 4-bit code that is three more than the associated BCD code. For example, 0_{10} is encoded in excess-3 as 0011_2, 1_{10} is encoded in excess-3 as 0100_2, and so on.
    a. List the illegal 4-bit excess-3 codes.
    b. Show that excess-3 correctly generates decimal digit carries by performing the BCD addition example given in Section 2.6.1 ($7_{10} + 5_{10}$) in excess-3. Use straight binary addition rules.

19. The Gray code ($g_{N-1}g_{N-2} \cdots g_1 g_0$) can be generated from the straight binary code ($b_{N-1}b_{N-2} \cdots b_1 b_0$). First, copy the most significant bit ($b_{N-1} = g_{N-1}$). Then, scan the straight binary code from the most significant bit to the least significant bit and compare pairs of bits. If $b_{N-1} \neq b_{N-2}$, $g_{N-2} = 1$; otherwise, $g_{N-2} = 0$. Similarly, if $b_{N-2} \neq b_{N-3}$, $g_{N-3} = 1$; otherwise, $g_{N-3} = 0$, and so on. Use this method to generate the Gray code for the following decimal numbers.
    a. 0_{10} to 31_{10} **b.** 81_{10} **c.** 200_{10} **d.** 44_{10}

20. Given that the ASCII code(s) for the lowercase letters a–z are respectively 61_{16}–$7A_{16}$, the uppercase letters A–Z are respectively 41_{16}–90_{16}, a period is $2E_{16}$, and a blank space is 20_{16}, translate the following sentence into ASCII.

    Learning does not always imply understanding.

REFERENCES AND READINGS

[1] Swartzlander, E., ed. *Computer Arithmetic II:* Los Alamitos, CA: IEEE Computer Society Press, 1990.

[2] Crowder, N. *The Arithmetic of Computers: An Introduction to Binary & Octal Mathematics.* Garden City, N Y: DoubleDay & Company, 1960.

[3] Gillie, A. *Binary Arithmetic and Boolean Algebra.* New York: McGraw-Hill, 1965.

[4] Glaser, A. *History of Binary & Other Nondecimal Numeration.* Southampton, PA: Glaser, 1971.

[5] Kapps, C., and S. Bergman. *Introduction to the Theory of Computing.* Columbus, OH: Merrill Publishing, 1975.

[6] Levine, S. *Computer Number Systems & Binary Arithmetic.* New York: J. F. Rider Publishing, 1965.

[7] Mackenzie, C. *Coded Character Sets: History & Development.* Reading, MA: Addison-Wesley, 1980.

[8] Menninger, K. *Number Words & Number Systems: A Cultural History of Numbers.* Cambridge, MA: MIT Press, 1969.

[9] Peterson, W., and E. Weldon. *Error-Correcting Codes.* Cambridge, MA: MIT Press, 1972.

[10] Rao, T., and E. Fujiwara. *Error-Control Coding for Computer Systems.* Englewood Cliffs, NJ: Prentice-Hall, 1989.

[11] Richards, R. *Digital Design.* New York: Wiley, 1971.

[12] Schmid, H. *Decimal Computation,* New York: Wiley, 1974.

[13] Spaniol, O. *Computer Arithmetic: Logic & Design.* New York: Wiley, 1981.

[14] Sterbenz, P. *Floating-Point Computation.* Englewood Cliffs, NJ: Prentice-Hall, 1974.

[15] Waser, S., and M. Flynn. *Introduction to Arithmetic for Digital Systems Designers.* New York: Holt, Rinehart & Winston, 1982.

Part I

DIGITAL ENGINEERING:

COMBINATIONAL

SYSTEMS

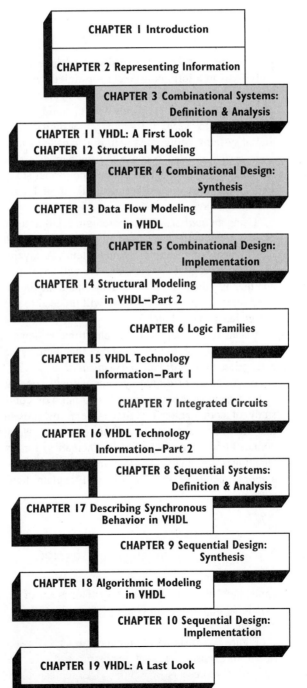

CHAPTER 1 Introduction

CHAPTER 2 Representing Information

CHAPTER 3 Combinational Systems:
Definition & Analysis

CHAPTER 11 VHDL: A First Look
CHAPTER 12 Structural Modeling

CHAPTER 4 Combinational Design:
Synthesis

CHAPTER 13 Data Flow Modeling
in VHDL

CHAPTER 5 Combinational Design:
Implementation

CHAPTER 14 Structural Modeling
in VHDL–Part 2

CHAPTER 6 Logic Families

CHAPTER 15 VHDL Technology
Information–Part 1

CHAPTER 7 Integrated Circuits

CHAPTER 16 VHDL Technology
Information–Part 2

CHAPTER 8 Sequential Systems:
Definition & Analysis

CHAPTER 17 Describing Synchronous
Behavior in VHDL

CHAPTER 9 Sequential Design:
Synthesis

CHAPTER 18 Algorithmic Modeling
in VHDL

CHAPTER 10 Sequential Design:
Implementation

CHAPTER 19 VHDL: A Last Look

Structural
Modeling

```
I1: NOT_OP
  port map (A, B);
```

Asynchronous
Data Flow
Modeling

```
B <= not A;
```

Technology
Modeling

```
B <= not A after 5 ns;
```

Synchronous
Data Flow
Modeling

```
B1: block (C = '1')
begin
  B <= guarded not A after 5 ns;
end block B1;
```

Algorithmic
Modeling

```
P1: process (PS)
begin
  NS <= STATE_TABLE(PS);
end process P1;
```

Combinational Logic
Schematics

Combinational Logic
Equations

$$Z = (A \cdot B) + C$$

Logic Families &
Integrated Circuits

Sequential Logic
Schematics

Sequential Logic
State Diagrams

00/0   11/0   01/0
01/1 (S0) (S1) 10/0
10/1   00/1   11/1

3
COMBINATIONAL SYSTEMS: DEFINITION AND ANALYSIS

In the previous chapter, we discussed how to represent information typically used on a daily basis in a form convenient for digital systems, namely the binary number system. The binary number system uses only two symbols, 0 and 1, which is compatible with digital electronic devices having only two operating states or conditions, "on" and "off." Having learned how to "speak the language" of digital systems, we are now ready to begin studying digital systems. A digital system is also called a *logic system* or a *switching system*. The term *logic system* is used because the mathematics that will be used to study digital systems, *switching algebra*, originated from the formal study of logic or human reasoning. The term *switching system* is used because early digital systems were constructed out of mechanical switching relays.

Using an *analysis/design* paradigm familiar to engineering, we will study digital systems by first defining the basic operating principles and illustrating the application of these principles by analyzing sample digital systems. Analyzing a sample digital system means finding out what the digital system does. We will develop a description of the input/output function and then use the description to understand the digital system's operation. Once we have mastered digital system *analysis*, we will proceed to the more challenging task of digital system *design* in Chapters 4 and 5.

Recall from Chapter 1 that electronic systems are broadly divided into two categories: analog and digital. Figure 3.1 shows that digital systems can be further divided into two subcategories: *combinational* and *sequential*.

In our study of digital systems we will concentrate first on combinational systems because sequential systems include and build on combinational systems. We will motivate the study of combinational digital systems by presenting an example combinational system that illustrates characteristic features. Then, we will examine *switching algebra*, which defines the fundamental rules (mathematics) for processing 1's and 0's. We will also learn techniques for analyzing combinational systems consisting of interconnected switching algebra operators. Finally, we will explain how to properly document combinational systems.

Figure 3.1

A taxonomy of digital systems

Electronic Systems

Analog Systems   Digital Systems

Combinational Systems   Sequential Systems

3.1 OVERVIEW OF COMBINATIONAL SYSTEMS

What is a combinational digital system? We will answer this question by looking at a typical application of a combinational system, given in Figure 3.2. Figure 3.2 shows how a combinational digital system could perform the task of controlling the ambient temperature of a building.

Figure 3.2(a) shows the inputs and outputs of the temperature controller. The inputs represent the sensed ambient temperature of the building encoded as a 3-bit Gray code, 000_2 (0_{10}) represents 0°C, 001_2 (1_{10}) represents 10°C, and so on. The three input signals representing the 3-bit Gray code are labeled *TEMP2*, *TEMP1*, and *TEMP0*. The output signal *HEAT* controls turning the building heating system on ("1") and off ("0"). Similarly, the output signal *COOL* controls turning the building cooling system on ("1") and off ("0"). Figure 3.2(b) defines how heating and cooling is controlled as a function of the sensed ambient temperature.

- If 30°C ≤ temperature ≤ 70°C, then turn the heat off and the cooling on.
- If temperature = 20°C, then turn both the heat and cooling off.
- If 0°C ≤ temperature ≤ 10°C, then turn the heat on and the cooling off.

This statement of the desired input/output function can be further refined by including the values of the individual signals. For example, the combinational system output in response to a temperature reading of 20°C can be restated as follows:

- If *TEMP2* = 0 and *TEMP1* = 1 and *TEMP0* = 1, then *HEAT* = 0 and *COOL* = 0.

Figure 3.2(c) shows the design of the temperature controller; the diagram is called a *logic schematic*. A detailed explanation of the combinational system design will be left for later discussions. For the present, note that a combinational digital system is composed of the interconnection of various symbols denoting *logic operators*. Each logic operator has a well-

Figure 3.2

A combinational digital system to control building temperature

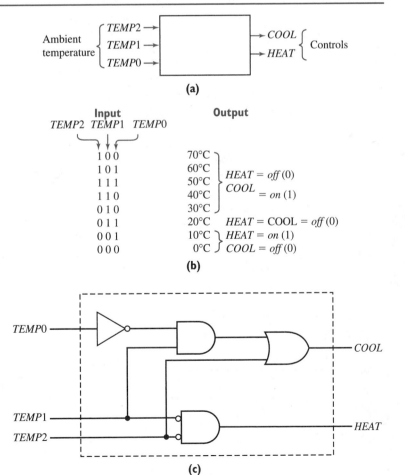

defined input/output transformation and collectively the logic operators generate the desired outputs in response to the inputs.[1]

A distinguishing characteristic of a combinational digital system is that at any given time the outputs depend only on the particular "combination" of values applied to the inputs, irrespective of any combination of inputs that might have been applied in the past. For the temperature controller example, the outputs *HEAT* and *COOL* are set to 0 whenever the input reads 20°C (*TEMP2* = 0, *TEMP1* = 1 , and *TEMP0* = 1), irrespective of any prior input temperature reading. Another way of stating this input/output characteristic is that a combinational digital system *has no memory*. Having no memory means that a combinational system has no way of storing information about previous inputs and using such information to affect the output. In contrast, we will find that a *sequential* digital system does have memory, so the output of a sequential system depends on both the present input and the *sequence* of past inputs.

3.2 SWITCHING ALGEBRA

The mathematics that define the rules for processing the binary values of 0 and 1 in a combinational digital system is called *switching algebra*, which is a subset of *Boolean algebra*. Boolean algebra is defined against a many-valued domain of {0, 1, 2, . . . , *N*}, whereas switching algebra is defined against a two-valued domain of {0, 1}. As we develop the theory of switching algebra, it is important not to get lost in the mathematics. Remember, switching algebra is simply a formal vehicle for (1) defining the basic building blocks of logic operations and (2) analyzing the behavior of a collection of these logic operations connected together to form a combinational digital system.

Boolean algebra was developed by an English mathematician named George Boole. In 1854, Boole published a treatise defining a mathematics or algebra for working with complex statements of propositional logic [1, 2]. The algebra provides a formal way to unambiguously describe logic statements and to determine the truth of such statements. Many years later, in 1938, a research assistant at Massachusetts Institute of Technology named Claude Shannon showed how Boole's algebra for logic can be applied to describe and analyze a network of switches comprising a combinational digital system [10].[2] It is worth emphasizing the importance of switching algebra as the foundation for digital system analysis and design. Without such a foundation, we would be left to manual, ad hoc techniques that would substantially limit our ability to construct complex digital systems.

3.2.1 Huntington's Postulates

You have probably been asked a series of questions by a child that get progressively more basic until you finally end up answering "Well, that is just how things are." Mathematics operates in a similar fashion: A series of questions that get progressively more basic can be asked about a mathematical theory until one arrives at a set of axioms or premises that are

Problem-Solving Skill

An important aspect in solving a problem is the ability to cast the problem in a mathematical framework. Such a framework provides a formal basis to define the problem and the set of operations that comprise the solution.

---

[1] The inquisitive reader should note that the combinational system design of the temperature controller can be further simplified if we are willing to make the assumption that the building temperature will never rise above 50°C (122°F). In this case, temperature readings of 60°C and 70°C will never happen and can be handled as "don't cares," which will be addressed in Chapter 4.

[2] Claude Shannon later developed many of the founding principles and concepts of information theory [11].

simply assumed to be true. Such axioms are called *postulates* or first-order principles that form a basic set of *laws* that define the mathematical theory. All other aspects of a mathematical theory are then derived from the fundamental postulates.

There are several possible sets of fundamental postulates that can be defined for switching algebra. A popular set of switching postulates proposed by E. Huntington in 1904 is given below [4, 5].

Huntington's Postulates for Switching Algebra

1. *Closure Properties*
   a. **Postulate 1a (P1a)**: *If* X *and* Y *are in the domain, that is, take on only the values {0, 1}, then* $(X + Y)$ *is also in the domain.*
   b. **Postulate 1b (P1b)**: *If* X *and* Y *are in the domain, that is, take on only the values {0, 1}, then* $(X \cdot Y)$ *is also in the domain.*
2. *Identity Properties*
   a. **Postulate 2a (P2a)**: $X + 0 = X$
   b. **Postulate 2b (P2b)**: $X \cdot 1 = X$
3. *Commutative Properties*
   a. **Postulate 3a (P3a)**: $X + Y = Y + X$
   b. **Postulate 3b (P3b)**: $X \cdot Y = Y \cdot X$
4. *Distributive Properties*
   a. **Postulate 4a (P4a)**: $X + (Y \cdot Z) = (X + Y) \cdot (X + Z)$
   b. **Postulate 4b (P4b)**: $X \cdot (Y + Z) = (X \cdot Y) + (X \cdot Z)$
5. *Complement Properties*
   a. **Postulate 5a (P5a)**: $X + \overline{X} = 1$
   b. **Postulate 5b (P5b)**: $X \cdot \overline{X} = 0$

These ten postulates define the basis for the entire theory of switching algebra! The symbols *X*, *Y*, and *Z* denote switching algebra variables that take on only the values 0 or 1. The three symbols—bullet "•", cross "+", and overbar "‾"—denote switching algebra operations, also called logic operations.

The arrangement of the Huntington's Postulates serves as an aide to understanding and learning them. The postulates are first grouped by *property*, such as closure, identity, commutative, et cetera. These properties are similar to the properties of the algebra of real numbers. For example, Postulates 1a and 1b state that the switching algebra operators "+" and "•" are *closed operators*. Closure means that the result of applying these operators to elements of the value set {0, 1} will always yield an element of the same value set {0, 1}. In other words, the domain *and* range of switching algebra is the set of values {0, 1}. Thus, $(X + Y)$ or $(X \cdot Y)$ will never yield a value like 3 or 506.

The switching algebra postulates are also grouped by *duality*. Notice the symmetry between the postulate pairs P1a/P1b, P2a/P2b, et cetera. The Pxb postulates can be generated from the associated Pxa postulates by (1) interchanging the 0's and 1's and (2) interchanging the + and • operations. This symmetrical property of switching algebra is called the *principle of duality*.

Postulates related by the principle of duality are called a *dual pair*. For instance, the Postulates P3a and P3b form a dual pair; Postulate P3a is called the *dual* of P3b and vice versa. The principle of duality is a useful and powerful property of switching algebra. Given any switching algebra expression, an alternate expression can be easily generated by invoking the principle of duality. We will see examples of exploiting this principle in later sections.

Principle of Duality

Switching algebra possesses a symmetry in that for any postulate or theorem, there is an associated (dual) postulate or theorem that is generated by performing the following actions:

- *interchanging 0's and 1's (0's are changed to 1's and 1's are changed to 0's) and*
- *interchanging + and • operations (+ operators are changed to • operators and • operators are changed to + operators).*

3.2.2 Switching Algebra Operations

The definition of the three logic operations (•, +, and ⁻) will be deduced from Huntington's Postulates and defined using *truth tables*. A truth table is a structured technique for systematically enumerating the output or result of a logic operation for every possible input. The term is borrowed from the early work on formalizing human logic or reasoning in which the validity (truth) of an argument set forth as a collection of propositions was analyzed via a truth table.

In addition to truth table definitions, commonly used *names* and *graphical symbols* will also be presented for each logic operation. The graphical symbols are grouped into two categories: *distinctive-shaped* and *rectangular-shaped*. The distinctive-shaped logic symbols, such as the symbols shown in Figure 3.2(c), have traditionally been more popular than the rectangular-shaped logic symbols. However, rectangular-shaped logic symbols are based on a more expressive, internationally agreed upon documentation standard that is growing in use [6]. Regardless of which convention is used, logic symbols provide an important capability for constructing a graphical representation of a combinational digital system to complement the mathematical representation provided by switching algebra.

Consider first the overbar ⁻ logic operation defined in Figure 3.3. The ⁻ operation is called the *complement*, *not*, or *invert* operation. We will generally use the terms **complement** and **not**, highlighted in bold type to denote their special meaning. The **not** operation is a *unary* logic operation, meaning that it takes one operand. The operand is the switching variable X and the result of applying the **not** operation to X is denoted as $\overline{X}$. The truth table in Figure 3.3(a) shows that the **not** operation changes a 0 to a 1 and a 1 to a 0. The input X is listed in the left column, and the output $\overline{X}$ is listed in the right column. We will adopt the convention of using a double vertical bar to separate truth table inputs from truth table outputs. Figure 3.3(b) and Figure 3.3(c) show that a small circle, commonly called a *bubble*, graphically denotes a **complement** operation.

Figure 3.4 defines the • logic operation and gives the associated graphical symbols. The • operation is called the *and* or *conjunction* operation; we will generally use the term **and**. The **and** operator is a binary logic operator, meaning that it takes two operands. The result of applying the **and** operator to the inputs X and Y is denoted as $X \cdot Y$. The truth table in Figure 3.4(a) has four entries because the two operands X and Y collectively have a total of $2^2 = 4$ possible values. Recall from Chapter 2 that a string of N bits has a total of 2^N possible unique values or bit combinations. The output of the **and** operator is 1 only when both input operands are equal to 1; that is, $X \cdot Y = 1$ only when $X = 1$ *and* $Y = 1$.

Finally, Figure 3.5 defines the + operation and gives the associated graphical symbols. The + operation is called the *or* or *disjunction* operation; we will generally use the term **or**. Like the **and** operator, the **or** operator is also a binary operator requiring two operands. The result of applying the **or** operator to the inputs X and Y is denoted as $X + Y$. Aptly named,

Figure 3.3

Complement switching algebra operation

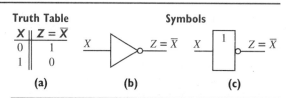

Truth Table	Symbols

X	Z = X̄
0	1
1	0

(a)   **(b)**   **(c)**

Figure (a) defines the **complement** operation via a truth table. Figures (b) and (c) give the associated graphical symbols.

Figure 3.4

Conjunction switching al-
gebra operation

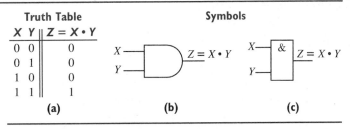

Truth Table		
X	Y	Z = X • Y
0	0	0
0	1	0
1	0	0
1	1	1

(a)   **(b)**   **(c)**

Figure (a) defines the **and** operation via a truth table. Figures (b) and
(c) give the associated graphical symbols.

Figure 3.5

Disjunction switching al-
gebra operation

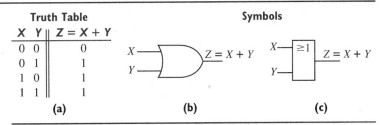

Truth Table		
X	Y	Z = X + Y
0	0	0
0	1	1
1	0	1
1	1	1

(a)   **(b)**   **(c)**

Figure (a) defines the **or** operation via a truth table. Figures (b) and (c) give
the associated graphical symbols.

the output of the **or** operator is 1 when either input operand is 1, including the case when
both inputs are 1; that is, $X + Y = 1$ when $X = 1$ *or* $Y = 1$ *or* $X = Y = 1$. To emphasize the
latter case, the **or** operator is also called the *inclusive or* operator.

3.3 ADDITIONAL LOGIC OPERATIONS

One might ask whether there are any other logic operators besides the one unary operator
(**not**) and two binary operators (**and** and **or**) discussed thus far. To answer this question, we
need to consider how many different operations can be defined against a set of independent
switching variables. In general, 2^{2^N} distinct operators can be defined against N independent
switching variables. Thus, there are $2^{2^1} = 4$ possible unary operations and $2^{2^2} = 16$ possible
binary operations.

3.3.1 Unary Operations

Figure 3.6 lists the four possible unary operations that can be defined for one switching variable. The first two truth tables in Figure 3.6(a) and Figure 3.6(b) define constant unary logic
operations that are rather uninteresting and not very useful for combinational systems. Regardless of the input value, the unary operation in Figure 3.6(a) always yields the value 0
and the unary operation in Figure 3.6(b) always yields the value 1.[3] Figure 3.6(c) defines the
not operation.

---

[3] To be fair, constant logic operations that yield either a "stuck-at-0" or "stuck-at-1" output regardless
of the input values are useful in defining fault models for testing combinational systems.

Figure 3.6

All possible unary switching algebra operations

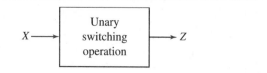

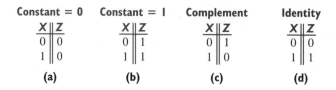

Constant = 0	Constant = I	Complement	Identity
X ‖ Z	X ‖ Z	X ‖ Z	X ‖ Z
0 ‖ 0	0 ‖ 1	0 ‖ 1	0 ‖ 0
1 ‖ 0	1 ‖ 1	1 ‖ 0	1 ‖ 1
(a)	(b)	(c)	(d)

The truth table in Figure 3.6(d) defines a unary logic operation called the *identity* or *buffer* operation. We will generally use the term **buffer**. Summarized in Figure 3.7, the **buffer** operation is straightforward; the output is the same as the input. Although the **buffer** operator does not seem to accomplish much, we will use this logic operator in later sections to restore signals and align circuit timings.

3.3.2 Binary Operations

Of the sixteen possible binary logic operators, only the following six operators are commonly used in combinational digital system design:

1. conjunction (**and**),
2. complement conjunction (**nand**),
3. disjunction (**or**),
4. complement disjunction (**nor**),
5. exclusive disjunction (**xor**), and
6. complement exclusive disjunction (**xnor**).

The **and** and **or** operations have already been defined; the remaining four logic operations (**nand**, **nor**, **xor**, and **xnor**) will be defined in the following paragraphs.

Figure 3.8 defines the **nand** logic operation (pronounced as it is spelled) and gives the associated graphical symbols. Comparing the **nand** definition in Figure 3.8(a) with the **and** definition in Figure 3.4(a) shows that the **nand** operation is the complement of the **and** operation, hence the name (**not and**). The output of the **nand** operator is 0 only when both input operands are equal to 1. Hence, the result of applying the **nand** operator to the inputs X and Y is denoted as $\overline{X \cdot Y}$. Figure 3.8(b) and Figure 3.8(c) show that the **nand** logic symbols are

Figure 3.7

Identity switching algebra operation

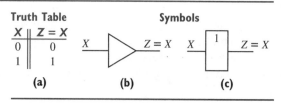

Truth Table		Symbols
X ‖ Z = X		
0 ‖ 0		
1 ‖ 1		
(a)	(b)	(c)

Figure (a) defines the **buffer** operation via a truth table. Figures (b) and (c) give the associated graphical symbols.

Figure 3.8

Complement conjunction switching algebra operation

Truth Table			Symbols

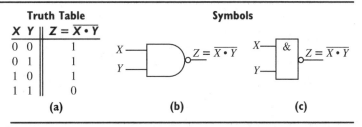

X	Y	Z = $\overline{X \cdot Y}$
0	0	1
0	1	1
1	0	1
1	1	0

(a)   (b)   (c)

Figure (a) defines the **nand** operation via a truth table. Figures (b) and (c) give the associated graphical symbols.

Figure 3.9

Complement disjunction switching algebra operation

Truth Table			Symbols

X	Y	Z = $\overline{X + Y}$
0	0	1
0	1	0
1	0	0
1	1	0

(a)   (b)   (c)

Figure (a) defines the **nor** operation via a truth table. Figures (b) and (c) give the associated graphical symbols.

derived from the **and** logic symbols by attaching bubbles to the outputs, which again illustrates the **and**/**nand** complement relationship.

Figure 3.9 defines the **nor** logic operation (pronounced as it is spelled) and gives the associated graphical symbols. Comparing the **nor** definition in Figure 3.9(a) with the **or** definition in Figure 3.5(a) shows that the **nor** operation is the complement of the **or** operation. The output of the **nor** operator is 0 when either input operand is 1, including when both inputs are 1. Hence, the result of applying the **nor** operator to the inputs X and Y is denoted as . The **nor** logic symbols are the **or** symbols with bubbles attached to the outputs.

Figure 3.10 defines the **xor** logic operation (pronounced as two words "*ex-or*") and gives the associated graphical symbols. The **xor** operation, also called the *exclusive-or* operation, is a variation of the **or** operation (which is also called the *inclusive-or* operation). Recall that the **or** operation yields a 1 when either input operand is 1 or when both inputs are 1; that is, its output is inclusive of the case when both inputs are 1. The truth table in Figure 3.10(a)

Figure 3.10

Exclusive disjunction switching algebra operation

Truth Table			Symbols

X	Y	Z = $X \oplus Y$
0	0	0
0	1	1
1	0	1
1	1	0

(a)   (b)   (c)

Figure (a) defines the **xor** operation via a truth table. Figures (b) and (c) give the associated graphical symbols.

Figure 3.11

Complement exclusive disjunction switching algebra operation

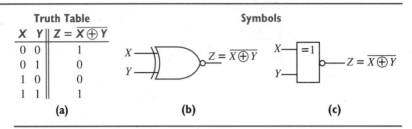

Truth Table		
X	Y	$Z = \overline{X \oplus Y}$
0	0	1
0	1	0
1	0	0
1	1	1

(a)   (b)   (c)

Figure (a) defines the **xnor** operation via a truth table. Figures (b) and (c) give the associated graphical symbols.

shows that the **xor** operation yields a 1 only when either input is 1; that is, its output is exclusive of the case when both inputs are 1. The result of applying the **xor** operator to the inputs X and Y is denoted as $X \oplus Y$.

Finally, Figure 3.11 defines the **xnor** logic operation (pronounced as two words "*exnor*") and gives the associated graphical symbols. The **xnor** operation is the complement of the **xor** operation. Hence, the **xnor** logic symbols are the **xor** logic symbols with bubbles attached to the outputs. The **xnor** operation is also called the *equivalence* operation because the output is 1 when both inputs are equal. The result of applying the **xnor** operator to the inputs X and Y is denoted as $\overline{X \oplus Y}$ or $X \odot Y$.

3.4 LOGIC MNEMONICS

Of the several popular mnemonics for representing and understanding the logic operations discussed in the previous sections, we will look at two: *switches* and *Venn Diagrams*. Readers with an engineering background may gravitate toward switches, whereas readers with a science background may gravitate toward Venn Diagrams.

Like a light switch in your home, a switch is either on or off. When a switch is on, its terminals are connected; when a switch is off, its terminals are disconnected. A switch can represent a switching algebra variable X using the following rules:

- $X = 0$ open switch, disable transmission
- $X = 1$ close switch, enable transmission

$X = 0$ $X = 1$

The **not** logic operation, $\overline{X}$, is easily denoted by inverting switch positions. The switch positions for X and $\overline{X}$ are mutually exclusive.

$\overline{X} = 0$ $\overline{X} = 1$

Binary logic operations can be represented by interconnecting switches. For example, Figure 3.12(a) shows that two switches in series represent the general **and**ing of two conditions denoted by the variables X and Y. These switches are used in Figure 3.12(b) to con-

Figure 3.12

Conjunction switching algebra operation

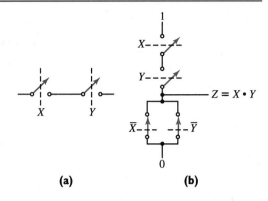

$$Z = X \cdot Y$$

(a)   **(b)**

struct an **and** operator. The bottom switches open and the top switches close to route a 1 to the output Z only when $X = 1$ and $Y = 1$.

As another example, Figure 3.13(a) shows that two switches in parallel represent the general **or**ing of two conditions denoted by the variables X and Y. These switches are used in Figure 3.13(b) to construct an **or** operator. The bottom switches open and the top switches close to route a 1 to the output Z only when $X = 1$ or $Y = 1$.

Venn Diagrams also graphically describe switching algebra operations. A Venn Diagram starts with a rectangle representing the two-valued switching algebra domain of $\{0, 1\}$. Then, circles are added to denote variables. For example, the circle below denotes the variable X; $X = 1$ inside the circle, and $X = 0$ outside the circle. These regions are reversed for .

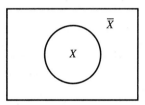

Figure 3.13

Disjunction switching algebra operation

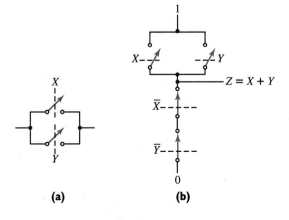

$$Z = X + Y$$

(a)   **(b)**

Figure 3.14

Conjunction and disjunction switching algebra operations

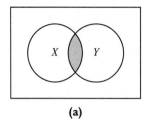

(a)

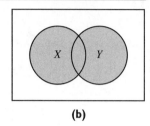

(b)

Figure 3.15

Exclusive disjunction switching algebra operations

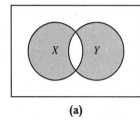

(a)

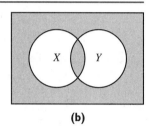

(b)

Figure 3.16

Logic operators

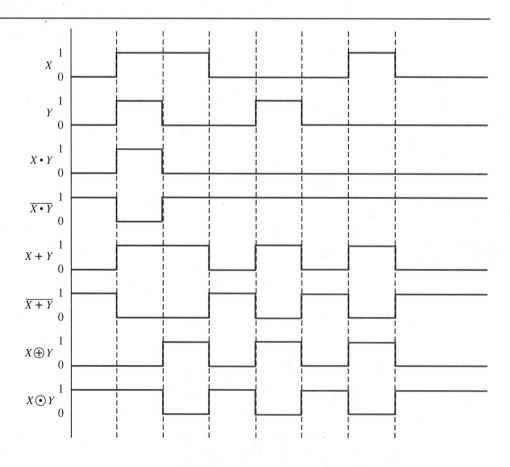

Binary logic operations can be represented by interconnecting circles and shading appropriate regions. For example, Figure 3.14(a) shows the Venn Diagram for the **and** operator; the shaded region denotes the condition that $X = 1$ and $Y = 1$. Figure 3.14(b) shows the Venn Diagram for the **or** operator; the shaded region denotes the condition that $X = 1$ or $Y = 1$.

As another example, Figure 3.15(a) shows the Venn Diagram for the **xor** operator; the shaded region denotes the condition that $X = 1$ or $Y = 1$, excluding the case that X and Y are both 1. The Venn Diagram for the **xnor** operator is shown in Figure 3.15(b)

To summarize our discussion thus far on switching algebra, Figure 3.16 presents a comparison of logic operators. Reference the respective truth tables to understand and verify the results.

3.5 MINIMAL SETS OF LOGIC OPERATORS

Having looked at the maximum number of unary/binary logic operators available, let us next look at the minimum number of logic operators needed to design a combinational digital system. The **and**, **or**, and **not** operators form a *minimal set* based on Huntington's Postulates. Two other minimal sets that are particularly interesting are the set composed of only the **nand** operator and the set composed of only the **nor** operator. These sets are interesting because using only one logic operator simplifies digital system design and manufacturing processes. We will investigate how digital systems are manufactured by realizing logic operators as an interconnection of electrical devices in Part 2.

To demonstrate that the **nand** operator forms a minimal and sufficient logic operator set, we show how the **nand** operator implements a variety of logic operations in Figure 3.17 (p. 52). We will examine Figure 3.17 in detail to introduce combinational system analysis concepts.

For example, consider the use of the **nand** operator to realize the **not** operator in Figure 3.17(a). To verify that connecting together the inputs of a **nand** operator forms an inverter, we need to prove that

$$\overline{X \bullet X} = \overline{X}$$

In other words, we need to prove the *equivalence* of the two *logic expressions* $\overline{X \bullet X}$ and $\overline{X}$. Two logic expressions are equivalent if they yield identical results for every possible combination of the input switching variables. In our example, the logic expressions $\overline{X \bullet X}$ and $\overline{X}$ are equivalent if they yield identical results for every possible value of the input switching variable X. One way of proving logic equivalence is by *complete enumeration* or *perfect induction*. Complete enumeration evaluates logic expressions for all possible value combinations of the input variables and tabulates the results in truth tables. If the truth tables yield identical outputs, the logic expressions are equivalent. The truth table for the **not** operator given in Figure 3.3(a) is repeated below.

X	$\overline{X}$
0	1
1	0

The truth table for the **nand** operator with connected inputs is given next.

X	$X \bullet X$	$\overline{(X \bullet X)} = \overline{X}$
0	0	1
1	1	0

Figure 3.17

A minimal logic operator set

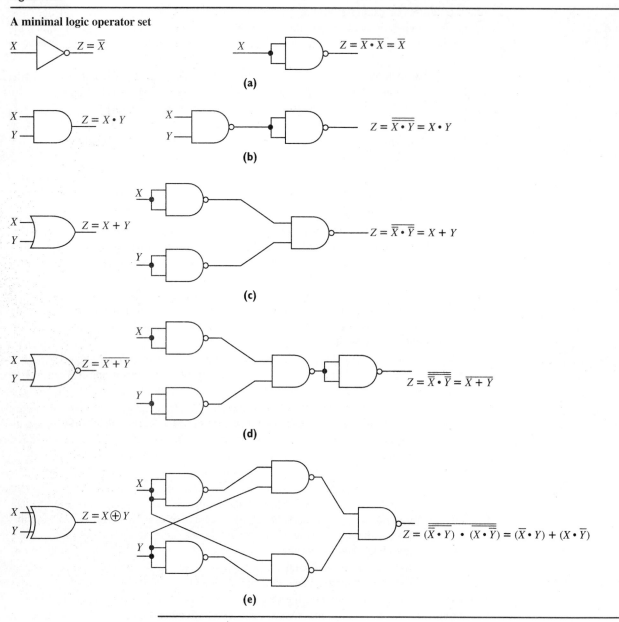

Figures (a)–(e) show how the **nand** logic operator respectively implements the **not, and, or, nor,** and **xor** logic operations.

The truth table is constructed in an incremental, two-step fashion by first constructing an entry for $X \cdot X$ and then an entry for $\overline{X \cdot X}$. Since the outputs of the two truth tables are identical, $\overline{X \cdot X} = \overline{X}$ and a **nand** operator with connected inputs indeed implements a **not** operator.

3.6 MULTI-INPUT LOGIC OPERATORS

As a final topic in our discussion of the fundamentals of switching algebra, we will consider multi-input logic operators. What is the output of a 3-input **and** operator, a 4-input **or** operator, or an n-input **nand** operator?

Figure 3.18 shows that the output of a multi-input logic operator is computed in a pairwise fashion by repeatedly using the definition of the associated binary-input logic operator. For example, Figure 3.18(a) shows that a 3-input **and** operator ($A \cdot B \cdot C$) is evaluated by first calculating the conjunction of the switching variables A and B to form ($A \cdot B$). Then this intermediate calculation is **and**ed with C to yield the output. The binary **and** logic operator extends in a straightforward fashion to handle multi-inputs; the output is 1 only when all inputs are 1. A similar analysis is shown in Figure 3.18(b) for a 4-input **or** operator; the output is 1 when any input is 1.

3.7 COMBINATIONAL SYSTEM ANALYSIS

Before we begin studying analysis techniques, let us summarize the salient aspects of combinational systems discussed thus far. Combinational designs are typically presented in a graphical form via *logic schematics*, also called *logic diagrams*. A logic schematic shows the system inputs, outputs, and how the inputs are processed by a set of logic operators to yield the outputs. Viewed from a mathematical perspective, the system inputs and outputs are switching algebra variables and the logic operators form a *switching algebra expression* that formally defines the relationships among the input and output variables. An important prop-

Figure 3.18

Multi-input logic operators

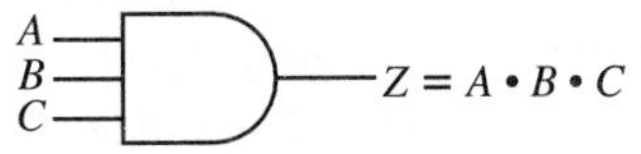

A B C	A $\cdot$ B	(A $\cdot$ B) $\cdot$ C
0 0 0	0	0
0 0 1	0	0
0 1 0	0	0
0 1 1	0	0
1 0 0	0	0
1 0 1	0	0
1 1 0	1	0
1 1 1	1	1

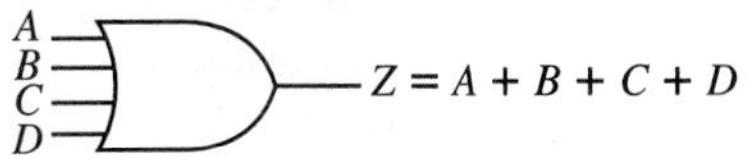

A B C D	(A + B)	(C + D)	(A + B) + (C + D)
0 0 0 0	0	0	0
0 0 0 1	0	1	1
0 0 1 0	0	1	1
0 0 1 1	0	1	1
0 1 0 0	1	0	1
0 1 0 1	1	1	1
0 1 1 0	1	1	1
0 1 1 1	1	1	1
1 0 0 0	1	0	1
1 0 0 1	1	1	1
1 0 1 0	1	1	1
1 0 1 1	1	1	1
1 1 0 0	1	0	1
1 1 0 1	1	1	1
1 1 1 0	1	1	1
1 1 1 1	1	1	1

(a)   (b)

Figure (a) presents the truth table for a 3-input **and** operator. Figure (b) presents the truth table for a 4-input **or** operator.

erty of a combinational digital system is that it exhibits no memory. At any given time, the outputs depend only on the particular combination of values applied to the inputs.

Let us now use our understanding of switching algebra fundamentals to analyze combinational systems. We will discuss two combinational digital system analysis techniques: *literal analysis* and *symbolic analysis*. Literal analysis generates an output *value* in response to a specific set of input *values*. Symbolic analysis is a more general technique that generates an output *logic expression* in response to a specific set of input *variables*.

3.7.1. Literal Analysis Technique

The term *literal* means actual value; 0 and 1 are actual binary values or literals. In combinational systems, literal analysis applies a set of literal or actual values to the inputs and propagates these values through the interconnection of logic operators to yield the output value(s). The output response of a logic operator is generated based on the truth table definitions given in previous sections. For example, what does the combinational system given in Figure 3.19 do? Figure 3.20 shows how to perform literal analysis to determine the combinational system output (Z) for the following three sample sets of inputs.

1. $A = 0, B = 1$, and $C = 0$
2. $A = 0, B = 1$, and $C = 1$
3. $A = 1, B = 1$, and $C = 1$

We use the truth table given in Figure 3.4(a) to determine the outputs of the **and** operators, and we use the truth table in Figure 3.5(a) and the explanation of multi-input logic operators to determine the output of the **or** operators. For instance, consider the input of all 1's ($A = 1, B = 1$, and $C = 1$). The three **and** operators yield three 1's because $1 \cdot 1 = 1$. The **or** operator yields the output $Z = 1$ because $1 + 1 + 1 = 1$. A thoughtful investigation of a few more input/output transforms reveals that the combinational system performs the majority function; the output is 1 whenever the majority (two or more) of inputs are 1.

Since propagating many sets of inputs through many logic operations is a laborious and possibly error-prone task, literal analysis is principally used to study the behavior of simple combinational systems in response to a small number of input stimuli. For more challenging analysis tasks, we will use symbolic analysis.

3.7.2 Symbolic Analysis Technique

Instead of propagating literal values through a combinational system, symbolic analysis propagates logic expressions through a combinational system. The resulting output logic expression defines the input/output behavior of the combinational digital system for *all* input value combinations.

Figure 3.21 shows how to perform symbolic analysis for the example combinational sys-

Figure 3.19

Example combinational system design

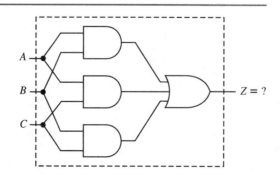

$Z = ?$

Figure 3.20

Literal analysis of the example combinational system design

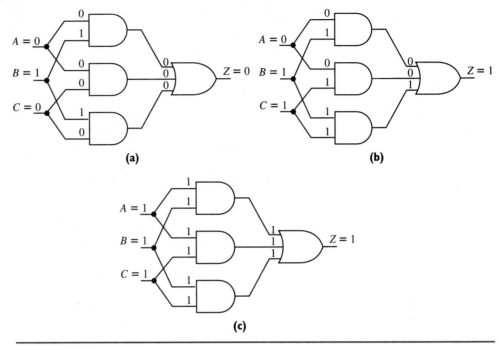

(a)

(b)

(c)

Figure (a) generates the output value in response to $A = 0$, $B = 1$, and $C = 0$. Figure (b) generates the output value in response to $A = 0$, $B = 1$, and $C = 1$. Figure (c) generates the output value in response to $A = 1$, $B = 1$, and $C = 1$.

Figure 3.21

Symbolic analysis of the example combinational system design

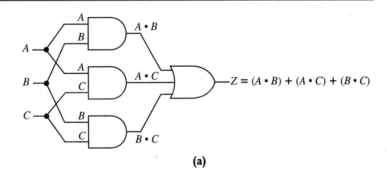

(a)

A B C	$(A \cdot B)$	$(A \cdot C)$	$(B \cdot C)$	$(A \cdot B) + (A \cdot C) + (B \cdot C)$
0 0 0	0	0	0	0
0 0 1	0	0	0	0
0 1 0	0	0	0	0
0 1 1	0	0	1	1
1 0 0	0	0	0	0
1 0 1	0	1	0	1
1 1 0	1	0	0	1
1 1 1	1	1	1	1

(b)

tem of Figure 3.19, which realizes the majority function. The input/output switching algebra expression is generated by tracing the flow of the input variables (A, B, and C) through the various logic operators. Moving from left to right, the three **and** operators yield the logic expressions $A \cdot B$, $A \cdot C$, and $B \cdot C$. These logic expressions serve as inputs to the **or** operator, which yields the output logic expression

$$Z = (A \cdot B) + (A \cdot C) + (B \cdot C)$$

The combinational system's response to a specific set of input values can be studied by assigning literal values to the input switching variables and evaluating the logic expression. The truth table shown in Figure 3.21(b) lists the results of evaluating the output logic expression for the eight possible combinations of the three input variables and verifies that the combinational digital system performs the majority function.

As another exercise, Figure 3.22 illustrates how to perform symbolic analysis to un-

Figure 3.22

Symbolic analysis of a modulo-8 incrementor

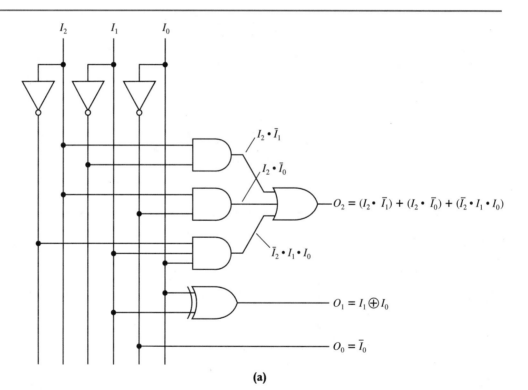

(a)

I_2 I_1 I_0	O_2	O_1	O_0
0 0 0	0	0	1
0 0 1	0	1	0
0 1 0	0	1	1
0 1 1	1	0	0
1 0 0	1	0	1
1 0 1	1	1	0
1 1 0	1	1	1
1 1 1	0	0	0

(b)

derstand the operation of a combinational system that realizes a modulo-8 increment function.

The combinational system in Figure 3.22(a) has multiple outputs. Each output logic expression is generated by propagating the input variables through the stages of logic operators. For the output O_2, the outputs of the three **and** operators, $(I_2 \cdot \overline{I_1})$, $(I_2 \cdot \overline{I_0})$, and $(\overline{I_2} \cdot I_1 \cdot I_0)$, drive the **or** operator to yield the logic expression

$$O_2 = (I_2 \cdot \overline{I_1}) + (I_2 \cdot \overline{I_0}) + (\overline{I_2} \cdot I_1 \cdot I_0)$$

For the output O_1, the **xor** operator yields the logic expression

$$O_1 = I_1 \oplus I_0$$

Finally, the output O_0 is simply $\overline{I_0}$.

The truth table shown in Figure 3.22(b) lists the results of evaluating the output logic expressions for the eight possible combinations of the three input variables. The input $I_2 I_1 I_0$ is a 3-bit binary number in the range 0 to 7. The output $O_2 O_1 O_0$ is also a 3-bit binary number in the range 0 to 7, formed by incrementing the input by 1—that is, $O_2 O_1 O_0 = I_2 I_1 I_0 + 1$.[4] The increment function wraps around when incrementing 7, making the operation modulo-8.

As a third exercise, Figure 3.23 illustrates how to perform symbolic analysis to understand the operation of a combinational system that detects even Gray codes. Again, moving from left to right, the input switching variables are propagated through the **not** operators and

Figure 3.23

Symbolic analysis of logic to detect even Gray codes

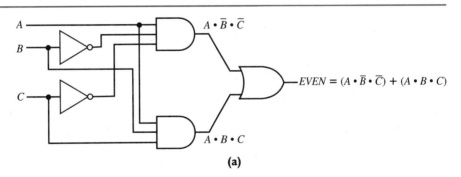

(a)

A B C	$(A \cdot \overline{B} \cdot \overline{C})$	$(A \cdot B \cdot C)$	$(A \cdot \overline{B} \cdot \overline{C}) + (A \cdot B \cdot C)$
0 0 0	0	0	0
0 0 1	0	0	0
0 1 0	0	0	0
0 1 1	0	0	0
1 0 0	1	0	1
1 0 1	0	0	0
1 1 0	0	0	0
1 1 1	0	1	1

(b)

Inputs signals B and C represent the numbers $0-3_{10}$, encoded using a 2-bit Gray code. Input signal A is a control signal.

[4] Somewhat like the ubiquitous $++$ operator in the programming language C.

the **and** operators to form the logic expressions $(A \cdot \overline{B} \cdot \overline{C})$ and $(A \cdot B \cdot C)$. These expressions are then combined via the **or** operator to yield the output expression.

$$EVEN = (A \cdot \overline{B} \cdot \overline{C}) + (A \cdot B \cdot C)$$

The truth table in Figure 3.23(b) shows that when $A = 0$, the combinational system is effectively shut off; the output $EVEN = 0$ regardless of the values of B or C. When $A = 1$, the output $EVEN = 1$ when the signals B and C encode an even Gray code number and $EVEN = 0$ when the signals B and C encode an odd Gray code number. For example, consider the last entry in the truth table with all inputs equal to 1. The output $EVEN = 1$ because $BC = 11$ is the two-bit Gray code for the even number 2_{10}.

Figure 3.23 illustrates an important aspect of digital systems: Input signals can generally be categorized as conveying either *data* or *control* information. For the even function, the signals B and C can be characterized as *data inputs* because they carry the Gray code numerical information, whereas signal A can be characterized as a *control input* because it effectively turns the even function on and off. The output $EVEN$ function also illustrates the *canonical sum* form of a logic expression, which will be discussed in the next section.

3.8 LOGIC EXPRESSIONS

Symbolic analysis emphasizes the central importance of *logic expressions*.

> Logic expressions define the behavior—the input/output transfer function—of combinational digital systems.

As we examine logic expressions in more detail, we will study rules for evaluating and representing logic expressions. To start, definitions of terms used with logic expressions are summarized below.

Term	Definition
Logic Expression	A mathematical formula consisting of logic operators and variables
Logic Operator	A function that, given a set of inputs, generates an output according to well-defined rules that are consistent with Huntington's Postulates defining switching algebra.
Logic Variable	A symbol representing the two possible switching algebra values of 0 and 1.
Logic Literal	The values 0 and 1.
	or
	A logic variable or the complement of a logic variable.

The multiple definitions for a logic literal deserve an explanation. This somewhat confusing situation exists because switching algebra and VHDL both use the term *literal*, but in different ways. VHDL uses the term *literal* to mean an actual value, whereas switching algebra uses the term *literal* to mean the appearance of a logic variable or its complement within a logic expression. To establish continuity between VHDL and switching algebra, we will tend to favor the former definition, as we have done thus far. We will always try to state context and meaning when using the term *literal* to avoid any potential confusion.

3.8.1 Operator Precedence

Operator precedence defines the order in which operators are evaluated in a logic expression. For example, what is the value of the logic expression at the top of p. 59? If the logic expression is evaluated left-to-right by evaluating the **and** operator $(0 \cdot 1 = 0)$ and then the

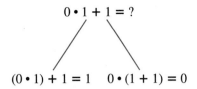

or operator $(0 + 1 = 1)$, the result is 1. Conversely, if the logic expression is evaluated right-to-left by evaluating the **or** operator $(1 + 1 = 1)$ and then the **and** operator $(0 \cdot 1 = 0)$, the result is 0. To avoid this type of ambiguity, we need to establish rules for evaluating logic expressions.

By convention, the **not** operator has higher precedence than the **and** operator, which, in turn, has higher precedence than the **or** operator. In other words, the **not** operator is applied before the **and** operator and the **and** operator is applied before the **or** operator. Hence, the correct evaluation of the logic expression $0 \cdot 1 + 1$ is left-to-right, yielding the value 1. Parentheses can be used to explicitly set the order of operator evaluation; expression evaluation starts with the innermost set of parentheses and expands outward. For example, the parentheses in the expression $0 \cdot (1 + 1) = 0$ explicitly force the **or** operator to be evaluated first. It is a good engineering practice to generously use parentheses in logic expressions whenever there might be confusion about how an expression should be correctly evaluated.

3.8.2 Canonical Forms

The term *canonical* means orthodox, adhering to the accepted or traditional. Logic expressions are so commonly used in digital design that conventions have been developed for writing logic expressions in standard forms; these standard forms are called *canonical forms*. There are two popular canonical forms for logic expressions: *canonical sum form* and *canonical product form*. Figure 3.24 shows the canonical forms for the majority function given in Figure 3.19.

The canonical sum form shown in Figure 3.24(a), also called the *disjunctive normal form* or *sum-of-products (SOP) form*, is composed of the *sum* of minterms. The term *sum* implies a logical sum and is used because the properties of the **or** operation for switching algebra are analogous to the properties of the addition or sum operation for real-number algebra. Each minterm is, in turn, composed of the *product* of the input variables or the complement of the input variables, that is, the literals. The term *product* implies a logical product and is used because the properties of the **and** operation for switching algebra are analogous to the properties of the multiplication or product operation for real-number algebra.

Figure 3.24

Standard canonical forms for logic expressions

Canonical sum form

$$MAJORITY = (\bar{A} \cdot B \cdot C) + (A \cdot \bar{B} \cdot C) + (A \cdot B \cdot \bar{C}) + (A \cdot B \cdot C)$$

Minterms

(a)

Canonical product form

$$MAJORITY = (A + B + C) \cdot (A + B + \bar{C}) \cdot (A + \bar{B} + C) \cdot (\bar{A} + B + C)$$

Maxterms

(b)

The basic idea behind the canonical sum form is that one of the minterms evaluates to 1 for each combination of inputs for which the output is 1. Minterms are easily generated from a truth table. Each input combination that yields a 1 output generates a minterm; the switching variables equal to 0 appear *complemented* in the associated minterm and the switching variables equal to 1 appear *uncomplemented* in the associated minterm. The truth table and minterms for the majority function are shown below.

Row #	A B C	MAJORITY	MINTERM	Minterm shorthand notation
0	0 0 0	0		
1	0 0 1	0		
2	0 1 0	0		
3	0 1 1	1	$\overline{A} \cdot B \cdot C$	m_3
4	1 0 0	0		
5	1 0 1	1	$A \cdot \overline{B} \cdot C$	m_5
6	1 1 0	1	$A \cdot B \cdot \overline{C}$	m_6
7	1 1 1	1	$A \cdot B \cdot C$	m_7

There is a minterm for every combination of inputs for which the majority function is a 1, namely the rows 3, 5, 6, and 7. The rightmost column shows a commonly used shorthand notation for minterms, m_x. The subscript x denotes the associated row number or, equivalently, the straight binary value of the input variable combination. This shorthand minterm notation is used in a shorthand canonical sum form notation called the *minterm list*. The minterm list for the majority function is shown here.

$$MAJORITY = m_3 + m_5 + m_6 + m_7$$
$$= \sum m(3, 5, 6, 7) \tag{3.1}$$

The canonical product form mirrors the canonical sum form due to the symmetrical properties of the principle of duality of switching algebra. The canonical product form shown in Figure 3.24(b), also called the *conjunctive normal form* or *product-of-sums (POS) form*, is composed of the product of *maxterms*. Each maxterm is, in turn, composed of the sum of the input variables or the complement of the input variables, that is, the literals.

The basic idea behind the canonical product form is that one of the maxterms evaluates to 0 for each combination of inputs for which the output is 0. Like minterms, maxterms are easily generated from a truth table. Each input combination that yields a 0 output generates a maxterm; the switching variables equal to 0 appear *uncomplemented* in the associated maxterm and the switching variables equal to 1 appear *complemented* in the associated maxterm. The truth table and maxterms for the majority function are shown below.

Row #	A B C	MAJORITY	MAXTERM	Maxterm shorthand notation
0	0 0 0	0	$A + B + C$	M_0
1	0 0 1	0	$A + B + \overline{C}$	M_1
2	0 1 0	0	$A + \overline{B} + C$	M_2
3	0 1 1	1		
4	1 0 0	0	$\overline{A} + B + C$	M_4
5	1 0 1	1		
6	1 1 0	1		
7	1 1 1	1		

There is a maxterm for every combination of inputs for which the majority function is a 0, namely the rows 0, 1, 2, and 4. Again, the rightmost column shows a commonly used shorthand notation for maxterms, M_x. The subscript x denotes the associated row number or, equivalently, the straight binary value of the input variable combination. This shorthand maxterm notation is used in a shorthand canonical product form notation called the *maxterm list*. The maxterm list for the majority function is shown here.

$$MAJORITY = M_0 \bullet M_1 \bullet M_2 \bullet M_4$$
$$= \prod M(0, 1, 2, 4) \tag{3.2}$$

It is helpful to note a few useful properties of the canonical sum and product forms. First, each minterm and maxterm should contain **all** input variables in either a complemented or uncomplemented form. For instance, $(\overline{A} \bullet B)$ is not a legal minterm for a combinational system with three inputs A, B, and C because the minterm does not contain all three input variables. Hence, a logic expression that is a sum of products is not necessarily a canonical sum and, similarly, a logic expression that is a product of sums is not necessarily a canonical product. Another property of canonical forms is that the minterm numbers (indices of the minterm list) and maxterm numbers (indices of the maxterm list) are mutually exclusive, and, collectively, account for all rows of the defining truth table.

3.9 DOCUMENTING COMBINATIONAL SYSTEMS

Combining the topics of digital design and the hardware description language VHDL underscores the point that properly documenting and conveying information about a digital system is as important as the process of designing the system. A digital system that is inadequately or incorrectly documented is of no use to anyone, regardless of the quality of the design.

Hence, this section will discuss several aspects of properly documenting a combinational logic design. The following documentation aspects are suggested *guidelines*, not inviolable *requirements* of all logic diagrams. A designer is free to devise whatever documentation yields the most straightforward description of a digital system.

3.9.1. Information Flow

Figure 3.25 shows that it is often a good practice to arrange logic diagrams so data flows left-to-right and control flows top-to-bottom. Differentiating data and control flow makes the design easier to understand.

For example, reconsider the even function in Figure 3.23(a). Using Huntington's Postulate 4b, the logic expression for *EVEN* can be rewritten from $EVEN = (A \bullet \overline{B} \bullet \overline{C}) + (A \bullet B \bullet C)$ to $EVEN = A \bullet ((\overline{B} \bullet \overline{C}) + (B \bullet C))$. The even function can now

Figure 3.25

Data and control flow for standard logic diagrams

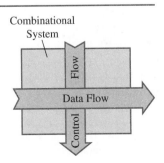

Figure 3.26

Redrawing the *EVEN* function to emphasize data and control

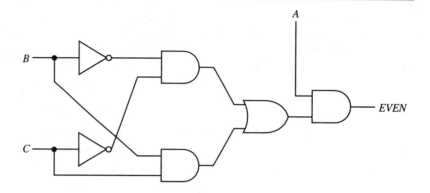

be redrawn to emphasize the data/control flow, as shown in Figure 3.26. The logic diagram here more clearly conveys the gating control function of signal *A*.

3.9.2 Connections

There are two popular conventions for denoting connected signal wires. One convention is that only signals that meet at a tee (⊤) are connected; signals that intersect in any manner other than a tee are not connected. Another convention is that a solid dot is placed over the intersection of signals that are connected. We will use the latter convention of denoting connectivity, as shown in Figure 3.26.

It is common practice to keep the size of a logic operator the same, regardless of the number of input connections. Logical operators with several inputs are drawn as follows:

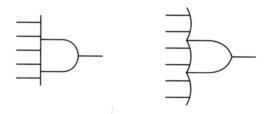

3.9.3 Active States and Bubbles

The temperature monitor design given at the outset of this chapter (Figure 3.2(c)) uses a logic operator that has not yet been discussed, namely, the logic operator generating the output signal *HEAT*. To understand the temperature monitor design and to complete our discussion of combinational system analysis, we need to understand the concept of *active states* and how bubbles are used to denote active states.

The term *state* refers to one of the logic states 0 or 1.[5] For a logic signal representing some condition or controlling some action, we need to know which logic state denotes the presence of the condition or the enabling of the action. For instance, consider a combinational system that controls an alarm; does a 0 output mean the alarm is enabled or does a 1 output mean the alarm is enabled? As the name implies, the term *active state* refers to a state

---

[5] The term *level* refers to one of the voltage levels "HIGH" or "LOW," which will be discussed as an implementation issue concerning mapping logic operators to physical gates.

that denotes the assertion of some condition or action, also called an *excitation state*. The term *active-0* means a 0 denotes the excitation state, whereas the term *active-1* means a 1 denotes the excitation state. Thus, if a 0 output means the alarm is enabled, the combinational system has an active-0 output; otherwise, a 1 means the alarm is enabled and the combinational system has an active-1 output. It is common practice to denote active states by the presence or absence of the logic negation symbol "○" (*bubble*). A bubble denotes an *active-0* state and the absence of a bubble denotes an *active-1* state. A logic signal is *asserted* when it takes on its active state.

Let us illustrate these concepts by taking a second look at the **not** operator shown below.

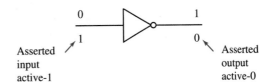

The bubble denotes a **not** or **complement** operation: a 0 input yields a 1 output and a 1 input yields a 0 output. The presence of a bubble on the output also denotes an active-0 output, which means that the condition or action represented by the inverter output is asserted when the output is 0. Active states also apply to inputs. The output is asserted active-0 when the input is asserted active-1, denoted by the absence of a bubble on the input.

Consider an alternate description of the **not** operator given next. Moving the bubble from

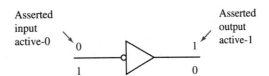

the output to the input does not affect the logic function; a 0 input yields a 1 output and a 1 input yields a 0 output. However, moving the bubble from the output to the input does affect the active state designation. The absence of a bubble on the output denotes an active-1 output, which means that the condition or action represented by the inverter output is asserted when the output is 1. The output is asserted active-1 when the input is asserted active-0, denoted by the presence of a bubble on the input.

It is important to recognize in this simple example that designating input/output active states does *not* change the input/output logic function, in other words, the input/output truth table. Rather, designating input/output active states augments the **not** logic function with additional information conveying the *intent* of the design by identifying input/output excitation states, in other words, the conditions under which the inputs and outputs are asserted.

3.9.4 Logic Operators and Diagrams

The active state notation can be exploited to develop an alternate set of logic operator symbols that can often be used to describe more clearly and concisely the behavior of combinational systems. The logic symbols shown in Figure 3.27 graphically illustrate the positive and negative alternate ways of stating switching algebra operations. The logic symbols in the left-hand column have active-0 outputs, while the logic symbols in the right-hand column have active-1 outputs. The standard set of logic symbols is shown in Figure 3.27(a.1), Figure

Figure 3.27

Alternate sets of symbols for logic operators

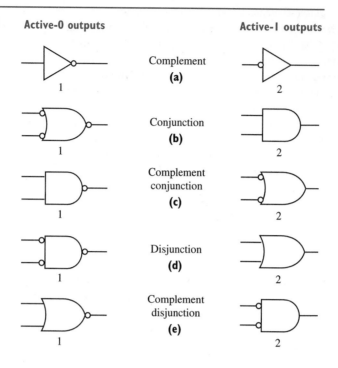

	Active-0 outputs		Active-1 outputs

Complement
(a)

Conjunction
(b)

Complement
conjunction
(c)

Disjunction
(d)

Complement
disjunction
(e)

3.27 (b.2), Figure 3.27(c.1), Figure 3.27 (d.2), and Figure 3.27(e.1); the remaining logic symbols form an alternate set.

The logic operators in Figure 3.27 illustrate that for every positive way of stating a logical proposition, there is an equivalent negative way of stating the same proposition. For example, consider the following logical proposition:

Digital combinational systems are successful when properly designed *and* documented.

An equivalent way of stating the same proposition in a negative way is as follows:

Digital combinational systems are *not* successful when *not* properly designed *or not* properly documented.

Applying these ideas to the **and** operation, one way to describe the **and** operation is that the output is asserted active-1 (takes on the value 1) when input X and input Y are asserted active-1 (take on the value 1). This is the conventional definition represented by the symbol in Figure 3.27(b.2) and the following truth table.

X Y	$X \cdot Y$
0 0	0
0 1	0
1 0	0
1 1	1

An equivalent, negative way of describing the **and** operation is that the output is asserted active-0 (takes on the value 0) when input X or input Y is asserted active-0 (takes on the value 0). This alternate definition is represented by the symbol shown in Figure 3.27(b.1) and the following truth table.

X Y	$\overline{X}$ $\overline{Y}$	$\overline{(X + Y)} = (\overline{X} \cdot \overline{Y})$
0 0	1 1	0
0 1	1 0	0
1 0	0 1	0
1 1	0 0	1

It is worth emphasizing that symbol pairs in Figure 3.27(a), Figure 3.27(b), et cetera, describe the same logic function from different perspectives. The symbols in the left-hand column view logic operations from an active-0 output perspective, whereas the symbols in the right-hand column view logic operations from an active-1 output perspective. A shorthand method for easily generating the logic symbol pairs is to first change all **and** operators to **or** operators and vice versa. Then, add a bubble to each input/output line that does not have a bubble and remove the bubble from each input/output line that does have a bubble. You are encouraged to study the logic symbols in Figure 3.27 to understand the alternate ways of describing logic operators.

We can use the alternate ways of describing logic operators to improve the documentation of combinational systems. For example, reconsider the logic diagram in Figure 3.17(e), which shows how the **nand** operator realizes the **xor** operator. Using the complete enumeration and symbolic analysis techniques discussed earlier, the following truth table verifies the combinational logic design, but the simple **xor** behavior is not obvious from the diagram.

X Y	$(\overline{X} \cdot Y)$	$(X \cdot \overline{Y})$	$\overline{(\overline{X} \cdot Y) \cdot (X \cdot \overline{Y})} = X \oplus Y$
0 0	1	1	0
0 1	0	1	1
1 0	1	0	1
1 1	1	1	0

A better choice of logic symbols is shown in Figure 3.28. The bubbles at the output of the second stage (**nand** operators labeled ③ and ④) cancel the bubbles at the input of the final stage (**nand** operator labeled ⑤) and can thus be ignored. Remembering that the **nand** operators labeled ① and ② with connected inputs implement **not** operators, we can read the output logic expression directly from the logic diagram as the **or** of $(\overline{X} \cdot Y)$ and $(X \cdot \overline{Y})$,

Figure 3.28

Improved logic diagram for exclusive disjunction design

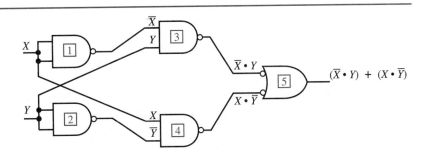

which is the **xor** function. Symbolic analysis is straightforward because it is immediately clear that the final logic operator is performing an **or** operation. Note that the behavior of the combinational system has not changed; we have only changed how the system is represented.

There are a few guidelines in selecting which symbol to use to represent a logic operator. First, logic operator symbols should reflect input/output active states. Active-0 inputs and outputs should have bubbles; active-1 input and outputs should not have bubbles. For example, the temperature monitor in Figure 3.2(c) enables cooling and heating when the respective outputs take on the value of 1. Consequently, symbols for the two logic operators driving the outputs should have active-1 outputs. Also, active states should be matched when connecting outputs to inputs. That is, bubbles should be connected to bubbles to cancel out complement operations.

3.9.5 Signal Names

We will conclude our discussion of documenting digital systems by studying common signal naming practices. First, we try to give a signal a name that reflects the information the signal carries. For example, a signal that controls a computer might be named INTERRUPT, or a signal that encodes voice might be named AUDIO.

A signal name should also convey active state information, in other words, whether the signal is active-0 or active-1. There are several common conventions for denoting active state information, such as adding a bar $^-$ over a signal name or a backslash / to the beginning of a signal name to denote active-0 excitation. Adding a bar over a signal name can be confusing because the overbar also denotes the **not** operation. Does $\overline{READY}$ denote an active-0 signal name or a logic expression of the **not** of a signal named $READY$? Mixing operator symbols with operand names is never a good practice. For example, does "A + B" denote an awkward variable name or the arithmetic expression of the addition of A and B?

Adding a backslash to the beginning of a signal name is easy and unambiguous, but is unfortunately not consistent with the signal naming practices in VHDL. Thus, we will adopt the practice of assuming a default case that a signal is active-1. If a signal is active-0, we will append a "_BAR" to the end of the signal name. This convention is consistent with VHDL signal naming practices and has the nice mnemonic property that $\overline{SIGNAL\_BAR} = SIGNAL$, which we will exploit in analyzing combinational systems.

To summarize active logic state notations and associated signal naming conventions, we will consider a simple combinational system monitoring part alignment in a manufacturing fixture. To be correctly aligned for manufacturing, the part in Figure 3.29 must touch the back alignment guide and at least one of the side alignment guides (left or right). The monitor receives from each alignment guide a status signal that is asserted if there is part contact. The monitor asserts its output signal when the part is properly aligned in the manufacturing fixture. Figure 3.29 shows four possible definitions of the monitor combinational system using various combinations of input and output active states. Examine the truth tables and logic diagrams to understand how active states relate to formulating logic expressions and selecting logic operator symbols.

3.10 COMBINATIONAL SYSTEM ANALYSIS: A SECOND LOOK

We will conclude this chapter by analyzing the temperature monitor combinational system presented at the outset (Figure 3.2) and repeated here (Figure 3.30). According to our signal naming convention, the inputs and outputs are by default active-1 signals.

Figure 3.29

A manufacturing alignment monitor

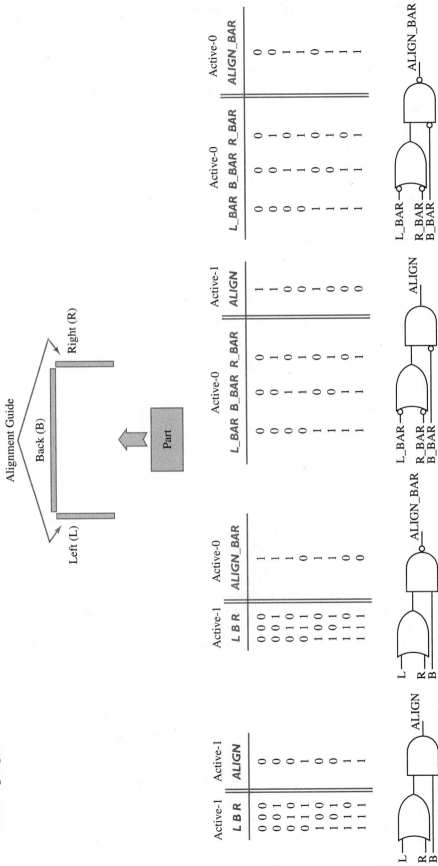

Figure 3.30

A temperature monitor

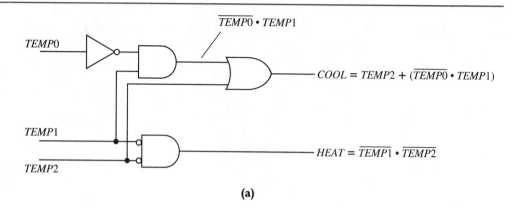

(a)

Degrees	TEMP2	TEMP1	TEMP0	COOL	HEAT
0°C	0	0	0	0	1
10°C	0	0	1	0	1
30°C	0	1	0	1	0
20°C	0	1	1	0	0
70°C	1	0	0	1	0
60°C	1	0	1	1	0
40°C	1	1	0	1	0
50°C	1	1	1	1	0

(b)

Figure 3.30(a) shows the formal symbolic analysis. An alternate way to analyze the behavior of the combinational system is to simply read the input/output transform directly from the logic diagram.

- The output *COOL* is asserted (*COOL* = 1) when *TEMP2* is asserted (*TEMP2* = 1) or when *TEMP1* is asserted (*TEMP1* = 1) and *TEMP0* is deasserted (*TEMP0* = 0).
- The output *HEAT* is asserted (*HEAT* = 1) when *TEMP2* is deasserted (*TEMP2* = 0) and *TEMP1* is deasserted (*TEMP1* = 0).

These statements can be used to fill in the entries in the truth table in Figure 3.30(b).

3.11 SUMMARY

In this chapter, we discussed how combinational digital systems are constructed and how to analyze and understand the behavior of a combinational logic design. We defined a set of mathematical postulates called Huntington's Postulates to present the fundamentals of switching algebra. Based on Huntington's Postulates, we then examined the following logic operators.

1. conjunction (**and**)
2. complement conjunction (**nand**)
3. disjunction (**or**)

4. complement disjunction (**nor**)
5. exclusive disjunction (**xor**)
6. complement exclusive disjunction (**xnor**)

We also discussed how to analyze a combinational system composed of the interconnection of several logic operators using literal and symbolic techniques. Symbolic combinational digital system analysis is summarized below.

Symbolic Analysis Technique

1. *Whenever possible, read the output expressions directly from the logic diagram.*
2. *If the complexity of the combinational system precludes reading the output expressions directly from the logic diagram, the output expressions can be incrementally generated as follows:*
   a. *Moving from the inputs to the outputs, construct the output logic expression for each operator as a function of its input logic expressions.*
   b. *Complement only when there is an assertion conflict, that is, an active-0 output drives an active-1 input or vice versa.*
3. *Use the output logic expressions to generate a truth table.*

The result of symbolic analysis is a logic expression that defines the input/output transfer function of a combinational system. Given the importance of logic expressions, we discussed two standard forms for representing logic expressions: canonical sum and canonical product.

Finally, we discussed several aspects of properly documenting combinational digital systems to effectively convey design information. Documentation includes data and control flow orientation, multi-input logic operator symbols, and signal-naming conventions. We also discussed how to describe a combinational system from an "active state" perspective and how these concepts could be judiciously used to improve documentation.

3.12 PROBLEMS

1. The touch-tone keypad of a telephone can be viewed as a digital system. The input is one of the buttons on the touch-tone keypad; the output is one or more signals activating the associated tone.
   a. Is the touch-tone keypad of a telephone a combinational or sequential digital system?
   b. Explain your answer.

2. The telephone network can be viewed as one large (very large) digital system. The input is the phone number entered one digit at a time by pressing the appropriate buttons on the touch-tone keypad. The output is one or more signals ringing a bell or sounding a buzzer at the receiving phone.
   a. Is a telephone network a combinational or sequential digital system?
   b. Explain your answer.

3. Show by perfect induction that the definitions of the **not**, **and**, and **or** switching algebra operations given respectively in Figure 3.3, Figure 3.4, and Figure 3.5 obey the following Huntington's Postulates.
   a. P2a/P2b   **b.** P3a/P3b   **c.** P4a/P4b   **d.** P5a/P5b

4. Use a truth table to describe the input/output behavior of the following logic operations.

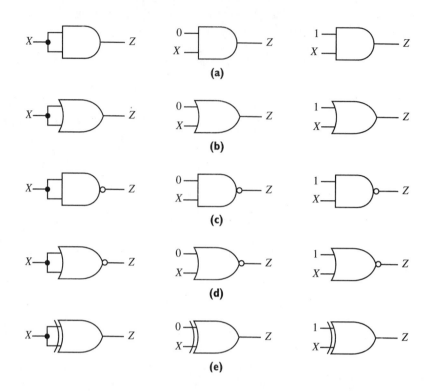

5. Show the truth tables for the $2^{2^2} = 16$ possible binary switching algebra operations.

6. If an input of a multi-input logic operator is not used, then it should be set to a known constant logic value (0 or 1) to ensure that the unused input does not float to an unknown or unwanted value. For the following multi-input logic operators, determine the logic values for the unused inputs so that the unused inputs do not interfere with the logic operations.

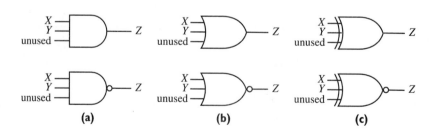

7. Using literal analysis, verify by complete enumeration that the three **nand** operators shown in Figure 3.17(c) correctly realize the **or** function.

8. Using literal analysis, verify by complete enumeration that the four **nand** operators shown in Figure 3.17(d) correctly realize the **nor** function.

9. Using literal analysis, show how the outputs *SUM* and *COUT* shown below are generated for the following sets of inputs.

a. $AUG = 0$, $ADD = 0$, and $CIN = 0$ **b.** $AUG = 1$, $ADD = 1$, and $CIN = 0$
c. $AUG = 0$, $ADD = 1$, and $CIN = 1$ **d.** $AUG = 1$, $ADD = 1$, and $CIN = 1$

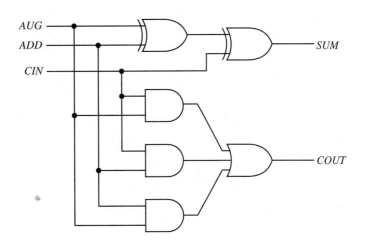

10. Using literal analysis, show how the output *SELECT* shown below is generated for the following sets of inputs.
  a. $C1 = 0$, $C2 = 0$, $I1 = 1$, $I2 = 0$, $I3 = 0$, and $I4 = 0$
  b. $C1 = 0$, $C2 = 1$, $I1 = 0$, $I2 = 1$, $I3 = 0$, and $I4 = 0$
  c. $C1 = 1$, $C2 = 1$, $I1 = 0$, $I2 = 0$, $I3 = 0$, and $I4 = 1$

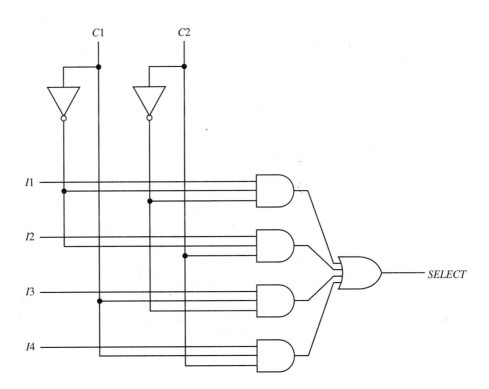

11. For the combinational schematic given in Problem 9, perform the following tasks.
    a. Generate the logic expressions for the outputs *SUM* and *COUT* using symbolic analysis. Show all intermediate logic expressions.
    b. Use the logic expressions to generate the truth tables for *SUM* and *COUT*.
    c. Describe the function of the combinational system.

12. For the combinational schematic given in Problem 10, perform the following tasks.
    a. Generate the logic expression for the output *SELECT* using symbolic analysis. Show all intermediate logic expressions.
    b. Use the logic expression to generate the truth table for *SELECT*.
    c. Describe the function of the combinational system.

13. Evaluate the following logic expressions.
    a. $\bar{0} + 1 \cdot 1 = $ _____
    b. $\overline{1 + 1 + 0} = $ _____
    c. $1 \cdot \bar{0} + \overline{(1 + 1)} = $ _____
    d. $\overline{(0 \cdot 1)} \cdot \overline{(1 + 0)} = $ _____

14. Perform the following tasks for the truth table shown below.
    a. List the minterms.
    b. List the maxterms.
    c. Generate the sum-of-products form of the output Z.
    d. Generate the minterm list form of the output Z.
    e. Generate the product-of-sums form of the output Z.
    f. Generate the maxterm list form of the output Z.

Row #	A B C	Z
0	0 0 0	1
1	0 0 1	0
2	0 1 0	1
3	0 1 1	1
4	1 0 0	0
5	1 0 1	0
6	1 1 0	0
7	1 1 1	1

15. Do the following switching algebra expressions describe the same logic function?

$$Z(A, B, C, D) = (\bar{A} \cdot B) + (\bar{C} \cdot D)$$

$$Z(A, B, C, D) = \prod M(0, 2, 3, 8, 10, 11, 12, 14, 15)$$

16. The sum-of-products logic expression form can be easily derived from the product-of-sums logic expression form and vice versa. Show the simple relationship between the sum-of-products and the product-of-sums forms for a given logic function.

17. Redraw the schematic of **nand** operators shown in Figure 3.17 to more clearly show the **or** function with an active-1 output.

18. Sometimes logic operators are drawn with active-1 and active-0 inputs. Show the truth tables for the following logic operators.

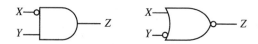

19. For the following combinational system, use symbolic analysis to generate the logic expression for the output *MAJORITY_BAR* and show that the system realizes the majority function with an active-0 output.

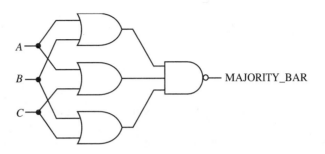

REFERENCES AND READINGS

[1] Boole, G. *The Laws of Thought,* Volume 2 in *George Boole's Logical Works.* Chicago, IL: Open Court Publishing, 1911. First published in 1854.

[2] Boole, G. *An Investigation of the Laws of Thought.* New York: Dover Publications, 1951. First published in 1854.

[3] Dershem, H. *Discrete Structures: Essential Computer Mathematics.* San Diego, CA: Harcourt Brace Jovanovich, 1988.

[4] Huntington, E. "Sets of Independent Postulates for the Algebra of Logic," *Transactions of the American Mathematical Society,* vol. 5, pp. 288–309, 1904.

[5] Huntington, E. "New Sets of Independent Postulates for the Algebra of Logic, with Special Reference to Whitehead and Russell's *Principia Mathematica*," *Transactions of the American Mathematical Society,* vol. 35, pp. 274–304, 1933.

[6] IEEE, Standard Graphic Symbols for Logic Functions, number ANSI/IEEE Std 91-1984, 1984.

[7] Mandl, M. *Introduction to Digital Logic Techniques & Systems.* Reston, VA: Reston Publishing, 1983.

[8] Pohl, I. *The Nature of Computation.* Potomac, MD: Computer Science Press, 1981.

[9] Schneeweiss, W. *Boolean Functions: With Engineering Applications & Computer Programs.* New York: Springer-Verlag, 1989.

[10] Shannon, C. "A Symbolic Analysis of Relay and Switching Circuits," *American Institute of Electrical Engineers (AIEE) Transactions,* vol. 57, pp. 713–723, 1933.

[11] Shannon, C. *The Mathematical Theory of Communication* in Weaver, W. *Recent Contributions to the Mathematical Theory of Communication.* University of Illinois (Urbana) Press, 1949. First published in Bell Systems Journal 1948.

[12] Skvarcius, R. *Discrete Mathematics with Computer Science Applications.* Menlo Park, CA: Benjamin/Cummings, 1986.

4
COMBINATIONAL DESIGN: SYNTHESIS

In Chapter 3, we discussed digital combinational system *analysis*. Starting with a combinational system composed of a set of interconnected switching algebra operators, we generated the input/output logic transform. We will now "turn the tables" and consider the more challenging task of combinational system *design*. Starting with a statement of the desired input/output logic transform, we will generate the corresponding set of interconnected switching algebra operators.

We will describe digital combinational system design in two stages: *synthesis* and *implementation*. Synthesis is addressed in this chapter; implementation is addressed in Chapter 5. Figure 4.1 shows that synthesis generates an optimized mathematical description of the desired function, whereas implementation generates an actual, physical system realizing the optimized mathematical description.

Figure 4.1

Digital combinational system synthesis

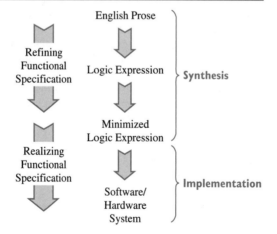

Figure 4.1 illustrates that synthesis often starts with the desired input/output function given as written prose, such as the following example.

> The tester output should indicate a warning notice when measured deviations exceed 5%, an error notice when measured deviations exceed 10%, and a system shutdown notice when measured deviations exceed 20%.

To design a combinational system realizing this monitoring function, we need to recast the written prose into the mathematics of switching algebra, in other words, a logic expression. Hence, we begin this chapter by studying how to derive logic expressions from functional specifications given via English descriptions.

After generating a logic expression, a variety of *minimization techniques* are used to further refine and improve the logic expression. The minimization techniques seek to minimize the complexity of the logic expression, which, in turn, tends to minimize the costs of the hardware that realizes the logic expression. We will study three general minimization techniques:

- algebraic reduction,
- Karnaugh maps, and,
- Quine-McCluskey tables.

4.1 GENERATING LOGIC EXPRESSIONS FROM PROSE

As we have noted, functional specifications for combinational systems are often initially formed as written statements, describing the desired outputs with respect to certain input conditions. Such written statements are helpful in conveying general objectives and ideas, but are generally too informal to support detailed logic synthesis. Consequently, written statements are translated into more precise and concise logic expressions. Recall from Chapter 3 that a logic expression defines the behavior—the input/output transfer function—of a combinational digital system. For simple functional specifications, the translation process is often obvious and straightforward. However, correctly translating an English description of a complex function into a logic expression can be difficult, especially if the description is incomplete or inconsistent. We will conduct English-to-logic translation using the following four steps.

1. Identify system inputs and outputs.
2. Define a logic variable for each system input and output.
3. Construct a truth table enumerating the outputs for each possible input.
4. Construct a canonical form logic expression from the truth table.

As an exercise in translating a written functional specification into a logic expression, consider the following specification.

> Design a printer controller that connects either user U1 or user U2 to printer P1 or printer P2. The digital system should also be capable of preventing all users from accessing printers.

The desired digital system is a simple example of a 2 × 2 crosspoint switch that allows any input to be connected to any output.

In the first steps of the translation process, we identify system inputs and outputs and define associated logic variables. The input and output signals for the printer controller are shown in Figure 4.2.

Figure 4.2

Input and output signals for a printer controller

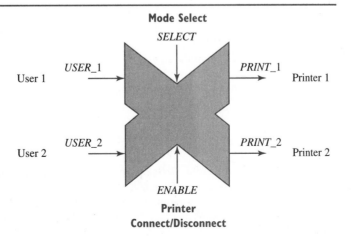

The input signals *USER*_1 and *USER*_2 carry the user data destined for the printers via the output signals *PRINT*_1 and *PRINT*_2. The input signal *SELECT* selects one of two possible input data ⇒ output printer routings:

- *USER*_1 ⇒ *PRINT*_1 and *USER*_2 ⇒ *PRINT*_2
- *USER*_1 ⇒ *PRINT*_2 and *USER*_2 ⇒ *PRINT*_1

The input signal *ENABLE* controls printer access. For convenience, we have assumed that both users cannot be routed to the same printer at the same time and that user data is carried by only one signal. In practice, we would recognize that the written functional specification is incomplete and ask for additional details to justify our assumptions.

The nature of the desired digital system begins to emerge and solidify as the English-to-logic translation process reformulates and refines the functional specification. Table 4.1 defines the system input/output logic variables in more detail. When defining the logic variables, we try to give a meaningful name to each one and, where appropriate, identify what the binary values of 0 and 1 represent. For example, for the logic variable *ENABLE*, 0 denotes disconnecting the users from the printers and 1 denotes connecting the users to the

Problem-Solving Skill

An important aspect of developing a solution is to carefully note all assumptions. Assumptions document the conditions under which the solution is valid.

Table 4.1

System input/output logic variables

Logic variable name	Description
*USER*_1	Input: User 1 Data
*USER*_2	Input: User 2 Data
SELECT	Input: Select Input ⇒ Output Routing Configuration
	"0" – *USER*_1 ⇒ *PRINT*_1 and *USER*_2 ⇒ *PRINT*_2
	"1" – *USER*_1 ⇒ *PRINT*_2 and *USER*_2 ⇒ *PRINT*_1
ENABLE	Input: Control User Access to Printers
	"0" – Disconnect Users from Printers
	"1" – Connect Users to Printers
*PRINT*_1	Output: Printer 1 Data
*PRINT*_2	Output: Printer 2 Data

printers. Actions and/or conditions associated with binary values may be specified by the designer or dictated by external interfaces and constraints.

After identifying system inputs/outputs and defining the associated logic variables, we construct a truth table that enumerates the output values in response to all possible input values. The truth table for the printer controller is shown on p. 77.

The truth table entries are based on the logic variable definitions given in Table 4.1. For example, consider row 13. *ENABLE* is asserted active-1, so the inputs are connected to the outputs. *SELECT* is also asserted active-1, so *USER*_1 is routed to *PRINT*_2 and *USER*_2 is routed to *PRINT*_1; in other words, *PRINT*_2 = *USER*_1 and *PRINT*_1 = *USER*_2. Review the other truth table entries to understand how each output value is generated.

In the final step of the translation process, we use the truth table to construct a canonical form logic expression for each output variable. Equations 4.1 and 4.2 give the sum-of-products (SOP) expressions for *PRINT*_1 and *PRINT*_2, respectively. (Refer to Chapter 3 for a discussion of the canonical sum-of-products form.)

$$
\begin{aligned}
PRINT\_1 = \ &(ENABLE \cdot \overline{SELECT} \cdot USER\_1 \cdot \overline{USER\_2}) \\
+ \ &(ENABLE \cdot \overline{SELECT} \cdot USER\_1 \cdot USER\_2) \\
+ \ &(ENABLE \cdot SELECT \cdot \overline{USER\_1} \cdot USER\_2) \\
+ \ &(ENABLE \cdot SELECT \cdot USER\_1 \cdot USER\_2)
\end{aligned}
$$

(4.1)

$$
\begin{aligned}
PRINT\_2 = {}& (ENABLE \cdot \overline{SELECT} \cdot \overline{USER\_1} \cdot USER\_2) \\
& + (ENABLE \cdot \overline{SELECT} \cdot USER\_1 \cdot USER\_2) \\
& + (ENABLE \cdot SELECT \cdot USER\_1 \cdot \overline{USER\_2} \\
& + (ENABLE \cdot SELECT \cdot USER\_1 \cdot USER\_2)
\end{aligned}
\tag{4.2}
$$

Row	ENABLE	SELECT	USER_1	USER_2	PRINT_1	PRINT_2
0	0	0	0	0	0	0
1	0	0	0	1	0	0
2	0	0	1	0	0	0
3	0	0	1	1	0	0
4	0	1	0	0	0	0
5	0	1	0	1	0	0
6	0	1	1	0	0	0
7	0	1	1	1	0	0
8	1	0	0	0	0	0
9	1	0	0	1	0	1
10	1	0	1	0	1	0
11	1	0	1	1	1	1
12	1	1	0	0	0	0
13	1	1	0	1	1	0
14	1	1	1	0	0	1
15	1	1	1	1	1	1

As another English-to-logic translation exercise, consider the following specification.

Design a combinational system that detects when a digitized sine wave equals or exceeds an upper limit of 6_{10} or a lower limit of -6_{10}. The sine wave is encoded as a 4-bit two's complement number in the symmetrical range -7_{10} to 7_{10}.

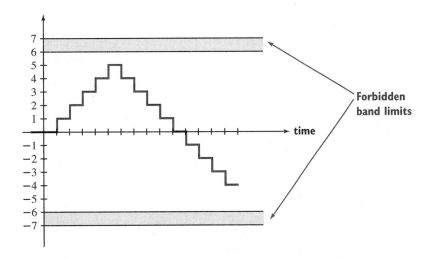

This exercise introduces *don't care* conditions.

Table 4.2

System input/output logic variables	Logic variable name	Description
	$S_3S_2S_1S_0$	Input: 4-bit two's complement number denoting sine wave magnitude in the symmetrical range -7_{10} to 7_{10}.
	LIMIT	Output: Assert active-1 when input equals or exceeds limits.
		0: $|S_3S_2S_1S_0| < 0110_2 = 6_{10}$
		1: $|S_3S_2S_1S_0| \geq 0110_2 = 6_{10}$

Again, we begin the translation process by identifying system inputs and outputs and defining associated logic variables, as shown in Table 4.2.

The truth table for the limit detector is given next.

Row	Decimal value	$S_3S_2S_1S_0$	*LIMIT*
0	0	0 0 0 0	0
1	1	0 0 0 1	0
2	2	0 0 1 0	0
3	3	0 0 1 1	0
4	4	0 1 0 0	0
5	5	0 1 0 1	0
6	6	0 1 1 0	1
7	7	0 1 1 1	1
8	-8	1 0 0 0	–
9	-7	1 0 0 1	1
10	-6	1 0 1 0	1
11	-5	1 0 1 1	0
12	-4	1 1 0 0	0
13	-3	1 1 0 1	0
14	-2	1 1 1 0	0
15	-1	1 1 1 1	0

The dash entry for the output *LIMIT* in response to the input $S_3S_2S_1S_0 = 1000_2 = -8_{10}$ in row 8 of the truth table illustrates a don't care condition.

When a particular combination of input values for a digital system should never occur because of external constraints, the condition is designated a *don't care* input. A don't care input literally means the designer does not care about the corresponding output because, in theory, the particular input condition will never occur. In the description of the sine wave limit detector, the input is defined to be a two's complement number in the *symmetrical* range -7_{10} to 7_{10}. Hence, the input $S_3S_2S_1S_0 = 1000_2 = -8_{10}$ should never occur and, consequently, the associated output value is designated a don't care. Switching algebra often designates a don't care by an X. However, VHDL designates a don't care by a dash, –, and uses X to designate an unknown value. To establish continuity between switching algebra and VHDL, we will use a dash to denote a don't care.

We complete the English-to-logic translation by using the truth table to generate the SOP expression for *LIMIT* given by Equation 4.3.

$$LIMIT(S_3, S_2, S_1, S_0) = \sum m(6, 7, 9, 10) + d(8) \qquad \textbf{(4.3)}$$

The notation "d(8)" denotes that minterm m_8 is a don't care.

Equation 4.3 assumes an active-1 output. We can easily modify the English-to-logic translation to accommodate an active-0 output by simply changing the output variable name from *LIMIT* to *LIMIT_BAR* in Table 4.2 and complementing the output variable values in the truth table. The resulting product-of-sums (POS) expression is given by Equation 4.4.

$$LIMIT\_BAR(S_3, S_2, S_1, S_0) = \prod M(6, 7, 9, 10) + d(8) \qquad \textbf{(4.4)}$$

4.2 MINIMIZATION TECHNIQUES

The English-to-logic translation yields a description of the desired combinational system as a logic expression. A logic diagram of the expressions given in Equations 4.1 and 4.2 is shown in Figure 4.3.

Figure 4.3

Logic diagram of a printer controller

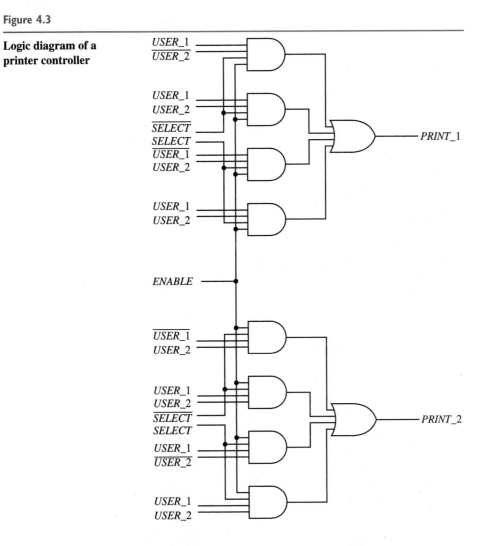

Later chapters will explain how logic operators are realized in hardware. For the present, however, note that the complexity and cost of digital system hardware is *generally* directly proportional to the complexity of the associated logic diagram. Consequently, we would like to minimize the number of logic operators and the inputs per operator in logic expressions. To that end, we will study three techniques for minimizing logic expressions:

1. algebraic reduction,
2. Karnaugh maps, and
3. Quine-McCluskey tables.

Before we begin discussing minimization techniques, we need to qualify the usefulness of such techniques. Sometimes, the number of logic operators is of secondary importance. For many digital designs, the physical area required for the *interconnect* of logic operators is more than the area required for the logic operators themselves. In such cases, one might opt for a logic design that minimizes interconnect rather than logic operators. Minimization techniques are clearly an important aspect of digital design; however, they may or may not be of primary importance, depending on the particular digital design.

4.3 ALGEBRAIC MINIMIZATION TECHNIQUE

Recall from Chapter 3 that Huntington's Postulates define the basic properties or laws of switching algebra. Several additional properties of switching algebra can be derived from Huntington's Postulates. Properties derived from fundamental postulates are called *theorems*. A set of commonly used switching algebra theorems are listed below.

- Involution Theorem
  1. **Theorem 1 (T1):** $\overline{\overline{X}} = X$
- Identity Theorems
  1. **Theorem 2a (T2a):** $X + 1 = 1$
  2. **Theorem 2b (T2b):** $X \cdot 0 = 0$
- Idempotency Theorems
  1. **Theorem 3a (T3a):** $X + X = X$
  2. **Theorem 3b (T3b):** $X \cdot X = X$
- Associative Theorems
  1. **Theorem 4a (T4a):** $X + (Y + Z) = (X + Y) + Z$
  2. **Theorem 4b (T4b):** $X \cdot (Y \cdot Z) = (X \cdot Y) \cdot Z$
- DeMorgan's Theorems
  1. **Theorem 5a (T5a):** $\overline{X + Y} = \overline{X} \cdot \overline{Y}$
  2. **Theorem 5b (T5b):** $\overline{X \cdot Y} = \overline{X} + \overline{Y}$
- Adjacency Theorems
  1. **Theorem 6a (T6a):** $X \cdot Y + X \cdot \overline{Y} = X$
  2. **Theorem 6b (T6b):** $(X + Y) \cdot (X + \overline{Y}) = X$
- Absorption Theorems
  1. **Theorem 7a (T7a):** $X + X \cdot Y = X$
  2. **Theorem 7b (T7b):** $X \cdot (X + Y) = X$
- Simplification Theorems
  1. **Theorem 8a (T8a):** $X + \overline{X} \cdot Y = X + Y$
  2. **Theorem 8b (T8b):** $X \cdot (\overline{X} + Y) = X \cdot Y$
- Consensus Theorems
  1. **Theorem 9a (T9a):** $X \cdot Y + \overline{X} \cdot Z + Y \cdot Z = X \cdot Y + \overline{X} \cdot Z$
  2. **Theorem 9b (T9b):** $(X + Y) \cdot (\overline{X} + Z) \cdot (Y + Z) = (X + Y) \cdot (\overline{X} + Z)$

Similar to the grouping of Huntington's Postulates, the above theorems are grouped in pairs, based on the principle of duality. Theorems 2a and 2b form a dual pair, Theorems 3a and 3b form a dual pair, and so on. We have already seen applications of some of these switching algebra theorems. For example, the alternate active-1/active-0 sets of logic symbols introduced in Chapter 3 are actually graphical representations of the DeMorgan's Theorems, T5a and T5b.

Recall the logical proposition example from Chapter 3:

Digital combinational systems are successful when properly designed and documented.

This logical proposition can be represented in switching algebra as

$$Z = X \cdot Y \tag{4.5}$$

The switching variable X denotes the condition "properly designed," Y denotes the condition "properly documented," and Z denotes the condition "successful digital combinational system." Using the Involution Theorem 1 and DeMorgan's Theorem 5b, Equation 4.5 can be reformulated as

$$Z = \overline{\overline{Z}} = \overline{(\overline{X \cdot Y})} = \overline{X} + \overline{Y}$$

which corresponds to the following equivalent way of stating the same logical proposition in a negative sense:

Digital combinational designs are not successful when not properly designed or not properly documented.

Although the theorems are given using simple switching variables, X, Y, and Z, the theorems hold true if the switching variables are replaced with arbitrarily complex logic expressions. For example, consider replacing the variable X with the expression $(A \cdot B)$ and the variable Y with the expression $(C \cdot D)$ in DeMorgan's Theorem 5a.

$$\overline{(A \cdot B) + (C \cdot D)} = \overline{(A \cdot B)} \cdot \overline{(C \cdot D)} \tag{4.6}$$

Equation 4.6 can be further expanded by applying DeMorgan's Theorem 5b to the terms $\overline{(A \cdot B)}$ and $\overline{(C \cdot D)}$, yielding Equation 4.7.

$$\overline{(A \cdot B) + (C \cdot D)} = \overline{(A \cdot B)} \cdot \overline{(C \cdot D)} \tag{4.7}$$
$$= (\overline{A} + \overline{B}) \cdot (\overline{C} + \overline{D})$$

Complementing a Logic Expression

1. Interchange 0's and 1's (a 0 is changed to a 1 and a 1 is changed to a 0).

2. Interchange + and • operations (a + operator is changed to a • operator and a • operator is changed to a + operator).

3. Interchange switching variables with their complements (X is changed to $\overline{X}$ and $\overline{X}$ is changed to X).

Equation 4.7 shows an easy way of generating the complement of any logic expression, based on a generalization of DeMorgan's Theorems (see left margin). Carefully note the similarities and differences between the principle of duality and DeMorgan's Theorems to avoid confusing these two properties: Both switching algebra properties interchange 0's/1's and +/• operations; however, only the DeMorgan's Theorems take the extra step of complementing variables.

Theorems can be verified using perfect induction. Recall from Chapter 3 that perfect induction or complete enumeration proves the equivalence of two logic expressions by constructing a truth table for each expression. If the two truth tables yield identical results for every possible combination of inputs, then the expressions are equivalent and the theorem is valid. For example, we can verify the Absorption Theorem 7b

$$X \cdot (X + Y) = X$$

by constructing the following truth table.

X Y	$(X + Y)$	$X \cdot (X + Y)$
0 0	0	0
0 1	1	0
1 0	1	1
1 1	1	1

The first and last columns are identical, so the Absorption Theorem 7b is valid.

Theorems can also be verified using the switch networks and Venn Diagrams introduced in Chapter 3. For example, Figure 4.4 verifies the Adjacency Theorem 6a. Figure 4.4(a) shows the switch network representation of the Adjacency Theorem 6a. One of Y or $\overline{Y}$ switches will always be closed, allowing the input-to-output transmission to be controlled by just X. Figure 4.4(b) shows the Venn Diagram representation of Adjacency Theorem 6a. The two shaded areas of the minterms $X \cdot Y$ and $X \cdot \overline{Y}$ combine to cover the variable X. In a similar manner, Figure 4.5 verifies the Simplification Theorem 8a.

Figure 4.4

The Adjacency Theorem

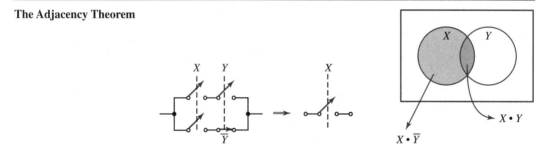

(a)   **(b)**

Figures (a) and (b) show switch network and Venn Diagram representations, respectively.

Figure 4.5

The Simplification Theorem

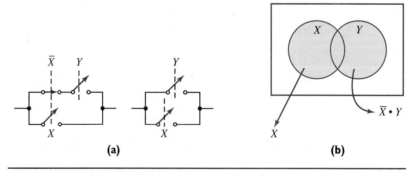

(a)   **(b)**

Figures (a) and (b) show switch network and Venn Diagram representations, respectively.

As a third option, theorems can be verified algebraically using Huntington's Postulates and previously proved theorems. For example, the proof of the Adjacency Theorem 6b is shown below.

$$(X + Y) \cdot (X + \overline{Y}) = X + (Y \cdot \overline{Y}) \quad \textit{Postulate 4a}$$
$$= X + 0 \quad \textit{Postulate 5b}$$
$$= X \quad \textit{Postulate 2a}$$

4.3.1 Simplifying Logic Expressions

The Adjacency, Absorption, Simplification, and Consensus Theorems are useful in algebraically minimizing logic expressions because these theorems state how to replace a particular form of a logic expression with an equivalent, simpler expression. For example, the logic expressions for the printer controller example given by Equations 4.1 and 4.2 can be algebraically minimized by repeated application of the Adjacency Theorem 6a. Consider, for instance, the first two minterms, m_{10} and m_{11}, for $PRINT\_1$ in Equation 4.1.

$$PRINT\_1 = (ENABLE \cdot \overline{SELECT} \cdot USER\_1 \cdot \overline{USER\_2})$$
$$+ (ENABLE \cdot \overline{SELECT} \cdot USER\_1 \cdot USER\_2) + \cdots$$

If X denotes $(ENABLE \cdot \overline{SELECT} \cdot USER\_1)$ and Y denotes $USER\_2$, the first two minterms of $PRINT\_1$ take the form

$$PRINT\_1 = (X \cdot \overline{Y}) + (X \cdot Y) + \cdots$$

and can be combined using Adjacency Theorem 6a to yield

$$PRINT\_1 = X + \cdots$$
$$= (ENABLE \cdot \overline{SELECT} \cdot USER\_1) + \cdots$$

Through repeated application of the Adjacency Theorem 6a, the logic expression for the 2×2 crosspoint switch can be simplified from the eight minterms in Equations 4.1 and 4.2 to the four minterms shown in Equations 4.8 and 4.9.

$$PRINT\_1 = (ENABLE \cdot \overline{SELECT} \cdot USER\_1)$$
$$+ (ENABLE \cdot SELECT \cdot USER\_2) \tag{4.8}$$

$$PRINT\_2 = (ENABLE \cdot \overline{SELECT} \cdot USER\_2)$$
$$+ (ENABLE \cdot SELECT \cdot USER\_1) \tag{4.9}$$

To gain proficiency with algebraic minimization techniques, we use the following example to illustrate the application of four theorems to simplify the logic expression

$$F = (\overline{A} \cdot D \cdot B) + (A \cdot \overline{B} \cdot C) + (\overline{B} \cdot C \cdot D) + (\overline{A} \cdot D \cdot \overline{B}) + A$$

$$F = (\overline{A} \cdot D \cdot B) + (A \cdot \overline{B} \cdot C) + (\overline{B} \cdot C \cdot D) + (\overline{A} \cdot D \cdot \overline{B}) + A$$
$$= (\overline{A} \cdot D \cdot B) + (\overline{A} \cdot D \cdot \overline{B}) + (A \cdot \overline{B} \cdot C) + (\overline{B} \cdot C \cdot D) + A$$
$$= (\overline{A} \cdot D) + (A \cdot \overline{B} \cdot C) + (\overline{B} \cdot C \cdot D) + A \qquad \textit{Adjacency T6a}$$
$$= A \cdot (\overline{B} \cdot C) + \overline{A} \cdot D + (\overline{B} \cdot C) \cdot D + A$$
$$= (A \cdot \overline{B} \cdot C) + (\overline{A} \cdot D) + A \qquad\qquad \textit{Consensus T9a}$$
$$= (A \cdot \overline{B} \cdot C) + A + D \qquad\qquad \textit{Simplification T8a}$$
$$= A + D \qquad\qquad\qquad\qquad \textit{Absorption T7a}$$

The second line is a reordering of terms in the first line. The third line is the result of applying the Adjacency Theorem 6a to the second line. The fourth line is a reordering of terms in the third line. The fifth line is the result of applying the Consensus Theorem 9a to the fourth line. The sixth line is the result of applying the Simplification Theorem 8a to the fifth line. The last line is the result of applying the Absorption Theorem 7a to the sixth line.

Sometimes the order in which algebraic transformations are applied is important in minimizing a logic expression. As an example, reconsider minimizing the logic expression for F given in the previous example by attempting to apply the Simplification Theorem 8a before the Consensus Theorem 9a.

$$
\begin{aligned}
F &= (\overline{A} \cdot D \cdot B) + (A \cdot \overline{B} \cdot C) + (\overline{B} \cdot C \cdot D) + (\overline{A} \cdot D \cdot \overline{B}) + A \\
&= (\overline{A} \cdot D) + (A \cdot \overline{B} \cdot C) + (\overline{B} \cdot C \cdot D) + A && \textit{Adjacency T6a} \\
&= (A \cdot \overline{B} \cdot C) + (\overline{B} \cdot C \cdot D) + A + D && \textit{Simplification T8a}
\end{aligned}
$$

Applying the Simplification Theorem 8a eliminates the $(\overline{A} \cdot D)$ product term needed for applying the Consensus Theorem 9a.

A final example illustrates that sometimes the minimization process involves an expansion step, in which an expression is temporarily made more complex to facilitate additional simplification steps.

$$
\begin{aligned}
F &= (A \cdot B) + (\overline{A} \cdot C) + (A \cdot \overline{B}) + (A \cdot \overline{C}) \\
&= (A \cdot B) + (\overline{A} \cdot C) + (A \cdot \overline{B}) + (A \cdot \overline{C}) + (B \cdot C) && \textit{Consensus T9a} \\
&= (A \cdot B) + (\overline{A} \cdot C) + (\overline{B} + \overline{C}) \cdot A + (B \cdot C) && \textit{Distributive P4b} \\
&= (A \cdot B) + (\overline{A} \cdot C) + \overline{(B \cdot C)} \cdot A + (B \cdot C) && \textit{DeMorgan T5b} \\
&= (A \cdot B) + (\overline{A} \cdot C) + A + (B \cdot C) && \textit{Simplification T8a} \\
&= (A \cdot B) + (\overline{A} \cdot C) + A && \textit{Consensus T9a} \\
&= (\overline{A} \cdot C) + A && \textit{Absorption T7a} \\
&= A + C && \textit{Simplification T8a}
\end{aligned}
$$

> **Problem-Solving Skill**
>
> *Sometimes in solving a problem it is necessary to take a step back to take two steps forward. Taking a step back may temporarily make the problem look more complex, but it may facilitate moving the solution out of a local optima toward the desired global optima.*

In the first step of this minimization process, the product term $(B \cdot C)$ is added to the logic expression. The resulting expression given in the second line is equivalent to the original expression because $(B \cdot C)$ is a redundant term, which can be seen by working the Consensus Theorem 9a "in reverse" with the product terms $(A \cdot B)$ and $(\overline{A} \cdot C)$. The extra term is removed after using it to invoke the other theorems.

4.3.2 Limitations

The above examples of algebraic minimization demonstrate that switching algebra postulates and theorems can be used to reduce the complexity of logic expressions. The examples also illustrate some disadvantages of using algebraic minimization techniques.

It is often difficult to know the best way to apply various algebraic transformations to simplify an expression. Moreover, it is often difficult to determine when to stop the minimization process, in other words, when a minimal form of the logic expression has been found. The following sections will describe more systematic minimization techniques that guarantee minimized logic expressions.

4.4 KARNAUGH MAP MINIMIZATION TECHNIQUE

In 1953, M. Karnaugh published an article describing a technique for geometrically representing a logic expression that facilitates minimizing the expression [7]. We will first explain how a *Karnaugh map* represents a logic expression. Next, we will study why a Karnaugh map facilitates logic minimization. Then, we will study sum-of-products (SOP) minimization, product-of-sums (POS) minimization, don't care minimization, and multiple output minimization, using Karnaugh maps.

4.4.1 Mapping Functions onto Karnaugh Maps

Karnaugh maps, also sometimes called *K-maps*, for logic functions of one, two, three, and four switching variables are shown in Figure 4.6 (p. 86). A Karnaugh map is composed of squares or cells. Each square represents a row in the truth table or, equivalently, a logic function minterm.

As an example, consider the Karnaugh map for the **xor** function shown in Figure 4.7. Moving from left to right across the Karnaugh map in Figure 4.7(c), the first square represents the minterm m_0, in other words, the $\overline{A} \cdot \overline{B}$ or $A = 0/B = 0$ truth table entry. Similarly, the second square represents the minterm m_1, in other words, the $\overline{A} \cdot B$ or $A = 0/B = 1$ truth table entry. The third square seems strangely out of order, representing the minterm m_3 instead of m_2. The rationale behind this geometric ordering of cells will be explained in the next section. Finally, the last cell represents the minterm m_2, in other words, the $A \cdot \overline{B}$ or $A = 1/B = 0$ truth table entry.

Figure 4.8 (p. 87) shows a Karnaugh map for the *HEAT* output of the temperature controller discussed in Chapter 3, Figure 3.2. The temperature controller inputs a 3-bit Gray code indicating the ambient building temperature and outputs signals controlling the heating and air conditioning systems. The output values listed in the truth table are mapped onto the Karnaugh map using the general three-variable Karnaugh map template given in Figure 4.6.

Karnaugh maps are sometimes presented as shown in Figure 4.9 (p. 87). This alternative form of a Karnaugh map highlights the regions of the map where each switching variable is

Figure 4.7

Three equivalent representations for the exclusive-or function

(a)

$$F = \Sigma m(1, 2)$$

(b)

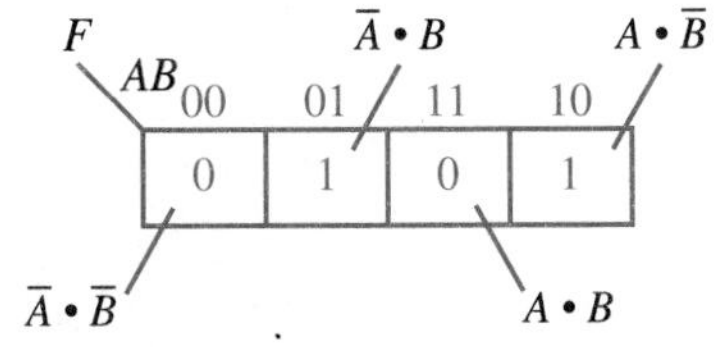

(c)

Figures (a)–(c) respectively show the truth table, minterm list, and Karnaugh map representations of the **xor** function.

Figure 4.6

Karnaugh maps for logic functions of one to four variables

Truth Table

Karnaugh Map

Minterm	A	F
0	0	m_0
1	1	m_1

F

A	0	1
	m_0	m_1

(a)

Minterm	A	B	F
0	0	0	m_0
1	0	1	m_1
2	1	0	m_2
3	1	1	m_3

F

AB	00	01	11	10
	m_0	m_1	m_3	m_2

(b)

Minterm	A	B	C	F
0	0	0	0	m_0
1	0	0	1	m_1
2	0	1	0	m_2
3	0	1	1	m_3
4	1	0	0	m_4
5	1	0	1	m_5
6	1	1	0	m_6
7	1	1	1	m_7

F

A \ BC	00	01	11	10
0	m_0	m_1	m_3	m_2
1	m_4	m_5	m_7	m_6

(c)

Minterm	A	B	C	D	F
0	0	0	0	0	m_0
1	0	0	0	1	m_1
2	0	0	1	0	m_2
3	0	0	1	1	m_3
4	0	1	0	0	m_4
5	0	1	0	1	m_5
6	0	1	1	0	m_6
7	0	1	1	1	m_7
8	1	0	0	0	m_8
9	1	0	0	1	m_9
10	1	0	1	0	m_{10}
11	1	0	1	1	m_{11}
12	1	1	0	0	m_{12}
13	1	1	0	1	m_{13}
14	1	1	1	0	m_{14}
15	1	1	1	1	m_{15}

F

AB \ CD	00	01	11	10
00	m_0	m_1	m_3	m_2
01	m_4	m_5	m_7	m_6
11	m_{12}	m_{13}	m_{15}	m_{14}
10	m_8	m_9	m_{11}	m_{10}

(d)

Figures (a)–(d) respectively show Karnaugh maps for logic functions of one to four variables.

Figure 4.8

Temperature controller Karnaugh map

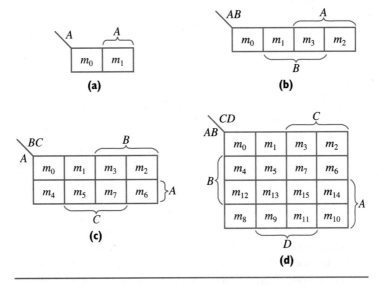

Degrees	TEMP2	TEMP1	TEMP0	HEAT
0°C	0	0	0	1
10°C	0	0	1	1
30°C	0	1	0	0
20°C	0	1	1	0
70°C	1	0	0	0
60°C	1	0	1	0
40°C	1	1	0	0
50°C	1	1	1	0

Figure 4.9

Alternate form of Karnaugh maps

Figures (a)–(d) respectively show alternate forms of Karnaugh maps for logic functions of one to four variables.

1 and is useful in representing logic expressions that are not in a canonical form. For example, consider mapping the following function onto a four-variable Karnaugh map, as shown in Figure 4.10.

$$F(A, B, C, D) = (A \cdot \overline{B}) + (\overline{A} \cdot \overline{B} \cdot C) + (\overline{A} \cdot B \cdot \overline{C} \cdot \overline{D}) + (\overline{A} \cdot B \cdot \overline{C} \cdot D)$$

Figure 4.10

Using the alternate form of Karnaugh maps

$$F = (A \cdot \overline{B}) + (\overline{A} \cdot \overline{B} \cdot C) + (\overline{A} \cdot B \cdot \overline{C} \cdot \overline{D}) + (\overline{A} \cdot B \cdot \overline{C} \cdot D)$$

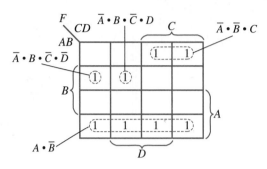

Figure 4.10 shows how each product term of F is mapped onto the Karnaugh map. For the product term $(A \cdot \overline{B})$, 1's are placed in the squares where $A = 1$ and $B = 0$, which are the four squares in the bottom row. For the product term $(\overline{A} \cdot \overline{B} \cdot C)$, 1's are placed in the squares where $A = 0$, $B = 0$, and $C = 1$, which are the two squares in the top right corner. The remaining two product terms are mapped in a similar fashion.

4.4.2 Minimizing Functions

A Karnaugh map can be used to quickly generate the minimal form of a logic expression because the squares are laid out so *the minterms of adjacent squares differ in only one variable.* Why would such an arrangement of squares facilitate minimization? Recall the Adjacency Theorem 6a, $X \cdot Y + X \cdot \overline{Y} = X$. If two product terms differ by only one variable, the variable can be eliminated and the product terms combined. For instance, the product terms $(\overline{A} \cdot B \cdot \overline{C} \cdot \overline{D})$ and $(\overline{A} \cdot B \cdot \overline{C} \cdot D)$ that are located in adjacent squares in Figure 4.10 differ in only the variable D and can thus be combined via the Adjacency Theorem to yield the product term $(\overline{A} \cdot B \cdot \overline{C})$. Hence, the Karnaugh map minimization technique involves looking for patterns of adjacent 1's that identify applications of the Adjacency Theorem, and the human brain is particularly good at pattern identification.

We need to be sure we understand the implications of the definition of "adjacent squares" in Karnaugh maps. Refer to Figure 4.11, which shows the cells that are adjacent to the minterm m_3 cell on a four-variable Karnaugh map. Each of the shaded cells is adjacent to the m_3 cell because each of the shaded minterms differ from the $m_3 = \overline{A} \cdot \overline{B} \cdot C \cdot D$ minterm by one variable; in particular, note that m_{11} is considered to be adjacent to m_3. More generally, the top row of a Karnaugh map is considered adjacent to the bottom row and the leftmost column is considered adjacent to the rightmost column.

The Karnaugh map minimization technique begins by identifying adjacent 1's in groups of $1(2^0)$, $2(2^1)$, $4(2^2)$, $8(2^3)$, and so on. We start with the *largest* possible groups and continue grouping until all 1's are included in a grouping, in other words, all 1's are *covered*. The largest groups of 1's are sought because such groups represent successive applications of the Adjacency Theorem.

For example, consider minimizing the logic function

$$F(A, B, C, D) = \sum m(0, 1, 5, 7, 8, 9, 15)$$

Figure 4.11

Example of adjacency on a Karnaugh map

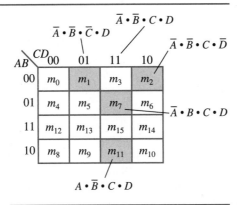

Cells adjacent to the minterm m_3 cell are shaded.

Figure 4.12 shows the Karnaugh map representation and groups of adjacent 1's. The two groupings given in dotted lines in Figure 4.12(a) show that the Adjacency Theorem can be used to combine the minterms m_0 and m_1

$$(\overline{A} \bullet \overline{B} \bullet \overline{C} \bullet \overline{D}) + (\overline{A} \bullet \overline{B} \bullet \overline{C} \bullet D) = \overline{A} \bullet \overline{B} \bullet \overline{C}$$

and the minterms m_8 and m_9.

$$(A \bullet \overline{B} \bullet \overline{C} \bullet \overline{D}) + (A \bullet \overline{B} \bullet \overline{C} \bullet D) = A \bullet \overline{B} \bullet \overline{C}$$

Moreover, the two resulting three-variable product terms can be combined into a single two-variable product term by again using the Adjacency Theorem.

$$(\overline{A} \bullet \overline{B} \bullet \overline{C}) + (A \bullet \overline{B} \bullet \overline{C}) = \overline{B} \bullet \overline{C}$$

Figure 4.12

Identifying the largest groupings of 1's

$$F(A,B,C,D) = \Sigma m(0, 1, 5, 7, 8, 9, 15)$$

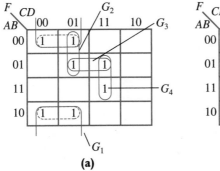

(a)

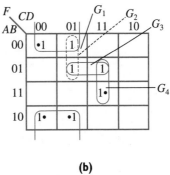

(b)

Figure (a) identifies all prime implicants. Figure (b) identifies essential prime implicants and secondary prime implicants.

In other words, the four groups of single 1's are combined into two groups of two 1's, which are, in turn, combined into a single group of four 1's. The resulting single group of four 1's is labeled G_1. The three groupings of two 1's (G_2, G_3, and G_4) cover the remaining minterms m_5, m_7, and m_{15}. When forming groups, it is permissible to include a minterm in more than one group, as happens with minterms m_1, m_5, and m_7.

The groups G_1–G_4 are called *prime implicants* of the logic function F. An *implicant* is any grouping, including a group of one 1. A prime implicant is a grouping that is not contained within a larger grouping, in other words, a product term that cannot be combined with any other product term via the Adjacency Theorem.

We do not need all the prime implicants in Figure 4.12(a) to represent the function F. The groups G_1, G_2, and G_4 or the groups G_1, G_3, and G_4 could be selected to cover all the 1's. Hence, the next step in Karnaugh map minimization is to select the prime implicants that form a *minimal cover* for the logic function. Prime implicants are selected by first identifying the *essential prime implicants*.

An essential prime implicant is a group that must be included because it covers a 1 (minterm) that is not covered by any other grouping. G_1 and G_4 are essential prime implicants because G_1 is the only group covering the minterms m_0, m_8, and m_9, and G_4 is the only group covering the minterm m_{15}. Figure 4.12(b) shows a common bookkeeping practice of identifying essential prime implicants by marking minterms covered by only one grouping with dots. The remaining prime implicants not identified as essential prime implicants are called *secondary prime implicants*.

The last step of Karnaugh map minimization selects secondary prime implicants to cover the 1's not covered by the essential prime implicants. There may be more than one acceptable minimum set of secondary prime implicants. In Figure 4.12(b), the remaining minterm m_5 can be "picked up" with either group G_2 or G_3. Both groups are of the same size, so we arbitrarily select G_3, yielding the following simplified expression for the logic function $F(A, B, C, D)$:

$$F(A, B, C, D) = (\overline{B} \bullet \overline{C}) + (\overline{A} \bullet B \bullet D) + (B \bullet C \bullet D) \qquad \textbf{(4.10)}$$

In selecting a minimum set of secondary prime implicants, it is again helpful to start by considering the largest groupings. However, a minimum set of secondary prime implicants may not always involve the largest groupings; an example is given in Figure 4.13. The Karnaugh map is called a *cyclic map* because the logic expression has *no* essential prime implicants; all prime implicants are secondary prime implicants. Note that the selected set of secondary prime implicants does not contain the group of four 1's denoted by the product term $A \bullet B$.

Figure 4.13

Selecting secondary prime implicants

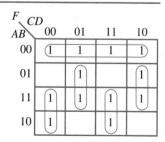

For small minimization problems, examining a few candidate selections of secondary prime implicants quickly reveals a minimum covering. For larger minimization problems, however, it may be impractical to attempt to determine the best selection of secondary prime implicants using a trial-and-error approach. A more systematic way of selecting secondary prime implicants is provided by the *Petrick algorithm*. The Petrick algorithm formulates the task of selecting secondary prime implicants as a logic expression and then minimizes the logic expression. Slick!

For instance, consider minimizing the logic function shown in Figure 4.14. The groups G_1–G_6 identify the prime implicants. There is only one essential prime implicant, G_1, consisting of the four corner minterms: m_0, m_2, m_8, and m_{10}. (Yes, the four corner minterms form a legitimate grouping.) The remaining groups, G_2–G_6, are secondary prime implicants that can be used to cover the remaining minterms: m_4, m_5, m_9, and m_{13}. The Petrick algorithm identifies a minimum set of secondary prime implicants by casting the covering task into the following equivalent product-of-sums expression.

$$m_4 \bullet m_5 \bullet m_9 \bullet m_{13} \tag{4.11}$$

$$(G_2 + G_3) \bullet (G_3 + G_4) \bullet (G_5 + G_6) \bullet (G_4 + G_5)$$

Equation 4.11 states that the minterm m_4 is covered by the groups G_2 *or* G_3 *and* the minterm m_5 is covered by the groups G_3 *or* G_4 *and* so on. Expanding Equation 4.11 by repeated applications of Distributive Postulate 4b and simplifying product terms by Idempotency Theorem 3b and Absorption Theorem 7a yields the equivalent sum-of-products expression given in Equation 4.12.

$$(G_2 \bullet G_4 \bullet G_5) + (G_2 \bullet G_4 \bullet G_6) + (G_3 \bullet G_4 \bullet G_6) + (G_3 \bullet G_5) \tag{4.12}$$

Figure 4.14

Petrick algorithm

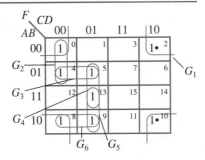

Equation 4.12 states that the remaining minterms can be covered using the following sets of secondary prime implicants:

- G_2, G_4, and G_5
- G_3, G_4, and G_6
- G_2, G_4, and G_6
- G_3 and G_5

G_2, G_4, G_5

(a)

G_2, G_4, G_6

(b)

G_3, G_4, G_6

(c)

G_3, G_5

(d)

The last set, containing the two secondary prime implicants G_3 and G_5, is a good choice because it covers the remaining minterms with the least amount of logic. Thus, Karnaugh map minimization yields the following simplified expression for the logic function of Figure 4.14.

$$F(A, B, C, D) = (\overline{B} \cdot \overline{D}) + (\overline{A} \cdot B \cdot \overline{C}) + (A \cdot \overline{C} \cdot D)$$

Let us summarize what we have learned thus far about Karnaugh map minimization.

Karnaugh Map Minimization Technique—SOP

1. *Identify all prime implicants by grouping adjacent 1's into groups $1(2^0)$, $2(2^1)$, $4(2^2)$, $8(2^3)$, and so on, starting with the largest groups.*
2. *Identify the prime implicants that form a minimal cover.*
   a. *Identify the essential prime implicants. An essential prime implicant contains minterms not contained in any other grouping.*
   b. *Select the minimum set of secondary prime implicants to cover the remaining minterms not covered by essential prime implicants. Secondary prime implicants can be selected informally by inspection or formally by the Petrick algorithm. It is permissible to reuse minterms covered by the essential prime implicants.*
3. *The resulting simplified logic expression is in sum-of-products form.*

As another Karnaugh map minimization technique exercise, reconsider the printer controller defined in Section 4.1. The Karnaugh maps for functions *PRINT*_1 and *PRINT*_2 are

Figure 4.15

Karnaugh maps for printer controller functions

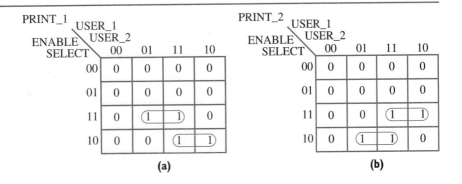

(a)   (b)

shown in Figure 4.15(a) and Figure 4.15(b), respectively. The groupings yield the following simplified expressions for *PRINT*_1 and *PRINT*_2.

$$PRINT\_1 = (ENABLE \bullet \overline{SELECT} \bullet USER\_1) \\ + (ENABLE \bullet SELECT \bullet USER\_2) \tag{4.13}$$

$$PRINT\_2 = (ENABLE \bullet \overline{SELECT} \bullet USER\_2) \\ + (ENABLE \bullet SELECT \bullet USER\_1) \tag{4.14}$$

Notice that the minimized expressions for *PRINT*_1 and *PRINT*_2 in Equations 4.13 and 4.14 are identical to the expressions in Equations 4.8 and 4.9 obtained by algebraic minimization.

4.4.3 Product-of-Sums (POS) Minimization

Thus far, we have used Karnaugh maps to generate minimized sum-of-products (SOP) forms. Karnaugh map logic minimization, however, can yield either a sum-of-products (SOP) form or product-of-sums (POS) form. Typically, one of the minimized logic forms yields a better, or less complex, result, depending on the particular logic expression.

The simple relationship between the SOP and POS forms

$$\text{minimal POS form of } F = \overline{\text{minimal SOP form of } \overline{F}}$$
$$\text{minimal SOP form of } F = \overline{\text{minimal POS form of } \overline{F}}$$

shows that POS Karnaugh map minimization is similar to SOP Karnaugh map minimization, except that 0's are grouped instead of 1's and there is an additional step of inverting the resulting expression. Grouping 0's yields the minimal SOP form of $\overline{F}$. Inverting the result using the generalized DeMorgan's Theorems yields the desired minimal POS form of F. The relationship between POS and SOP logic minimization is shown below.

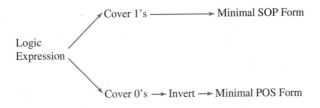

As an example, we will minimize the following logic function, given in canonical POS form.

$$F(A, B, C, D) = \prod M(0, 1, 3, 7, 8, 10, 11, 15)$$

The logic function F is easily mapped onto a Karnaugh map by entering 0's in the squares denoted by the maxterms M_0, M_1, M_3, M_7, M_8, M_{10}, M_{11}, and M_{15}. All other squares receive 1's.

The minimal SOP form of F given in Equation 4.15 is generated by grouping the 1's shown in Figure 4.16(a).

$$F(A, B, C, D) = (B \cdot \overline{C}) + (B \cdot \overline{D}) + (A \cdot \overline{C} \cdot D) + (\overline{A} \cdot C \cdot \overline{D}) \quad \textbf{(4.15)}$$

The minimal POS form of F is generated by first grouping the 0's shown in Figure 4.16(b), which yields $\overline{F}$.

$$\overline{F(A, B, C, D)} = (C \cdot D) + (A \cdot \overline{B} \cdot \overline{D}) + (\overline{A} \cdot \overline{B} \cdot \overline{C})$$

Then, $\overline{F}$ is inverted using the generalized DeMorgan Theorem to yield the minimal POS form of F.

$$F(A, B, C, D) = (\overline{C} + \overline{D}) \cdot (\overline{A} + B + D) \cdot (A + B + C) \quad \textbf{(4.16)}$$

Compare the minimal SOP form of F (Equation 4.15) and the minimal POS form of F (Equation 4.16) shown in Figure 4.17. Using the simple complexity metric of the total number of operators and associated inputs (ignoring complement operators) shows that the SOP form requires 5 logic operators having a total of 14 inputs, while the POS form requires only 4 logic operators having a total of 11 inputs. Hence, in this case, the minimal POS form yields a less complex result.

Product-of-sums (POS) Karnaugh map minimization is summarized below.

Karnaugh Map Minimization Technique—POS

1. *Identify all prime implicants by grouping adjacent 0's into groups $1(2^0)$, $2(2^1)$, $4(2^2)$, $8(2^3)$, and so on, starting with the <u>largest</u> groups.*
2. *Identify the prime implicants that form a minimal cover.*
   a. *Identify the essential prime implicants. An essential prime implicant contains maxterms not contained in any other grouping.*
   b. *Select the minimum set of secondary prime implicants to cover the remaining maxterms not covered by essential prime implicants. Secondary prime implicants can be selected informally by inspection or formally by the Petrick algorithm. It is permissible to reuse maxterms covered by the essential prime implicants.*
3. *Invert the resulting simplified sum-of-products logic expression using DeMorgan's Theorems.*
4. *The resulting logic expression is in product-of-sums form.*

Figure 4.16

Sum-of-Products (SOP) and Product-of-Sums (POS) groupings

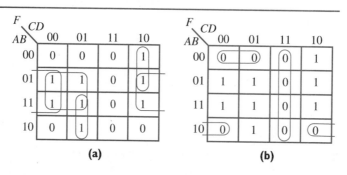

Figure 4.17

**Sum-of-Products (SOP)
and Product-of-Sums
(POS) minimizations**

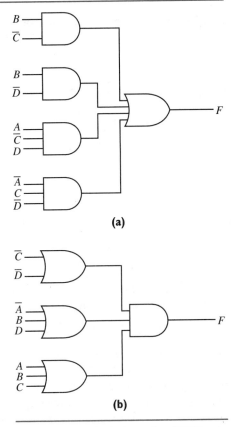

Figures (a) and (b) show results of SOP and
POS logic minimizations, respectively.

There is no easy way to determine if a particular logic expression is best synthesized in a product-of-sums or sum-of-products form. If time permits, it is often wise to check both approaches, especially on exams!

4.4.4 Don't Care Input Values

Recall from our earlier discussion on translating English descriptions of combinational digital systems into logic expressions that *don't cares* denote unspecified input/output transformations. If an output does not matter for a combination of inputs, the combination of inputs is called a *don't care condition* and the associated output value is designated by a *don't care value*, "–." A don't care condition can arise when the environment constrains a particular combination of inputs to never occur or when the environment does not monitor the output for a particular combination of inputs.

A don't care input condition allows additional flexibility during logic minimization because we are free to select the output to be either 0 or 1, depending on which value best simplifies the logic. Though don't care conditions can serve as convenient "wildcards" for minimization, we must be careful about assuming that an input condition can never occur. It is difficult to forsee all possible eventualities.

As an example of don't care conditions, reconsider the temperature controller shown in Figure 4.8. If we are willing to make the assumption that the building temperature will never rise above 50°C (122°F), the input temperature readings of 60°C and 70°C will never happen and can be handled as don't cares.

The truth table for the *HEAT* output of the temperature controller just described is shown below.

Temperature (°C)	TEMP2	TEMP1	TEMP0	HEAT
0	0	0	0	1
10	0	0	1	1
30	0	1	0	0
20	0	1	1	0
70	1	0	0	–
60	1	0	1	–
40	1	1	0	0
50	1	1	1	0

The input temperature readings of 60°C (101_2) and 70°C (100_2) are don't care input conditions and the corresponding values of the output *HEAT* are denoted by a dash, –. The minterm list representation for the new version of the *HEAT* logic expression is given as

$$HEAT(TEMP2, TEMP1, TEMP0) = \sum m(0, 1) + d(4, 5) \qquad \textbf{(4.17)}$$

The notation $d(4, 5)$ denotes that minterms m_4 and m_5 are don't cares.

The Karnaugh map minimization technique for expressions with don't care conditions follows the same steps discussed in the previous sections, with the addition that we are free to use a don't care minterm as either a 1 or a 0 to facilitate the grouping process. In Figure 4.18, we have opted to let the don't care minterms m_4 and m_5 represent 1's, yielding $HEAT = \overline{TEMP1}$.

A don't care minterm can serve as *either* a 0 or a 1, but not *both* a 0 and a 1. In other words, a don't care minterm cannot serve as a 1 for one grouping and a 0 for another grouping. Once the value of a don't care minterm is chosen, it cannot be changed.

Admittedly, for this simple digital system, the reduction in logic complexity is relatively minor; recall the original definition of *HEAT*, $HEAT = \overline{(TEMP1 + TEMP2)}$. However, significant gains can be made by exploiting don't care input conditions for larger systems. As a final point, it is important to recognize the potential implications of assuming don't care inputs. If the input temperature readings of 60°C or 70°C ever did occur, the heat would be turned on, which could cause serious damage. Remember the old saying, "Never say never."

As another example of Karnaugh map minimization with don't cares, Figure 4.19 shows

Figure 4.18

Karnaugh map for a temperature controller with don't care inputs

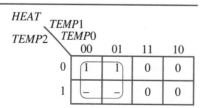

Figure 4.19

Karnaugh map for a limit detector with don't care inputs

the Karnaugh map groupings for the sine wave limit detector defined in Table 4.2. The don't care minterm m_8 is treated as a 1 to yield

$$LIMIT(S_3, S_2, S_1, S_0) = (\overline{S}_3 \cdot S_2 \cdot S_1) + (S_3 \cdot \overline{S}_2 \cdot \overline{S}_1) + (S_3 \cdot \overline{S}_2 \cdot \overline{S}_0)$$

4.4.5 Multiple Outputs

Thus far, we have focused attention on minimizing a combinational system design "output-by-output." That is, the logic for each output is minimized independent of any other output logic. Such an approach *locally* optimizes each output expression, but may not *globally* optimize the entire combinational system because it does not exploit the possibility of *sharing* logic operators between outputs. Duplicate implicants or subexpressions that appear in multiple outputs can be combined to further reduce the overall complexity of a combinational system.

Karnaugh map minimization for a single logic function can be extended in a straightforward fashion to incorporate simultaneously minimizing multiple logic functions. The basic idea is to consider the implicants that can be shared between expressions along with the implicants for each expression in minimizing logic.

Since the power of Karnaugh maps is based on pattern-matching skills, perhaps the easiest way to introduce multiple-output (function) Karnaugh map minimization is to examine several typical groupings that represent shared implicants. Two such groupings are shown in Figure 4.20 (p. 98). The logic functions F_1 and F_2 are minimized individually in Figure 4.20(a) and collectively in Figure 4.20(b). Figure 4.20(a) uses eight logic operators, whereas Figure 4.20(b) uses only six.

Duplicate prime implicants are an obvious example of Karnaugh map groupings that represent shared implicants. The duplicate **and** operators for the prime implicants $A \cdot C \cdot D$ in Figure 4.20(a) can be combined into a single **and** operator in Figure 4.20(b) to eliminate one logic operator.

Duplicate implicants can also indicate opportunities for multiple-output optimization. Consider the implicant m_5 for both functions F_1 and F_2. In Figure 4.20(a), the **and** operator for the prime implicant $\overline{A} \cdot B \cdot \overline{C}$ covering the minterm m_5 for F_1 and the **and** operator for the prime implicant $\overline{A} \cdot \overline{C} \cdot D$ covering the minterm m_5 for F_2 can be combined to eliminate another logic operator. The resulting single **and** operator for the implicant $\overline{A} \cdot B \cdot \overline{C} \cdot D$ covers the minterm m_5 for both F_1 and F_2 in Figure 4.20(b).

Figure 4.21 (p. 99) shows another set of Karnaugh map groupings that can represent shared implicants. Recognizing that the prime implicants $\overline{A} \cdot \overline{C} \cdot \overline{D}$ in F_2 and $A \cdot \overline{C} \cdot \overline{D}$ in F_3 can collectively cover the left-hand column of 1's (minterms m_0, m_4, m_8, and m_{12}) for F_1 eliminates the **and** logic operator for $\overline{C} \cdot \overline{D}$.

Figure 4.20

Example groupings for Karnaugh map multiple-output minimization

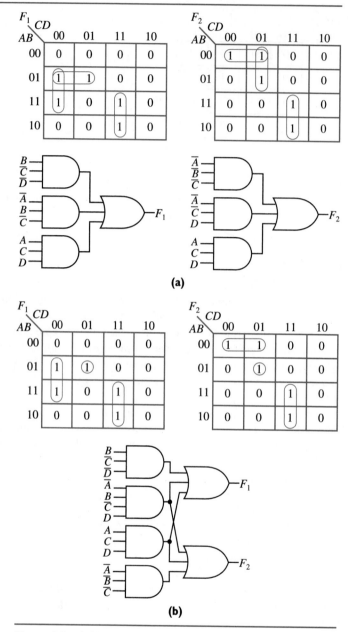

(a)

(b)

Figures (a) and (b) respectively consider minimizing F_1 and F_2 individually and collectively.

4.4.6 Limitations

Karnaugh map logic minimization has several advantages over algebraic logic minimization. Karnaugh map logic minimization exploits human pattern recognition abilities and is more procedurally oriented, detailing a specific set of steps to obtain a minimized expression.

However, Karnaugh map logic minimization also has disadvantages that restrict its general utility. Most notably, Karnaugh maps are typically limited to logic expressions involving

Figure 4.21

Example groupings for Karnaugh map multiple-output minimization

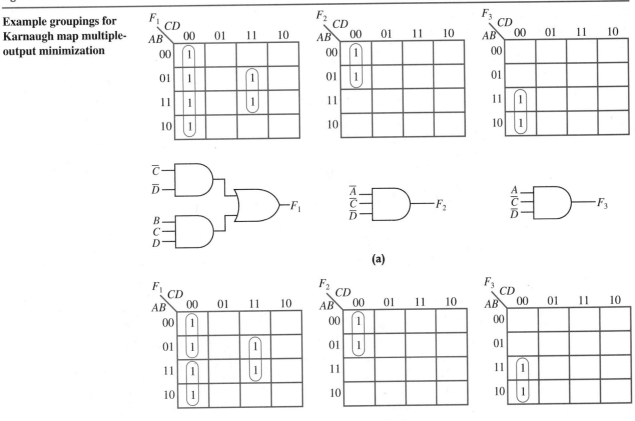

Figures (a) and (b) respectively consider minimizing F_1, F_2, and F_3 individually and collectively.

at most four or five switching variables because maps of six or more variables become unwieldy. Figure 4.22 (p. 100) shows a Karnaugh map for a logic expression of five switching variables, $F(A, B, C, D, E)$. The process of grouping adjacent 1's or 0's is the same as was described for a four-variable Karnaugh map, with the addition that minterms in corresponding positions on the $A = 0$ and $A = 1$ planes are also considered adjacent. If we imagine the $A = 0$ plane lying on top of the $A = 1$ plane, vertically aligned cells are considered adjacent (for example, m_1 is adjacent to m_{17}, m_{13} is adjacent to m_{29}, and so on).

As will be explained in the next section, the Quine-McCluskey minimization technique uses tables instead of maps, which makes the Quine-McCluskey minimization technique more amenable to simplifying more complex logic expressions.

Figure 4.22

Karnaugh map for a logic function of five switching variables

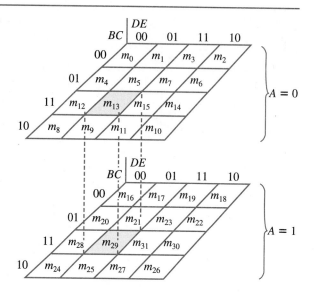

4.5 QUINE-McCLUSKEY MINIMIZATION TECHNIQUE

Quine-McCluskey table logic minimization is based on the same basic principles as Karnaugh map logic minimization. First, the prime implicants are found using successive applications of the Adjacency Theorem. Then, the minimum cover of prime implicants is found by identifying the essential prime implicants and the necessary secondary prime implicants. The Quine-McCluskey minimization technique will be introduced by again simplifying the sum-of-products logic expressions for the *PRINT*_1 and *PRINT*_2 outputs of the printer controller example respectively given in Equations 4.1 and 4.2. We will then follow the same format used for Karnaugh map minimization and discuss sum-of-products (SOP) minimization, product-of-sums (POS) minimization, don't care minimization, and multiple-output minimization, using the Quine-McCluskey table technique.

First, an *implicant table* is constructed to systematically and exhaustively apply the Adjacency Theorem 6a. Table 4.3 shows an implicant table for the *PRINT*_1 logic expression. We start with the implicants that are the SOP minterms of *PRINT*_1, listed in the second column of the table. The four minterms equate to four groups of single 1's on a corresponding Karnaugh map. The implicants are arranged into three rows according to the number of *uncomplemented* variables contained in each implicant. Arranging the implicants in such a manner limits the number of possible applications of the Adjacency Theorem that need to be investigated. Since the Adjacency Theorem applies to only implicants that differ by one variable,

Table 4.3

Quine-McCluskey mini-mization technique	**Number of uncomplemented variables**	**Implicants (SOP minterms)**	**Implicants (Adjacency Theorem #1)**
	2	$ENABLE \cdot \overline{SELECT} \cdot USER\_1 \cdot \overline{USER\_2}$	
	3	$ENABLE \cdot \overline{SELECT} \cdot USER\_1 \cdot USER\_2$	
		$ENABLE \cdot SELECT \cdot \overline{USER\_1} \cdot USER\_2$	
	4	$ENABLE \cdot SELECT \cdot USER\_1 \cdot USER\_2$	

the Adjacency Theorem potentially only applies between implicants in consecutive rows, for example, between rows 1 and 2, rows 2 and 3, et cetera.

To determine if the four groups of single 1's can be combined into larger groupings of two 1's, we look for applications of the Adjacency Theorem. We start by placing a check mark beside all possible combinations of implicants between the first and second rows. We find one application of the Adjacency Theorem given by

$$ENABLE \cdot \overline{SELECT} \cdot USER\_1 = (ENABLE \cdot \overline{SELECT} \cdot USER\_1 \cdot \overline{USER\_2})$$
$$+ (ENABLE \cdot \overline{SELECT} \cdot USER\_1 \cdot USER\_2)$$

Table 4.4 shows that the implicants used in the reduction are checked (column 2) and the resulting implicant is listed in the next column (column 3), again according to the number of uncomplemented variables.

Continuing with the implicants in column 2, there are two applications of the Adjacency Theorem between the implicants in the second and third rows.

$$ENABLE \cdot USER\_1 \cdot USER\_2 = (ENABLE \cdot \overline{SELECT} \cdot USER\_1 \cdot USER\_2)$$
$$+ (ENABLE \cdot SELECT \cdot USER\_1 \cdot USER\_2)$$

$$ENABLE \cdot SELECT \cdot USER\_2 = (ENABLE \cdot SELECT \cdot \overline{USER\_1} \cdot USER\_2)$$
$$+ (ENABLE \cdot SELECT \cdot USER\_1 \cdot USER\_2)$$

The implicants used in the reduction are checked (column 2) and the resulting implicants are added to column 3, as shown in Table 4.5. The implicants listed in the third column of Table 4.5 represent all the possible groupings of two 1's on a Karnaugh map.

Next, we look for all possible groupings of four 1's by trying to successively apply the Adjacency Theorem to the implicants in the newly formed third column. Since the Adjacency Theorem cannot be applied to any combination of these implicants, we are finished applying the Adjacency Theorem. The prime implicants of $PRINT\_1$ are the implicants

Table 4.4

Quine-McCluskey minimization technique	Number of uncomplemented variables	Implicants (SOP minterms)	Implicants (Adjacency Theorem #1)
	2	√ $ENABLE \cdot \overline{SELECT} \cdot USER\_1 \cdot \overline{USER\_2}$	$ENABLE \cdot \overline{SELECT} \cdot USER\_1$
	3	√ $ENABLE \cdot \overline{SELECT} \cdot USER\_1 \cdot USER\_2$	
		$ENABLE \cdot SELECT \cdot \overline{USER\_1} \cdot USER\_2$	
	4	$ENABLE \cdot SELECT \cdot USER\_1 \cdot USER\_2$	

Table 4.5

Quine-McCluskey minimization technique	Number of uncomplemented variables	Implicants (SOP minterms)	Implicants (Adjacency Theorem #1)
	2	√ $ENABLE \cdot \overline{SELECT} \cdot USER\_1 \cdot \overline{USER\_2}$	$ENABLE \cdot \overline{SELECT} \cdot USER\_1$
	3	√ $ENABLE \cdot \overline{SELECT} \cdot USER\_1 \cdot USER\_2$	$ENABLE \cdot USER\_1 \cdot USER\_2$
		√ $ENABLE \cdot SELECT \cdot \overline{USER\_1} \cdot USER\_2$	$ENABLE \cdot SELECT \cdot USER\_2$
	4	√ $ENABLE \cdot SELECT \cdot USER\_1 \cdot USER\_2$	

that have *not* been checked off, namely, the three implicants listed in the third column of Table 4.5. Remember, a check means that the implicant is contained within a larger grouping and thus cannot be a prime implicant.

We next find the minimum cover of prime implicants by identifying the essential prime implicants and the necessary secondary prime implicants by constructing a *covering table*. The covering table for *PRINT*_1 is shown in Table 4.6. Minterms are listed along the top of the covering table and prime implicants are listed down the left side of the covering table. Each row lists the minterms of *PRINT*_1 covered by each prime implicant; covered minterms are marked by an X.

For Karnaugh maps, essential prime implicants are identified by minterm squares covered by only one grouping. For Quine-McCluskey tables, essential prime implicants are identified by minterms covered by only one prime implicant, that is, the columns of the covering table that have only one X. The first and third columns of Table 4.6 each have only one X, so *ENABLE* • $\overline{SELECT}$ • *USER*_1 and *ENABLE* • *SELECT* • *USER*_2 are respectively essential prime implicants. Table 4.7 shows that singularly covered minterms are identified by circling the associated X.

Since the two essential prime implicants cover all the minterms of *PRINT*_1, there is no need to select any secondary prime implicants. The Quine-McCluskey minimization technique is finished and the resulting simplified logic expression for *PRINT*_1 is given by

$$PRINT\_1 = (ENABLE • \overline{SELECT} • USER\_1) + (ENABLE • SELECT • USER\_2)$$

Notice that the Quine-McCluskey minimization technique yields the same logic expression obtained via the algebraic (Equation 4.8) and Karnaugh map (Equation 4.13) minimization techniques.

4.5.1 Cellular Notation

The Quine-McCluskey minimization example given in the previous section illustrates that it is somewhat cumbersome to use the full switching variable names to denote implicants, particularly for long names. To make Quine-McCluskey tables more compact and easier to generate, we will introduce a shorthand notation called *cellular notation*. Cellular notation is based on using minterm numbers.

For example, consider the logic expression

$$F(A, B, C, D) = \sum m(0, 1, 4, 5, 7, 12, 14, 15)$$

given in minterm list notation; the corresponding implicant table using cellular notation is shown in Table 4.8 (p. 104). The first column of implicants (second column of the table) are the SOP minterms, denoted by minterm numbers and grouped according to the number of uncomplemented variables per minterm. The minterm number 0 denotes $m_0 = (\overline{A} • \overline{B} • \overline{C} • \overline{D})$, which has no uncomplemented variables; the minterm number 1 denotes $m_1 = (\overline{A} • \overline{B} • \overline{C} • D)$, which has one uncomplemented variable; and so on.

Two implicants in column 2 in adjacent rows can be combined via the Adjacency Theorem if the difference of their respective minterm numbers is a positive power of two. In subtracting minterm numbers, the minterm number of the implicant in the higher row is subtracted from the minterm number of the implicant in the lower row. For example, consider the implicants in the first row (0) and second row (1 and 4). The minterms m_0 and m_1 can be combined because $1 - 0 = 1 = 2^0$.

$$(\overline{A} • \overline{B} • \overline{C}) = (\overline{A} • \overline{B} • \overline{C} • \overline{D}) + (\overline{A} • \overline{B} • \overline{C} • D)$$

Similarly, the minterms m_0 and m_4 can be combined because $4 - 0 = 4 = 2^2$.

$$(\overline{A} • \overline{C} • \overline{D}) = (\overline{A} • \overline{B} • \overline{C} • \overline{D}) + (\overline{A} • B • \overline{C} • \overline{D})$$

Table 4.6

Covering table for *PRINT_*1 logic expression

	Minterms				
Prime implicants	*ENABLE* · $\overline{SELECT}$ · *USER_1* · $\overline{USER\_2}$	*ENABLE* · $\overline{SELECT}$ · *USER_1* · *USER_2*	*ENABLE* · *SELECT* · *USER_1* · *USER_2*	*ENABLE* · *SELECT* · $\overline{USER\_1}$ · *USER_2*	*ENABLE* · *SELECT* · *USER_1* · *USER_2*
ENABLE · $\overline{SELECT}$ · *USER_1*	X		X		
ENABLE · *USER_1* · *USER_2*			X		X
ENABLE · *SELECT* · *USER_2*				X	X

Table 4.7

Covering table for *PRINT_*1 logic expression

	Minterms				
Prime implicants	*ENABLE* · $\overline{SELECT}$ · *USER_1* · $\overline{USER\_2}$	*ENABLE* · $\overline{SELECT}$ · *USER_1* · *USER_2*	*ENABLE* · *SELECT* · *USER_1* · *USER_2*	*ENABLE* · *SELECT* · $\overline{USER\_1}$ · *USER_2*	*ENABLE* · *SELECT* · *USER_1* · *USER_2*
ENABLE · $\overline{SELECT}$ · *USER_1*	⊗		X		
ENABLE · *USER_1* · *USER_2*			X		X
ENABLE · *SELECT* · *USER_2*				⊗	X

Table 4.8

Quine-McCluskey mini-mization using cellular notation	Number of uncomplemented variables	Implicants (SOP minterms)	Implicants (Adjacency Theorem #1)	Implicants (Adjacency Theorem #2)
	0	0		
	1	1		
		4		
	2	5		
		12		
	3	7		
		14		
	4	15		

These two logic reductions are recorded in Table 4.9. The implicant $(\overline{A} \cdot \overline{B} \cdot \overline{C})$ is denoted by the cellular notation 0,1(1) in column 3: the first two numbers, 0,1, record the minterms that form the implicant and the last number, (1), is the minterm difference that records the weight of the deleted switching variable D. The switching variables A, B, C, and D are respectively of weight $2^3 = 8$, $2^2 = 4$, $2^1 = 2$, and $2^0 = 1$. In a similar fashion, the implicant $(\overline{A} \cdot \overline{C} \cdot \overline{D})$ is denoted by 0,4(4), recording that minterms 0 and 4 form the implicant and variable B is deleted. Following the same practice explained earlier, each time an implicant is involved in an application of the Adjacency Theorem, the implicant is checked off.

Table 4.10 shows the results of the first pass of the Adjacency Theorem, listing all four-variable implicants in column 2 and all three-variable implicants in column 3. Review column 3 to understand the logic reductions and the cellular notations.

The Quine-McCluskey logic reduction process continues by repeatedly applying the Adjacency Theorem to column 3, combining three-variable implicants into two-variable implicants. Two implicants in column 3 in adjacent rows can be combined via the Adjacency Theorem if the difference between the first two minterms is a positive power of two *and* the weights given in parentheses are the same.

For example, implicants 0,1(1) and 4,5(1) can be combined because $4 - 0 = 4 = 2^2$ and both implicants are missing the D switching variable of weight 1.

$$(\overline{A} \cdot \overline{C}) = (\overline{A} \cdot \overline{B} \cdot \overline{C}) + (\overline{A} \cdot B \cdot \overline{C})$$

Table 4.9

Combining four-variable implicants to form three-variable implicants	Number of uncomplemented variables	Implicants (SOP minterms)	Implicants (Adjacency Theorem #1)	Implicants (Adjacency Theorem #2)
	0	√ 0	0,1(1) 0,4(4)	
	1	√ 1		
		√ 4		
	2	5		
		12		
	3	7		
		14		
	4	15		

Table 4.10

Three- and four-variable implicants	Number of uncomplemented variables	Implicants (SOP minterms)	Implicants (Adjacency Theorem #1)	Implicants (Adjacency Theorem #2)
	0	$\checkmark$ 0	0,1(1) 0,4(4)	
	1	$\checkmark$ 1 $\checkmark$ 4	1,5(4) 4,5(1) 4,12(8)	
	2	$\checkmark$ 5 $\checkmark$ 12	5,7(2) 12,14(2)	
	3	$\checkmark$ 7 $\checkmark$ 14	7,15(8) 14,15(1)	
	4	$\checkmark$ 15		

The resulting implicant ($\overline{A} \cdot \overline{C}$) is denoted by the cellular notation 0,1,4,5(1,4) in column 4 in Table 4.11, recording that minterms 0, 1, 4, and 5 form the implicant and the switching variables B of weight 4 and D of weight 1 are deleted. For completeness, Table 4.11 also shows that the implicants 0,4(4) and 1,5(4) can be combined to yield the implicant 0,4,1,5(1,4) because $1 - 0 = 1 = 2^0$ and both implicants are missing the B switching variable of weight 4. However, since 0,1,4,5(1,4) and 0,4,1,5(1,4) denote the *same* implicant, we will arbitrarily elect to use 0,1,4,5(1,4) and ignore 0,4,1,5(1,4).

The implicant table in Table 4.11 is complete because no further logic reduction can be performed. All implicants that are *not* checked are prime implicants. To identify essential and secondary prime implicants, the prime implicants are listed in the covering table shown in Table 4.12. The circled X's identify minterms covered by only one prime implicant, making the associated prime implicant an essential prime implicant. The implicant 0,1,4,5(1,4) is the only essential prime implicant.

The last step in the Quine-McCluskey minimization technique is to cover the remaining minterms by selecting secondary prime implicants. To that end, a *reduced covering table* for the minterms m_7, m_{12}, m_{14}, and m_{15} is shown in Table 4.13. There are several possible combinations of secondary prime implicants that can be selected to cover the minterms because each column of the reduced covering table has more than one X. A covering table that has at

Table 4.11

Quine-McCluskey minimization — completed implicant table	Number of uncomplemented variables	Implicants (SOP minterms)	Implicants (Adjacency Theorem #1)	Implicants (Adjacency Theorem #2)
	0	$\checkmark$ 0	$\checkmark$ 0,1(1) $\checkmark$ 0,4(4)	0,1,4,5(1,4) 0,4,1,5(1,4)
	1	$\checkmark$ 1 $\checkmark$ 4	$\checkmark$ 1,5(4) $\checkmark$ 4,5(1) 4,12(8)	
	2	$\checkmark$ 5 $\checkmark$ 12	5,7(2) 12,14(2)	
	3	$\checkmark$ 7 $\checkmark$ 14	7,15(8) 14,15(1)	
	4	$\checkmark$ 15		

Table 4.12

Covering table using cellular notation

Prime implicants	Minterms							
	0	1	4	5	7	12	14	15
0,1,4,5(1,4)	⊗	⊗	X	X				
5,7(2)				X	X			
12,14(2)						X	X	
7,15(8)					X			X
14,15(1)							X	X
4,12(8)			X			X		

least two X's in each column is called a *cyclic table*. Since Table 4.13 is relatively small, examining a few candidate selections quickly reveals that the secondary prime implicants 7,15(8) and 12,14(2) provide an efficient covering.

The Petrick algorithm introduced for Karnaugh map minimization can also be used for Quine-McCluskey table minimization to provide a more systematic way of selecting secondary prime implicants. The steps of the Petrick algorithm are the same, except we will use cellular notation for the groups. The Petrick algorithm translates the reduced covering table into the following equivalent product-of-sums expression.

$$m_7 \bullet m_{12} \bullet m_{14} \bullet m_{15} \qquad (4.18)$$

$$(5,7(2) + 7,15(8)) \bullet (12,14(2) + 4,12(8)) \bullet (14,15(1) + 12,14(2)) \bullet (7,15(8) + 14,15(1))$$

Equation 4.18 states that the minterm m_7 is covered by the implicants 5,7(2) *or* 7,15(8) *and* the minterm m_{12} is covered by the implicants 12,14(2) *or* 4,12(8) *and* so on. Expanding Equation 4.18 by repeated applications of Distributive Postulate 4b and simplifying product terms by the Idempotency Theorem 3b and Absorption Theorem 7a yields the equivalent sum-of-products expression given in Equation 4.19.

$$(5,7(2) \bullet 12,14(2) \bullet 14,15(1)) + (5,7(2) \bullet 4,12(8) \bullet 14,15(1))$$
$$(7,15(8) \bullet 4,12(8) \bullet 14,15(1)) + (7,15(8) \bullet 12,14(2)) \qquad (4.19)$$

Equation 4.19 states that the remaining minterms can be covered using the following sets of secondary prime implicants:
- 5,7(2), 12,14(2), and 14,15(1)
- 5,7(2), 4,12(8), and 14,15(1)
- 7,15(8), 4,12(8), and 14,15(1)
- 7,15(8) and 12,14(2)

Table 4.13

Reduced covering table using cellular notation

Prime implicants	Minterms			
	7	12	14	15
5,7(2)	X			
7,15(8)	X			X
14,15(1)			X	X
12,14(2)		X	X	
4,12(8)		X		

The last set, containing the two implicants 7,15(8) and 12,14(2), is a good choice of secondary prime implicants because it covers the remaining minterms with the least amount of logic, thus verifying our earlier by-inspection selection of secondary prime implicants.

We have completed the Quine-McCluskey minimization technique; the resulting simplified expression for $F(A, B, C, D)$ is given below.

$$F(A, B, C, D) = 0,1,4,5(1,4) + 7,15(8) + 12,14(2)$$
$$= (\overline{A} \cdot \overline{C}) + (B \cdot C \cdot D) + (A \cdot B \cdot \overline{D})$$

Let us summarize what we have learned thus far about Quine-McCluskey minimization.

Quine-McCluskey Minimization Technique—SOP

1. *Construct the implicant table.*
   a. *Form the first column of implicants by grouping the minterms according to the number of uncomplemented variables contained per minterm.*
   b. *Form each succeeding column of implicants by applying the Adjacency Theorem to pairs of implicants contained in the previous column.*
      - *Check only pairs of implicants in adjacent rows.*
      - *Two implicants of adjacent rows can be combined if the difference of their respective minterm numbers is a positive power of two and they contain the same switching variables.*
   c. *Mark all implicants that are used in successfully applying the Adjacency Theorem.*
2. *Construct the covering table.*
   a. *The implicants that are not marked in the implicant table are prime implicants. List the prime implicants vertically and the minterms horizontally.*
   b. *For each prime implicant, fill in the corresponding row by marking the columns of the minterms covered by the prime implicant.*
   c. *Mark the minterms that are covered by only one prime implicant to identify essential prime implicants. If all minterms are covered by the essential prime implicants, the minimization process is finished.*
3. *Construct the reduced covering table.*
   a. *The implicants not marked as essential prime implicants are secondary prime implicants. List the secondary prime implicants vertically and the remaining minterms horizontally.*
   b. *For each secondary prime implicant, fill in the corresponding row by marking the columns of the minterms covered by the secondary prime implicant.*
   c. *Identify a minimum number of secondary prime implicants that best cover the remaining minterms by inspection or by the Petrick algorithm.*
4. *Combine the essential prime implicants with the selected secondary prime implicants to generate the resulting simplified logic expression in a sum-of-products form.*

4.5.2 Product-of-Sums (POS) Minimization

Recall from the earlier Karnaugh map discussions that sometimes 0's group better than 1's, in which case, a product-of-sums (POS) form yields a better minimized logic expression than a sum-of-products (SOP) form. The relationships that exist between SOP and POS Karnaugh map minimization techniques also exist between SOP and POS Quine-McCluskey table minimization techniques, where we group 0's instead of 1's and invert the resulting expression. Grouping 0's yields the minimal SOP form of $\overline{F}$. Inverting $\overline{F}$ using the generalized DeMorgan's Theorems yields the desired minimal POS form of F.

As an example, let us reconsider simplifying the logic function discussed in Section 4.4.3.

$$F(A, B, C, D) = \prod M(0, 1, 3, 7, 8, 10, 11, 15)$$

Table 4.14

POS Quine-McCluskey minimization

Number of Complemented variables	Implicants (POS maxterms)	Implicants (Adjacency Theorem #1)	Implicants (Adjacency Theorem #2)
0	$\sqrt{}$ 0	0,1(1) 0,8(8)	
1	$\sqrt{}$ 1 $\sqrt{}$ 8	1,3(2) 8,10(2)	
2	$\sqrt{}$ 3 $\sqrt{}$ 10	$\sqrt{}$ 3,7(4) $\sqrt{}$ 3,11(8) 10,11(1)	3,7,11,15(4,8) 3,11,7,15(4,8)
3	$\sqrt{}$ 7 $\sqrt{}$ 11	$\sqrt{}$ 7,15(8) $\sqrt{}$ 11,15(4)	
4	$\sqrt{}$ 15		

The logic function F is given in canonical POS form. In POS Quine-McCluskey table minimization, we start by listing the *maxterms* in the second column of Table 4.14. We order the maxterms by number of *complemented* variables per maxterm. The maxterm number 0 denotes $M_0 = (A + B + C + D)$, which has no complemented variables; the maxterm number 1 denotes $M_1 = (A + B + C + \overline{D})$, which has one complemented variable; and so on. All groups of two 0's are listed in the third column. For example, the maxterms M_0 and M_1 can be combined using the Adjacency Theorem 6b because $1 - 0 = 1 = 2^0$.

$$(A + B + C) = (A + B + C + D) + (A + B + C + \overline{D})$$

The one group of four 0's is listed in the fourth column. The unchecked implicants are prime implicants and are listed in the covering table in Table 4.15.

The implicant 3,7,11,15(4,8) is the only essential prime implicant. The remaining secondary prime implicants are listed in the cyclic reduced covering table in Table 4.16. The secondary prime implicants 0,1(1) and 8,10(2) form a minimal cover for the remaining maxterms, yielding the complement of F.

$$\overline{F(A, B, C, D)} = 3,7,11,15(4,8) + 8,10(2) + 0,1(1)$$
$$= (C \cdot D) + (A \cdot \overline{B} \cdot \overline{D}) + (\overline{A} \cdot \overline{B} \cdot \overline{C})$$

Inverting the complement of F using the generalized DeMorgan Theorems quickly yields the minimal POS form of F.

$$F(A, B, C, D) = (\overline{C} + \overline{D}) \cdot (\overline{A} + B + D) \cdot (A + B + C) \tag{4.20}$$

Table 4.15

POS covering table using cellular notation

Prime implicants	Maxterms							
	0	1	3	7	8	10	11	15
0,8(8)	X				X			
0,1(1)	X	X						
1,3(2)		X	X					
8,10(2)					X	X		
10,11(1)						X	X	
3,7,11,15(4,8)			X	Ⓧ			X	Ⓧ

Table 4.16

POS reduced covering table using cellular notation

Prime implicants	Maxterms			
	0	1	8	10
0,8(8)	X		X	
0,1(1)	X	X		
1,3(2)		X		
8,10(2)			X	X
10,11(1)				X

The simplified form of the function F in Equation 4.20, generated using POS Quine-McCluskey table minimization, is equal to the simplified form of the function F in Equation 4.16, generated using POS Karnaugh map minimization.

4.5.3 Don't Care Input Values

In Section 4.4.4, we discussed how don't care conditions can be exploited to improve the results of Karnaugh map minimization. Don't care input conditions can be similarly exploited to improve the results of Quine-McCluskey minimization. The basic Quine-McCluskey algorithm remains the same, with the additions that don't care implicants are ignored in the implicant table and don't care minterms are *not* included in covering tables.

The following example illustrates Quine-McCluskey minimization with don't care input conditions.

$$F(A, B, C, D) = \sum m(2, 3, 5, 7) + d(6, 8, 9)$$

The implicant table is shown in Table 4.17. Don't care minterms d_6, d_8, and d_9 are used with the 1 minterms m_2, m_3, m_5, and m_7 to help form successively larger groupings. For example, the don't care minterm d_6 is used to form the 2,3,6,7(1,4) grouping. The implicant table shows that there are two prime implicants: 2,3,6,7(1,4) and 5,7(2). Although the product term 8,9(1) is not checked in the implicant table, 8,9(1) is ignored and not considered as a prime implicant because it is composed entirely of don't care minterms. Also, the product terms 2,3,6,7(1,4) and 2,6,3,7(1,4) denote the same prime implicant, so we arbitrarily elect to use 2,3,6,7(1,4) and ignore 2,6,3,7(1,4).

The covering table is shown in Table 4.18. Since we are only concerned with covering

Table 4.17

Quine-McCluskey minimization using don't cares

Number of uncomplemented variables	Implicants (SOP minterms)	Implicants (Adjacency Theorem #1)	Implicants (Adjacency Theorem #2)
1	√ 2 √ 8	√ 2,3(1) √ 2,6(4) 8,9(1)	2,3,6,7(1,4) 2,6,3,7(1,4)
2	√ 3 √ 5 √ 6 √ 9	√ 3,7(4) 5,7(2) √ 6,7(1)	
3	√ 7		

Table 4.18

Covering table using don't cares

Prime implicants	Minterms			
	2	3	5	7
2,3,6,7(1,4)	Ⓧ	Ⓧ		X
5,7(2)			Ⓧ	X

the 1 minterms m_2, m_3, m_5, and m_7, the don't care minterms d_6, d_8, and d_9 are not included as columns in the covering table. The covering table reveals that both prime implicants are essential prime implicants. The resulting minimized expression for $F(A, B, C, D)$ is given by

$$F(A, B, C, D) = 2,3,6,7(1,4) + 5,7(2)$$
$$= (\overline{A} \cdot C) + (\overline{A} \cdot B \cdot D)$$

As another example of don't care Quine-McCluskey table minimization, Table 4.19 shows the implicant table for the *HEAT* function of our familiar temperature controller.

$$HEAT(TEMP2, TEMP1, TEMP0) = \sum m(0,1) + d(4,5)$$

The result

$$HEAT(TEMP2, TEMP1, TEMP0) = 0,1,4,5(1,4)$$
$$= \overline{TEMP1}$$

is equal to the simplified result obtained in Section 4.4.4.

4.5.4 Multiple Outputs

Quine-McCluskey table minimization can be extended to accommodate simultaneous consideration of multiple outputs by adding an implicant table to identify common prime implicants and extending covering tables to include shareable implicants. Consider a combinational system having two outputs, $F_1(A, B, C, D)$ and $F_2(A, B, C, D)$, given in minterm list notation as

$$F_1(A, B, C, D) = \sum m(0, 1, 2, 3, 8, 9)$$
$$F_2(A, B, C, D) = \sum m(2, 3, 5, 13)$$

We start the multiple-output Quine-McCluskey minimization by finding the prime implicants for the individual function F_1 and F_2. We will also need to find the prime implicants for a third function $F_{1,2}(A, B, C, D) = \sum m(2, 3)$, which represents the minterms common to

Table 4.19

Quine-McCluskey minimization using don't cares

Number of uncomplemented variables	Implicants (SOP minterms)	Implicants (Adjacency Theorem #1)	Implicants (Adjacency Theorem #2)
0	√ 0	√ 0,4(4) √ 0,1(1)	0,1,4,5(1,4) 0,4,1,5(1,4)
1	√ 1 √ 4	√ 1,5(4) √ 4,5(1)	
2	√ 5		

both F_1 and F_2. The prime implicants of $F_{1,2}(A, B, C, D)$ can be shared between F_1 and F_2. Using the process described in the previous section, we construct three implicant tables:

1. an implicant table for F_1 given in Table 4.20,
2. an implicant table for F_2 given in Table 4.21, and
3. an implicant table for $F_{1,2}$ given in Table 4.22.

Then, we use the prime implicants from the implicant tables to construct the covering table shown in Table 4.23 (p. 112). The circled X's show that the term 0,1,8,9(1,8) is an essential prime implicant for F_1 and the term 5,13(8) is an essential prime implicant for F_2. The reduced covering table is shown in Table 4.24.

The Petrick algorithm can be used to list all possible combinations of secondary implicants and identify the best choice. Alternatively, we can observe that the secondary prime implicant 2,3(1) of $F_{1,2}$ is the best choice because it covers all the minterms that 0,1,2,3(1,2) of F_1 and 2,3(1) of F_2 cover. In other words, the secondary prime implicant 2,3(1) of $F_{1,2}$ includes or *dominates* the other two secondary prime implicants.

Table 4.20

Multiple-output Quine-McCluskey minimization	Number of uncomplemented variables	Implicants (SOP minterms)	Implicants (Adjacency Theorem #1)	Implicants (Adjacency Theorem #2)
	0	√ 0	√ 0,1(1) √ 0,2(2) √ 0,8(8)	0,1,2,3(1,2) 0,2,1,3(1,2) 0,1,8,9(1,8) 0,8,1,9(1,8)
	1	√ 1 √ 2 √ 8	√ 1,3(2) √ 2,3(1) √ 1,9(8) √ 8,9(1)	
	2	√ 3 √ 9		

Table 4.21

Multiple-output Quine-McCluskey minimization	Number of uncomplemented variables	Implicants (SOP minterms)	Implicants (Adjacency Theorem #1)
	1	√ 2	2,3(1)
	2	√ 3 √ 5	5,13(8)
	3	√ 13	

Table 4.22

Multiple-output Quine-McCluskey minimization	Number of uncomplemented variables	Implicants (SOP minterms)	Implicants (Adjacency Theorem #1)
	1	√ 2	2,3(1)
	2	√ 3	

Table 4.23

Covering table for multiple outputs

Function	Prime implicants	Minterms F_1						Minterms F_2			
		0	1	2	3	8	9	2	3	5	13
F_1	0,1,2,3(1,2)	X	X	X	X						
	0,1,8,9(1,8)	X	X			⊗	⊗				
F_2	2,3(1)							X	X		
	5,13(8)									⊗	⊗
$F_{1,2}$	2,3(1)			X	X			X	X		

Table 4.24

Reduced covering table for multiple outputs

Function	Prime implicants	Minterms F_1		Minterms F_2	
		2	3	2	3
F_1	0,1,2,3(1,2)	X	X		
F_2	2,3(1)			X	X
$F_{1,2}$	2,3(1)	X	X	X	X

The results of minimizing F_1 and F_2 individually and collectively are shown in Figure 4.23(a) and Figure 4.23(b), respectively. Notice two important points. First, simultaneously minimizing the two output functions results in an overall simpler design. Second, although the overall design in Figure 4.23(b) is simpler, the complexity of F_1 is actually *increased*.

Figure 4.23

Minimizing combinational systems with multiple outputs

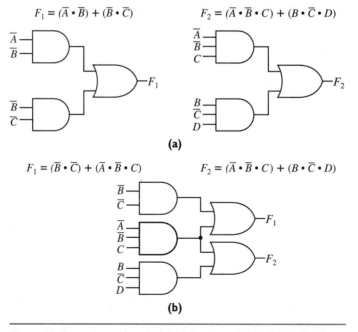

(a)

(b)

> **Problem-Solving Skill**
> *Locally optimizing the parts of a solution does not always ensure that the overall solution is globally optimized. Sometimes, sacrificing local optimality can lead to a better global optimality for the overall system.*

Figure (a) shows the results of minimizing F_1 and F_2 individually. Figure (b) shows the results of minimizing F_1 and F_2 collectively.

4.6 EXERCISE: WEATHER VANE

To conclude this chapter and summarize what we have learned, we will consider the following digital combinational system.

Design a combinational system that accepts four inputs indicating north, south, east, and west wind direction and generates an output whenever the wind direction is northeast (north and east directions asserted) or northwest (north and west directions asserted).

We first define the system input/output logic variables in Table 4.25. Defining logic variables reveals an ambiguity in the English functional specification: The system specification does not define input/output active states. For convenience, we have assumed active-1 inputs and outputs.

The truth table for the weather vane is shown below.

Row	N S E W	N_EW
0	0 0 0 0	0
1	0 0 0 1	0
2	0 0 1 0	0
3	0 0 1 1	–
4	0 1 0 0	0
5	0 1 0 1	0
6	0 1 1 0	0
7	0 1 1 1	–
8	1 0 0 0	0
9	1 0 0 1	1
10	1 0 1 0	1
11	1 0 1 1	–
12	1 1 0 0	–
13	1 1 0 1	–
14	1 1 1 0	–
15	1 1 1 1	–

We have taken advantage of the physical constraints on wind direction to introduce don't care conditions. That is, it is physically impossible for wind to simultaneously blow in opposing directions; thus, any entry asserting both north and south or east and west will never occur and can be considered a don't care input.

Figure 4.24 (p. 114) shows that the Karnaugh map and Quine-McCluskey table logic minimizations both yield the same answer. A product-of-sums minimization yields a slightly better design than a sum-of-products minimization.

$$N\_EW = N \cdot (E + W)$$

Table 4.25

System input/output logic variables	Logic variable name	Description
	N	Input: Assert active-1 if North wind
	S	Input: Assert active-1 if South wind
	E	Input: Assert active-1 if East wind
	W	Input: Assert active-1 if West wind
	N_EW	Output: Assert active-1 if Northeast or Northwest wind

Figure 4.24

Karnaugh map and Quine-McCluskey table minimization of weather vane

NS \ EW	00	01	11	10
00	0	0	–	0
01	0	0	–	0
11	–	–	–	–
10	0	1	–	1

0	✓0,1(1)	✓0,4(4)	✓0,1,2,3(1,2) ✓0,2,4,6(2,4)	0,1,2,3,4,5,6,7(1,2,4)
	✓0,2(2)	✓0,8(8)	✓0,1,4,5(1,4) 0,4,8,12(4,8)	
			✓1,3,5,7(2,4)	
1	✓1,3(2)	✓2,6(4)	✓2,3,6,7(1,4)	
2	✓2,3(1)	✓4,6(2)	✓4,5,6,7(1,2)	4,5,6,7,12,13,14,15(1,2,8)
4	✓1,5(4)	✓4,12(8)	✓4,5,12,13(1,8)	
8	✓4,5(1)	✓8,12(4)	✓4,6,12,14(2,8)	
3	✓3,7(4)	✓5,13(8)	✓3,7,11,15(4,8)	
5	✓5,7(2)	✓12,13(1)	✓5,7,13,15(2,8)	
6	✓6,7(1)	✓6,14(8)	✓6,7,14,15(1,8)	
12	✓3,11(8)	✓12,14(2)	✓12,13,14,15(1,2)	
7				
11	7,15(8)	13,15(2)		
13	11,15(4)	14,15(1)		
14				
15				

Prime Implicants	Maxterms						
	0	1	2	4	5	6	8
0,1,2,3,4,5,6,7(1,2,4)	x	(x)	(x)	x	x	x	
4,5,6,7,12,13,14,15(1,2,8)				x	x	x	
0,4,8,12(4,8)	x			x			(x)

It is instructive to note that the logic expression could have been easily generated directly from the initial functional specification: The *N_EW* is asserted when *N* is asserted and either *E* or *W* is also asserted. This observation serves as a useful "sanity check" on our solution.

4.7 SUMMARY

In this chapter, we introduced combinational digital synthesis. Synthesis is the first stage in design, whereby an informal system specification is transformed into a formal, optimized mathematical expression. We examined how a combinational system is often initially defined via an English description of desired input/output function and then discussed how to translate such a description into switching algebra—the mathematics of combinational logic. Then we saw how the switching algebra expression is minimized with respect to the number of logic operators and the inputs per operator to minimize the actual physical hardware that will be used to realize the mathematical expression in the second stage of design, implementation.

We discussed three minimization techniques:

- algebraic transformations,

- Karnaugh maps, and
- Quine-McCluskey tables.

Algebraic techniques apply postulates and theorems to simplify logic expressions. Algebraic techniques can sometimes be the easiest minimization approach, but are often difficult to use for complex expressions. For logic expressions involving one to four switching variables, Karnaugh maps offer a simple, systematic procedure for obtaining a minimized logic expression. Karnaugh maps rely on the pattern-matching strengths of humans to recognize graphical applications of the Adjacency Theorem. For logic expressions involving more than four switching variables, Quine-McCluskey tables offer a structured procedure for obtaining a minimized logic expression. Similar to Karnaugh maps, Quine-McCluskey tables use successive applications of the Adjacency Theorem.

We also discussed the options of generating either a minimal product-of-sums (POS) or sum-of-products (SOP) form for a logic expression. SOP minimization focuses on the conditions when the logic expression yields a 1, whereas POS minimization focuses on the conditions when the logic expression yields a 0. Typically, one of the forms will be more conducive to minimization than the other.

Finally, we learned how to exploit don't care input conditions and commonalities between multiple-output logic functions during the minimization process. A designer is typically not concerned about the system output in response to a don't care input condition because the output is not being monitored when the particular input combination occurs or the input combination should never occur. In either case, we have the added flexibility to assign output values for don't care inputs in a manner that simplifies the logic design. Further gains in logic reduction can be obtained by considering multiple-output logic expressions simultaneously during the minimization process to identify and take advantage of common product terms.

4.8 PROBLEMS

1. An error-checking system accepts three binary signals, A, B, and C. The output is asserted when
   - $A = B$ and $A \neq C$ or
   - $A = C$ and $A \neq B$.

   a. Generate a truth table, a canonical sum-of-products logic expression, and a canonical product-of-sums logic expression for the proposed combinational system, assuming an active-1 output.

   b. Generate a truth table, a canonical sum-of-products logic expression, and a cononical product-of-sums logic expression for the proposed combinational system, assuming an active-0 output.

2. As part of a local high school science fair program, you have been asked to design a digital system to detect even and odd numbers. The input is a 4-bit binary number (no special encoding) and the output must drive the following liquid crystal display. If the input is an even number (consider 0 an even number), then segments 1, 2, 4, and 5 must be asserted to display the letter "E." If the input is an odd number, then segments 1, 2, 3, and 5 must be asserted to display the letter "O."

   a. Generate a truth table, sum-of-products logic expressions, and product-of-sums logic expressions for the proposed combinational system, assuming active-1 outputs.

   b. Generate a truth table, sum-of-products logic expressions, and product-of-sums logic expressions for the proposed combinational system, assuming active-0 outputs.

3. A digital system compares two 2-bit binary numbers, A and B. The result of the comparison is given by three output signals, LT, EQ, and GT. The output LT is asserted only if $A < B$; the output EQ is asserted only if $A = B$; and the output GT is asserted only if $A > B$.

   a. Generate a truth table, canonical sum-of-products logic expressions, and canonical product-of-sums logic expressions for the proposed combinational system, assuming active-1 outputs.

   b. Generate a truth table, canonical sum-of-products logic expressions, and canonical product-of-sums logic expressions for the proposed combinational system, assuming active-0 outputs.

4. Prove the Absorption Theorem 7a

$$X + X \cdot Y = X$$

   a. Using perfect induction.
   b. Using algebraic transformations.

5. Prove the Simplification Theorem 8a

$$X + \overline{X} \cdot Y = X + Y$$

   a. Using perfect induction.
   b. Using algebraic transformations. (*Hint:* $X + Y \cdot Z = (X + Y) \cdot (X + Z)$)

6. For each logic equivalence given below, show how the logic expression on the left of the equals sign can be reduced to the logic expression on the right of the equals sign using switching algebraic transformations. State the postulates and/or theorems used.

   a. $(X + Z) \cdot (X + \overline{X} \cdot Y) \cdot (X + \overline{Z}) = X$
   b. $(A \cdot B \cdot C \cdot D) + (A \cdot D) + (\overline{A} \cdot B \cdot C) + (\overline{A} \cdot D) + (B \cdot C \cdot D) = (\overline{A} \cdot B \cdot C) + D$
   c. $(A \cdot B) + (A \cdot B \cdot C) + (\overline{A} \cdot D) + (\overline{B} \cdot D) = (A \cdot B) + D$

7. Find the result of the following logic expression.

$$\overline{(A \cdot B + \overline{C}) \cdot (\overline{D} + E \cdot \overline{F}) \cdot (G + \overline{H})}$$

8. Show how the following logic expressions can be represented by a Karnaugh map.

   a. $(A \cdot B \cdot C) + (\overline{A} \cdot \overline{B}) + (B \cdot \overline{C})$
   b. $C + (A \cdot B \cdot \overline{C} \cdot D) + (A \cdot \overline{B} \cdot \overline{C})$
   c. $(\overline{A} \cdot \overline{B} \cdot \overline{C} \cdot \overline{D}) + (\overline{A} \cdot B \cdot C \cdot \overline{D}) + (A \cdot B \cdot C \cdot D) + (A \cdot \overline{B} \cdot C \cdot \overline{D})$

9. For the combinational system defined in Problem 1,

   a. Map the active-1 output onto a Karnaugh map.
   b. Identify all implicants, prime implicants, essential prime implicants, and secondary prime implicants.
   c. Obtain the minimized logic expression.

10. For the combinational system defined in Problem 3,

   a. Map the active-1 outputs onto Karnaugh maps.
   b. Identify all implicants, prime implicants, essential prime implicants, and secondary prime implicants.
   c. Obtain the minimized logic expressions.

11. For the combinational system defined by the following minterm list,

$$F(A, B, C, D) = \sum m\ (0, 2, 6, 7, 8, 10, 13)$$

   a. Map the function onto a Karnaugh map.
   b. Identify all implicants, prime implicants, essential prime implicants, and secondary prime implicants.
   c. Obtain the minimized logic expression.

12. For the combinational system defined by the following minterm list,

$$F(A, B, C, D) = \sum m\ (5, 6, 7, 12, 13, 14)$$

   a. Map the function onto a Karnaugh map.

   b. Identify all implicants, prime implicants, essential prime implicants, and secondary prime implicants.

   c. Obtain the minimized logic expression.

13. For the following logic expression,

$$F(A, B, C, D, E) = \sum m(2, 4, 5, 7, 10, 11, 21, 28) + d(3, 18)$$

   a. Map the logic expression onto a Karnaugh map.

   b. Identify all implicants, prime implicants, essential prime implicants, and secondary prime implicants.

   c. Obtain the minimized logic expression.

14. For the following logic expression,

$$F(A, B, C, D) = \sum m(4, 5, 7, 8, 11, 12, 15)$$

   a. Generate the minimized sum-of-products expression.

   b. Generate the minimized product-of-sums expression.

   c. Is one form better than the other form? Explain your answer.

15. Minimize the following logic expressions individually using Quine-McCluskey tables. Show the implicant table. Show the covering table and, if necessary, the reduced covering table.

   a. $F_1(A, B, C, D) = \sum m(0, 2, 10, 13)$

   b. $F_2(A, B, C, D) = \sum m(5, 7, 13, 15)$

   c. $F_3(A, B, C, D) = \sum m(1, 8, 9, 10, 12, 14)$

16. Minimize the logic expression given in Problem 13 using Quine-McCluskey tables.

   a. Show the implicant table.

   b. Show the covering table and, if necessary, the reduced covering table.

17. Minimize the following logic expressions individually and collectively using Quine-McCluskey tables.

$$F_1(A, B, C, D) = \sum m(2, 3, 4, 8, 9, 10, 11)$$
$$F_2(A, B, C, D) = \sum m(3, 4, 7, 8, 9)$$

   a. Show the implicant tables.

   b. Show the covering tables and, if necessary, the reduced covering tables.

   c. Do the individual and collective minimizations yield different results?

18. Many logic minimization algorithms used by design automation tools use the concept of *Boolean cubes*. A truth table for an *n*-input logic expression can be represented as an *n*-dimensional Boolean cube. As an example, the Boolean cube representation of the function $F = \sum m(4, 5, 6)$ is shown below. The eight nodes or vertices of the Boolean cube represent the eight input condi-

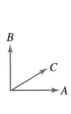
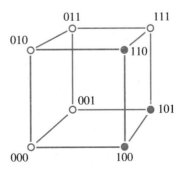

A	B	C	F
0	0	0	0
0	0	1	0
0	1	0	0
0	1	1	0
1	0	0	1
1	0	1	1
1	1	0	1
1	1	1	0

tions; darkened nodes, called the *on-set*, denote minterms. Adjacent minterms linked by a single Boolean cube edge can be combined using the Adjacency Theorem.

a. Draw a Boolean cube representation for $F = \sum m(0, 2, 4, 6)$.

b. Identify all adjacent minterms and applications of the Adjacency Theorem. (*Hint*: Minterms linked by a Boolean cube plane can be combined using successive applications of the Adjacency Theorem.)

c. Generate a minimized logic expression.

d. Repeat part (c) using a Karnaugh map.

REFERENCES AND READINGS

[1] Newton, R., ed. *Selected Papers on Logic Synthesis for Integrated Circuit Design.* IEEE Press, 1987.

[2] Edwards, M., ed. *Automatic Logic Synthesis Techniques for Digital Systems.* New York: Mc-Graw-Hill, 1992.

[3] Brayton, R. "New Directions in Logic Synthesis," *Proc. VLSI Logic Synthesis and Design Conf.,* Kyoto, Japan, 1991.

[4] Brayton, R., G. Hachtel, and C. McMullen. *Espresso-IIC: Logic Minimization Algorithms for VLSI Synthesis.* Boston MA: Kluwer Academic Publishers, 1984.

[5] Fabricius, E. *Modern Digital Design and Switching Theory.* Ann Arbor, MI: CRC Press, 1992.

[6] Hill, F. and G. Peterson. *Introduction to Switching Theory and Logical Design.* New York: Wiley, 1981.

[7] Karnaugh, M. "A Map for Synthesis of Combinational Logic," *Trans. AIEE, Comm. and Electron.,* vol. 27, pp. 593–599, November 1953.

[8] McCluskey, E. "Minimization of Boolean Functions," *Bell System Technical Journal,* vol. 35, no. 5, pp. 1417–1444, November 1956.

[9] Peatman, J. *Digital Hardware Design.* New York: McGraw-Hill, 1980.

[10] Pilsl, M. *Logic Level Synthesis.* Boston, MA: Kluwer Academic Publishers, 1992.

[11] Quine, W. "A Way to Simplify Truth Functions," *American Mathematics Monthly,* vol. 62, no. 9, pp. 627–631, 1955.

[12] Winkel, D. *The Art of Digital Design: An Introduction to Top-Down Design.* Englewood Cliffs, NJ: Prentice-Hall, 1980.

5
COMBINATIONAL DESIGN: IMPLEMENTATION

In Chapter 4, we discussed the first stage of combinational digital systems design: synthesis. Starting from a statement of the desired input/output function given in English, we learned how to develop a truth table. From the truth table, we then developed a logic expression using one of three minimization techniques: algebraic manipulations, Karnaugh maps, or Quine-McCluskey tables. The minimization techniques optimize a logic expression with respect to complexity, that is, the number of logic operators and the number of inputs per operator.

In this chapter we will discuss the second stage of combinational digital system design: *implementation*. Figure 5.1 shows that implementation involves translating or mapping a logic expression into actual hardware.

We will examine several approaches for implementing logic expressions, such as using hardware components called *gates, multiplexers, decoders, memories,* and *programmable logic devices* (PLDs). We will conclude this chapter by discussing situations where we might want to deviate from formal logic design principles for practical reasons; such situations are called *design practices*. We will look at two examples of design practices: *macro logic design* and *iterative circuits.*

Figure 5.1

Digital combinational system design

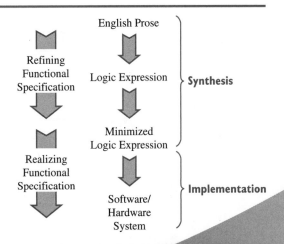

5.1 SPECIFICATION VERSUS IMPLEMENTATION

We will first make the distinction between *synthesis* and *implementation* to emphasize that a combinational digital system specification, that is, a logic expression, may have *several* different possible realizations. For example, consider the design of the 4-input majority function shown in Figure 5.2. The output of the 4-input majority function is 1 whenever the majority, that is, three or more, of the inputs are 1. Figure 5.2(a) shows the synthesis of the majority function. The schematic uses **and** and **or** logic *operators* to define a minimized logic expression in a sum-of-products (SOP) form.

> **Problem-Solving Skill**
>
> *Just as a problem may have many possible solutions, a functional specification may have many possible realizations.*

$$Z = (A \bullet B \bullet C) + (A \bullet B \bullet D) + (A \bullet C \bullet D) + (B \bullet C \bullet D)$$

Figure 5.2(b) and Figure 5.2(c) show two possible implementations of the majority function. Figure 5.2(b) uses **nand** *gates* in a two-level network, while Figure 5.2(c) uses **and** and **or** *gates* in a programmable logic array (PLA). These implementations will be explained in more detail in later sections.

For the present, understand that a specification is a mathematical abstraction, whereas a realization is a physical piece of hardware. More specifically, a logic specification is a switching algebra expression of an input/output function, whereas a logic realization is a device that actually executes the input/output function. To further distinguish synthesis and implementation, we will use the term *logic operator* when describing a specification and the term *logic gate* when describing an implementation. A logic operator denotes a switching algebra function, whereas a logic gate is a physical device that realizes a logic operator. Sometimes the terms logic operators and logic gates are used interchangeably when a particular implementation is implied or assumed. For example, Figure 5.2(a) could denote both a specification and a realization by simply replacing the logic operators with corresponding discrete logic gates. To avoid such confusion, we will try to be careful about how the terms are used. If a schematic refers to an actual hardware implementation, the logic elements will be called gates; otherwise, the logic elements will be called operators.

Figure 5.2

A 4-input majority function

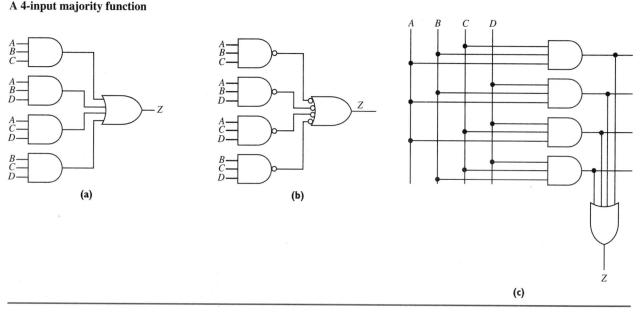

Figure (a) shows the specification in a sum-of-products form. Figures (b) and (c) show two possible realizations using **nand** gates and a programmable logic array (PLA), respectively.

Although we are decomposing design into synthesis and implementation processes, it is important to recognize the interdependencies between synthesis and implementation. A particular synthesis approach may target a certain implementation or a particular implementation may imply a certain synthesis approach. We will examine these interdependencies in more detail as we discuss combinational implementation strategies.

5.2 TWO-LEVEL NETWORKS

In Chapter 4, we generated two-level logic expressions in either a sum-of-products or product-of-sums form. The term *two-level logic* means that there are two stages of logic between the inputs and the outputs. For the sum-of-products form, the first stage of logic is **and** operators and the second stage is **or** operators. The stages are reversed for the product-of-sums form.

A two-level logic expression can be realized using a variety of *two-level gate networks.* The **and-or** and **or-and** two-level gate networks are two obvious choices, obtained by simply mapping the logic operators comprising SOP and POS expressions directly into the corresponding discrete logic gates. There are, however, reasons why the **and-or** and **or-and** gate networks may not yield the best implementation. For instance, inverting gates, such as **nand** and **nor** gates, are often faster and easier to build than noninverting gates, such as **and** and **or** gates. Hence, using inverting gates might yield a more economical implementation having better performance. In addition, for certain logic expressions, the **xor** gate can be exploited to reduce circuit complexity (that is, gate count). The following sections will explore several commonly used two-level gate networks.

5.2.1 **Nand-Nand** Networks

Recall from Chapter 3 that the **nand** logic operator forms a minimal set, meaning that any logic expression can be described using only **nand** operators. Consequently, any logic expression can be implemented using only **nand** gates. Figure 5.3 shows the active-0 and active-1 symbols for a **nand** gate, introduced in Figure 3.27.

The **nand** gate in Figure 5.3(a) is viewed as performing the conjunction operation with active-1 inputs and an active-0 output. The **nand** gate in Figure 5.3(b) is viewed in its alternate role of performing the disjunction operation with active-0 inputs and an active-1 output.

Given the dual conjunction/disjunction roles of the **nand** gate, Figure 5.4(a) (p. 122) shows how a **nand-nand** network can implement a sum-of-products logic expression with active-1 inputs and outputs. A **nand-nand** network can also implement a product-of-sums logic expression with active-0 inputs and outputs, as shown in Figure 5.4(b).

For example, the sum-of-products logic expression for the majority function given in Figure 5.2(a) is implemented by the **nand-nand** two-level network shown in Figure 5.2(b)

Figure 5.3

Alternative logic symbols

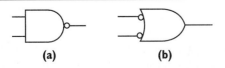

(a)   (b)

Figures (a) and (b) give the standard and alternative symbols for the **nand** gate, respectively.

Figure 5.4

Two-level network

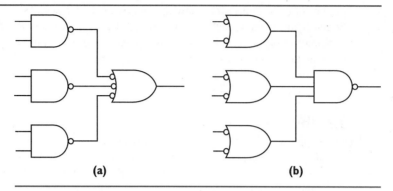

(a)   **(b)**

A **nand-nand** network can implement either a sum-of-products logic expression with active-1 inputs and outputs (Figure (a)) or a product-of-sums logic expression with active-0 inputs and outputs (Figure (b)).

and repeated below. Notice the advantage of using the active state notation to clarify the op-

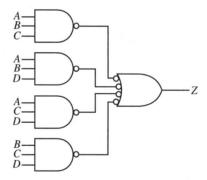

eration of the **nand-nand** network. The output bubbles of the first stage cancel the input bubbles of the second stage, immediately revealing the sum-of-products function. Remember that each symbol represents a physical **nand** gate.

As another example, we will design a combinational digital system realizing the truth table to the left using a **nand-nand** network. The output *EQUAL* is asserted active-1 when $A_1 A_0 = B_1 B_0$.

$A_1 A_0$	$B_1 B_0$	EQUAL
0 0	0 0	1
0 0	0 1	0
0 0	1 0	0
0 0	1 1	0
0 1	0 0	0
0 1	0 1	1
0 1	1 0	0
0 1	1 1	0
1 0	0 0	0
1 0	0 1	0
1 0	1 0	1
1 0	1 1	0
1 1	0 0	0
1 1	0 1	0
1 1	1 0	0
1 1	1 1	1

The minimized sum-of-products expression for *EQUAL*,

$$EQUAL = (\overline{A_1} \cdot \overline{A_0} \cdot \overline{B_1} \cdot \overline{B_0}) + (\overline{A_1} \cdot A_0 \cdot \overline{B_1} \cdot B_0) + (A_1 \cdot \overline{A_0} \cdot B_1 \cdot \overline{B_0})$$
$$+ (A_1 \cdot A_0 \cdot B_1 \cdot B_0)$$

is implemented as a **nand-nand** network in Figure 5.5.

5.2.2 **Nor-Nor** Networks

Similar to the **nand** gate, any logic expression can be implemented using only **nor** gates because the **nor** operator also forms a minimal set. Figure 5.6 shows the active-0 and active-1 alternate symbols for a **nor** gate, repeated from Figure 3.27. The **nor** gate in Figure 5.6(a) is viewed as performing the traditional disjunction operation with active-1 inputs and an active-0 output. The **nor** gate in Figure 5.6(b) is viewed in its

Figure 5.5

An *EQUAL* two-level network

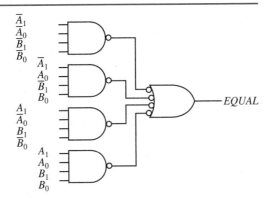

Figure 5.6

Alternative logic symbols

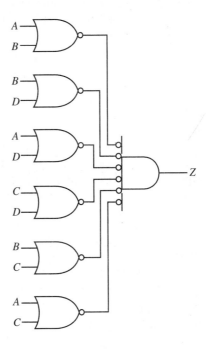

Figures (a) and (b) give the standard and alternate symbols for a **nor** gate, respectively.

dual role of performing the conjunction operation with active-0 inputs and an active-1 output.

Figure 5.7(a), p. 124, shows how a **nor-nor** network can implement a sum-of-products logic expression with active-0 inputs and outputs. A **nor-nor** network can also implement a product-of-sums logic expression with active-1 inputs and outputs, as shown in Figure 5.7(b).

As an example, the product-of-sums logic expression for the 4-input majority function

$$Z = (A + B) \cdot (B + D) \cdot (A + D) \cdot (C + D) \cdot (B + C) \cdot (A + C)$$

is implemented by the **nor-nor** two-level network shown below. Again, remember that each

symbol represents a physical **nor** gate. Also, notice that for the 4-input majority function, the product-of-sums implementation has more gates than the sum-of-products implementation.

As a counter example, the product-of-sums implementation for the *EQUAL* function

$$EQUAL = (\overline{A_0} + B_0) \cdot (A_0 + \overline{B_0}) \cdot (\overline{A_1} + B_1) \cdot (A_1 + \overline{B_1})$$

Figure 5.7

A two-level network

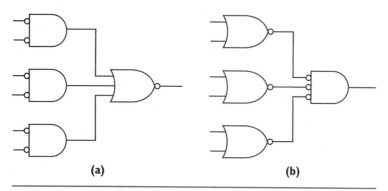

(a)   (b)

A **nor-nor** network can implement either a sum-of-products logic expression with active-0 inputs and outputs (Figure (a)) or a product-of-sums logic expression with active-1 inputs and outputs (Figure b)).

implemented as a **nor-nor** network in Figure 5.8 is less complex (has fewer inputs) than the sum-of-products implementation given earlier in Figure 5.5. As pointed out in Chapter 4, either the sum-of-products or product-of-sums form typically yields a more efficient implementation, depending on the logic expression.

5.2.3 Exclusive-Or and Equivalence Gates

Since exclusive-or (**xor**) and equivalence (**xnor**) logic operators are defined in terms of **and** and **or** logic operators,

$$X \oplus Y = (X \cdot \overline{Y}) + (\overline{X} \cdot Y)$$
$$X \odot Y = (X \cdot Y) + (\overline{X} \cdot \overline{Y})$$

the **xor** and **xnor** gates are often viewed as more "complex" gates, representing more "expressive" logic functions. This expressive power can be exploited to implement certain logic expressions.

The easiest way to explain how to detect possible applications of **xor** and **xnor** gates is to study a few examples. Figure 5.9 illustrates Karnaugh maps for several three-variable logic expressions and associated implementations using **xor** and **xnor** gates.

Figure 5.8

An EQUAL two-level network

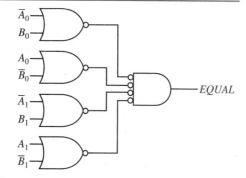

$EQUAL$

Figure 5.9

Karnaugh maps for logic expressions using xor/xnor gates

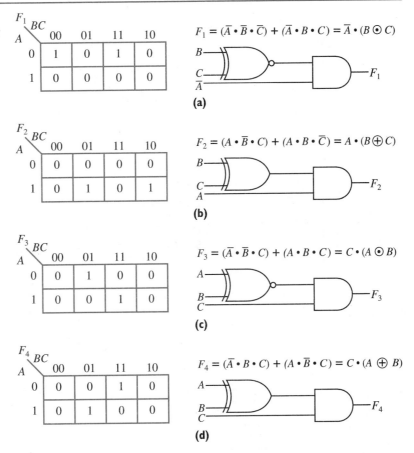

$F_1 = (\overline{A} \cdot \overline{B} \cdot \overline{C}) + (\overline{A} \cdot B \cdot C) = \overline{A} \cdot (B \odot C)$

(a)

$F_2 = (A \cdot \overline{B} \cdot C) + (A \cdot B \cdot \overline{C}) = A \cdot (B \oplus C)$

(b)

$F_3 = (\overline{A} \cdot \overline{B} \cdot C) + (A \cdot B \cdot C) = C \cdot (A \odot B)$

(c)

$F_4 = (\overline{A} \cdot B \cdot C) + (A \cdot \overline{B} \cdot C) = C \cdot (A \oplus B)$

(d)

Notice the pattern of 1's that suggest the possible use of **xor** or **xnor** gates. In Figure 5.9(a) and Figure 5.9(b), the 1's are separated by one cell. In Figure 5.9(c) and Figure 5.9(d), the 1's are in diagonally adjacent cells. Figure 5.10 (p. 126) shows that these checkerboard Karnaugh map patterns of single 1's can be extended to groups of 1's. In Figure 5.10(a), the two groupings are separated by a column of cells. In Figure 5.10(b), the two groupings are diagonally adjacent.

To show the utility of **xor** and **xnor** gates, we can exploit the "suspicious" checkerboard pattern of 1's on the Karnaugh map of the *EQUAL* logic function shown in Figure 5.11 to yield the following expression:

$$EQUAL = (A_0 \odot B_0) \cdot (A_1 \odot B_1)$$

Thus, the **xnor** implementation in Figure 5.12 yields an even simpler realization than the **nor-nor** network in Figure 5.8.

The implementations of the *EQUAL* function using a **nand-nand** network in Figure 5.5, a **nor-nor** network in Figure 5.8, and **xnor** gates in Figure 5.12 demonstrate the strengths and weaknesses of various two-level gate networks. Certain two-level gate networks are good at implementing certain types of logic expressions. A good digital designer judiciously matches a two-level network to the particular nature of a logic expression to generate an efficient implementation.

Figure 5.10

Groupings using xor/xnor gates

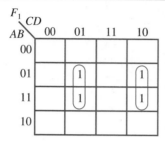

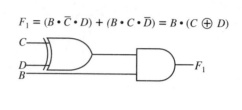

$$F_1 = (B \cdot \bar{C} \cdot D) + (B \cdot C \cdot \bar{D}) = B \cdot (C \oplus D)$$

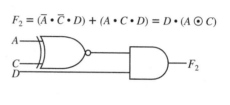

(a)

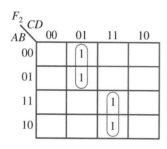

$$F_2 = (\bar{A} \cdot \bar{C} \cdot D) + (A \cdot C \cdot D) = D \cdot (A \odot C)$$

(b)

Figure 5.11

Karnaugh map for the EQUAL logic expression

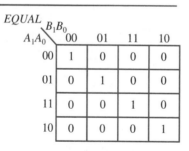

Figure 5.12

EQUAL implementation using xnor gates

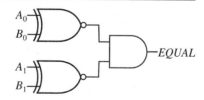

5.2.4 A-O-I Networks

In addition to providing devices containing discrete gates, manufacturers also often provide devices containing collections of gates wired together into networks. These predefined or "canned" gate networks serve as larger "building blocks," also called packages, for implementing logic expressions faster and easier.

One such network is the **and-nor** network, commonly called an *A-O-I* network, where A-O-I stands for "**and-or-invert**." Figure 5.13(a) shows how an A-O-I network can implement a sum-of-products logic expression with active-1 inputs and an active-0 output. An

Figure 5.13

A-O-I network

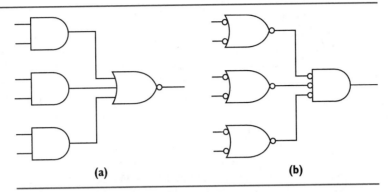

(a)   (b)

An A-O-I network can implement either a sum-of-products logic expression
with active-1 inputs and active-0 outputs (Figure (a)) or a product-of-sums
logic expression with active-0 inputs and active-1 outputs (Figure (b)).

A-O-I network can also implement a product-of-sums logic expression with active-0 inputs
and an active-1 output, as shown in Figure 5.13(b).

Any logic function F can be easily implemented using an A-O-I network by simply generating the sum-of-products expression for the inverse of F, that is, $\overline{F}$. In other words, we
group 0's on a Karnaugh map or maxterms in a Quine-McCluskey table. For instance, consider an A-O-I network implementation of the manufacturing alignment monitor combinational system discussed in Chapter 3. The truth table, given in Figure 3.29(b) and repeated
below, asserts the output *ALIGN_BAR* active-0 when a part is correctly aligned by touch-
ing the back alignment guide (B) and at least one of the side alignment guides, left (L) or
right (R).

L B R	ALIGN_BAR	Condition
0 0 0	1	
0 0 1	1	
0 1 0	1	
0 1 1	0	Part aligned back and right
1 0 0	1	
1 0 1	1	
1 1 0	0	Part aligned left and back
1 1 1	0	Part aligned left, back, and right

Grouping 0's yields

$$\overline{ALIGN\_BAR} = (L \cdot B) + (R \cdot B)$$

Inverting both sides of the logic expression directly yields the A-O-I form

$$ALIGN\_BAR = \overline{(L \cdot B) + (R \cdot B)}$$

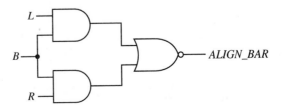

5.3 MULTILEVEL NETWORKS

Since the standard sum-of-products and product-of-sums logic forms denote two levels of logic, two-level networks are straightforward implementation techniques. However, there may be situations where a two-level network implementation is not preferred or not feasible. One factor, for example, that may preclude a two-level network implementation is that gates typically have limited *fan-in* and *fan-out*. Fan-in refers to the maximum number of inputs a gate can accept, for example, there are no 25-input **nand** gates. Fan-out refers to the maximum number of inputs a gate can drive. Under such situations, we may want to consider *multilevel* networks, that is, networks involving more than two levels or stages of logic between the inputs and the outputs.

For instance, the **and-or** two-level sum-of-products gate network of the 4-input majority function shown in Figure 5.14(a) uses 3-input **and** gates and a 4-input **or** gate. Alternatively, Figure 5.14(b) gives a three-level network implementation for the majority function that uses only 2-input **and** and **or** gates.

A two-level sum-of-products network can be transformed into a multilevel network by *factoring* the logic expression, as shown below.

$$Z = (A \cdot B \cdot C) + (A \cdot B \cdot D) + (A \cdot C \cdot D) + (B \cdot C \cdot D)$$
$$= [(A \cdot B) \cdot (C + D)] + [(C \cdot D) \cdot (A + B)]$$

Using the Distributive Postulate 4b, the term $A \cdot B$ is factored out of the first two product terms and the term $C \cdot D$ is factored out of the second two product terms.

Similar manipulations can be performed on product-of-sums networks by *collapsing* instead of factoring the logic expression. Collapsing multiplies out product terms. For example, the **or-and** two-level product-of-sums gate network of the 4-input majority function shown in Figure 5.15(a) uses 2-input **or** gates and a 6-input **and** gate. Alternatively, Figure 5.15(b) gives a four-level network implementation for the majority function that uses only 2-input **and** and **or** gates.

The majority function two-level product-of-sums network can be transformed into a multilevel network by collapsing the logic expression, as shown below.

$$Z = (B + D) \cdot (C + D) \cdot (B + C) \cdot (A + C) \cdot (A + D) \cdot (A + B)$$
$$= [(B \cdot C) + (B \cdot D) + (D \cdot C) + (D \cdot D)] \cdot [(B \cdot A) + (B \cdot C) + (C \cdot A) + (C \cdot C)]$$
$$\cdot [(A \cdot A) + (A \cdot B) + (D \cdot A) + (D \cdot B)]$$

Figure 5.14

Multilevel implementation for sum-of-products majority function

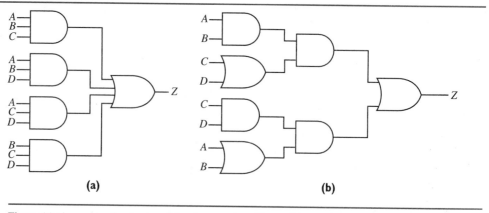

(a)   (b)

Figure (a) gives a two-level network implementation. Figure (b) gives a three-level network implementation.

$$=[(B \bullet C) + (B \bullet D) + (D \bullet C) + D] \bullet [(B \bullet A) + (B \bullet C) + (C \bullet A) + C]$$
$$\bullet [(A \bullet B) + (D \bullet A) + (D \bullet B) + A]$$
$$= [(B \bullet C) + D] \bullet [(A \bullet B) + C] \bullet [(B \bullet D) + A]$$

Using the Distributive Postulate 4b, the term $(B + D)$ is **and**ed with the term $(C + D)$ to yield $(B \bullet C) + (B \bullet D) + (C \bullet D) + (D \bullet D)$. Similarly, the terms $(B + C)/(A + C)$ and $(A + D)/(A + B)$ are also **and**ed together. The results form the second equation. Then, $(D \bullet D)$ is replaced by D, $(C \bullet C)$ is replaced by C, and $(A \bullet A)$ is replaced by A, using the Idempotency Theorem 3b. The results form the third equation. Finally, $(B \bullet D) + (C \bullet D) + D$ is replaced by D using the Absorption Theorem 7a. Other terms are collapsed in a similar fashion to form the fourth and last equation.

Reformulating logic expressions into multilevels can also sometimes aid logic minimization. For instance, factoring can identify common pieces of logic within a combinational system and reuse a single copy of the common logic to realize all instances. As an example, consider the minimized two-level sum-of-products logic expression in Equation 5.1.

$$Z = (A \bullet C) + (A \bullet D) + (A \bullet E) + (B \bullet C) + (B \bullet D) + (B \bullet E) + (\overline{C} \bullet D) \quad \textbf{(5.1)}$$

The variable A can be factored out of the first three product terms and the variable B can be factored out of the next three product terms, yielding Equation 5.2.

$$Z = A \bullet (C + D + E) + B \bullet (C + D + E) + (\overline{C} \bullet D) \quad \textbf{(5.2)}$$

Next, the subexpression $(C + D + E)$ can be factored out of the first two terms to yield the three-level logic expression in Equation 5.3.

$$Z = (C + D + E) \bullet (A + B) + (\overline{C} \bullet D) \quad \textbf{(5.3)}$$

Figure 5.15

Multilevel implementation for product-of-sums majority function

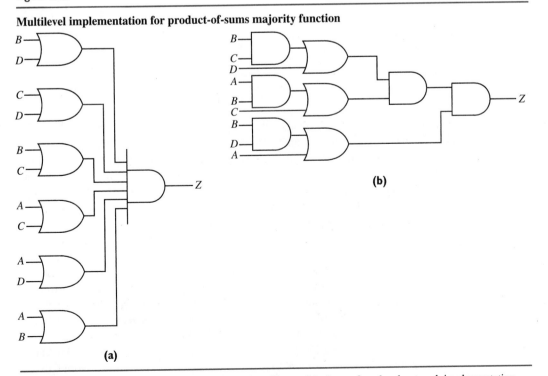

(a)

(b)

Figure (a) gives a two-level network implementation. Figure (b) gives a four-level network implementation.

Figure 5.16

Multistage logic minimization

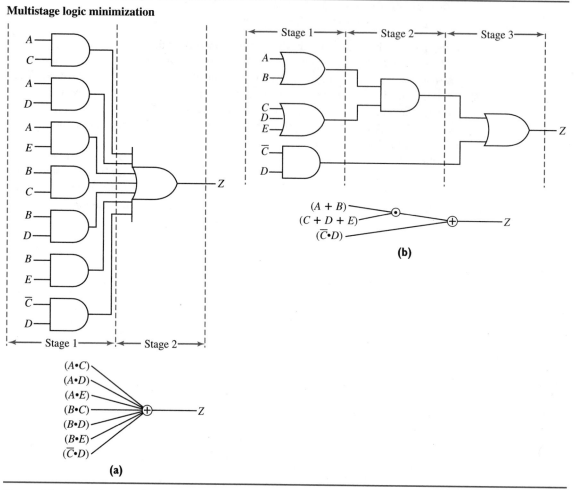

(a)

(b)

Figures (a) and (b) show two-level and three-level implementations, respectively.

The two-level and three-level logic implementations are compared in Figure 5.16. The two-level logic implementation uses eight gates, whereas the three-level logic implementation uses only five gates. To keep the analysis simple, inverters are ignored, which is a reasonable approximation because logic implementations often routinely provide input variables and their complements, as will be seen in later discussions on programmable logic devices in this chapter and on sequential systems in Chapter 8. The implementations for the logic function shown in Figure 5.16(a) and Figure 5.16(b) illustrate a general logic complexity versus logic delay trade-off often associated with alternative implementations (see Figure 5.17, p. 131).

In Part 2 of the text, we will discuss in detail how logic gates are fabricated and their associated physical properties, such as area and speed. For the present, it is sufficient to know that a gate consumes a certain amount of *area*, that is, physical reality. Also, a gate has a *propagation delay*, meaning that there is a finite amount of time between the input stimulus and the output response. Increased gate count implies increased area; increased logic levels implies increased delay. Figure 5.17 shows that, in general, input/output logic delay can be improved, but at the expense of increased logic complexity. Similarly, logic complexity can be improved, but at the expense of increased logic delay.

Figure 5.17

General logic complexity/delay trade-off

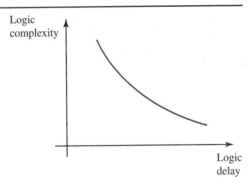

5.4 MULTIPLEXERS

In addition to two-level and multilevel gate networks, combinational logic can also be implemented using *multiplexers*. First, we will examine what a multiplexer does, and then we will learn how to use a multiplexer to implement logic.

A multiplexer, also called a *mux*, serves as a rotary switch, routing or connecting 1 of N input signals to an output signal.[1] Figure 5.18 shows that multiplexers are sometimes combined with *demultiplexers* to route several communication lines over a single, shared channel. The multiplexer serves as a "traffic cop," deciding which input signal is routed to the output. In a typical communication scenario, the multiplexer and demultiplexer switches are first both set to the top position, allowing signal S_1 to use the shared wire or channel for a period of time. Then, the multiplexer and demultiplexer switches are set to connect S_2 to the channel for a period of time, and so on.

Figure 5.18 also shows the distinctive trapezoidal shape often associated with a mux. The control signals determine which input is connected to the output. The number of control lines determines the number of inputs. Since n bits can denote 2^n unique encodings, n control lines can select one of 2^n multiplexer inputs. Thus, manufacturers typically offer 4-input multiplexers (having two control lines), 8-input multiplexers (having three control lines), and so on.

Figure 5.18

Using multiplexers and demultiplexers to share a communication channel

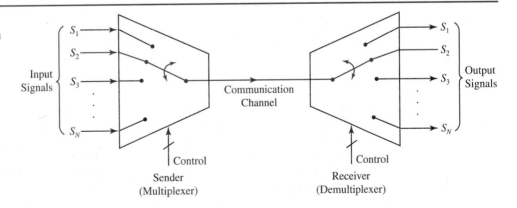

---

[1] **multiplex**—Designating a simultaneous communication of two or more messages on the same wire or radio frequency. The American Heritage Dictionary, Houghton Mifflin Company. Reprinted by permission.

Figure 5.19

A 4-input multiplexer

$CNTRL_1$	$CNTRL_0$	I_3	I_2	I_1	I_0	Z	Z_BAR
0	0	–	–	–	0	0	1
0	0	–	–	–	1	1	0
0	1	–	–	0	–	0	1
0	1	–	–	1	–	1	0
1	0	–	0	–	–	0	1
1	0	–	1	–	–	1	0
1	1	0	–	–	–	0	1
1	1	1	–	–	–	1	0

(a)

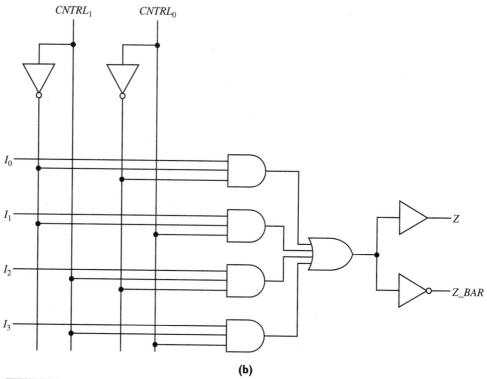

(b)

Figures (a) and (b) give the truth table and logic design for a 4-input multiplexer, respectively.

Figure 5.19 shows the truth table and logic design for a 4-input multiplexer. It is common practice to provide both active-1 and active-0 multiplexer outputs. Figure 5.19(a) shows a shorthand notation for constructing a truth table. Instead of listing every one of the $2^6 = 64$ possible input combinations, *don't care* values, –, are used to build a more compact truth table that more clearly shows the multiplexing function. For example, the first two entries in the multiplexer truth table state that when $CNTRL_1 = 0$ and $CNTRL_0 = 0$, the output $Z = I_1$, regardless of the values on I_1, I_2, and I_3. Recall from Chapter 4 that a don't care dash represents a 0 or a 1.

5.4.1 Implementing Logic Expressions
A multiplexer can also be used to implement logic expressions, in addition to its role as "traffic cop." For example, Figure 5.20 shows how a 16-input mux can implement the

Figure 5.20

Using a 16-input multiplexer to implement a majority logic function

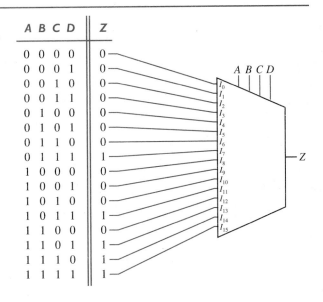

A B C D	Z
0 0 0 0	0
0 0 0 1	0
0 0 1 0	0
0 0 1 1	0
0 1 0 0	0
0 1 0 1	0
0 1 1 0	0
0 1 1 1	1
1 0 0 0	0
1 0 0 1	0
1 0 1 0	0
1 0 1 1	1
1 1 0 0	0
1 1 0 1	1
1 1 1 0	1
1 1 1 1	1

4-input majority function. The majority function inputs (A, B, C, and D) act as the multiplexer control signals, selecting one of the 16 multiplexer inputs. The values for the 16 multiplexer inputs are taken directly from the truth table. Any four-variable logic expression can be implemented in this manner by simply changing the values of the 16 multiplexer inputs.

Figure 5.21 shows a better way to implement a 4-variable logic expression, using only an 8-input multiplexer. We select any three of the four majority function inputs to control the 8-input mux; then, we use the remaining majority function input to generate the values for the 8 mux inputs. Figure 5.21 uses the three inputs A, B, and C to control the 8-input mux and the fourth input D to generate the 8 mux inputs. For instance, when $A = 0$, $B = 1$, and $C = 1$, the seventh and eighth entries in the truth table show that the majority function

Figure 5.21

Using an 8-input multiplexer to implement a majority logic function

A B C	D	Z	
0 0 0	0 / 1	0 / 0	Z = 0
0 0 1	0 / 1	0 / 0	Z = 0
0 1 0	0 / 1	0 / 0	Z = 0
0 1 1	0 / 1	0 / 1	Z = D
1 0 0	0 / 1	0 / 0	Z = 0
1 0 1	0 / 1	0 / 1	Z = D
1 1 0	0 / 1	0 / 1	Z = D
1 1 1	0 / 1	1 / 1	Z = 1

Figure 5.22

Using a 4-input multiplexer to implement a majority logic function

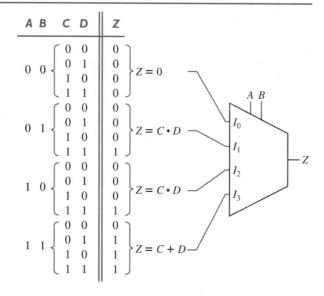

output $Z = D$. This condition is encoded in the mux implementation by connecting the signal D to the mux input I_3. Thus, when $A = 0$, $B = 1$, and $C = 1$, the mux routes input I_3 to the output, making $Z = D$.

This general strategy of partitioning logic inputs between mux control and data inputs can be used to map a logic expression onto a variety of different sized multiplexers. To generate a 4-input multiplexer implementation of the majority function, we select two of the four majority function inputs to control the 4-input mux, say A and B. Then, we use the remaining two majority function inputs (C and D) to generate the values for the four mux data inputs. (See Figure 5.22.)

For instance, when $A = 0$ and $B = 1$, the truth table entries 5–8 show that the majority function output $Z = C \cdot D$. This condition is encoded in the mux implementation by connecting an **and** gate to the mux input I_1. Thus, when $A = 0$ and $B = 1$, the mux routes input I_1 to the output, making $Z = C \cdot D$.

To see an example of the interdependencies between synthesis and implementation, notice that a multiplexer implementation may or may not take advantage of logic minimization. Using a 4-input multiplexer to realize a four-variable function requires two-variable logic minimization. However, using a 16-input multiplexer to realize a four-variable logic function requires *no* logic minimization; the 16 data inputs are taken directly from the defining truth table.

5.5 DECODERS

Another component that can be used to implement combinational logic is a *decoder*. Following the same format used to discuss multiplexers in the previous section, we will first examine what a decoder does, and then we will learn how to use a decoder to implement logic.

We will discuss a particular class of decoders called *binary decoders*. Figure 5.23 shows that a binary decoder transforms one n-bit binary input signal into 2^n 1-bit output signals. Consequently, binary decoders are also called n-to-2^n decoders; common binary decoders are 2-to-4, 3-to-8, et cetera.

Figure 5.23

General binary decoder

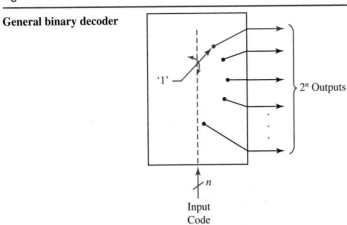

The important property of a binary decoder is that only *one* of the 2^n output signals is asserted, depending on the value of the n-bit binary input signal. Figure 5.23 assumes active-1 outputs, showing that the input signal controls a switch to assert one of the outputs.

Binary decoders are often used in address decoding applications, as illustrated in Figure 5.24. The processor communicates with the four input/output (I/O) devices. To select a particular input/output device, the processor sends the device address to the 2-to-4 binary decoder. The binary decoder, in turn, decodes the address and selects or activates the appropriate I/O device by asserting the associated enable signal, *EN*.

Figure 5.25 (p. 136) shows the truth table and logic design for a 2-to-4 binary decoder.

Figure 5.24

Address decoding application of binary decoders

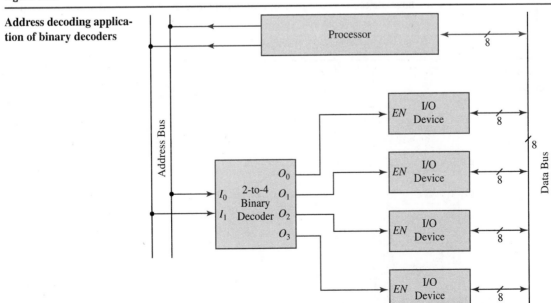

Figure 5.25

A 2-to-4 binary decoder

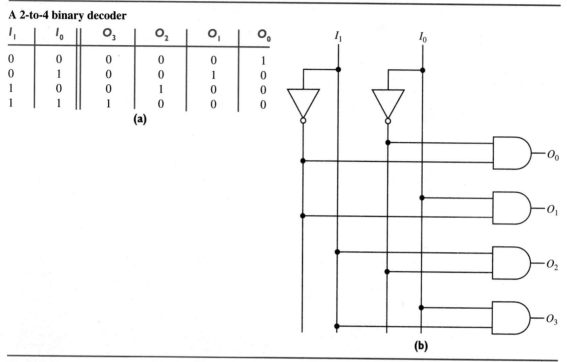

I_1	I_0	O_3	O_2	O_1	O_0
0	0	0	0	0	1
0	1	0	0	1	0
1	0	0	1	0	0
1	1	1	0	0	0

(a)

(b)

Figures (a) and (b) give the truth table and logic design for a 2-to-4 binary decoder, respectively.

5.5.1 Implementing Logic Expressions

A binary decoder is also called a *function generator* or a *minterm generator* because it can implement a general logic sum-of-products expression. Figure 5.26 shows how the 4-input majority function can be implemented using a 4-to-16 binary decoder. We just "pick-off" the minterms and connect them through an **or** gate to the output. Anytime the inputs (A, B, C, and D) have three or more 1's, one and only one of the selected minterms will be

Figure 5.26

Using a decoder to implement a 4-input majority logic function

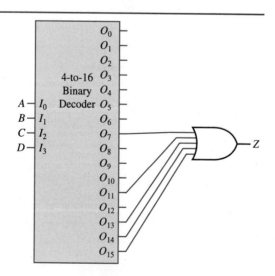

Figure 5.27

Implementing multiple-output logic functions using decoders

TEMPERATURE	TEMP2	TEMP1	TEMP0	HEAT	COOL
0°C	0	0	0	1	0
10°C	0	0	1	1	0
30°C	0	1	0	0	1
20°C	0	1	1	0	0
70°C	1	0	0	0	1
60°C	1	0	1	0	1
40°C	1	1	0	0	1
50°C	1	1	1	0	1

(a)

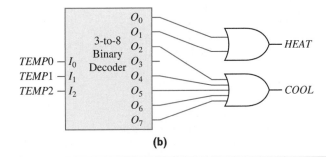

(b)

The controller inputs the ambient temperature encoded as a 3-bit Gray code and output signals to turn off/on heating and cooling. Figure (a) and Figure (b) respectively give the truth table and logic implementation using a 3-to-8 binary decoder for a temperature controller.

asserted, which will assert the output. For example, if $A = 1$, $B = 0$, $C = 1$, and $D = 1$, O_{11} will be asserted active-1. All other outputs will be deasserted. The **or** of the selected minterms equals 1, which asserts the output Z.

Binary decoders are typically not used to implement large combinational systems because the number of outputs (2^n) becomes unwieldy. However, binary decoders can be effectively used to implement small, multiple-output combinational systems. Figure 5.27 shows how a 3-to-8 binary decoder can implement the temperature controller introduced in Chapter 3.

Figure 5.28 (p.138) takes advantage of the multiple function implementation capability of decoders to realize a sine wave inverter. Figure 5.28(a) shows that the sine wave inverter accepts a 4-bit, two's complement number denoting the waveform magnitude in the symmetrical range -7_{10} to 7_{10} and inverts the input.

The decoder outputs that drive each system output are taken directly from the truth table. For instance, the minterms $m_1 - m_7$ assert Z_3. Hence, the corresponding decoder outputs $O_1 - O_7$ are connected as inputs to the **or** gate that drives Z_3.

Again we see an example of the interrelationships between synthesis and implementation: Decoders, as minterm generators, do not exploit logic minimization. Thus, the logic minimization techniques discussed in Chapter 4 are not required during synthesis for a decoder implementation strategy.

5.6 MEMORY

A memory stores digital information, 1's and 0's. Recall that one of the most important characteristics of a combinational digital system is that it has no memory, meaning that a

Figure 5.28

Using a decoder to implement a sine wave inverter

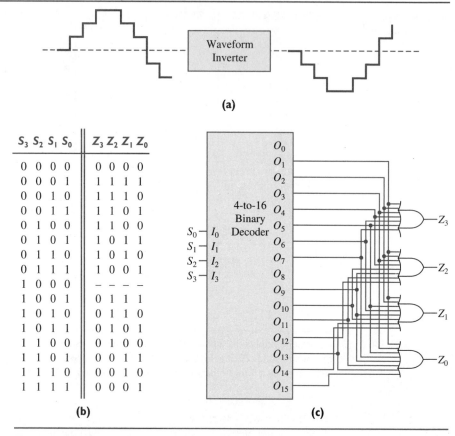

(a)

S_3 S_2 S_1 S_0	Z_3 Z_2 Z_1 Z_0
0 0 0 0	0 0 0 0
0 0 0 1	1 1 1 1
0 0 1 0	1 1 1 0
0 0 1 1	1 1 0 1
0 1 0 0	1 1 0 0
0 1 0 1	1 0 1 1
0 1 1 0	1 0 1 0
0 1 1 1	1 0 0 1
1 0 0 0	– – – –
1 0 0 1	0 1 1 1
1 0 1 0	0 1 1 0
1 0 1 1	0 1 0 1
1 1 0 0	0 1 0 0
1 1 0 1	0 0 1 1
1 1 1 0	0 0 1 0
1 1 1 1	0 0 0 1

(b)

(c)

Figure (a) shows the input/output diagram. Figures (b) and (c) respectively show the truth table specification and decoder implementation.

combinational digital system cannot store information about previous inputs. In Part 3, we will combine combinational logic with memory to build *sequential systems*.

Figure 5.29 shows that a memory can be viewed as a table; each entry in the table is a place for storing information. The n-bit address selects one of the 2^n entries in the table. The address decoding is performed by a n-to-2^n binary decoder. Each table entry is a *word*, that is, a group of m bits. The number n of address signals (rows) and the number m of data signals (columns) determines the size of the table, that is, how many bits the $n \times m$ memory stores.

There are many different kinds of memories that use a variety of ways to read and write the stored data. Unfortunately, naming conventions can be a little confusing and do not always reflect the associated function. *RAM (Random Access Memory)* allows the user to electronically read and write any memory entry in any order, hence the name "random access." A read operation outputs the contents of the addressed memory entry as the value of the *DATA* signal; a write operation updates the contents of the addressed memory entry with the value of the *DATA* signal. Although not entirely descriptive, the term RAM is used for historical reasons, differentiating this type of storage medium from sequential storage media such as magnetic tape.

ROM (Read-Only Memory) allows the user to electronically read, but *not* write, any memory entry in any order. The data stored in ROM is set once, at the time the ROM is

Figure 5.29

Viewing memory as a table

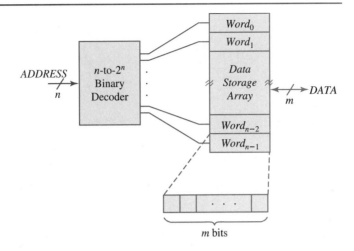

manufactured, and thereafter cannot be changed. Another type of read-only memory, called a *PROM (Programmable Read-Only Memory),* allows the user to set or "program" the stored data by selectively applying voltages to an array of small fuses inside the PROM. The data can be programmed only once; the stored data cannot be changed once the fuses are blown.

This programming restriction is removed with *EPROM (Electrically Programmable Read-Only Memory), EEPROM (Electrically Erasable Programmable Read-Only Memory),* and *flash* read-only memories, which can be programmed by the user. EPROM is programmed electrically and erased or reset via ultraviolet irradiation. Information is stored in an EPROM by applying voltages to certain input pins, which selectively assert individual locations or *cells.* Information is erased in an EPROM by exposing the EPROM to ultraviolet light, which deasserts or resets all memory cells. EPROM devices are easily recognized because they have a quartz window over the memory to allow for ultraviolet light exposure.

EEPROM is programmed *and* erased electronically. Information is stored/erased in an EEPROM by applying voltages to certain input pins, which selectively assert/deassert individual locations or cells. Similar to EEPROMs, flash memories are programmed and erased electronically. However, the memory is erased in bulk; individual memory locations cannot be selectively erased.

5.6.1 Implementing Logic Expressions

Regardless of how memory is programmed, we can use memory to implement general logic expressions. For example, consider the simple 16×1 memory shown in Figure 5.30 (p. 140), which implements the 4-input majority function we have been examining. The decoder activates one of the $2^4 = 16$ table entries, based on the 4-bit input address. The 16 1-bit table entries hold the truth table for the majority function. Anytime the inputs (A, B, C, and D) have three or more 1's, the decoder selects a memory location that stores a 1 and the value is routed to the output, Z. Simply changing the contents of memory changes the logic function.

If the storage array contains more than one bit per table entry, multiple logic expressions can be implemented. Figure 5.31 uses a 16×4 memory to implement the sine wave inverter discussed in the previous section.

To further understand how memory realizes a general logic expression and as a lead into the next section on programmable logic devices, we will develop an alternative representation for memory that focuses on its logic function rather than its storage function. In its logic-function role, a memory can be viewed as implementing a canonical sum-of-products

Figure 5.30

Using memory to implement a 4-input majority function

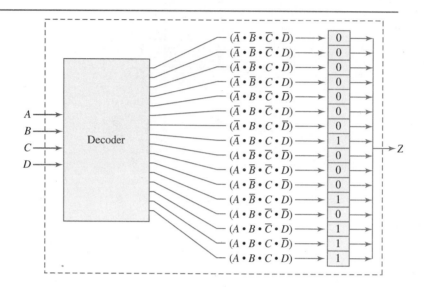

expression. The decoder generates the minterms, forming the product, or **and**, of all combinations of the input variables. The contents of the storage array determines which minterms contribute to the sum-of-products expression, forming the sum, or **or**, of the desired minterms. Thus, memory can be viewed as shown in Figure 5.32.

The decoder forms an **and** *plane*, generating the minterms. The **and** plane is called *fixed* or *nonprogrammable* because the generated minterms are hardwired and cannot be changed. The storage array forms an **or** plane, generating the sum of the minterms. The **or** plane is programmable because the minterms that contribute to the output sum-of-products logic expression can be changed by simply changing the contents of the storage array. A stored 1 includes a minterm; a stored 0 excludes a minterm.

Figure 5.31

Using memory to implement a sine wave inverter

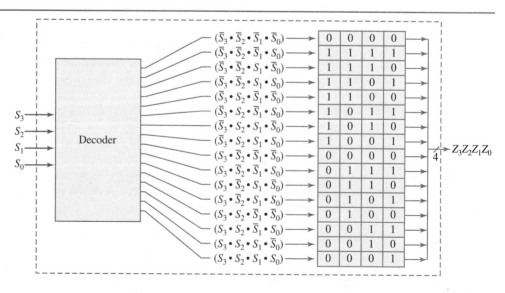

Figure 5.32

Logic-function view of memory

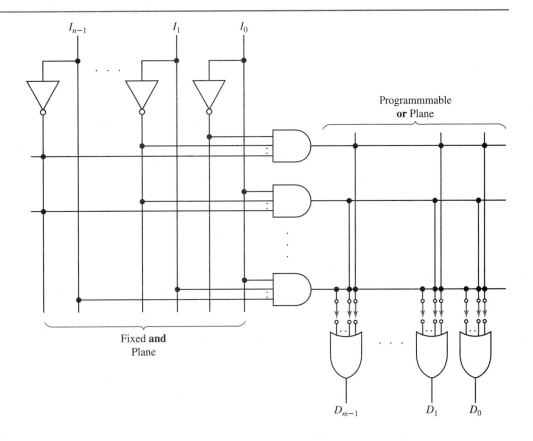

Figure 5.33 (p. 142) shows a shorthand notation commonly used to avoid drawing lots of wires on logic schematics. A single wire conceptually represents the combined input signals for a logic operator. Figure 5.33(a) and Figure 5.33(b) respectively show that fixed connections are denoted by • and programmable connections are denoted by X. To illustrate the utility of this shorthand notation, Figure 5.32 is redrawn in Figure 5.34 (p. 142). Compare and contrast Figure 5.29 and Figure 5.34 to understand the storage/logic dual roles of memory.

5.7 PROGRAMMABLE LOGIC DEVICES

One disadvantage of using memory to implement combinational logic is that memory often "wastes" logic. Memory does not take advantage of any minimization techniques because memory directly implements the truth table. Also, when a combinational system has fewer outputs than the memory word (table entry) size, portions of the memory are not fully utilized. Thus, in the mid-1970s, manufacturers started providing hardware components better suited than memories for implementing logic; these components are collectively called *programmable logic devices* (PLDs). Programmable logic devices offer the convenience of *structured logic design* associated with memories with the added benefit of improved logic utilization and performance.

Manufacturers offer a broad selection of programmable logic devices and the selection is

Figure 5.33

Shorthand notation for structured logic schematics

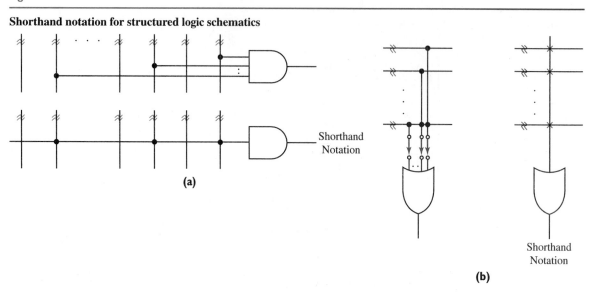

Shorthand
Notation

(a)

Shorthand
Notation

(b)

Figure 5.34

Logic-function view of memory using shorthand notation

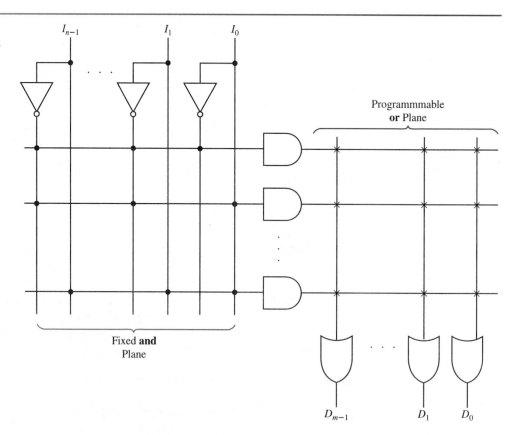

continually changing. We will investigate the following representative sampling of PLDs that illustrate general architectural features:

- programmable logic arrays (PLAs),
- programmable array logic (PAL),[2]
- programmable multilevel logic arrays (PMLs), and
- field-programmable gate arrays (FPGAs).

5.7.1 Programmable Logic Arrays (PLAs)

Programmable logic arrays, also called *field-programmable logic arrays* (FPLAs), were introduced in the mid-1970s. Using the shorthand notation described in the previous section (Figure 5.33), Figure 5.35 shows the general logic architecture of a PLA.

The principal difference between the logic architecture of a memory and a PLA is that the **and** plane of a PLA is programmable, whereas the **and** plane of a memory is fixed. Having a programmable **and** plane allows the user to implement minimized product terms. For example, the 4-input majority function is implemented via the simple PLA, shown in Figure 5.36 (p. 144). The notation **X** denotes a programmed connection. Compare the majority function implementations given in Figure 5.30 and Figure 5.36: Figure 5.30 uses a memory and implements the canonical logic expression, whereas Figure 5.36 uses a PLA and implements the minimized logic expression.

Figure 5.35

General logic structure of a programmable logic array (PLA)

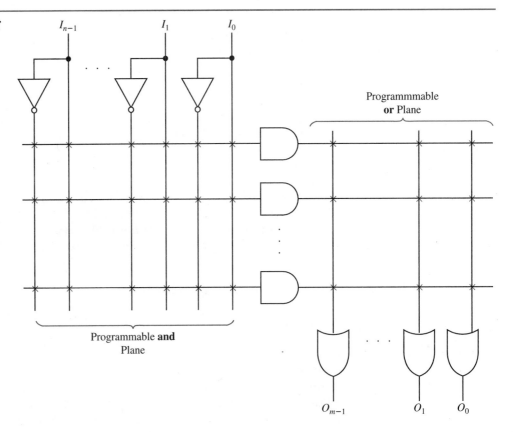

[2] PAL® is a registered trademark of Advanced Micro Devices, Inc.

Figure 5.36

Programmable logic array (PLA) implementation of a majority function

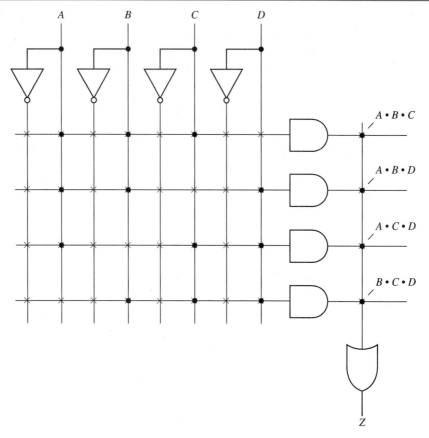

Figure 5.37 shows a PLA implementation of the *EQUAL* function

$$EQUAL = \sum m(0, 5, 10, 15)$$

that was discussed in previous sections.

Similar to memories, PLAs can be programmed in a variety of ways, including PROM, EPROM, or EEPROM technologies. Devices using EPROM-based programming are often called *erasable programmable logic devices* (EPLDs). Devices using EEPROM-based programming are often called *electrically erasable programmable logic devices* (E²PLDs).

5.7.2 Programmable Array Logic (PAL)

There are situations where the flexibility of a programmable **or** plane is largely unnecessary, such as with logic expressions that do not share minterms. Hence, the increased component complexity required to provide a programmable **or** plane is not justified in many cases. In addition, programmable **and** and **or** planes contribute to longer input/output signal propagation delays, which limits the application of PLAs in high-speed digital systems.

To address these deficiencies, programmable array logic (PAL) was introduced in the late 1970s. Again using the shorthand notation, Figure 5.38 (p. 146) shows the general logic architecture of a PAL. A PAL has a programmable **and** plane and a fixed **or** plane. Each output **or** gate is connected to a set number of product lines coming out of the **and** plane. For simplicity, Figure 5.38 shows each output **or** gate connected to only two product lines. Typi-

Problem-Solving Skill

Always try to address the most important aspects of a problem. Do not waste time developing a technically superior solution that is of little practical use.

Figure 5.37

Programmable logic array
(PLA) implementation of
the *EQUAL* function

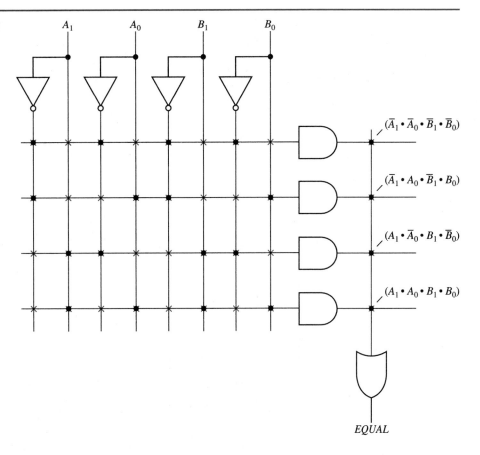

$$Z = [(A + \overline{B}) \bullet C] + [(\overline{C} + D) \bullet A]$$

cally, an output **or** gate is connected to eight or more product lines. PALs trade off reduced function associated with the fixed **or** plane for increased performance. Similar to PLAs, PALs can be programmed in a variety of ways, including PROM, EPROM, or EEPROM technologies.

5.7.3 Programmable Multilevel Logic Arrays (PMLs)

Programmable multilevel logic arrays (PMLs) employ variations of the conventional **and/or**, two-plane structured logic architecture to realize multilevel logic expressions. For example, Figure 5.39 (p. 147) shows a programmable multilevel logic array using only a single array or plane of **nand** gates. The outputs of some of the **nand** gates are fed back as inputs to the programmable array. Complex logic expressions involving several stages of logic can be implemented by judiciously programming the feedback connections.

For instance, Figure 5.40 (p. 148) shows the three-level logic implementation of

$$Z = [(A + \overline{B}) \bullet C] + [(\overline{C} + D) \bullet A]$$

Study the **nand** gate transformation and implementation. First, the logic operators in Figure 5.40(a) are replaced by **nand** gates in Figure 5.40(b). Active state notation is used to clearly convey logic functions. Next, the bubbles in the new schematic in Figure 5.40(b) are checked and inverters are added where unwanted logic inversions have been introduced. The bubbles at the outputs of gates ② and ④ and the inputs of gate ⑤ cancel each other out. The bubbles at the inputs to gates ① and ③, however, are not canceled out, so we need to

Figure 5.38

General logic structure of programmable array logic (PAL)

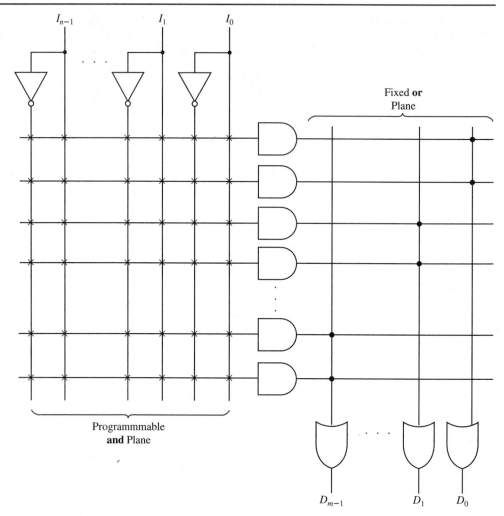

add inverters to keep the overall logic expression for Z unchanged. Thus, in Figure 5.40(b), **nand** gate ① accepts $\overline{A}$ and B instead of A and $\overline{B}$. Similarly, **nand** gate ③ accepts C and $\overline{D}$ instead of $\overline{C}$ and D. Finally, the **nand** gates in Figure 5.40(b) are mapped onto the programmable multilevel logic array in Figure 5.40(c).

5.7.4 Field-Programmable Gate Arrays (FPGAs)

PLAs, PALs, and PMLs have similar logic architectures, involving one or more large blocks or planes of logic. Field-programmable gate arrays (FPGAs) offer a different approach to programmable logic devices, involving distributed structures of smaller blocks of logic. Figure 5.41 (p. 149) shows several popular field-programmable gate array architectures. A logic expression is implemented with a FPGA by decomposing the logic expression into smaller functions; each smaller function is implemented by a logic block. Each logic block is programmed, and the wiring channels are programmed to interconnect the logic blocks to yield the desired implementation.

Typically, FPGA logic blocks are all identical and each block can be individually programmed to realize a "chunk" of logic. The complexity of a logic block varies considerably

Figure 5.39

Programmable multilevel logic array using a single nand plane

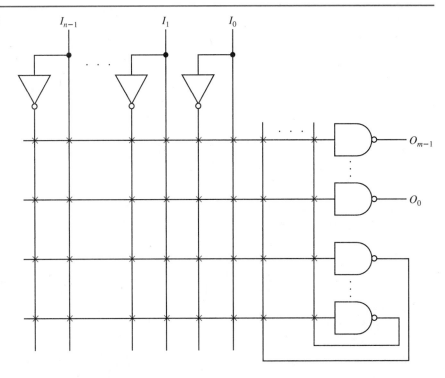

between different FPGAs, ranging from a "fine-grain" block realizing one or two logic functions of two to four variables to a "coarse-grain" block realizing several logic functions of four to ten variables. The implementation strategy of a logic block also varies among different FPGAs, including combinations of discrete gates, programmable multiplexers, memories, PLAs, PALs, and PMLs. FPGAs containing PLA, PAL, or PML logic blocks are commonly called *segmented PLDs*.

Programmable interconnect adds complexity that complicates using FPGAs, but it also adds a degree of freedom that can be exploited to improve implementation performance. Programmable interconnect consists of a set of wires and programmable switches. Figure 5.42 (p. 150) shows that the programmable switches are typically located between logic blocks and at the intersection of horizontal and vertical wiring channels. The switches are programmed to connect or disconnect wire segments to establish desired communication paths. The darkened path in Figure 5.42 illustrates how the switches might be set to route an output signal from logic block *A* to a system output and to an input of logic block *D*.

5.8 DESIGN PRACTICES

We will conclude this chapter on combinational logic implementation techniques and this part on combinational systems by examining situations where we might want to deviate from formal design principles and use informal design techniques, often called *design practices*. If judiciously used, design practices can be a valuable part of a good engineer's digital design skills. However, the abuse of design practices reduces digital design to a collection of ad hoc, inane methods. The two design practices we will examine are *iterative circuits* and *macro logic design*.

Figure 5.40

Implementation of three-level logic array using a single nand plane

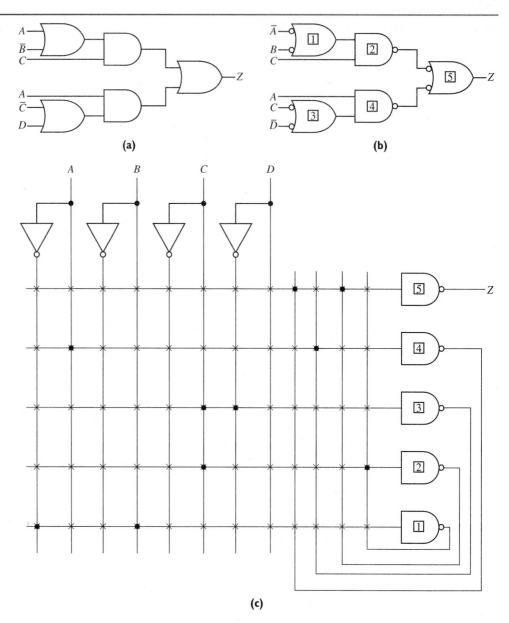

(a)

(b)

(c)

5.8.1 Iterative Circuits

Figure 5.43 (p. 150) shows that the basic idea behind *iterative circuits* is to decompose a function into a set of simpler functions that can be cascaded together. The chain of cascaded circuits incrementally computes the desired function.

To illustrate an iterative circuit implementation, we will consider the design of an arithmetic logic function: addition. The adder accepts two N-bit operands, A and B, and yields an $(N + 1)$-bit output, SUM, where $SUM = A + B$. We could design the adder using one of the formal minimization techniques discussed in Chapter 4; the truth table would have 2^{2N} entries. Alternatively, we could design the adder as an iterative circuit, as shown in Figure 5.44 (p. 150). The addition is performed column by column with a cascade of "bit-wise adders," propagating carries where appropriate. The iterative adder is more commonly called

Figure 5.41

**Field-programmable gate
array architectures
(FPGAs)**

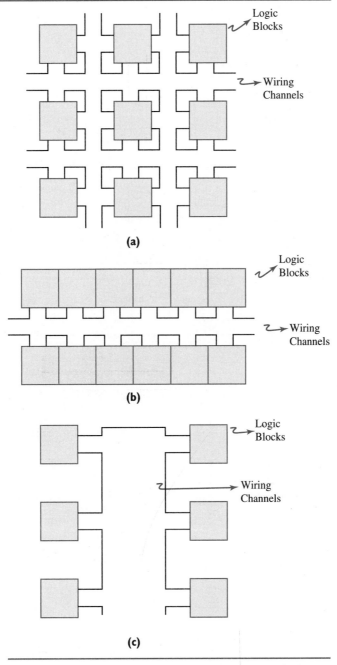

Figures (a), (b), and (c) show matrix, row, and column architectures,
respectively.

a *ripple carry adder*, or simply a *ripple adder*, because the carries "ripple" through the iterative structure.

Figure 5.45 (p. 151) shows a more detailed look at an individual bit-wise adder, more
commonly called a *full adder*. The block diagram in Figure 5.45(a) shows that a full adder
accepts corresponding operand bits (a_i and b_i) and a carry-in bit (c_i) from the preceding,

Figure 5.42

Programmable interconnect

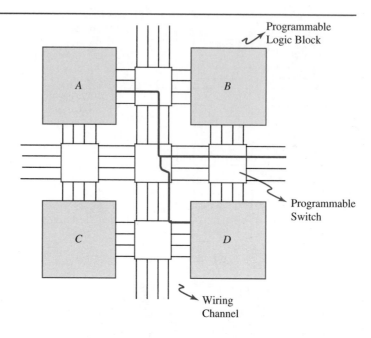

Figure 5.43

General architecture of iterative circuits

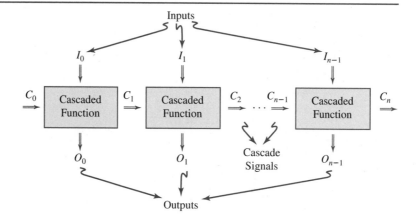

Figure 5.44

Iterative implementation of an adder

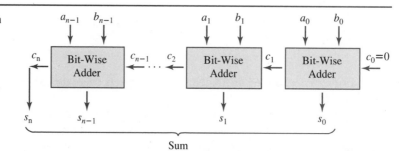

Figure 5.45

A bit-wise, or full, adder

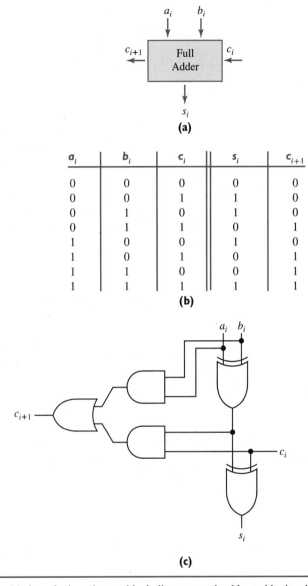

(a)

a_i	b_i	c_i	s_i	c_{i+1}
0	0	0	0	0
0	0	1	1	0
0	1	0	1	0
0	1	1	0	1
1	0	0	1	0
1	0	1	0	1
1	1	0	0	1
1	1	1	1	1

(b)

(c)

Figures (a)–(c) show the input/output block diagram, truth table, and logic schematic of a full adder.

lower-order full adder stage and yields a carry-out bit (c_{i+1}) for the succeeding, higher-order full adder stage and a sum bit (s_i). The truth table for the full adder given in Figure 5.45(b) is taken from the discussion of binary addition given in Chapter 2.

As another example of iterative circuits, Figure 5.46 (p. 152) shows a comparator that accepts two N-bit binary numbers, A and B, as operands and yields a result that shows their relative magnitudes. Again, we could design the comparator using one of the formal minimization techniques discussed in Chapter 4; the truth table would have 2^{2N} entries. Alternatively, we could generate the N-bit comparison by cascading the successive applications of 1-bit or bit-wise comparators. Each bit-wise comparator accepts the results of the previous, higher-order bit-wise comparator, adds in its results, and passes the resulting information along to the next, lower-order bit-wise comparator.

Figure 5.46

Iterative implementation of a comparator

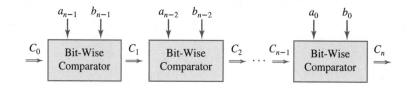

The design of the bit-wise comparator is shown in Figure 5.47. There are two cascade signals, x_i and y_i, because at least two bits are needed to encode the three possible results:
1. $A = B$,
2. $A < B$, or
3. $A > B$.

Using the encodings given in Table 5.1, the truth table for the bit-wise comparator is given in Figure 5.47(b). If the previous higher-order bits of A and B are equal ($x_i y_i = 00$),

Figure 5.47

A bit-wise comparator

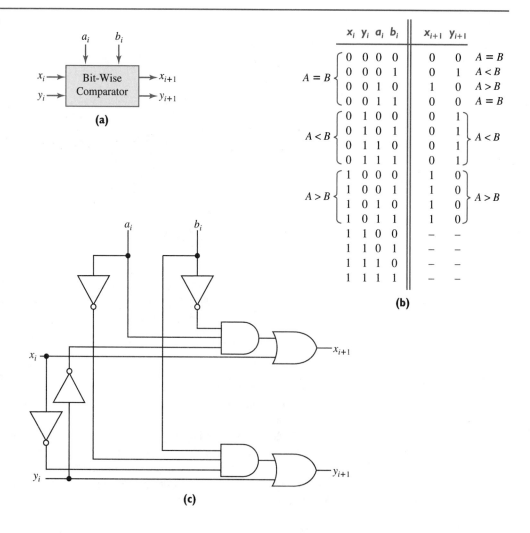

	x_i	y_i	a_i	b_i	x_{i+1}	y_{i+1}	
$A = B$	0	0	0	0	0	0	$A = B$
	0	0	0	1	0	1	$A < B$
	0	0	1	0	1	0	$A > B$
	0	0	1	1	0	0	$A = B$
$A < B$	0	1	0	0	0	1	
	0	1	0	1	0	1	$A < B$
	0	1	1	0	0	1	
	0	1	1	1	0	1	
$A > B$	1	0	0	0	1	0	
	1	0	0	1	1	0	$A > B$
	1	0	1	0	1	0	
	1	0	1	1	1	0	
	1	1	0	0	–	–	
	1	1	0	1	–	–	
	1	1	1	0	–	–	
	1	1	1	1	–	–	

(b)

Table 5.1

	$x_i y_i$	Encoded result
Cascade signals encoding for a bit-wise comparator	00	$A = B$
	01	$A < B$
	10	$A > B$
	11	Don't care

then we set the outputs x_{i+1} and y_{i+1} according to the relative values of a_i and b_i. If the previous, higher-order comparisons have determined that $A < B$ or $A > B$, ($x_i y_i = 01$ or $x_i y_i = 10$), then we simply pass the result along irrespective of the values of a_i and b_i. Finally, the output values for the last four truth table entries are don't cares because the $x_i y_i = 11$ encoding is not used. Figure 5.47(c) shows the minimized logic schematic for the bit-wise comparator.

Iterative circuits use a serial-oriented implementation, computing a desired logic function as the successive application of more primitive logic functions. Not all combinational functions are amenable to an iterative implementation. However, logic functions involving multi-bit operands are often good candidates for an iterative circuit, where bit-wise computations are accumulated to yield the desired function.

5.8.2 Macro Logic Design

Macro logic design implements a desired logic expression using a collection of predefined logic functions, typically without executing a formal logic synthesis process. In other words, predefined logic functions, commonly called *macros* or *commercial off-the-shelf components* (COTS), are used as building blocks to compose a system quickly. The multiplexers, binary decoders, and memories discussed in the previous sections can be viewed as examples of macros. Manufacturers offer a wide selection of macros or logic functions, detailed in *data books*.

For example, Figure 5.48 shows an alternate implementation for the sine wave limit detector combinational system examined in Chapter 4. The sine wave limit detector accepts a digitized sine wave; sine wave magnitude is represented by a 4-bit two's

Figure 5.48

Macro logic design of a sine wave limit detector

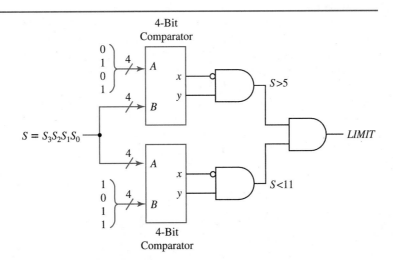

Table 5.2

Out-of-range binary numbers

Unsigned value	Binary input	Two's complement value
6	0110	6
7	0111	7
8	1000	−8
9	1001	−7
10	1010	−6

complement number, $S_3 S_2 S_1 S_0$. The sine wave limit detector asserts its output if $|S_3 S_2 S_1 S_0| \geq 0110_2 = 6_{10}$.

In Chapter 4, we performed the conventional combinational synthesis approach: We enumerated the input/output transform of the limit detector in a truth table having $2^4 = 16$ entries and generated a minimized logic expression using Karnaugh maps. Alternatively, we can recognize a few elementary functions within the sine wave limit detector—a couple of comparators and a few **and** gates—that can be easily assembled to quickly realize the combinational system.

The comparator studied in the previous section performs the numerical comparison of the input sine wave magnitude. Since the 4-bit comparators do not recognize two's complement numbers, the comparators are augmented with additional logic to recognize the out-of-range binary numbers listed in Table 5.2. If the top comparator asserts that $S > 5$ and the bottom comparator asserts that $S < 11$, then S is an out-of-range number and the last **and** gate asserts the output *LIMIT*.

The macro design of the sine wave limit detector uses much more logic than the custombuilt, minimized logic expressions. However, the macro design required far less time; we easily generated the schematic in a few minutes.

A macro can even be a logic gate. For example, consider the design of an *identity comparator* that accepts two 8-bit signals and asserts its output if the two signals are equal. We could design the identity comparator using one of the formal minimization techniques discussed in Chapter 4; the truth table would have $2^{16} = 65,536$ entries. Executing such an approach will eventually yield a solution, given enough patience and persistence or a good synthesis computer-aided design (CAD) tool. However, the design shown in Figure 5.49 can be easily generated by recognizing the following:

1. An **xnor** gate provides an equivalence function that checks bit-for-bit equality.
2. An **and** gate provides a sum function that combines the individual bit-for-bit checks to yield the desired byte-to-byte identity comparison.

The output is asserted active-1 if $A_0 = B_0$, $A_1 = B_1$, and so on.

The basic idea behind macro logic design is that *sometimes* insight can be gained into how best to design a logic expression by studying how the new function can be built out of existing functions. There is no well-defined algorithm for recognizing applications of macro logic functions. However, some designers are very good at this "bottom-up" style of logic design and find the design practice quite useful.

Problem-Solving Skill
Never blindly execute a mathematical analysis and/or design technique to obtain a solution to a problem. The solution can sometimes be readily obtained with a little practical thought and understanding of the problem.

Figure 5.49

An 8-bit identity comparator

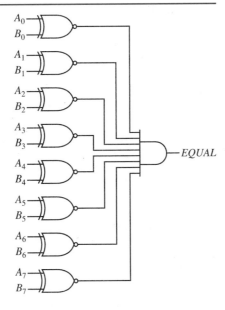

5.9 SUMMARY

In this chapter, we discussed the second stage of combinational logic design: implementation. Implementation realizes a logic expression in actual, physical hardware. Just as a specification can have many possible implementations, a logic expression can have many possible realizations. To avoid possible confusion in interpreting a schematic, we defined some terminology for logic elements. If a schematic refers to a specification, the logic elements are called *operators*. If a schematic refers to an implementation, the logic elements are called *gates*. Operators are mathematical functions; gates are physical devices.

Since standard minimization techniques generate two-level sum-of-products or product-of-sums logic expressions, we discussed a variety of two-level gate network implementation schemes, such as **nand-nand** and **nor-nor**. We also learned how **xor** and **xnor** gates can be exploited to yield efficient implementations of certain logic functions having a distinctive checkerboard pattern of 1's on a Karnaugh map.

We then studied multilevel gate networks, involving more than two levels or stages of logic. Two-level sum-of-products and product-of-sums networks can be stretched out into multilevel networks using factoring and collapsing switching algebra manipulations, respectively. Stretching out logic expressions into multilevel networks tends to serialize logic functions, which can reduce logic complexity but typically increases logic delay.

We also discussed logic implementation strategies using multiplexers, decoders, memories, and programmable logic devices (PLDs). There is a wide variety of memories, including random access memory (RAM) and many variants of read-only memory (ROM). Likewise, there is a wide variety of programmable logic devices, including coarse-grain and fine-grain architectures.

Finally, we discussed two design practices: iterative circuits and macro logic design. A design practice is largely an ill-defined, heuristic implementation strategy that takes advantage of designer ingenuity and creativity to generate a design without necessarily using formal design principles. Iterative circuits decompose a logic function into an iterative or incremental series of simpler computations. Macro logic design exploits common logic components, like multiplexers and decoders, to quickly implement a complex logic function.

The resulting design may not be as good as a custom-built design using formal synthesis principles, but it may be good enough.

Since we have described several ways to implement a logic expression, you may be wondering how to select the best implementation for a particular design. Unfortunately, there is no algorithm for selecting the best implementation approach; the choice is typically made based on good *engineering judgment*. Engineering judgment takes into account the particular design task and the general merits of each implementation approach. As we study digital design further, we will try to build the experience base required to make good engineering judgments.

5.10 PROBLEMS

1. Implement the minimized form of the following logic expression.

$$F(A, B, C, D) = \sum m(1, 5, 6, 9, 11, 13)$$

   a. Use a two-level **nand-nand** gate network.
   b. Use a two-level **nor-nor** gate network.
   c. Which implementation is better?

2. Implement the minimized form of the following logic expression.

$$F(A, B, C, D) = \sum m(3, 6, 7, 9, 11, 12, 13) + d(14, 15)$$

   a. Use a two-level **nand-nand** gate network.
   b. Use a two-level **nor-nor** gate network.
   c. Which implementation is better?

3. Implement the minimized form of the following logic expression.

$$F(A, B, C, D) = \sum m(3, 7, 11, 12, 13, 14)$$

   a. Use **and** and **or** gates.
   b. Use **xor** and **and** gates.
   c. Which implementation is better?

4. Implement the logic expression given in Problem 3 using only 2-input **and** and **or** gates by factoring the minimized sum-of-products function.

5. Consider the simple circuit shown below.
   a. Draw the timing diagram for the outputs of the two 2-input multiplexers.
   b. What is a possible application of the combinational system?

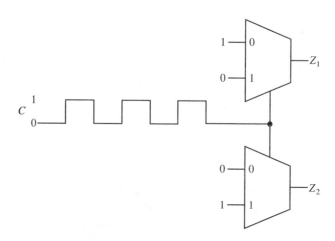

6. Draw a two-level **and-or** gate network of an 8-input multiplexer having active-1 inputs and both active-0/active-1 outputs.

7. Multiplexers often have an *enable* (EN) input that enables/disables the data/control inputs. When the enable input is deasserted, the output is deasserted, regardless of the value of the data/control inputs. With the enable input, multiplexers can be connected together to form larger multiplexers. Use the 4-input multiplexer having an active-0 enable shown in the truth table below to construct an 8-input multiplexer.

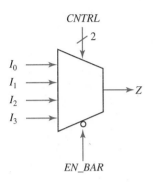

EN_BAR	CNTRL$_1$	CNTRL$_0$	I$_3$	I$_2$	I$_1$	I$_0$	Z
0	0	0	–	–	–	0	0
0	0	0	–	–	–	1	1
0	0	1	–	–	0	–	0
0	0	1	–	–	1	–	1
0	1	0	–	0	–	–	0
0	1	0	–	1	–	–	1
0	1	1	0	–	–	–	0
0	1	1	1	–	–	–	1
1	–	–	–	–	–	–	0

8. Recall from Chapter 2 that the *excess-3 code* is a variation of the binary coded decimal (BCD) code. Each decimal digit is represented by a 4-bit code that is three more than the associated BCD code. For example, 0_{10} is encoded in excess-3 as 0011_2, 1_{10} is encoded in excess-3 as 0100_2, and so on. Design a combinational logic system that detects (active-1) prime numbers represented by a 4-bit excess-3 code using each of the following:
   a. one 16-input multiplexer
   b. one 8-input multiplexer
   c. one 4-input multiplexer

9. If a logic expression has a special property called *simple disjoint decomposition,* then the logic expression can be implemented using a series of multiplexers. Simple disjoint decomposition means that a logic expression can be factored or decomposed into a series of embedded logic expressions, as shown below.

$$F(A, B, C, D) = F(A, B, G(C, D))$$

   a. Show that $F(A, B, C, D) = \sum m(5, 6, 8, 11)$ possesses the simple disjoint decomposition property.
   b. Implement $F(A, B, C, D)$ using only two 4-input multiplexers in cascade.

10. For the logic expression given in Equation 5.1,
   a. Verify that the expression is given in a minimal sum-of-products form using Quine-McCluskey tables.
   b. Implement the logic expression using one 8-input multiplexer.

11. Draw a logic diagram of a 3-to-8 decoder having active-1 inputs and outputs.

12. Show how a 3-to-8 decoder can control four devices; each device receives two signals: start and stop.

13. Decoders often have an *enable* (EN) input that enables/disables the inputs. When the enable input is deasserted, all outputs are deasserted, regardless of the value of the inputs. With the enable input, decoders can be connected together to form larger decoding functions. Use the 3-to-8 decoder having an active-0 enable shown in the truth table below to construct a 4-to-16 decoder.

EN_BAR	I_2	I_1	I_0	O_7	O_6	O_5	O_4	O_3	O_2	O_1	O_0
0	0	0	0	0	0	0	0	0	0	0	1
0	0	0	1	0	0	0	0	0	0	1	0
0	0	1	0	0	0	0	0	0	1	0	0
0	0	1	1	0	0	0	0	1	0	0	0
0	1	0	0	0	0	0	1	0	0	0	0
0	1	0	1	0	0	1	0	0	0	0	0
0	1	1	0	0	1	0	0	0	0	0	0
0	1	1	1	1	0	0	0	0	0	0	0
1	–	–	–	0	0	0	0	0	0	0	0

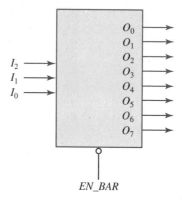

EN_BAR

14. Use the 3-to-8 decoder described in Problem 13 and a 2-to-4 decoder to construct a 5-to-32 decoder.

Table 5.3

A 2-out-of-5 code

Decimal digit	2-out-of-5 code
0	11000
1	00011
2	00101
3	00110
4	01001
5	01010
6	01100
7	10001
8	10010
9	10100

15. In addition to binary coded decimal and excess-3, the *2-out-of-5* code is another code used to represent decimal digits. Table 5.3 lists the *2-out-of-5* representations for the decimal digits 0–9. The 2-out-of-5 code has the property that a single bit error can be detected. If a single bit inadvertently changes from 0 to 1 or vice versa for any 2-out-of-5 code, the resulting bit string is not a legal 2-out-of-5 code.

a. Use a 4-to-16 decoder and **or** gates to implement a combinational system that accepts a 4-bit binary number and yields the corresponding 5-bit 2-out-of-5 code.

16. A *binary encoder* performs the reverse function of a binary decoder. For example, a 4-to-2 binary encoder accepts four inputs and yields a 2-bit binary number indicating the asserted input, assuming only one input is asserted at any time. A *priority encoder* allows more than one input to be asserted at any time and outputs a binary number indicating the asserted input having the highest priority. Shown below is the table for a 2-to-4 priority encoder; input priorities are assigned in descending order, starting with I_3 and ending with I_0.

I_3	I_2	I_1	I_0	O_1	O_0
0	0	0	1	0	0
0	0	1	–	0	1
0	1	–	–	1	0
1	–	–	–	1	1

a. Use a 4-to-16 decoder and **or** gates to implement a 4-to-2 priority encoder.

17. Implement the 4-to-2 priority encoder described in Problem 16 using memory.

a. Use a 256-bit memory configured as 64 × 4. Identify unused memory locations as sized blocks.

b. A 256-bit memory is *very* small by today's standards. If we implemented the 4-to-2 priority encoder using a 16K memory configured as 4K $\times$ 4, how much memory would be unused?

18. Memory provides a convenient way to implement "table lookup" kinds of combinational functions. Show how two 16×8 memories can be used to implement a combinational system that converts a 4-bit binary number into its corresponding 2-digit, ASCII-encoded character string. For example,

$$3_{10} = 0011_2 = 30 \quad 33_{16}$$

19. Design an 8-to-3 priority encoder.

   a. Obtain a minimized logic expression. (*Hint*: Instead of processing a truth table having $2^8 = 256$ entries, try using a design practice and generate the three output logic expressions based on an understanding of the 8-to-3 priority encoding function.)
   b. Implement the minimized logic expressions using a programmable logic array.

20. Design a full adder having an equal number of stages of gates, that is, levels of logic, between input and output for both the sum and carry functions.

21. Show a gate-level schematic for a 4-bit ripple carry adder, that is, a ripple carry adder that computes the sum of two 4-bit operands.

22. An iterative binary subtracter can be defined like an iterative binary adder. The truth table for a *full subtracter* is given in Table 5.4. A full subtracter performs the arithmetic operation

$$\text{Difference} = \text{minuend} - \text{subtrahend} - \text{borrow-in}$$

and generates a borrow-out bit if

$$\text{Subtrahend} + \text{borrow-in} > \text{minuend}$$

   a. Generate minimized logic expressions for the difference and borrow-out bits.
   b. Show the gate-level schematic for a 4-bit ripple borrow subtracter, that is, a ripple borrow subtracter that computes the difference of two 4-bit operands.
   c. Use the literal analysis technique discussed in Chapter 3 to show that the 4-bit ripple borrow subtracter works for $9_{10} - 7_{10}$ and $15_{10} - 8_{10}$.

Table 5.4

Rules for subtracting two binary numbers. These rules describe how to generate the difference and borrow-out bits on a column-by-column basis when subtracting two binary numbers.	Borrow-in bit from previous column	Minuend bit	Subtrahend bit	Difference bit	Borrow-out bit to next column
	0	0	0	0	0
	0	0	1	1	1
	0	1	0	1	0
	0	1	1	0	0
	1	0	0	1	1
	1	0	1	0	1
	1	1	0	0	0
	1	1	1	1	1

23. An easy way to build a digital multiplier is to use ripple carry adders to sum the partial products, as shown on p. 160. Notice that an n-bit multiplier generates a $2n$-bit product.

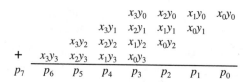

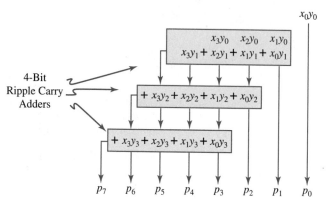

4-Bit
Ripple Carry
Adders

a. Using the 4-bit ripple carry adder designed in Problem 21, show the gate-level schematic of a 4-bit multiplier.

b. Use the literal analysis technique discussed in Chapter 3 to show that the 4-bit multiplier works for $5_{10} \times 6_{10}$ and $13_{10} \times 11_{10}$.

REFERENCES AND READINGS

[1] Goto, S., ed. *Design Methodologies,* Volume 6 in Goto, S., *Advances in CAD in VLSI.* New York: North-Holland, 1986.

[2] Alford, R. *Programmable Logic Designer's Guide.* Indianapolis, IN: H. W. Sams, 1989.

[3] Almaini, A. *Electronic Logic Systems.* Englewood Cliffs, NJ: Prentice-Hall, 1989.

[4] Blakeslee, T. *Digital Design with Standard MSI & SSI,* New York: Wiley, 1979.

[5] Bostock, G. *Programmable Logic Devices: Technology & Applications.* New York: McGraw-Hill, 1988.

[6] Brock, L. *Designing with MSI.* Sunnyvale, CA: Signetics, 1970.

[7] Brown, S. *Field-Programmable Gate Arrays.* Norwell, MA: Kluwer Academic Publishers, 1992.

[8] Burton, V. *The Programmable Logic Device Handbook.* Blue Ridge Summit, PA: Tab Books, 1990.

[9] Dempsey, J. *Basic Digital Electronics with MSI Applications.* Reading, MA: Addison-Wesley, 1977.

[10] Dietmeyer, D. *Logic Design of Digital Systems.* Boston: Allyn & Bacon, 1988.

[11] Lala, P. *Digital System Design Using Programmable Logic Devices.* Englewood Cliffs, NJ: Prentice-Hall, 1990.

[12] Parr, E. *The Logic Designer's Guidebook.* New York: McGraw-Hill, 1984.

[13] Pellerin, D. *Practical Design Using Programmable Logic.* Englewood Cliffs, NJ: Prentice-Hall, 1991.

Part 2 DIGITAL ENGINEERING: MANUFACTURING TECHNOLOGIES

**Structural
Modeling**

```
I1: NOT_OP
   port map (A, B);
```

**Asynchronous
Data Flow
Modeling**

```
B <= not A;
```

**Technology
Modeling**

```
B <= not A after 5 ns;
```

**Synchronous
Data Flow
Modeling**

```
B1: block (C = '1')
begin
 B <= guarded not A after 5 ns;
end block B1;
```

**Algorithmic
Modeling**

```
P1: process (PS)
begin
 NS <= STATE_TABLE(PS);
end process P1;
```

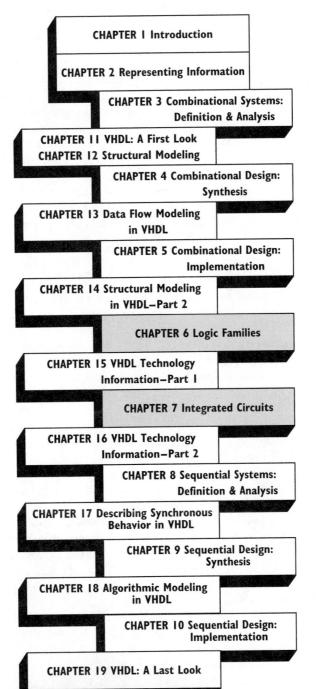

CHAPTER 1 Introduction

CHAPTER 2 Representing Information

CHAPTER 3 Combinational Systems: Definition & Analysis

CHAPTER 11 VHDL: A First Look
CHAPTER 12 Structural Modeling

CHAPTER 4 Combinational Design: Synthesis

CHAPTER 13 Data Flow Modeling in VHDL

CHAPTER 5 Combinational Design: Implementation

CHAPTER 14 Structural Modeling in VHDL–Part 2

CHAPTER 6 Logic Families

CHAPTER 15 VHDL Technology Information–Part 1

CHAPTER 7 Integrated Circuits

CHAPTER 16 VHDL Technology Information–Part 2

CHAPTER 8 Sequential Systems: Definition & Analysis

CHAPTER 17 Describing Synchronous Behavior in VHDL

CHAPTER 9 Sequential Design: Synthesis

CHAPTER 18 Algorithmic Modeling in VHDL

CHAPTER 10 Sequential Design: Implementation

CHAPTER 19 VHDL: A Last Look

**Combinational Logic
Schematics**

**Combinational Logic
Equations**

$Z = (A \cdot B) + C$

**Logic Families &
Integrated Circuits**

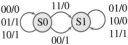

**Sequential Logic
Schematics**

**Sequential Logic
State Diagrams**

6
LOGIC FAMILIES

Congratulations! You have completed the study of the analysis and design of combinational digital systems. Before we move on to sequential digital systems, we need to take time out to examine how digital systems are manufactured. Thus far, the logical aspects of gates have been emphasized, implicitly assuming *ideal* gates having no power dissipation, input/output delay, or connectivity restrictions. However, in practice, gates are made of devices that have physical limitations; it is important to understand such *nonidealities* when implementing a digital system.

Various techniques for realizing logic gates are called *logic families*. This chapter will focus on the basic operation of each logic family and associated advantages and disadvantages. Chapter 7 will address in more detail device physics and manufacturing processes.

6.1 ELECTRICAL SIGNALS AND LOGIC CONVENTIONS

The objective in realizing a logic gate is to devise a scheme for (1) using a physical medium to denote the binary logic states of 0 and 1 and (2) changing the physical medium to denote switching algebra operations. Examples of physical media include mechanical, electrical, magnetic, and optical. Presently, the most popular scheme is to denote the binary logic states with the presence or absence of electrical charge, or *voltage*.

Two voltages, traditionally labeled *high* (H) and *low* (L), are required to denote the two logic states 0 and 1. Table 6.1 lists the two possible mappings of logic states onto voltage levels or polarities. The *positive logic convention* represents the logic state 0 by a *low* voltage and logic state 1 by a *high* voltage. The *negative logic convention* represents the logic state 0 by a *high* voltage and logic state 1 by a *low* voltage.

Figure 6.1 shows that a single physical gate can be viewed as performing two different logic operations, depending on the logic convention. The truth table in Figure 6.1(a) given in terms of voltage levels can denote either a logic **and** or a logic **or** operation, depending on how logic states are mapped onto voltage levels. Mapping L's into 0's and H's into 1's using the positive logic convention transforms the physical truth table in Figure 6.1(a) into the **and** logic truth table in Figure 6.1(b). Conversely, mapping L's into 1's and H's into 0's using the negative logic convention transforms the physical truth table in Figure 6.1(a) into the **or** logic truth table in Figure 6.1(c). To avoid confusion, manufacturers often supply truth tables in terms of voltage levels; the associated logic operation depends on which logic convention the designer elects to use.

Some digital design techniques, called *mixed logic techniques*, use both positive and negative logic conventions in a single design. Such techniques use logic *state* negation symbols "○," logic *level* polarity symbols "▷," and various signal naming conventions to denote active logic states and levels throughout a digital design. Although mixed logic techniques can exploit certain optimizations, the added complexity of such design techniques is often not

Table 6.1

	Convention	Logic state	Voltage level
Logic conventions	Positive logic	0	L
		1	H
	Negative logic	0	H
		1	L

warranted. Consequently, we will use a simpler approach of adopting a *single* logic convention throughout a digital design. Unless otherwise stated, we will assume a positive logic convention: A 0 logic state will be represented by a low (L) voltage level and a 1 logic state will be represented by a high (H) voltage level.

One final note before we begin discussing logic families. When explaining circuit operation, we will assume *conventional* current flow. By convention, current flows from the positive potential to the negative potential (even though electrons actually flow in the opposite direction, from the negative potential to the positive potential).

Figure 6.1

Logic function depends on logic-to-voltage mapping

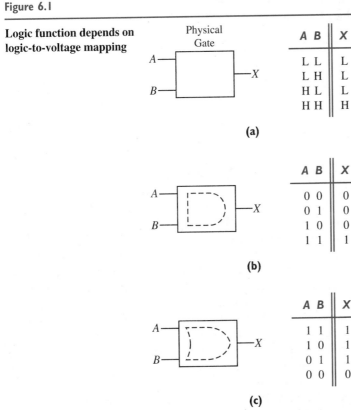

(a)

(b)

(c)

Figure (a) presents a physical truth table using voltage levels. Figures (b) and (c) present logic truth tables obtained from Figure (a) using the positive and negative logic conventions, respectively.

6.2 METAL-OXIDE-SEMICONDUCTOR LOGIC FAMILIES

Having devised a scheme for using the physical medium of electrical charge to denote logic states, we now need the means of changing the physical medium to denote switching algebra operations. We need an electronic switch to charge a circuit node to a high voltage or discharge a circuit node to a low voltage. A popular electronic switch is a *metal-oxide-semiconductor field-effect transistor* (MOSFET), also called an *insulated-gate field-effect transistor* (IGFET). The following sections will explain the MOSFET and show various ways in which the transistor is used to construct logic families.

6.2.1 MOSFET Switch

MOSFETs are generally classified as either *N-channel*, commonly called *NMOS transistors*, or *P-channel*, commonly called *PMOS transistors*. In addition, NMOS and PMOS transistors are further classified as either *enhancement-mode* or *depletion-mode* devices.[1] Combining these two classifications yields the following four types of MOSFETs.

1. NMOS enhancement
2. NMOS depletion
3. PMOS enhancement
4. PMOS depletion

Like the term *metal-oxide-semiconductor*, the terms *N-channel/P-channel* and *enhancement/depletion* refer to how the devices are fabricated, which will be discussed in Chapter 7.

Figure 6.2 gives a listing of various symbols used to denote MOSFETs. A MOSFET is a four-terminal device described as follows:

Terminal Symbol	Terminal Name
G	Gate
D	Drain
S	Source
B	Body or Bulk

The *drain* (D) and *source* (S) are the switch terminals. The *gate* (G) is the controlling terminal, controlling whether the drain and source are connected or disconnected. The *bulk* or *body* (B) is usually not an active terminal; it is typically connected to a low voltage for NMOS or a high voltage for PMOS. The MOSFET symbols showing only three terminals in Figure 6.2 assume such connections for the body (B) terminal.

Figure 6.3 (p. 166) shows the switching action of each type of MOSFET when a low, zero, and high voltage is applied to the gate with respect to the source. When a MOSFET is acting like a closed switch, the drain and source terminals are connected and current can flow between the switch terminals. In this state, the device is *turned on* or *saturated*. When a MOSFET is acting like an open switch, the drain and source terminals are disconnected and no current flows between the switch terminals. In this state, the device is *turned off* or *cut off*.

Enhancement-mode devices act like normally turned off or open switches. Figure 6.3(a) shows that a high voltage must be applied to the gate with respect to the source of an NMOS enhancement transistor to turn the device on, in other words, close the normally open switch. In a complementary fashion, Figure 6.3(b) shows that a low voltage must be applied to the gate with respect to the source of a PMOS enhancement transistor to turn the device on; a high or no gate voltage leaves a PMOS enhancement transistor in its normally off condition.

Depletion-mode devices act as normally turned on or closed switches. Figure 6.3(c) shows that a low voltage must be applied to the gate with respect to the source of an NMOS

---

[1] Unfortunately, half the battle of learning any technology is conquering the terminology.

Figure 6.2

MOSFET symbols

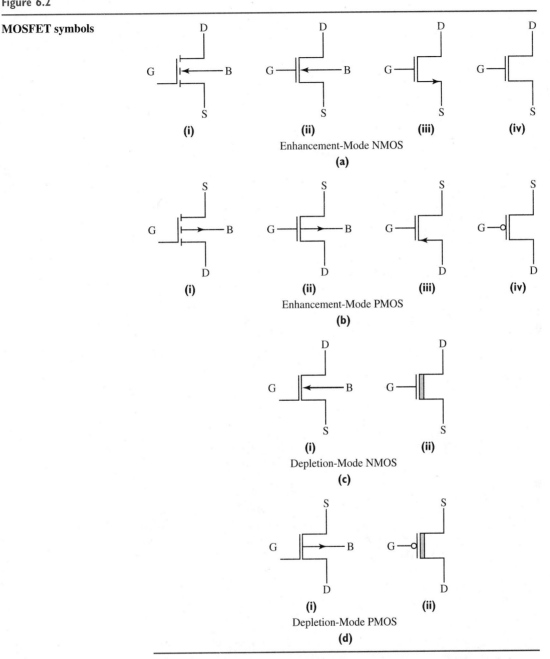

Figures (a) and (b) list the NMOS and PMOS enhancement transistor symbols, respectively. Figures (c) and (d) list the NMOS and PMOS depletion transistor symbols, respectively.

depletion transistor to turn the device off, in other words, open the normally closed switch. The PMOS depletion transistor shown in Figure 6.3(d) again acts in a complementary fashion. A high voltage between gate and source is needed to turn the device off; a low or no gate voltage leaves a PMOS depletion transistor in its normally on condition.

Although it is convenient to think of MOSFETs as simple electronic switches, MOSFETs are, unfortunately, not ideal switches. A MOSFET is not a perfect closed switch because the

Figure 6.3

MOSFETs as switches

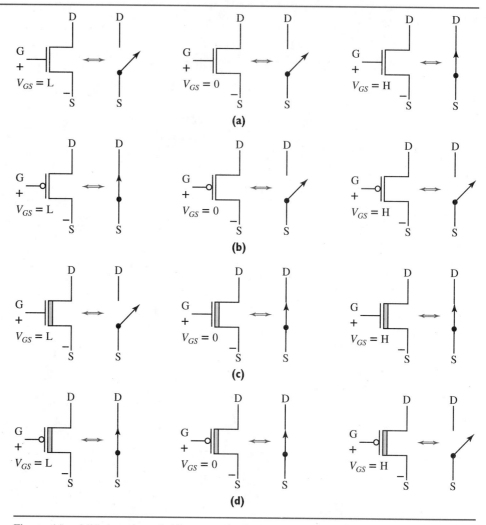

Figures (a) and (b) show the switching properties for NMOS and PMOS enhancement transistors, respectively. Figures (c) and (d) show the switching properties of NMOS and PMOS depletion transistors, respectively.

drain-to-source resistance of a saturated MOSFET (*on-resistance*) is low, but not zero. The *on-resistance* of a MOSFET is typically on the order of tens to thousands of ohms, which approximates a closed connection. Similarly, a MOSFET is not a perfect open switch because the drain-to-source resistance of a cut off MOSFET (*off-resistance*) is high, but not infinite. The *off-resistance* of a MOSFET is typically on the order of megaohms, which approximates an open connection. In addition, a MOSFET requires a minimum threshold voltage (V_t) to turn on; $V_t \simeq 1$ V for NMOS enhancement transistors and $V_t \simeq -1$ V for PMOS enhancement transistors.

6.2.2 Static NMOS

Historically, PMOS transistors were the first MOSFETs to be widely used in building logic gates. However, NMOS logic families have supplanted PMOS logic families because NMOS transistors can turn on and off faster than PMOS transistors. NMOS logic families can be classified into two categories: *static* and *dynamic*. The output of a static NMOS logic gate is a "static" function of the inputs, meaning that the output will persist as long as the inputs persist. In contrast, the output of a dynamic logic gate must be periodically reset or refreshed regardless of how long the inputs persist. The difference between static and dynamic NMOS logic will become more apparent as we examine the associated circuit details.

To introduce NMOS logic families, we show in Figure 6.4 how an NMOS depletion-mode transistor and an NMOS enhancement-mode transistor connected in series can implement a logic inverter. The depletion-mode transistor is the *pull-up* device and is responsible for charging the output up to a high voltage. Notice that the NMOS depletion-mode transistor is always turned on because $V_{GS} = 0$; its drain-to-source on-resistance is denoted by R_{ON}^{PU}. The enhancement-mode transistor is the *pull-down* device and is responsible for discharging the output down to a low voltage. The input turns the pull-down device on and off. The drain-to-source on-resistance and off-resistance of the pull-down device are denoted by R_{ON}^{PD} and R_{OFF}^{PD}, respectively.

Figure 6.4(b) shows the state of the pull-up and pull-down devices with a high input. A high (H) input that is greater than the threshold voltage V_t turns on the NMOS enhancement-

Figure 6.4

A static NMOS inverter

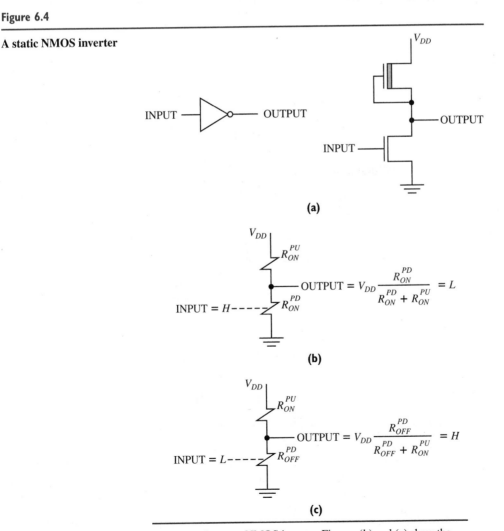

Figure (a) shows an NMOS inverter. Figures (b) and (c) show the state of the transistors with a high and low input, respectively.

mode device, effectively closing the switch and connecting the output to a low potential via the voltage divider resistor network. Hence, we have achieved half of our desired inverter or complement function; a high (H) input yields a low (L) output.

Figure 6.4(c) shows the state of the pull-up and pull-down devices with a low input. A low (L) input that is less than the threshold voltage V_t turns off the NMOS enhancement-mode device, effectively opening the switch and connecting the output to a high potential via the voltage divider resistor network. Hence, we have achieved the remaining aspect of our desired inverter or complement function; a low (L) input yields a high (H) output.

6.2.3 Dynamic NMOS

The basic idea behind dynamic NMOS logic is that "the output is high unless the inputs are such to cause the output to be low." Gate evaluation is divided into two phases. The first phase *precharges* the output to a high voltage. Then, the second phase conditionally discharges the output depending on the inputs and the logic function.

Figure 6.5 shows a dynamic NMOS inverter. Figure 6.5(b) shows that during the precharge phase, ϕ_1 turns on the enhancement NMOS pull-up transistor Q_1 and the output is

Figure 6.5

A dynamic NMOS inverter

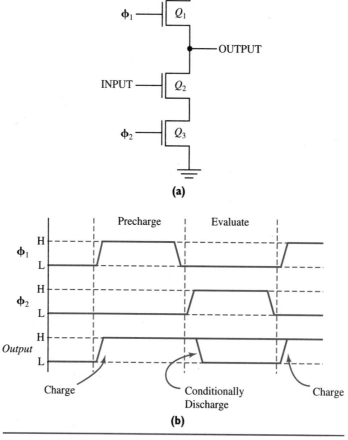

Figure (a) shows a dynamic NMOS inverter. Figure (b) shows nonoverlapping, antiphase clocks controlling the two-phase evaluation.

Problem-Solving Skill
Two-phased evaluation offers an interesting perspective on problem solving. During the first phase, the output is preset to a default value. During the second phase, the output either stays at the default value or conditionally changes, depending on the inputs and the desired function.

charged to a high voltage. Also, during the precharge phase, the input value sets up. During the evaluation phase, ϕ_2 turns on the enhancement NMOS pull-down transistor Q_3. If the input is low, Q_2 is cut off and the precharged output stays at a high voltage. If the input is high, Q_2 turns on and the output is discharged to a low voltage through Q_2 and Q_3. Again, we have realized an inverter; a low (L) input yields a high (H) output and vice versa. Dynamic logic is called *precharged* or *clocked* logic because of the *two-phased* clocked evaluation scheme.

6.2.4 Static CMOS

On June 18, 1963, Frank Wanlass applied for a patent for a "low stand-by power *complementary* field effect circuit," which formed the basis for *complementary metal-oxide-semiconductor* (CMOS) logic [12]. The term *complementary* is used because CMOS uses both N-channel and P-channel enhancement-mode MOSFETs.

Figure 6.6 shows examples of static CMOS logic. The PMOS transistors serve as pull-up circuits; the NMOS transistors serve as pull-down circuits. The NMOS and PMOS transis-

Figure 6.6

CMOS logic functions

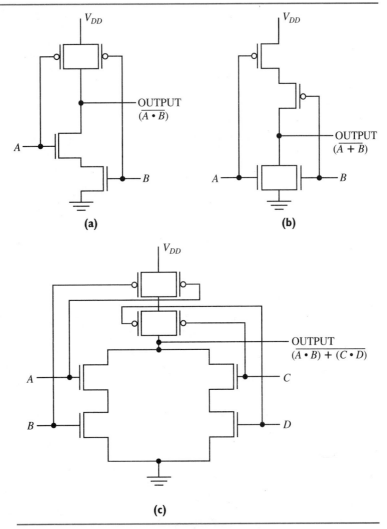

Figure (a) shows a CMOS **nand** gate. Figure (b) shows a CMOS **nor** gate.
Figure (c) shows a CMOS **and-or-invert** structure.

tors operate antiphase; when the PMOS pull-up transistors are turned on, the NMOS pull-down transistors are turned off and vice versa.

Consider, for example, the **nand** gate. The output is low only when both inputs are high, which turns on the series-connected NMOS pull-down transistors and turns off the parallel-connected PMOS pull-up transistors. The CMOS gates are drawn to emphasize the series/parallel connections. The series/parallel MOSFET structures can be combined to build complex logic functions, such as the **and-or-invert** function shown in Figure 6.6(c).

6.2.5 Dynamic CMOS

A dynamic CMOS logic family, called *domino logic* [7], is shown in Figure 6.7. Taking advantage of the complementary switching properties of PMOS and NMOS, the two clocks used with dynamic NMOS logic can be replaced by a single clock, ϕ. The PMOS logic used with static CMOS is replaced by a single PMOS transistor Q_1, which controls precharging. An inverter structure, Q_5 and Q_6, drives the output of each gate. The inverter controls the sequence in which cascaded or multistage logic gates are evaluated.

To understand domino logic operation, examine Figure 6.8. During the precharge phase ($\phi = L$), the input to each inverter driving each dynamic gate output is set to a high voltage, which drives all gate outputs to a low voltage. Thus, all inputs to each stage of logic, except the first stage, will be low, effectively deactivating the associated NMOS logic blocks. During the evaluation phase ($\phi = H$), the inputs will activate the NMOS logic blocks associated with the first stage, which, in turn, will activate the outputs of the first stage. This action activates the NMOS logic blocks associated with the second stage, which, in turn activates the outputs of the second stage, and so on. Logic evaluation ripples through the logic stages in strict sequence, similar to the action of knocking over a string of dominos.

Problem-Solving Skill
Sometimes the inspiration for the solution to a problem can come from simple observances of natural actions or behaviors.

Figure 6.7

Domino logic

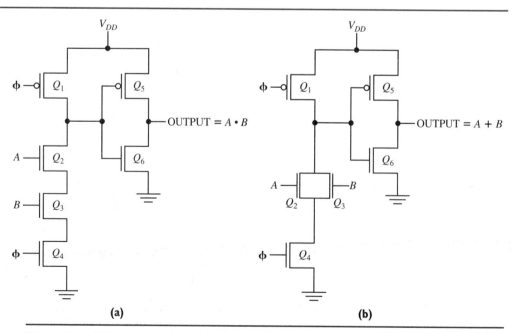

(a)   (b)

Figures (a) and (b) show an **and** and **or** gate, respectively.

Figure 6.8

Domino logic operation

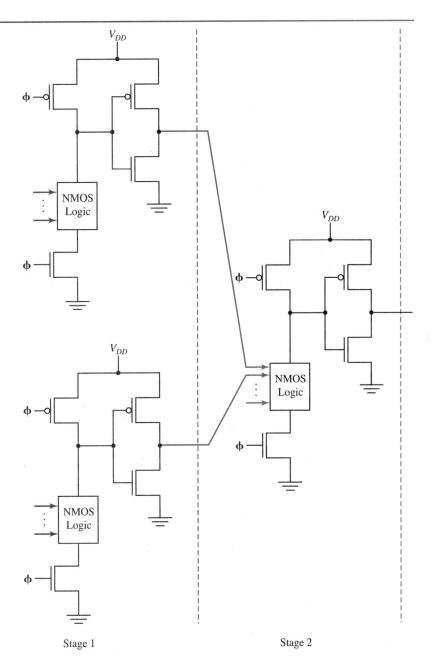

Stage 1   Stage 2

6.3 BIPOLAR LOGIC FAMILIES

Another type of transistor commonly used as an electronic switch is a *bipolar* transistor, also called a *bipolar junction transistor* or BJT. The development of bipolar logic families predates the development of MOS logic families. We discussed MOS logic families first because MOS logic is presently more prevalent than bipolar logic and the switching actions of a MOSFET are conceptually simpler than the switching actions of a BJT. The following sections will explain the bipolar transistor and show various ways in which the transistor is used to construct logic functions.

6.3.1 Bipolar Switch

Bipolar transistors are generally classified as either NPN or PNP. NPN and PNP refer to the detailed physics and fabrication of bipolar transistors, which we will discuss in Chapter 7. Figure 6.9 lists the symbols used to denote NPN and PNP BJTs.

A bipolar transistor is a three-terminal device described as follows:

Terminal Symbol	Terminal Name
B	Base
E	Emitter
C	Collector

An NPN BJT, shown in Figure 6.9(a), is denoted by an arrow pointing out of the emitter terminal. A PNP BJT, shown in Figure 6.9(b), is denoted by an arrow pointing into the emitter terminal. The arrow denotes the direction of positive emitter current, I_E.

The collector (C) and emitter (E) are typically the switch terminals and the base (B) is the controlling terminal. However, as mentioned earlier, BJT switching action is a little trickier than MOSFET switching action. MOSFET switching is controlled by one parameter, V_{GS}, whereas BJT switching action is controlled by *three* parameters:

1. base voltage with respect to the emitter V_{BE},
2. base voltage with respect to the collector V_{BC}, and
3. base current, I_B.

The effect of these parameters on BJT switching is summarized in Figure 6.10. A bipolar transistor has four modes of operation: *cutoff, forward active, saturation,* and *reverse active.* Each mode of bipolar transistor operation exhibits different switching properties.

Figure 6.10(a) shows that a BJT is in *cutoff* mode when a low voltage is applied to the base with respect to the emitter and collector of an NPN BJT or when a high voltage is applied to the base with respect to the emitter and collector of a PNP BJT. In *cutoff* mode, a bipolar transistor approximates an open connection between the emitter and collector switch terminals.

Figure 6.10(b) and Figure 6.10(c) respectively show that an NPN BJT is in *forward active* mode when $V_{BE} = H$ and $V_{BC} = L$ and *saturation* mode when $V_{BE} = H$ and $V_{BC} = H$. Voltage polarities are reversed for a PNP BJT. In *forward active* and *saturation* modes, a bipolar transistor approximates a closed switch between the collector and emitter switch terminals. The principal difference between forward active and saturation modes is that a BJT is typically conducting more collector/emitter current in saturation mode than in forward active mode. Moreover, Figure 6.10(b) shows that in forward active mode, the current between

Figure 6.9

Bipolar transistor symbols

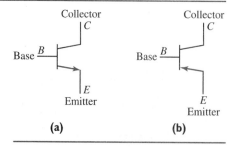

(a)   **(b)**

Figures (a) and (b) list the NPN and PNP transistor symbols, respectively.

Figure 6.10

Bipolar junction transistor (BJT) as an electrical switch

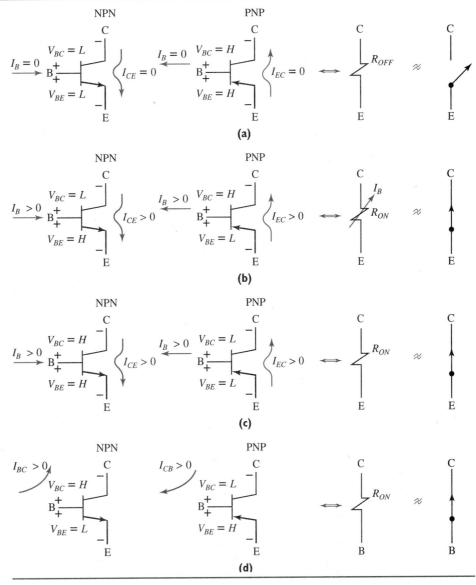

Figures (a), (b), (c), and (d) show the switching properties in *cutoff* mode, *forward active* mode, *saturation* mode, and *reverse active* mode, respectively.

the emitter and collector, or equivalently the on-resistance, is a function of the base current (I_B).

Finally, *reverse active* mode, shown in Figure 6.10(d), is somewhat of a rogue mode of operation, where the base and collector terminals effectively become the switch terminals. Although the reverse active mode is seldom used, we will see how transistor-transistor logic (TTL) uses this mode of operation to build logic gates.

Like MOSFETs, BJTs are not ideal electronic switches. A BJT exhibits a low, but not zero, on-resistance as a closed switch and a high, but not infinite, off-resistance as an open switch. Also, a bipolar transistor requires a minimum threshold voltage (V_t) to turn on and conduct current.

6.3.2 Diodes

Before we explain how to use the switching action of a bipolar transistor to construct logic gates, we need to discuss another electronic device called a *diode*. The symbol of a diode is given below.

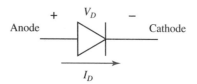

A diode conducts current in only one direction. When the terminal voltage V_D is greater than or equal to the diode's threshold voltage V_t, the diode conducts current, I_D. In this condition, the diode is *forward-biased* and the triangle serves as a reminder of the direction of positive or forward current flow. When the polarity of the terminal voltage is reversed such that the cathode is positive with respect to the anode, the diode effectively conducts no current.[2] In this condition, the diode is *reverse-biased* or *cut off*.

It is no coincidence that the behavior of a diode is similar to many aspects of the behavior of a transistor. In Chapter 7, we will learn that transistor operation and fabrication are largely based on the physical principles of diodes.

6.3.3 Transistor-Transistor Logic (TTL)

Of the long lineage of bipolar logic families some of the early ones, such as resistor-transistor logic (RTL) and diode-transistor logic (DTL), are no longer widely used and will not be covered here. We will begin our look at bipolar logic families by discussing *transistor-transistor logic* (TTL), which was introduced in 1965. As an introduction to TTL, we show a TTL **nand** gate in Figure 6.11.

To understand the operation of the TTL **nand** gate, it is instructive to split the gate into two stages: input and output. Figure 6.12 shows the output stage. Q_3 and Q_4 are connected in a structure called a *push-pull* or *totem pole* configuration. Q_2 is called a *phase splitter*. Q_2 ensures that the output transistors Q_3 and Q_4 operate in antiphase. Either Q_4 is turned off and Q_3 is turned on or vice versa. Figure 6.12(a) shows that Q_2 is cut off when a low voltage is applied to its base. With Q_2 cut off, a low voltage is applied to the base of Q_3 and a high voltage is applied to the base of Q_4. This action turns off Q_3 and turns on Q_4, pulling the output high. Figure 6.12(b) shows that Q_2 saturates when a high voltage is applied to its base. With Q_2 saturated, a high voltage is applied to the base of Q_3 and a low voltage is applied to the base of Q_4. This action turns on Q_3 and turns off Q_4, pulling the output low. Thus, if the base of $Q_2 = L$, the output is H; if the base of $Q_2 = H$, the output is L. The output diode D_1 is called a *hold-off* diode. The diode is used to ensure that Q_2 is able to pull the base of Q_4 low enough to cut off Q_4, in other words, $V_{BE}^4 < V_t$.

The purpose of the input stage shown in Figure 6.13 (p. 176) is to drive the base of Q_2 high or low, such that the output is the logical **nand** of the two inputs. Q_1 is a *multiple-emitter* bipolar transistor. The number of inputs can be increased by adding more emitters to Q_1. Figure 6.13(a) shows that when either $INPUT_1$ or $INPUT_2$ is low, Q_1 saturates.[3] Acting like a closed switch, Q_1 forces the base of Q_2 low, which, as explained above, forces the output

---

[2] Actually, the diode conducts a very small amount of current, called *leakage current*.
[3] Taking the emitter of an NPN BJT low with respect to the base is the same as taking the base high with respect to the emitter. Both actions turn the transistor on, provided $V_{BE} \geq V_t$.

Figure 6.11

Transistor-transistor logic (TTL) nand gate

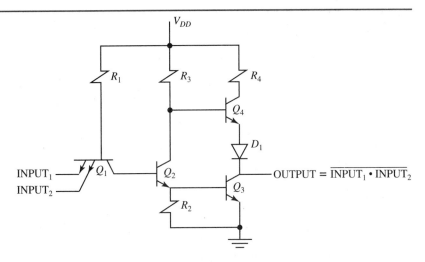

high. Figure 6.13(b) shows that when both inputs are high, Q_1 is in the *reverse active* mode, connecting the base of Q_2 to a high potential through R_1. Forcing the base of Q_2 high forces the output low. Thus, the output is low only when all inputs are high, otherwise the output is high. In other words, the output is the logical **nand** of the inputs, assuming the positive logic convention.

Figure 6.12

Output stage of transistor-transistor logic (TTL) nand gate

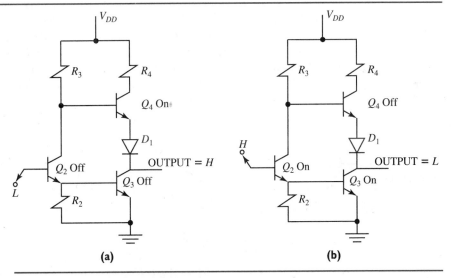

Figure (a) shows the state of the transistors with the base of Q_2 low. Figure (b) shows the state of the transistors with the base of Q_2 high.

6.3.4 Schottky TTL (STTL)

Since the introduction of the basic TTL logic family in 1965, a series of several follow-on generations of TTL-related logic families have been introduced, offering improved performance. *Schottky TTL logic* (STTL), introduced in the early 1970s, is an example of a high-speed variant of basic TTL.

Figure 6.13

Input stage of transistor-transistor logic (TTL) nand gate

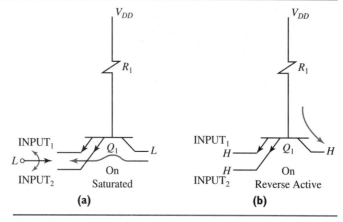

(a)   (b)

Figure (a) shows the state of Q_1 when a low voltage is applied to one or both inputs. Figure (b) shows the state of Q_1 when a high voltage is applied to both inputs.

Schottky TTL is derived from basic TTL by replacing bipolar transistors with *Schottky-clamped* bipolar transistors, which have faster switching properties. Figure 6.14 shows that a *Schottky-clamped* BJT is formed by connecting a *Schottky-barrier diode* between the base and collector terminals. A Schottky-barrier diode is a special kind of diode having a threshold voltage lower than the conventional diode described in Section 6.3.2.

The Schottky diode increases bipolar transistor switching speed by preventing the transistor from going into saturation. Since a BJT is conducting "harder" in saturation mode than in forward active mode, it takes longer to turn off a BJT in saturation mode than a BJT in forward active mode. As base current increases to drive the transistor from forward active mode into saturation mode, the Schottky diode turns on. Further increases in base current are shunted away from the transistor by the Schottky diode, thereby preventing the transistor from being driven into full saturation.

Figure 6.14

Schottky-clamped bipolar transistor

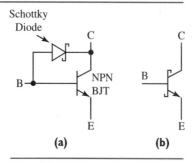

(a)   (b)

Figure (a) shows the symbol for a Schottky-barrier diode and how the diode is connected between the base and collector. Figure (b) shows a shorthand notation for denoting a Schottky-clamped transistor.

6.3.5 Merged Transistor Logic (MTL)

The relatively high number of components (resistors, diodes, and bipolar transistors) involved in TTL logic translates into digital systems that dissipate a lot of power and take up a lot of space. *Merged transistor logic* is a stripped down version of TTL logic that is used to implement large digital systems, where power dissipation and area are serious concerns.

Figure 6.15 shows one type of merged transistor logic called *integrated injection logic* (I^2L). In I^2L, an NPN BJT and a PNP BJT are merged together to form a multiple-output inverter. Q_2 is a *multiple-collector* NPN bipolar transistor. When the input is low, Q_2 is cut off and the outputs are high. Conversely, when the input is high, Q_2 saturates and the outputs are low. Q_1 is a PNP bipolar transistor that supplies the base current needed to operate Q_2. Figure 6.15(b) highlights Q_1's function by replacing the PNP BJT by a current source. Figure 6.15(c) gives a commonly used shorthand notation for the I^2L gate in Figure 6.15(a), emphasizing the multiple-output inverter function.

The I^2L inverter can be used to build other logic functions by taking advantage of *wired logic*. Figure 6.16 (p. 178) shows how I^2L inverters can be connected to realize the **nor** and **nand** functions. Both Figure 6.16(a) and Figure 6.16(b) illustrate that an implicit **and** function is realized by simply connecting the inverter outputs together. Such a connection is called a *wired-**and** connection*. Figure 6.16(a) shows an I^2L **nor** gate, which is easily seen by transforming

$$\overline{A} \cdot \overline{B}$$

into

$$\overline{A + B}$$

using DeMorgan's Theorems (see Chapter 4). Figure 6.16(b) shows an I^2L **nand** gate, which

Figure 6.15

Integrated injection logic

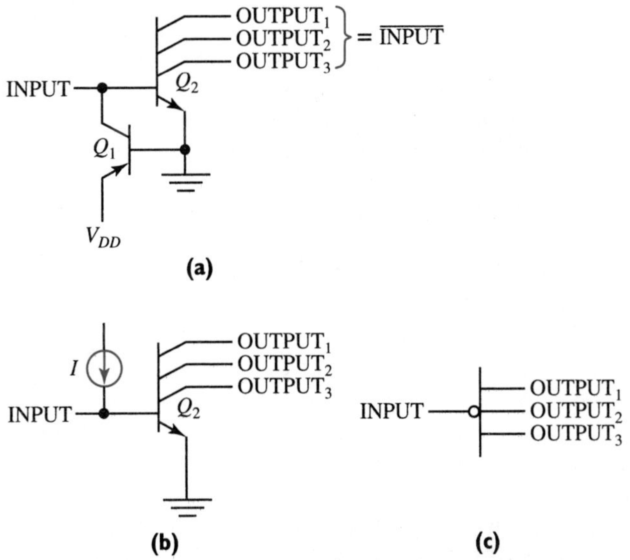

Figure (a) shows the transistor circuit for the basic inverter gate.
Figure (b) shows Q_1 redrawn as a current source. Figure (c)
shows a common shorthand notation for the I^2L gate.

Figure 6.16

Building logic functions in integrated injection logic

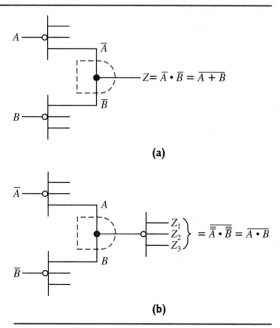

(a)

(b)

Figures (a) and (b) show how I^2L inverters can be used to realize the **nor** and **nand** logic functions, respectively.

is easily seen by transforming

$$\overline{\overline{\overline{A}}} \cdot \overline{\overline{\overline{B}}}$$

into

$$\overline{A \cdot B}$$

Wired logic will be discussed in more detail in Section 6.5.5.

Like TTL logic, several new generations of merged transistor logic have been introduced that improve the performance of I^2L, such as *integrated Schottky logic* (ISL) and *Schottky transistor logic* (STL). Both ISL and STL improve speed performance by preventing transistor saturation.

6.3.6 Emitter-Coupled Logic (ECL)

Emitter-coupled logic (ECL) is another bipolar logic family, but it operates quite differently than the TTL-related logic families studied thus far. TTL-related logic families propagate *voltage* levels throughout a circuit to affect the desired logic function. ECL logic propagates (or "steers") *current* levels throughout a circuit to affect the desired logic function. ECL is also called a *nonsaturating* logic family because the circuits are designed so conducting transistors do not go into saturation.

As an example, Figure 6.17 shows an ECL structure that is both an **or** and **nor** gate. An ECL gate typically offers a logic function and its complement. Similar to TTL, it is instructive to split the ECL gate into input and output stages. The input stage performs the **or/nor** logic; the output stage serves as a buffer.

Consider first the output stage composed of Q_5 and Q_6. The output stage buffers or

Figure 6.17

Emitter-coupled logic (ECL)

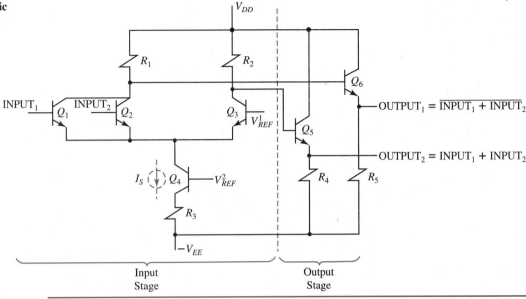

ECL structure yields **or** and **nor** functions. V^1_{REF} and V^2_{REF} are reference voltages that are preset by a network of resistors and/or diodes not shown.

"strengthens" the signals generated by the input stage, which improves the performance of the logic gate. Q_5 and Q_6 are connected in an *emitter-follower* configuration (hence the name emitter-coupled logic). The emitter voltages of Q_5 and Q_6 "follow" their respective base voltages.

The base voltages of Q_5 and Q_6 are generated by the input stage shown in Figure 6.18 (p. 180). The input stage used without the output stage is a common variant of ECL called *current-mode logic* (CML). The input stage is basically a *differential amplifier*. Q_4 and R_3 form a simple current source, having value I_s.

Figure 6.18(a) shows that when $INPUT_1$ and/or $INPUT_2$ are high (H), the circuit is designed so either Q_1 and/or Q_2 turn on. With Q_1 and/or Q_2 conducting, the emitter of Q_3 is forced high enough to turn off Q_3. I_S is routed down through the left-hand circuit path. V^6_B is pulled low through Q_1 and/or Q_2; V^5_B is pulled high through R_2.

Figure 6.18(b) shows that when both inputs are low, Q_1 and Q_2 are turned off. Q_3 conducts, routing I_S down through the right-hand circuit path. V^5_B is pulled low through Q_3; V^6_B is pulled high through R_1.

Putting the input and output stages together, $OUTPUT_1 = H$ when both inputs are low, otherwise $OUTPUT_1 = L$; in other words, $OUTPUT_1 = \overline{INPUT_1 + INPUT_2}$. In a complementary fashion, $OUTPUT_2 = L$ when both inputs are low, otherwise $OUTPUT_2 = H$; in other words, $OUTPUT_2 = INPUT_1 + INPUT_2$.

Complex logic functions can be realized by *stacking* the basic ECL differential transistor circuit structure, as shown in Figure 6.19 (p. 181). The bottom differential configuration composed of Q_1, Q_2, and Q_3 provides $A + B$. The top differential configuration composed of Q_4, Q_5, and Q_6 provides $C + D$. Stacking the two differential configurations **and**s the two expressions, yielding $Z = (A + B) \cdot (C + D)$. Power supply levels and circuit tolerances limit the degree of stacking.

Figure 6.18

Current steering action of emitter-coupled logic (ECL)

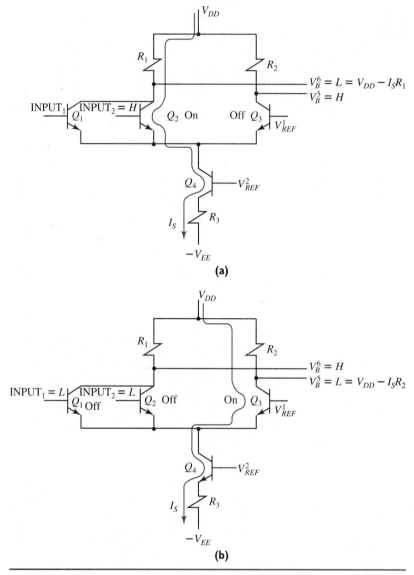

(a)

(b)

Figures (a) and (b) show the state of transistors when the inputs are high and low, respectively.

6.4 BiCMOS LOGIC FAMILY

Recently, the merits of CMOS and bipolar logic families have been combined to yield *BiCMOS*. Figure 6.20 shows a BiCMOS **nand** gate [5, 6]. The MOSFETs turn on and off the totem pole configured output NPN bipolar transistors by turning on and off the associated base currents.

Transistors Q_1, Q_2, Q_3, and Q_4 form the basic CMOS **nand** structure shown in Figure 6.6(a). The MOSFETs Q_5, Q_6, and Q_7 are called *base discharge* transistors and help turn off Q_8/Q_9.

When inputs A and B are high, Q_1 and Q_2 turn off and Q_5 and Q_6 turn on, which respectively cuts off Q_8's base current and forces Q_8's base to a low potential. These actions

Figure 6.19

Stacking ECL circuit structures to build complex logic functions

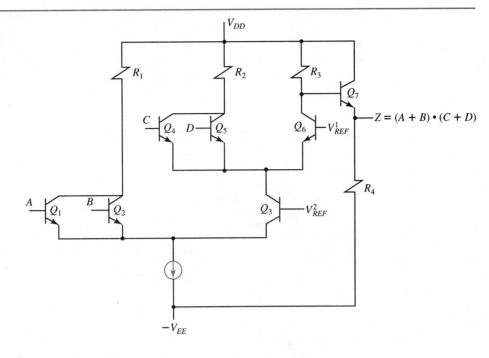

$Z = (A + B) \cdot (C + D)$

Figure 6.20

BiCMOS nand gate

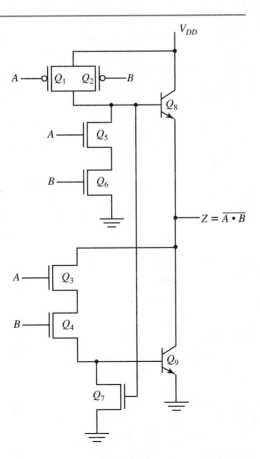

$Z = \overline{A \cdot B}$

combine to turn off Q_8. At the same time, Q_3 and Q_4 turn on, which supplies a current path for Q_9's base current. Q_9 turns on, connecting the output Z to a low potential.

When inputs A and/or B are low, Q_3 and/or Q_4 turn off and Q_7 turns on, which respectively cuts off Q_9's base current and forces Q_9's base to a low potential. These actions combine to turn off Q_9. At the same time, Q_1 and/or Q_2 turn on, which supplies a current path for Q_8's base current. Q_8 turns on, connecting the output Z to a high potential.

6.5 ELECTRICAL CHARACTERISTICS

In the previous sections we introduced a diverse collection of MOS and bipolar logic families. We studied the basic operation of each logic family by adopting a simplified view of transistors as electronic switches and analyzing circuit currents/voltages. With this functional understanding of MOS and bipolar logic families, we can next discuss electrical characteristics that define the physical properties and limitations of logic families. As a digital systems designer, you will need to be aware of these electrical characteristics because a design must be logically *and* electrically correct to ensure proper operation. Even if a design is logically correct, it may fail to operate correctly if the design violates electrical limitations and usage rules of a particular logic family.

6.5.1 Noise Margins

Thus far, circuit voltage levels have generally been referred to in symbolic terms as either a high (H) or low (L) voltage. Different logic families have different definitions for what is considered a high and low voltage. In this section, we will discuss the voltage levels of various logic families and an associated electrical characteristic called *noise margin*.

Figure 6.21 shows that logic family voltage levels are, in general, specified using eight voltages. The eight voltages are divided into two sets: *input voltage levels* and *output voltage levels*. Each set of voltage levels is further divided into two ranges: the range of voltages representing a high value and the range of voltages representing a low value.

Any voltage in the range $V_{IH\,min} \leq V_I \leq V_{IH\,max}$ at the input of the inverter is guaranteed to be considered a high level, which will generate a low output guaranteed to be in the range $V_{OL\,min} \leq V_O \leq V_{OL\,max}$. Conversely, any voltage in the range $V_{IL\,min} \leq V_I \leq V_{IL\,max}$ at the input of the inverter is guaranteed to be considered a low level, which will generate a high output guaranteed to be in the range $V_{OH\,min} \leq V_O \leq V_{OH\,max}$.

Typically, a designer is concerned with only four of the eight voltages shown in Figure 6.21 because power supply values set the extreme upper ($V_{IH\,max}$ and $V_{OH\,max}$) and lower ($V_{IL\,min}$ and $V_{OL\,min}$) voltages. Hence, manufacturers sometimes specify logic family voltage

Figure 6.21

Input and output voltage levels

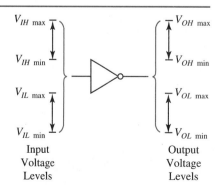

Input Voltage Levels   Output Voltage Levels

Table 6.2

Typical voltage levels for logic families. All values are in volts.

Logic family	$V_{IH\,max} \Leftrightarrow V_{IH\,min}$	$V_{IL\,max} \Leftrightarrow V_{IL\,min}$	$V_{OH\,max} \Leftrightarrow V_{OH\,min}$	$V_{OL\,max} \Leftrightarrow V_{OL\,min}$
TTL †	$5 \Leftrightarrow 2$	$0.8 \Leftrightarrow 0$	$5 \Leftrightarrow 2.5$	$0.4 \Leftrightarrow 0$
STTL †	$5 \Leftrightarrow 2$	$0.8 \Leftrightarrow 0$	$5 \Leftrightarrow 2.7$	$0.5 \Leftrightarrow 0$
CMOS †	$5 \Leftrightarrow 3.5$	$1.0 \Leftrightarrow 0$	$5 \Leftrightarrow 4.9$	$0.1 \Leftrightarrow 0$
ECL ††	$-0.81 \Leftrightarrow -1.13$	$-1.48 \Leftrightarrow -1.95$	$-0.81 \Leftrightarrow -0.98$	$-1.63 \Leftrightarrow -1.95$

Note:

$\dagger : V_{DD} = 5\,\text{V}$

$\dagger\dagger : V_{DD} = 0\,\text{V and} -V_{EE} = -5.2\,\text{V}$

levels using four voltages: $V_{IH} \Rightarrow V_{IH\,min}$, $V_{IL} \Rightarrow V_{IL\,max}$, $V_{OH} \Rightarrow V_{OH\,min}$, and $V_{OL} \Rightarrow V_{OL\,max}$. In the simplest case, logic family voltage levels may also be specified using only *two* voltages (V_L and V_H), which represent average or nominal low and high voltages, respectively. Table 6.2 shows typical voltage levels for several logic families.

Can the definition of the voltage levels of one logic family be "better than" another logic family? The answer is yes, if we consider *noise margins*. Noise is an unwanted, yet common, electrical signal that originates from many sources, such as atmospheric radiation and power supplies. Any circuit signal is susceptible to being corrupted by noise, whereby noise voltage adds to and/or subtracts from the circuit signal voltage. If the noise signal is strong enough, a circuit signal can erroneously change voltage levels from a high to a low or vice versa. This voltage error translates into a logic error. To avoid such inadvertent logic errors, logic family input/output voltage levels are designed to minimize noise interference or, equivalently, to maximize *noise immunity*. Noise margins are electrical characteristics that indicate the noise immunity of a particular logic family; the higher the noise margins, the better the noise immunity.

There are two noise margins:

$$NM_{LOW} = V_{IL\,max} - V_{OL\,max}$$
$$NM_{HIGH} = V_{OH\,min} - V_{IH\,min}$$

The noise margin NM_{LOW} defines the maximum additional noise voltage that can be added to an output low voltage without corrupting the signal. That is, an output low voltage can tolerate being increased by a noise voltage up to NM_{LOW} and still be considered a low input signal by succeeding logic gates. Similarly, the noise margin NM_{HIGH} defines the maximum additional noise voltage that can be subtracted from an output high voltage without corrupting the signal. That is, an output high voltage can tolerate being decreased by a noise voltage up to NM_{HIGH} and still be considered a high input signal by succeeding logic gates. Based on Table 6.2, CMOS logic families typically offer the largest noise margins, while ECL logic families typically offer the smallest noise margins.

6.5.2 Speed

Speed is another important electrical characteristic of a logic family. *Speed* is the reciprocal of *propagation delay*. Propagation delay is the time delay between the input stimuli and the output response. Since electrons travel at a finite speed, it takes a finite amount of time to propagate changes in input signal voltages throughout a circuit to yield appropriate changes in output signal voltages. Charge has to move around to turn on and off transistors to charge and discharge circuit nodes.

To discuss propagation delay, we need to define some common timing terms. First, since every event in an electrical circuit takes a finite amount of time, signal value transitions do not occur instantaneously or in zero time. Figure 6.22(a) shows a signal having finite transition times, characterized by the parameters *rise time* (t_r) and *fall time* (t_f).

The transition from a low voltage to a high voltage is a *rising edge*. Rise time is defined to be the time required to transition from $(0.1 \times high)$ to $(0.9 \times high)$. The transition from a high voltage to a low voltage is a *falling edge*. Fall time is defined to be the time required to transition from $(0.9 \times high)$ to $(0.1 \times high)$.

Figure 6.22(b) shows that propagation delay is often given as two numbers $(t_{PLH}$ and $t_{PHL})$, relative to rising and falling output transitions. The propagation delay t_{PLH} denotes output propagation delay for a low to high transition. The propagation delay t_{PHL} denotes output propagation delay for a high to low transition.

Two propagation delays are given because the input/output delay may be asymmetrical, in other words, not the same for a rising output transition versus a falling output transition. Asymmetrical propagation delay is due to asymmetrical charge and discharge circuit paths. The **nand** gate in Figure 6.6(a), for example, has asymmetrical charge and discharge circuit paths. The charge path involves two PMOS transistors in parallel, whereas the discharge path involves two NMOS transistors in series. For simplicity, propagation delay is also given as a single number, which typically represents a nominal input/output delay or the average of t_{PLH} and t_{PHL}.

It is difficult to give representative propagation delays for various logic families because the numbers are always changing, which explains why every textbook gives a different set of logic family speeds. However, it is useful to discuss relative speed performance trends and comparisons. Bipolar logic families are generally faster than MOS logic families. Within

Figure 6.22

Timing parameters

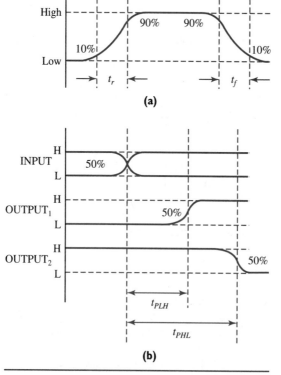

(a)

(b)

Figure (a) shows a signal having finite transition times (t_r) and (t_f). Figure (b) shows propagation times.

bipolar logic families, nonsaturating families, such as Schottky-based logic and current-mode logic, are generally faster than saturating logic, such as basic TTL. Finally, within the nonsaturating bipolar logic families, ECL tends to consistently be the fastest logic family.

6.5.3 Power Dissipation

High speed is gained at the expense of high power dissipation. High speed implies that circuit nodes are charging and discharging at fast rates. Thus, more charge is being moved around the circuit, which implies higher currents. Increased currents result in increased power dissipation because *power = voltage × current*.

The power dissipation of a logic gate consists of two components: *static* and *dynamic*.

$$P_{TOTAL} = P_{STATIC} + P_{DYNAMIC}$$

Static power dissipation, also called *quiescent power dissipation*, is the power dissipated by a logic gate when signals are not changing value; in other words, the logic gate is in a quiescent state. Dynamic power dissipation is the power dissipated by a logic gate when input/output signals change values. Dynamic power dissipation accounts for the additional movement of charge associated with signals changing voltage levels.

The power dissipation of some logic families is dominated principally by the static power dissipation component, such as the bipolar logic families discussed in Section 6.3. For these logic families, the power dissipation per logic gate is typically given as a nominal value in units of milliwatts per gate (mW/gate).

In contrast, the power dissipation of CMOS, BiCMOS, and most dynamic logic families is dominated principally by the dynamic power dissipation component. The nonoverlapping clocks used in dynamic NMOS and the complementary switching properties of NMOS and PMOS transistors used in CMOS and BiCMOS ensure that there is effectively no static current path between power and ground. For these logic families, the power dissipation per logic gate is typically given as a nominal value in units of microwatts per gate per megahertz (switching rate) per picofarad (node capacitance) (μW/Gate • MHz • pF). Depending on the switching rate, complementary and dynamic logic can offer substantial power savings.

6.5.4 Fan-in and Fan-out

Figure 6.23 illustrates two electrical characteristics that define restrictions on interconnecting logic gates: *fan-in* and *fan-out*.

Figure 6.23

Fan-in and fan-out electrical characteristics

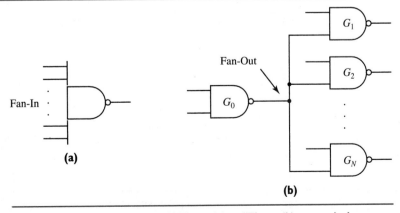

Fan-in and fan-out are illustrated in Figure (a) and Figure (b), respectively.

Fan-in specifies the maximum number of inputs per logic gate. The fan-in capability of a logic family is influenced by several factors, such as noise margins, current sourcing/sinking, and switching rate. Consider, for example, switching rate and noise margins. Adding more inputs to a static CMOS gate (Figure 6.6) increases the number of series NMOS pull-down transistors for the **nand** gate or PMOS pull-up transistors for the **nor** gate. Increasing the number of series-connected MOSFETs increases the effective pull-up/pull-down resistance because each MOSFET exhibits a finite on-resistance. Increasing the pull-up/pull-down resistance degrades the switching speed of the gate, in other words, the ability to charge or discharge the output node. Moreover, increasing pull-up/pull-down resistance can degrade noise margins to the point where the gate fails to operate correctly because the output voltage cannot be pulled high or low enough. Hence, there is a limit to the number of inputs a logic gate can support, specified by the fan-in.

Fan-out, shown in Figure 6.23(b), specifies the maximum number of gate inputs that can be connected to a gate output. Many of the factors influencing fan-in also influence fan-out. Consider, for example, current sourcing/sinking. Recall for electrical-based logic families, that the "flow" of binary 0's and 1's through stages of logic implies the movement of charge, in other words, current. Current sourcing/sinking refers to the ability of a gate output to supply the current required by the connected gate inputs and is typically defined by the following four currents:

1. $I_{IL\,max}$,
2. $I_{IH\,max}$,
3. $I_{OL\,max}$, and
4. $I_{OH\,max}$.

These currents, shown in Figure 6.24, can be positive or negative. By definition, positive current flows into a terminal; negative current flows out of a terminal.

In Figure 6.24(a), the **nand** gate G_0 is supplying current, also called *sourcing current*, to the *load* **nand** gates $G_1 - G_N$. $I_{IH\,max}$ defines the maximum current that an input requires when the associated voltage is high. Similarly, $I_{OH\,max}$ defines the maximum current that an output can supply when the associated voltage is high. For correct operation, the combined input sink current demand cannot exceed the maximum output source current capability. That is,

$$\sum_{i=1}^{N} I_{IH\,max_i} \leq I_{OH\,max} \tag{6.1}$$

Figure 6.24

Current sourcing and sinking requirements

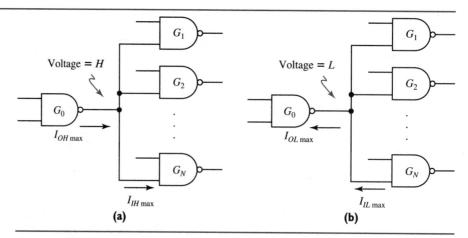

Figures (a) and (b) show a gate output driving the connected gate inputs high and low, respectively.

In Figure 6.24(b), the **nand** gate G_0 is receiving current, also called *sinking current*, from the *load* **nand** gates G_1-G_N. $I_{IL\,max}$ defines the maximum current that an input supplies when the associated voltage is low. Similarly, $I_{OL\,max}$ defines the maximum current that an output can receive when the associated voltage is low. Again, for correct operation, the combined input source current demand cannot exceed the maximum output sink current capability. That is,

$$\sum_{i=1}^{N} I_{IL\,max_i} \leq I_{OL\,max} \qquad\qquad (6.2)$$

6.5.5 Wired Logic

Another aspect of interconnecting logic gates concerns *wired logic*. Sometimes, connecting the outputs of several logic gates together generates an implicit logic function, without the need for explicitly adding a logic gate. Such implicit logic functions are called *wired logic* or *dotted logic* functions. The characteristics of a logic family determine the properties of wired logic and whether it is supported. For example, recall from Figure 6.16 that connecting the outputs of several I²L gates together forms an implicit **and** function, called a *wired-***and** function. I²L actually depends on wired logic to implement Boolean expressions. In contrast, ECL supports *wired-***or** logic. That is, the outputs of several **nor** gates, shown in Figure 6.17, can be **or**ed by simply connecting the outputs together.

Be aware that some logic families *do not* support wired logic. For example, the totem pole configured outputs of the TTL logic family shown in Figure 6.11 should not be connected. To understand this restriction, consider the situation where the totem pole outputs of two **nand** gates are connected to a common node. What happens when one TTL **nand** gate tries to drive the common node high, while the other TTL **nand** gate tries to drive the common node low? The gates are *fighting* and neither gate will win. The output will be at some illegal, intermediate voltage that is neither high nor low. Moreover, the pull-up/pull-down actions of the totem pole output structures act in concert to yield high currents that can damage the logic gates.

To take advantage of wired logic, manufacturers often supply special versions of a logic family having output structures that can be connected. For example, Figure 6.25 (p. 188) shows a version of TTL having an *open-collector* output, rather than a totem pole output.

The fighting problem associated with connecting gate outputs having an active pull-up and pull-down totem pole structure is resolved by removing the active pull-up circuit. Comparing Figure 6.11 and Figure 6.25(a) reveals that the elements R_4, Q_4, and D_1, which perform the active pull-up action, have been removed, leaving the collector of Q_3 open. For proper operation, the open collector output must be connected to power (V_{DD}) through an externally supplied resistor. The extra resistor forms a passive pull-up. Figure 6.25(b) shows that the passive pull-up supports a wired-**and** function. For MOS-related logic families, the fighting problem is solved in a similar fashion with *open-drain* outputs.

6.5.6 Tri-State

The final logic family electrical characteristic that we will discuss is *tri-state* outputs.[4] Tri-state outputs can be viewed as another approach to solving the fighting problem associated with connecting gate outputs having an active pull-up and pull-down. As the name implies, a logic gate having a tri-state output can be in one of *three* states: low, high, and *high impedance*. In the high impedance state, both the active pull-up and pull-down devices are turned

---

[4] Tri-state® is a registered trademark of National Semiconductor Corporation.

Figure 6.25

TTL with open-collector outputs

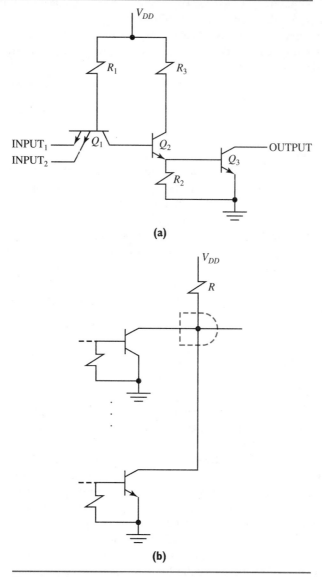

Figure (a) shows the schematic for a TTL **nand** gate with an *open-collector* output stage. Figure (b) shows how TTL with open-collector outputs supports wired-**and** logic.

off, effectively disconnecting the gate from its output node. That is, in the high impedance state, the gate is not driving its output. The high impedance state is also called a *high-Z* state, pronounced "high zee."

Manufacturers typically provide versions of both MOS and bipolar logic families that support tri-state outputs. When gates having tri-state outputs are connected together, all but one of the gates are placed in the high impedance state. The gate not placed in the high impedance state controls the output node.

Figure 6.26

Electronic-based logic families

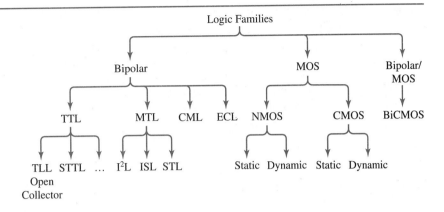

6.6 SUMMARY

In this chapter, we studied how logic gates are actually built, focusing on electrical-based logic families. The logic level of abstraction was peeled away, revealing the circuit level of abstraction. We examined circuit devices and how their current/voltage characteristics realize logic functions. Figure 6.26 shows the wide spectrum of logic families that we discussed.

Logic families are broadly classified into three groups, based on the type of transistor: (1) bipolar, (2) MOS, and (3) bipolar/MOS. Bipolar logic families use the bipolar junction transistor (BJT); MOS logic families use the metal-oxide-semiconductor field-effect transistor (MOSFET); and bipolar/MOS logic families use both BJTs and MOSFETs. Transistors act as electronic switches to connect the output node to a high or low potential based on the input voltages and the desired switching algebra function.

Having an understanding of the function of each logic family, we then discussed several important performance aspects, such as noise margin, speed, power dissipation, fan-in/fan-out, and wired logic. Noise is an unwanted spurious electrical signal that can corrupt electrical signals carrying logic information. Noise margin is a figure-of-merit that determines how much noise a logic family can tolerate and still operate correctly. Speed describes how fast a gate can respond to input stimuli, and power dissipation describes how much energy is expended. High speed and low power dissipation is desirable but usually not obtainable because high switching rates typically imply high switching energies. Fan-in/fan-out and wired logic address connectivity limitations and properties of logic families.

A digital system designer should have an understanding of the general properties of logic families because they influence system performance and cost. Each logic family has advantages and disadvantages, which should be comparatively analyzed to select the logic family that best supports design objectives and requirements.

6.7 PROBLEMS

1. A gate has the following truth table given in terms of physical voltages.

A	B	Z
L	L	L
L	H	H
H	L	H
H	H	L

    a. Using the positive logic convention, what is the logic function of this gate?

    b. Using the negative logic convention, what is the logic function of this gate?

2. A problem with many dynamic logic families that do not possess active pull-up and pull-down circuits that operate concurrently is that transient signal transitions, called *glitches*, can cause erroneous behavior. A gate may start to discharge its output node based on an initial set of inputs. However, the initial set of inputs may change, perhaps in response to signal values propagating through several stages of logic. If the final set of inputs require a high gate output, then the initial discharge action was in error and unfortunately cannot be undone. Consider the following CMOS dynamic logic circuit.

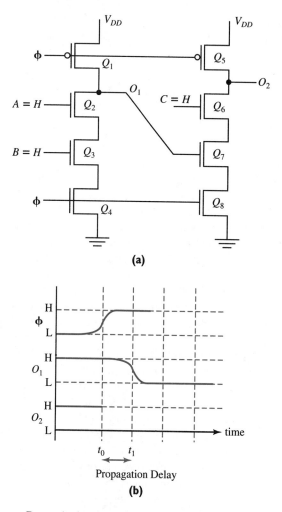

    a. Draw a logic schematic of the circuit.

    b. Describe the desired behavior of the cascaded gates and what the final value of O_2 *should* be.

    c. Complete the timing diagram, describing how an erroneous final value of O_2 could be generated.

3. Draw the circuit schematic of a static CMOS inverter and describe its operation.

4. Describe the switching actions of the NMOS and PMOS transistors comprising the **nor** gate given in Figure 6.6(b) for each possible combination of inputs.

5. Draw the circuit schematic of a static CMOS logic structure implementing the following the logic expression.

$$F(A, B, C, D, E, F) = (A \cdot B \cdot (C + D)) + E + F$$

6. What logic function is realized by the circuit given below?

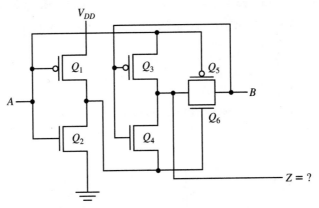

The PMOS/NMOS transistor pair Q_5/Q_6 is called a *transmission gate*. The complementary gate signals for Q_5 and Q_6, A and $\bar{A}$ respectively, turn both transistors simultaneously on or off. When Q_5 and Q_6 are both on, current can flow, so information can be transmitted between source and drain.

7. Consider the TTL gates shown below.

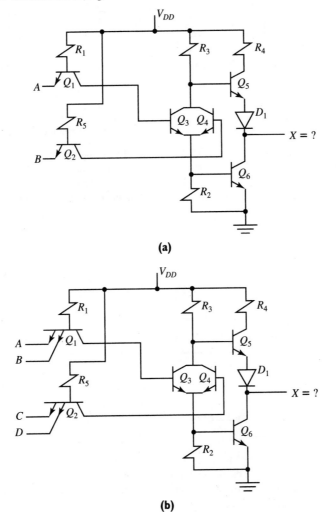

(a)

(b)

a. Construct truth tables and identify logic operations.

b. Describe the state or mode of each transistor for each possible combination of inputs.

8. For standard TTL, the fan-in/fan-out currents are listed below.

- $I_{IL\,max} = 1.6\,\text{mA}$
- $I_{OL\,max} = 16\,\text{mA}$
- $I_{IH\,max} = 40\,\mu\text{A}$
- $I_{OH\,max} = 400\,\mu\text{A}$

Examine the cascaded TTL **nand** gates given below.

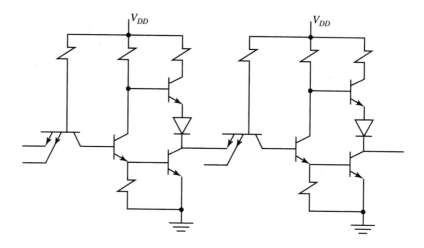

a. Show $I_{IL\,max}$ and $I_{OL\,max}$ when the output of the first **nand** gate is low.

b. Show $I_{IH\,max}$ and $I_{OH\,max}$ when the output of the first **nand** gate is high.

9. For open-collector TTL logic, the value of the external, pull-up resistor must be selected so that current and voltage operational specifications are not violated. The resistor must be large enough to prevent exceeding the output sink capacity $I_{OL\,max}$, but small enough to prevent violating the minimum output voltage $V_{OH\,min}$. Use the fan-in/fan-out currents given in Problem 8, the voltages given in Table 6.2, and the open-collector TTL given below to do the following tasks.

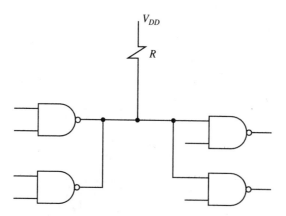

a. Compute the minimum value of the pull-up resistor, R.

b. Compute the maximum value of the pull-up resistor, R. (*Hint:* $I_{OH\,max} = 250\,\mu\text{A}$ for open-collector TTL logic.)

   c. Discuss the advantages and disadvantages of selecting a particular resistor value within the computed range.

10. Implement the following 3-input majority function using I^2L logic.

$$F(A, B, C) = (A \cdot B) + (A \cdot C) + (B \cdot C)$$

11. Describe the logic function and transistor operation of the circuit given below.

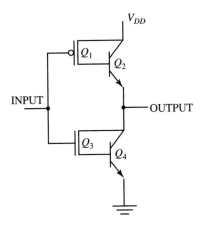

12. Can TTL logic interface directly with CMOS logic? In other words, can the output of a TTL gate be directly connected to the input of a CMOS gate and vice versa? For TTL, use the voltages given in Table 6.2 and the fan-in/fan-out currents given in Problem 8. For CMOS, use the voltages given in Table 6.2 and assume the following fan-in/fan-out currents:

   • $I_{IL\,max} = 1\ \mu A$
   • $I_{OL\,max} = 4\ mA$
   • $I_{IH\,max} = 1\ \mu A$
   • $I_{OH\,max} = 4\ mA$

   Explain your answer and note any restrictions.

13. Compute the noise margins NM_{LOW} and NM_{HIGH} for the four logic families given in Table 6.2. Rank the logic families based on the noise margin metric from best to worst and explain your rationale.

14. List the factors that contribute to high-speed logic circuits.

15. Manufacturers often specify CMOS gate power dissipation using the equation

$$P_{DYNAMIC} \simeq C \times V^2 \times F$$

   where,
   • C is the circuit node capacitance being charged and/or discharged, typically given in picofarads (pF) ($1\ pF = 1 \times 10^{-12}$ farads);
   • V is the supply voltage, typically given in volts; and
   • F is the switching rate, typically given in megahertz (MHz) ($1\ MHz = 1 \times 10^6$ hertz).

   Compute the power dissipation of a 240,000 gate design operating at 100 KHz, 500 KHz, and 1 MHz, assuming $C = 15\ pF$ and $V = 5$ V. Also, assume that only one-third of the gates switch per input.

16. Can the static CMOS logic family discussed in this chapter support wired logic? (*Hint:* Think about what would happen if the outputs of two static CMOS inverters described in Problem 3 were connected to a common node.)

17. Describe the logic function and transistor operation of the circuit given below.

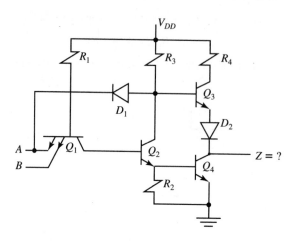

Describe a useful application of this circuit. (*Hint*: Think about using a decoder [see Chapter 5].)

REFERENCES AND READINGS

[1] Carr, W. and J. Mize. *MOS/LSI Design and Application*. New York: McGraw-Hill, 1972.

[2] Glaser, A. and C. Searle. *Integrated Circuit Engineering*. Reading, MA: Addison-Wesley, 1979.

[3] Glasser, L. and D. Dobberpuhl. *The Design and Analysis of VLSI Circuits*. Reading, MA: Addison-Wesley, 1985.

[4] Hodges, D. and H. Jackson. *Analysis and Design of Digital Circuits*. New York: McGraw-Hill, 1983.

[5] Horiuchi, T., I. Sakai, K. Yamazaki, and O. Kudoh. "A New BiCMOS Gate Array Cell with Diode Connected Bipolar Driver," *Proceedings of the 1990 Bipolar Circuits and Technology Meeting*, pp. 124–126, September 1990.

[6] Imai, K., T. Yamazaki, and T. Tashiro. "A Cryo-BiCMOS Technology with Si/SiGe Heterojunction Bipolar Transistors," *Proceedings of the 1990 Bipolar Circuits and Technology Meeting*, pp. 90–93, September 1990.

[7] Krambeck, R., C. Lee, and H. Law. "High-Speed Compact Circuits with CMOS," *IEEE Journal of Solid-State Circuits*, vol. SC-17, no. 3, pp. 614–619, June 1982.

[8] Mavor, J., M. Jack, and P. Denyer. *Introduction to MOS LSI Design*. Reading, MA: Addison-Wesley, 1983.

[9] Mukherjee, A. *Introduction to NMOS and CMOS VLSI Systems Design*. Englewood Cliffs, NJ: Prentice-Hall, 1986.

[10] Ong, D. *Modern MOS Technology: Processes, Devices, & Design*. New York: McGraw-Hill, 1984.

[11] Penney, W. and L. Lau. *MOS Integrated Circuits*. New York: Van Nostrand Reinhold, 1972.

[12] Riezenman, M. "Wanlass's CMOS Circuit," *IEEE Spectrum*, vol. 28, no. 5, p. 44, May 1991.

[13] Weste, N. and K. Eshraghian. *Principles of CMOS VLSI Design: A Systems Perspective*. Reading, MA: Addison-Wesley, 1985.

7
INTEGRATED CIRCUITS

In Chapter 6, we learned how logic gates are realized using electronic circuit components, such as resistors, diodes, and transistors. In this chapter, we will "peel off" the circuit layer of abstraction to investigate further how individual electronic circuit components are realized. Electronic components are typically realized as *semiconductor devices* on a single, *monolithic* substrate. The resulting circuit, in other words, digital design, is called an *integrated circuit* (IC), a *chip*, or a *die*.

Although integrated circuits are several levels of abstraction removed from logic schematics, we include an introduction in this text because integrated circuits are a powerful and pervasive technology. Also, introducing integrated circuits will provide you with a perspective on how basic digital system engineering relates to state-of-the-art digital products. Understanding the general capabilities of integrated circuit technology will provide valuable insight into how the simple digital system examples presented in this text scale to the complex digital systems being developed by industry.

Integrated circuit technology is also called *microelectronic* technology because the semiconductor devices are small—very small. Leading integrated circuits containing several *million* devices are smaller than a typical postage stamp. The information contained in 12 reams (a 2-foot stack) of 8.5-inch $\times$ 11-inch sheets of double-spaced text can be stored in an integrated circuit occupying an area of about 0.5 inch $\times$ 0.5 inch! Jack Kilby and Robert Noyce share the credit for inventing the integrated circuit. In 1958, working at Texas Instruments, Jack Kilby built an integrated circuit realizing a phase-shift oscillator (analog circuit) consisting of several transistors, resistors, and capacitors. At the same time, Robert Noyce independently developed an integrated circuit at Fairchild Camera and Instrument.

In the following sections, we will present an overview of integrated circuit technology. First, we will provide a brief introduction to the operation of semiconductor diodes and transistors. Then, we will discuss the integrated circuit manufacturing process, examining aspects of *lithography*, *testing*, and *packaging*. The chapter will conclude with an overview of *application-specific integrated circuit* (ASIC) design techniques and integrated circuit economics.

7.1 DIODE: THE pn JUNCTION

Recall from Chapter 6 that a diode conducts current in only one direction, from anode to cathode. The triangle serves as a reminder of the direction of diode current flow, as shown here.

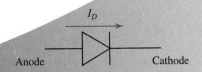

We will now learn how diodes operate, based on the properties of materials called *semiconductors*. As the name implies, semiconductors have moderate conductivity, more than insulators such as glass, but less than conductors such as metals. Although there are many semiconductors, we will focus on the semiconductor *silicon* because it is widely used and its properties are well understood. For an introduction to semiconductors and semiconductor physics, you are encouraged to study Appendix C.

Figure 7.1 shows that a semiconductor diode is formed by joining pieces of p-type and n-type silicon together; the union is called a *pn junction*. The abstract view of n-type and p-type silicon emphasizes the factors affecting diode conduction. For n-type silicon, the large positive charges are called *donor* ions (Group V elements) because they donate *electrons*, represented as small negative charges. For p-type silicon, the large negative charges are called *acceptor* ions (Group III elements) because they accept (some might say steal) electrons from neighboring atoms and thereby create vacancies called *holes*, represented as small positive charges. The donor and acceptor ions are considered immobile, fixed at various locations within the silicon crystal lattice. The electrons and holes are considered mobile, free to move around the silicon crystal lattice.

Two principal forces determine the physics of a pn junction: *diffusion* and *drift*. When a piece of p-type silicon is joined to a piece of n-type silicon, holes diffuse across the junction from the p-type silicon into the n-type silicon because the p-type silicon has a much higher concentration of holes than the n-type silicon. Likewise, electrons diffuse from the n-type silicon into the p-type silicon because the n-type silicon has a much higher concentration of free electrons than the p-type silicon. In other words, a diffusion process occurs because of the hole and electron *concentration gradients* across the junction, similar to the process of the smell of a strong perfume spreading throughout a room.

Holes that diffuse into the n-type silicon leave behind unneutralized, negatively charged acceptor ions. Likewise, electrons that diffuse into the p-type silicon leave behind unneutralized, positively charged donor ions. This region of "uncovered" or unneutralized charge surrounding the pn junction is called the *depletion region* or *space-charge region*. The layers of negative acceptor and positive donor ion charge form a dipole that sets up an electric field (E). The electric field exerts a *drift* force that opposes the diffusion force. Consider, for example, a hole placed exactly at the pn junction in Figure 7.1. The diffusion force due to the concentration gradient pushes the hole to the right, while the drift force due to the electric

Figure 7.1

The pn junction forms a diode

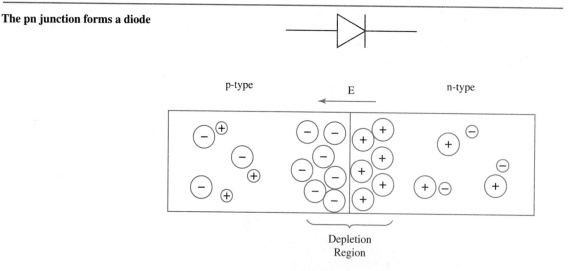

p-type

E

n-type

Depletion
Region

Problem-Solving Skill

The steady-state behavior of physical systems is often the result of several contributing dynamic processes.

field pushes the hole to the left. Electrons and holes continue to diffuse across the junction and the depletion region continues to expand until the drift force due to the electric field counterbalances the diffusion force due to the concentration gradient. Dynamic thermal equilibrium is established and no net current flows across the pn junction.

7.1.1 Forward and Reverse Biasing

How does a pn junction perform the rectifying action of a diode; in other words, how does it allow current to flow in only one direction? Consider first the *forward bias* condition shown in Figure 7.2(a), where a voltage is applied across the pn junction, making the anode positive with respect to the cathode.

Under forward bias conditions, electrons flow out of the negative terminal of the voltage

Figure 7.2

Forward and reverse biasing a diode

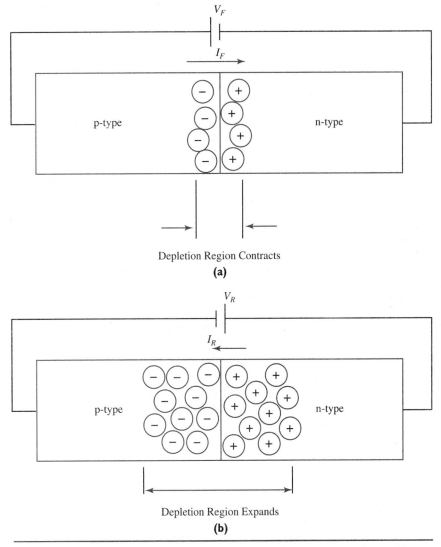

Figures (a) and (b) show the conditions of a pn junction under forward and reverse bias, respectively.

supply, V_F, into the cathode and neutralize some of the positively charged donor ions comprising the depletion region in the n-type silicon. At the same time, electron/hole pair generation occurs in the p-type silicon to supply electrons that flow out of the anode and into the positive terminal of the voltage supply. The generated holes drift toward the junction and neutralize some of the negatively charged acceptor ions comprising the depletion region in the p-type silicon. Thus, a capacitive charging current *decreases* the width of the depletion region, which reduces the associated electric field, which reduces the drift force. The drift force no longer counterbalances the diffusion force. The diffusion of holes into the n-type silicon and free electrons into the p-type silicon creates a net current flow (I_F) from the anode to the cathode.

Figure 7.2(b) shows that under reverse bias conditions, a capacitive charging current *increases* the width of the depletion layer. Electrons flow out of the n-type silicon toward the positive terminal of the voltage supply, V_R, exposing additional positively charged donor ions comprising the depletion region in the n-type silicon. At the same time, electrons flow into the p-type silicon from the negative terminal of the voltage supply. These electrons recombine with holes, exposing additional negatively charged acceptor ions comprising the depletion region in the p-type silicon. Thus, a capacitive charging current expands the depletion region, which increases the associated electric field. The counterbalance between drift and diffusion forces at the pn junction is again upset, but this time favoring the drift force. The increased drift force supports the flow of holes from the n-type side to the p-type side and free electrons from the p-type side to the n-type side. Since there are relatively few holes in n-type silicon and relatively few free electrons in p-type silicon, the net current flow from cathode to anode is very small compared to the forward-biased current.

Figure 7.3(a) and Figure 7.3(b) respectively show ideal and typical diode current (I)/voltage (V) plots. An ideal diode behaves like a switch, freely conducting current when forward-biased and conducting no current when reverse-biased. However, an actual diode only approximates this ideal switching behavior due to the properties of a semiconductor pn junction. Instead of turning on at $V = 0$, an actual diode has a *threshold voltage* (V_t). The forward bias potential must be greater than this threshold voltage before a diode begins to

Figure 7.3

Ideal and typical diode current (I)/voltage (V) characteristics

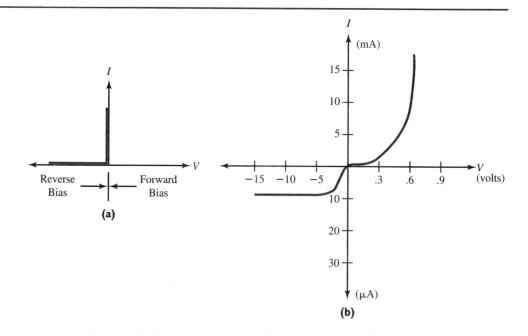

conduct appreciable forward current. Figure 7.3(b) shows that a typical threshold voltage for silicon is ≈ 0.6 V. In addition, Figure 7.3(b) shows that an actual semiconductor diode conducts a small amount of current under reverse bias. The magnitude of the reverse diode current, also called *leakage current*, is typically on the order of tens of microamperes. Note the current and voltage scale changes for forward and reverse bias conditions.

7.2 METAL-OXIDE-SEMICONDUCTOR TRANSISTORS

Having an understanding of pn junctions, we are now ready to study semiconductor MOSFETs. In Chapter 6, we introduced the *metal-oxide-semiconductor field-effect transistor*, MOSFET, as a voltage-controlled electronic switch. The gate-to-source voltage controls the switching action between the source and drain terminals. In this section, we will explain in more detail the solid-state physics of a MOSFET.

Recall that there are four principal types of MOSFETs, classified by the mode of operation (enhancement/depletion) and type of channel (N-channel/P-channel). Figure 7.4(a) shows a simplified diagram of the construction of an enhancement-mode NMOS transistor.

An enhancement-mode NMOS transistor is constructed by forming two n-type or donor-

Figure 7.4

Construction of an enhancement-mode NMOS transistor

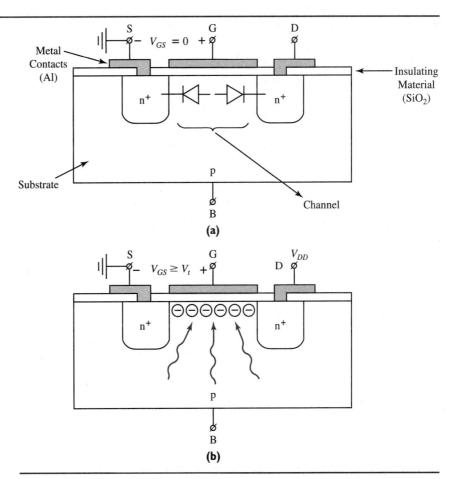

(a)

(b)

Figure (a) shows an enhancement NMOS transistor when $V_{GS} = 0$. Figure (b) shows an enhancement NMOS transistor when $V_{GS} \geq V_t$.

doped regions within a piece of p-type or acceptor-doped silicon. The two n-type regions form the source and drain; the p-type silicon forms the body or substrate. By convention, a superscript of "+" or "−" is added to a doping concentration symbol of "n" and "p" to denote relative doping magnitudes. The symbol n^+ denotes that the n-type source and drain regions are more heavily doped than the p-type substrate. The region between the source and drain is called the *channel*. The gate electrode lies above the channel, separated from the substrate by a thin layer of insulating material. The insulating material is typically silicon dioxide (SiO_2). The gate-insulator-channel structure forms a parallel plate capacitor.

With no voltage applied to the controlling gate terminal with respect to the source, the two back-to-back drain-to-substrate and substrate-to-source diodes prevent any appreciable current flow between source and drain. In this condition, an enhancement-mode NMOS transistor acts like an open switch between source and drain. As the gate potential is increased, more positive charge accumulates on the gate terminal, which capacitively attracts negative charge into the channel. Figure 7.4(b) shows that at a point defined to be the threshold voltage, V_t, the gate potential has drawn enough free electrons into the channel to form an n-type region connecting the source and drain. This condition is called *inversion* because the p-type channel is inverted or changed into an n-type channel, providing a conducting path between source and drain. When $V_{GS} \geq V_t$, an enhancement-mode NMOS transistor acts like a closed switch between source and drain.

An enhancement-mode PMOS transistor works similarly to the enhancement-mode NMOS transistor except that the doping concentrations, terminal voltage polarities, and current directions are reversed. That is, the source and drain regions for an enhancement-mode PMOS transistor are p^+ regions in an n-type substrate. Applying a negative potential (L) to the gate with respect to the source attracts positive holes to form a channel connecting the source and drain.

Figure 7.5 shows the construction on a depletion-mode NMOS transistor. The construction of a depletion-mode NMOS transistor is similar to an enhancement-mode NMOS transistor, except the depletion NMOS transistor has a thin n-type channel region joining the source and drain. Thus, with no potential applied to the gate terminal there is a conducting path between source and drain and a depletion-mode NMOS transistor behaves like a closed switch. Accumulating negative charge on the gate, in other words, applying a negative voltage, capacitively attracts positive charge into the channel, which depletes the channel of free electrons. At a point defined to be the threshold voltage, V_t, the gate potential has forced enough free electrons out of the channel to destroy the connection between the source and drain. In this condition, a depletion-mode NMOS transistor acts like an open switch between source and drain.

Figure 7.5

Construction of a depletion-mode NMOS transistor

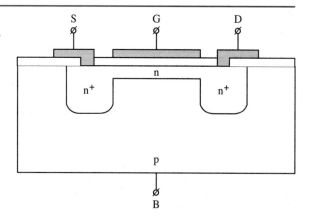

In summary, a MOSFET serves as a voltage-controlled switch for digital logic applications. The voltage on the gate capacitively controls the charge in the channel connecting the source and drain switch terminals. The channel charge, in turn, controls source/drain conductivity.

7.3 BIPOLAR TRANSISTORS

Chapter 6 also introduced the bipolar junction transistor (BJT) as a voltage-controlled switch. The base-to-emitter voltage controls the switching action between the collector and emitter terminals. There are two types of BJTs, classified by the type of base material (NPN or PNP). Figure 7.6 shows a simplified diagram of the construction of an NPN bipolar junction transistor (BJT). An NPN biopolar transistor is constructed by sandwiching a piece of p-type silicon between two pieces of n-type silicon.

Recall from Chapter 6 that an NPN BJT is in *cutoff* mode when a low voltage is applied to the base with respect to the emitter and collector. In this condition, the base-collector and base-emitter pn junctions are reverse-biased. No current can flow between the collector and emitter, and the bipolar transistor acts like an open switch.

Now, consider the *forward active* mode where a high voltage is applied to the base with respect to the emitter and the base-to-collector bias remains at a low voltage. In this

Figure 7.6

Construction of an NPN bipolar transistor

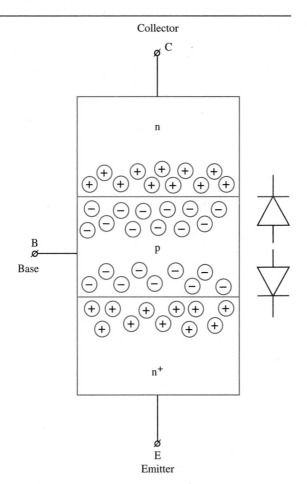

condition, the base-emitter pn junction is forward-biased and the base-collector pn junction is reversed-biased. Normally, we would expect the reversed-biased base-collector pn junction to prevent any current flow, yet a BJT in forward active mode behaves like a closed switch between the collector and emitter. Why is there substantial current flow across the reverse-biased base-collector pn junction in the forward active mode? The answer reveals the key principles of bipolar transistor behavior.

Recall from Figure 7.2(b) that a reverse-biased pn junction supports a drift current flow from cathode to anode. This reverse current flow is normally very small compared to the magnitude of the forward-biased current flow because the reverse current depends on *minority* carriers, in other words, holes in the n-type silicon and free electrons in the p-type silicon. But, the forward-biased base-emitter junction is supplying minority carriers, lots of minority carriers.

Figure 7.7 shows that once electrons cross the forward-biased base-emitter junction, they become minority carriers in the p-type base. The base is constructed to be very thin, so most of the injected electrons from the emitter are caught up by the drift force and swept across the reverse-biased base-collector junction into the collector. The relative magnitude of the free electron and hole flow across the base-emitter junction is indicated by the size of the respective arrows. Many more electrons than holes cross the base-emitter junction because the emitter has a greater doping concentration (n^+) than the base (p). That is, asymmetrical p-type and n-type doping concentrations cause asymmetrical carrier flow across the junction. Thus, a small base current affecting the biasing of the base-emitter junction controls the conductivity between the collector and emitter switch terminals. The term *bipolar* is used because both holes and electrons participate in the transistor conduction process. Note that

Figure 7.7

Bipolar transistor carrier flow in forward active mode

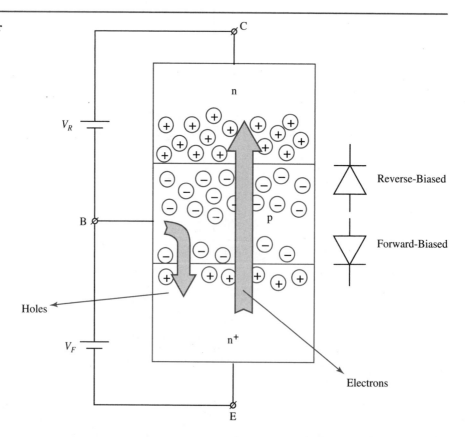

MOSFETs are *unipolar* devices because only electrons (NMOS) or holes (PMOS) participate in the transistor conduction process.

7.4 FABRICATION AND PACKAGING

Having explained the basic physical principles of semiconductor diodes and transistors, we will now investigate how these semiconductor devices are fabricated on a single substrate to form an integrated circuit. The integrated circuit fabrication process can be divided into the following three phases:

- wafer preparation,
- wafer processing, and
- testing and packaging.

Although many different types of semiconductors may be used to construct integrated circuits, we will continue to focus on silicon in the following discussions.

7.4.1 Wafer Preparation

The integrated circuit fabrication process starts with *sand*. Yes, common sand is basically silicon dioxide (SiO_2), and the earth's abundant supply of sand is one reason for silicon's popularity. Raw sources of silicon-based materials are refined to produce a cylindrical rod of silicon, called an *ingot*. Special care is taken to minimize impurities and crystal lattice imperfections. Advances in crystal growth technology over the past two decades have enabled the production of larger ingots. For example, in the early 1960s, silicon ingots were approximately 1 to 2 inches in diameter. In contrast, silicon ingots produced today are routinely 5 to 6 inches in diameter. The advantages of growing larger ingots will be apparent as we explain the IC fabrication process.

After a silicon ingot has been grown, it is cut into thin (typically 0.2 to 0.5 mm thick) slices, called *wafers*. Each wafer is then mechanically and chemically polished to remove any surface damage caused by the cutting process. The silicon wafer is now ready to be used as the substrate for fabricating integrated circuits in the wafer-processing phase.

7.4.2 Wafer Processing

The wafer-processing phase defines the electronic devices and their interconnect that realize a digital system. The electronic devices are fabricated by a series of *patterning* steps that define the semiconductor structures discussed earlier. We will explain these patterning steps by examining a basic fabrication process for the enhancement-mode NMOS transistor shown in Figures 7.8(a) through 7.8(i) (pp. 204–205).

The NMOS transistor fabrication process begins by defining the area that will contain the device, called the *active area*. An insulating ring, called the *field oxide*, is constructed around the active area to isolate the NMOS transistor from neighboring devices and prevent unwanted parasitic interactions. Figures 7.8(a) through 7.8(d) show the definition of the active area. The p-type silicon substrate is first coated with a layer of silicon nitride (Si_3N_4). The silicon nitride will be selectively removed in the areas where we wish to form the insulating field oxide. Thus, we need to define a *pattern* on the silicon nitride that specifies areas that will remain on the silicon surface and areas that will be removed. The silicon nitride pattern is defined by a process called *lithography*, which transfers geometric shapes onto the surface of an integrated circuit.

Lithography defines the silicon nitride pattern by (1) coating the silicon nitride with a material called *resist*, (2) defining a resist pattern, and (3) transferring the resist pattern to the silicon nitride. The resist pattern is defined by selectively subjecting the resist to some form

Figure 7.8

**Simplified illustration of the
NMOS fabrication process**

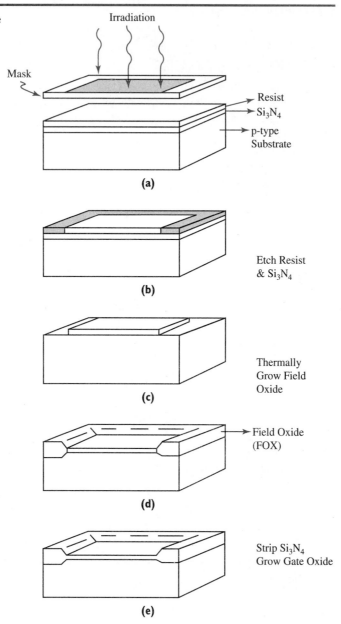

of irradiation. Ultraviolet light is presently the most common form of irradiation, but other forms, such as X ray or charged-particle (electron or ion) beam, are being investigated because smaller devices can be created with higher frequency irradiation. When ultraviolet light is used as irradiation, the resist is called *photoresist* and the lithographic process is called *photolithography*. Photoresist has a solubility that varies with exposure to ultraviolet light. *Positive* photoresist has a solubility that *increases* with exposure to ultraviolet light.

In Figure 7.8(a), the resist pattern is defined on the glass plate, called a *mask*. The mask has darkened areas where we wish to block the irradiation from reaching the resist. After exposure, the shaded areas in Figure 7.8(b) denote where the photoresist is more soluble because of exposure to light. In Figure 7.8(c), the resist pattern is transferred to the silicon ni-

Figure 7.8 (continued)

Simplified Illustration of the NMOS fabrication process

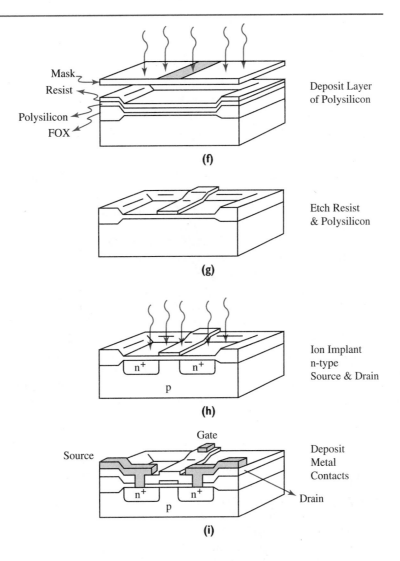

tride by first applying a solvent to dissolve the softened photoresist and then dissolving away the exposed areas of silicon nitride. The hardened photoresist is then stripped away, leaving the patterned silicon nitride. Next, the silicon substrate is exposed to oxygen at high temperatures to thermally grow silicon dioxide (SiO_2), which forms the field oxide (FOX). Figure 7.8(d) shows that field oxide only grows where the silicon nitride has been stripped away.

In Figure 7.8(e), the silicon nitride mask has been stripped away and a very thin layer of silicon dioxide ($\simeq 500\,\text{Å}$)[1] has been grown by *chemical vapor deposition* (CVD) to form the NMOS gate oxide. The gate oxide is made as thin as possible to obtain a gate-to-channel capacitance that is as large as possible. Since capacitance is the proportionality factor that determines the amount of stored charge for a given applied voltage,

$$charge = capacitance \times voltage \tag{7.1}$$

a higher gate-to-channel capacitance increases the amount of charge induced in the channel

---

[1] An angstrom (Å) is defined to be 1×10^{-10} meters.

for a given gate voltage. In other words, you get more "bang (charge) for your buck (voltage)." The gate oxide is deposited instead of thermally grown because it is easier to control the thickness of the oxide when it is deposited.

Figure 7.8(f) and Figure 7.8(g) again use photolithography to define the gate. First, the substrate is covered with a layer of polycrystaline silicon, commonly called *polysilicon*. (Polysilicon does not have a pure, single crystal structure.) Next, a pattern is defined in the polysilicon, following the same lithographic steps described earlier. A layer of photoresist is put down on top of the polysilicon and selectively exposed. The softened areas of photoresist are stripped away and the exposed polysilicon is chemically etched away, leaving the polysilicon gate structure shown in Figure 7.8(g).

Next, the source and drain regions are defined in Figure 7.8(h). N-type donor atoms, such as phosphorous (P) or arsenic (Ar), are injected into the substrate by *ion implantation* and diffused to the desired depth by a thermal *drive-in* process. The energy of the implanted donor atoms is set so the atoms penetrate the thin gate oxide areas, but not the polysilicon gate or field oxide areas.

To complete the enhancement-mode NMOS transistor fabrication process, a final lithographic process is performed to define the pattern of metal contacts for the source, drain, and gate terminals shown in Figure 7.8(i).

The basic device-level wafer-processing procedure is scaled up and applied to the entire integrated circuit, whereby a series of masks and lithographic steps define all the circuit devices and their interconnect. Then, the integrated circuit wafer-processing procedure is again scaled up and applied to an entire wafer. Figure 7.9 shows that the result of the wafer-processing phase is many integrated circuits replicated across the wafer's surface.

Around the outside of an integrated circuit are input/output (I/O) "pads" that provide access to the chip's input/output signals. I/O pads are also called *bonding pads* because wires are bonded to these pads to connect the chip's input/output signals to terminals or pins on the enclosing package.

7.4.3 Testing and Packaging

Following the wafer-processing phase, testing verifies that an integrated circuit works properly. Then, packaging "wraps the integrated circuit in a larger container" for general use.

Testing is an important part of the integrated circuit manufacturing process because disturbances associated with the wafer-processing steps described in the previous section can cause circuit defects. Examples of such disturbances include variations in ion implantation energies, oxidation times, or ambient temperature and pressure. Even a small particle of dust can cause an integrated circuit to fail to operate correctly. Testing advanced integrated circuits is a challenging task because the designs are complex, making exhaustive verification impractical. Also, the I/O pads typically provide the only "portholes" for "seeing" into a chip and exercising its logic; this limited access to individual nodes makes it difficult to check devices and isolate faults. (For more information about testing, see references [3], [11], and [12].)

After testing has identified the good and bad integrated circuits on a wafer, a diamond saw cuts the wafer up into individual chips; this process is called *scribing*. The bad chips are discarded and the good chips proceed to final assembly, also called *packaging*. There are several reasons for enclosing an integrated circuit into a larger housing or package:

- It protects the integrated circuit from physical damage.
- It scales up the integrated circuit input/output signals to dimensions consistent with printed circuit boards.
- It provides for power dissipation.
- It provides for easy socket insertion and removal.

Figure 7.9

A wafer containing many integrated circuits

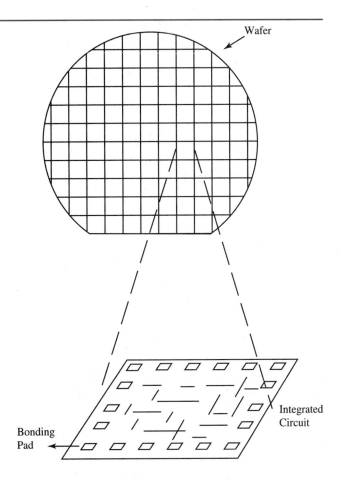

There are a variety of packaging schemes to accommodate a wide range of integrated circuit and application requirements, such as number of input/output signals, product price, and package area. Packaging technology has evolved from dual-in-line packages to more sophisticated packages, such as the pin-grid array or chip carrier. The newer packaging technologies support more input/output signals in smaller "footprints," in other words, package sizes. Figure 7.10 (p. 208) shows a sampling of common chip packages.

Presently, the densest packaging is achieved with *multichip module* (MCM) technology. MCM technology places *multiple* integrated circuits onto a common substrate, which is then enclosed in a single package. The substrate can be composed of a variety of materials, including silicon or ceramic compounds.

7.5 APPLICATION-SPECIFIC INTEGRATED CIRCUITS (ASICs)

Now that we have seen an overview of semiconductor device and integrated circuit manufacturing, we will next discuss integrated circuit design methodologies, also commonly called *application-specific integrated circuit* (ASIC) design methodologies. Figure 7.11 (p. 208) shows several popular ASIC (pronounced "$\bar{a}$-*sick*") design methodologies and their relative characteristics.

A variety of ASIC design methodologies have been developed to provide a broad range of manufacturing capabilities. These ASIC design methodologies are often comparatively

Figure 7.10

Sampling of common integrated circuit packages

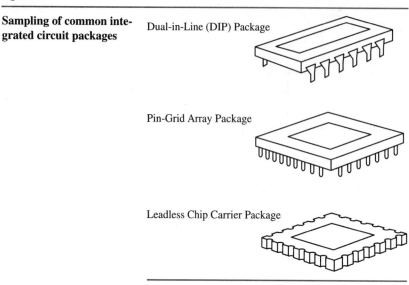

Dual-in-Line (DIP) Package

Pin-Grid Array Package

Leadless Chip Carrier Package

Figure (a) shows a dual-in-line (DIP) package. Figure (b) shows a pin-grid array (PGA) package. Figure (c) shows a leadless chip carrier (LCC) package.

positioned using the metrics of design performance, development time, and development cost. In general, design methodologies supporting higher performance digital systems incur more time and cost.

At one end of the range are *standard products* and *programmable logic devices*, which were introduced in Chapter 5. Standard products are "off-the-shelf" integrated circuits that contain predefined logic functions, such as multiplexers, decoders, or several individual logic gates. Exploiting standard products minimizes development time and costs, but often at the expense of design performance. There may not be an exact match between desired logic function and predefined logic functions, so the standard products may introduce additional stages of logic and/or gates. As the name implies, programmable logic devices (PLDs) contain general logic structures that can be programmed or customized by the user to realize

Figure 7.11

Comparison of digital design methodologies

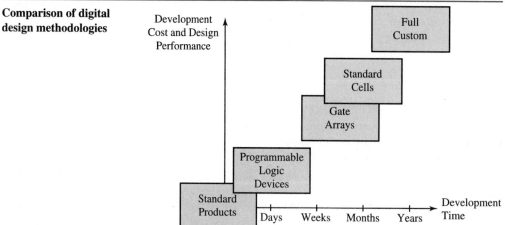

specific logic functions. PLDs can offer better system performance than standard products because PLD programmability allows greater flexibility in matching desired logic functions. However, the task of programming PLDs adds to development time.

At the other extreme are *full custom* integrated circuits. With these, the user is responsible for the entire design and fabrication. The full custom ASIC design methodology offers high system performance because special attention can be given to critical devices and interconnections. System performance can be optimized by judiciously controlling factors such as device location, transistor sizing, and interconnect length. However, the detailed work involved in full custom ASIC design takes a long time and requires many engineers, which equates to high development costs. Thus, full custom ASIC digital design is generally pursued only in situations where performance is critical or the product is expected to have high volume sales that can recoup the high initial investment.

Between the two extremes, standard cell ASIC design imposes a more structured design discipline, restricting the generality of full custom ASIC design. Standard cell design is similar to standard product design in that both methodologies seek to compose a desired function quickly, using a library of predefined functions. However, a standard product is an integrated circuit, whereas a standard cell is a *piece* of an integrated circuit. A standard product design typically yields a printed circuit board containing several integrated circuits; a standard cell design typically yields a single integrated circuit containing several cells. Composing an integrated circuit design out of predesigned pieces of logic reduces development time and costs; however, these advantages are gained at the expense of optimizing individual devices. Consequently, standard cell ASIC designs often do not achieve the system performance attainable with full custom.

Finally, gate array ASIC design imposes even more logical and physical structure than standard cell design to gain further reductions in development time and costs. Similar to standard cell design, gate array design composes or "builds up" a digital design from logic cells. However, gate array logic cells are generally less complex than standard logic cells, typically only a few gates. Moreover, the gate array logic cells are predesigned *and* prefabricated up to the final metalization steps. The cells are located in a regular two-dimensional pattern across the surface of the integrated circuit, hence the name "gate array." A digital design is realized by defining the logic function of the individual cells and their interconnect via the final metalization step in the fabrication process. Compared to standard cell design, gate array design generally offers the advantage of reduced development time and costs, but at the expense of degraded system performance.

7.6 IC ECONOMICS

Advancements in microelectronic technology have been one of the most significant scientific and engineering accomplishments in recent history. Producing an integrated circuit no bigger than a thumbnail having many times the processing throughput of computers that occupied entire rooms is truly an awesome feat. Integrated circuits are being incorporated into products from a wide range of leading industrial markets, including information processing, automotive, communication, entertainment, and aerospace. Compared to older discrete component technologies, integrated circuits occupy less space, weigh less, and dissipate less power. Thus, integrated circuits can realize complex digital systems that provide new functions and capabilities at unprecedented performance levels.

Figure 7.12 (p. 210) illustrates the impressive rate of advancement of integrated circuit technology. Since dynamic random access memories (DRAMs) have traditionally paced the advancement of integrated circuit technology, several generations of DRAMs are plotted on a complexity versus time chart to show the dramatic development progress. Several terms

Problem-Solving Skill
Do not waste time reinventing solutions. Always take time to survey what work has already been done on solving a particular problem. Where appropriate, try to build off previous work.

Figure 7.12

Historical development of
integrated circuits

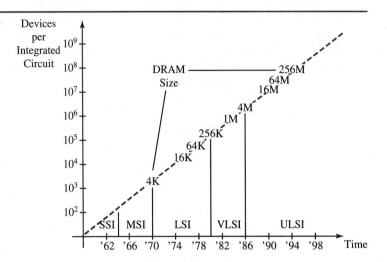

are customarily used to characterize the stages of integrated circuit technology advancement
and associated level of integration or chip complexity. Although there are no commonly
agreed-upon definitions for these complexity terms, a typical set of definitions is presented
below.

Term	Definition
SSI	Small-Scale Integration: $0-10^2$ devices
MSI	Medium-Scale Integration: 10^2-10^3 devices
LSI	Large-Scale Integration: 10^3-10^5 devices
VLSI	Very Large-Scale Integration: 10^5-10^7 devices
ULSI	Ultra Large-Scale Integration: $\geq 10^7$ devices

Starting from just a few devices in the early 1960s, state-of-the-art digital integrated circuits presently contain over 1 million devices. To appreciate the magnitude of this level of
complexity, imagine scaling up the dimensions of an integrated circuit such that circuit wires
equate to highways and a chip equates to a street map. The complexity of 1 million devices
would thus equate to an urban map the size of California and Nevada!

Advances in chip complexity are achieved by making the devices smaller and the chips
bigger. For example, from the early 1970s to the present, the nominal die size has increased
from less than 1 mm² to over 100 mm². At the same time, minimum feature size has shrunk
from 10 μm to 1 μm. Again, to put these accomplishments in perspective, the diameter of a
human hair is $\simeq 70$ μm!

7.7 SUMMARY

In Chapter 6 we studied how logic gates are realized using circuit-level devices, such as resistors, transistors, and diodes. In this chapter, we studied how these circuit devices are realized using integrated circuit technology.

We investigated how the properties of semiconductors can be exploited to implement circuit devices, such as diodes and transistors. We saw that although field-effect and bipolar
transistors have different semiconductor structures and operating principles, they are both
charge control devices. Charge inserted on the gate terminal of a MOSFET controls the conductivity between source and drain terminals. Similarly, charge inserted on the base terminal
of a BJT controls the conductivity between emitter and collector terminals.

After gaining an understanding of semiconductor circuit device structures, we then examined an overview of the integrated circuit fabrication process. We learned that through a series of lithographic steps, patterns are defined on the semiconductor surface to fabricate active areas, device structures, and interconnect. By working at a very small scale where dimensions are in terms of microns, integrated circuits can be produced that contain over 1 million devices.

Since the impact of microelectronics is so pervasive and influential, it is important to recognize and understand the significance and implications of integrated circuits. The ability to manufacture a highly complex digital system is an integrated or monolithic fashion on a small piece of semiconductor is a powerful technology, the full potential of which has yet to be tapped.

7.8 PROBLEMS

1. The pn-junction diode current/voltage characteristic is typically given by the following equation:

$$I = I_s(e^{V/V_T} - 1)$$

where,

- I_s is the *saturation current*, given as 10 fA and
- $V_T = kT/q$.
  - k is Boltzman's Constant $= 1.38 \times 10^{-23}$ J/K.
  - T is temperature in degrees Kelvin.
  - q is the charge of an electron in coulombs, 1.6×10^{-19} C.

  a. Generate an I/V plot for a pn-junction diode at room temperature $T = 300$ K, showing forward and reverse bias conditions.

  b. Estimate the threshold voltage V_t, where the diode begins to conduct substantial current and acts like a closed switch.

  c. What is the significance of I_s?

2. Using the pn-junction I/V equation given in Problem 1,

  a. Show the effect of increasing the room temperature by 20°F on pn-junction I/V characteristics. (*Note*: I_s is proportional to temperature. Assume I_s doubles with every 10°C rise in temperature above nominal room temperature.)

  b. Why might it be important to consider heat dissipation techniques when packaging digital integrated circuits?

3. Using the pn-junction I/V equation given in Problem 1, find the forward bias voltage required to yield a diode current of 1 mA. Assume $T = 300$ K.

4. Draw a cross section of an enhancement-mode PMOS transistor under the following conditions.

  a. No bias.

  b. Forward bias. Describe operation.

5. Draw a cross section of a depletion-mode PMOS transistor under the following conditions.

  a. No bias.

  b. Reverse bias. Describe operation.

6. The current/voltage characteristic for an enhancement-mode NMOS in saturation is typically given by the equation

$$I_{DS} = \left(\frac{\mu_e \varepsilon_{ox}}{2t_{ox}}\right)\left(\frac{W}{L}\right)(V_{GS} - V_t)^2$$

where,

- μ_e is the electron mobility;
- ε_{ox} is the dielectric constant of the gate oxide;

- t_{ox} is the thickness of the gate oxide;
- W is the gate width, which is the dimension perpendicular to the plane of the paper in Figure 7.8(g);
- L is the gate length, which is the dimension horizontally aligned with the plane of the paper in Figure 7.8(g); and
- V_t is the threshold voltage.
  a. Generate an I/V plot showing the "on" or saturation current for an enhancement-mode NMOS transistor $0V \leq V_{GS} \leq 5V$. Assume $\mu_e = 500$ cm^2/V • sec, $\varepsilon_{ox} = 3.54 \times 10^{-13}$ F/cm, $t_{ox} = 1200$ Å, $W = 5$ μm, $L = 1$ μm, and $V_t = 0.5$ V.
  b. Faster circuits can be obtained by generating higher currents to charge/discharge nodes in less time. Using the I/V equation given above, describe an NMOS enhancement device fabrication process optimized for fast logic circuits.

7. Resistors are generally not easy to realize in integrated circuit technology. Large resistances often require long conducting paths arranged in a serpentine layout to conserve chip area. The resistance of a conducting path can be computed by the equation

$$R = R_s \left(\frac{L}{W} \right)$$

where,
- R_s is called the *sheet resistance* and is characteristic of the conducting material;
- L is the length of the conducting path; and
- W is the width of the conducting path.
  a. Calculate the length of a 5-μm-wide n^+ diffusion region to realize a 300-Ω resistor, assuming $R_s = 25$ Ω.
  b. Calculate the length of a 5-μm-wide n^+ polysilicon region required to realize a 300-Ω resistor, assuming $R_s = 50$ Ω.

8. For a PNP bipolar transistor,
  a. Draw the semiconductor cross section, similar to Figure 7.6.
  b. Describe terminal voltages, junction biases, junction conduction mechanisms, and carrier flow for the forward active mode.

9. Describe an integrated circuit fabrication process for making a CMOS inverter. Starting with an n-type wafer, describe the general steps and lithography patterns. Also, draw a series of wafer cross sections showing device structures.

10. Identify and draw a circuit schematic of the following device and discuss its possible uses in logic families.

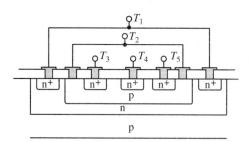

11. Repeat Problem 10 with terminals T_1 and T_2 connected together.

12. Draw a circuit schematic of the following devices. Identify the logic family and discuss the circuit operation.

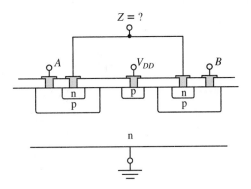

13. Draw a circuit schematic of the following devices. Identify the logic family and discuss the circuit operation.

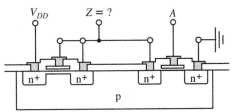

14. Describe the advantages and disadvantages of the following ASIC design methodologies.
    a. standard products
    b. programmable logic devices
    c. gate arrays
    d. full custom

REFERENCES AND READINGS

[1] Yang, E., ed. *Fundamentals of Semiconductor Devices.* New York: McGraw-Hill, 1978.

[2] Sze, M., ed. *Physics of Semiconductor Devices.* New York: Wiley, 1981.

[3] Bennetts, R. *Design of Testable Logic Circuits.* Reading, MA: Addison-Wesley, 1984.

[4] Einspruch, N. *VLSI Handbook.* Orlando, FL: Academic Press, 1985.

[5] Embabi, S., A. Bellaovar, and M. Elmasry. *Digital BiCMOS Integrated Circuit Design.* Boston, MA: Kluwer Academic Publishers, 1993.

[6] Grove, A. *Physics and Technology of Semiconductor Devices.* New York: Wiley, 1963.

[7] Hollis, E. *Design of VLSI Gate Array ICs.* Englewood Cliffs, NJ: Prentice-Hall, 1987.

[8] Hurst, S. *Custom-Specific Integrated Circuits.* New York: Marcel Dekker, 1985.

[9] Keyes, R. *The Physics of VLSI Systems.* Reading, MA: Addison-Wesley, 1987.

[10] Lee, H. *Fundamentals of Microelectronic Processing.* New York: McGraw-Hill, 1990.

[11] McCluskey, E. "A Survey of Design for Testability Scan Techniques," *VLSI Design Magazine,* pp. 36–61, December 1984.

[12] Miczo, A. *Digital Logic Testing and Simulation.* New York: Harper Row, 1986.

[13] Milnes, A. *Semiconductor Devices and Integrated Electronics.* New York: Van Nostrand Reinhold, 1980.

[14] Neudeck, G. *The PN Junction Diode,* in *Modular Series in Solid State Devices.* Reading, MA: Addison-Wesley, 1983.

[15] Pierret, R. *Field Effect Devices,* in *Modular Series in Solid State Devices.* Reading, MA: Addison-Wesley, 1983.

[16] Reinhard, D. *Introduction to Integrated Circuit Engineering.* Boston, MA: Houghton Mifflin Co., 1987.

[17] Rice, R. *VLSI: The Coming Revolution in Applications and Design.* New York: IEEE Press, 1980.

[18] Runyan, W. and K. Bean. *Semiconductor Integrated Circuit Processing Technology.* Reading, MA: Addison-Wesley, 1990.

[19] Zanger, H. *Semiconductor Devices and Circuits.* New York: Wiley, 1984.

Part 3   DIGITAL ENGINEERING:

SEQUENTIAL

SYSTEMS

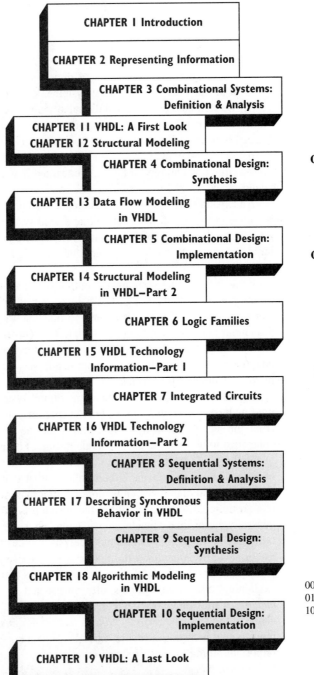

CHAPTER 1 Introduction

CHAPTER 2 Representing Information

CHAPTER 3 Combinational Systems: Definition & Analysis

CHAPTER 11 VHDL: A First Look
CHAPTER 12 Structural Modeling

CHAPTER 4 Combinational Design: Synthesis

CHAPTER 13 Data Flow Modeling in VHDL

CHAPTER 5 Combinational Design: Implementation

CHAPTER 14 Structural Modeling in VHDL–Part 2

CHAPTER 6 Logic Families

CHAPTER 15 VHDL Technology Information–Part 1

CHAPTER 7 Integrated Circuits

CHAPTER 16 VHDL Technology Information–Part 2

CHAPTER 8 Sequential Systems: Definition & Analysis

CHAPTER 17 Describing Synchronous Behavior in VHDL

CHAPTER 9 Sequential Design: Synthesis

CHAPTER 18 Algorithmic Modeling in VHDL

CHAPTER 10 Sequential Design: Implementation

CHAPTER 19 VHDL: A Last Look

Structural Modeling

```
I1: NOT_OP
  port map (A, B);
```

Asynchronous Data Flow Modeling

```
B <= not A;
```

Technology Modeling

```
B <= not A after 5 ns;
```

Synchronous Data Flow Modeling

```
B1: block (C = '1')
begin
 B <= guarded not A after 5 ns;
end block B1;
```

Algorithmic Modeling

```
P1: process (PS)
begin
 NS <= STATE_TABLE(PS);
end process P1;
```

Combinational Logic Schematics

Combinational Logic Equations

$$Z = (A \cdot B) + C$$

Logic Families & Integrated Circuits

Sequential Logic Schematics

Sequential Logic State Diagrams

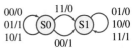

8

SEQUENTIAL SYSTEMS: DEFINITION AND ANALYSIS

We have completed the study of digital system manufacturing technology. Before we press on to study sequential digital systems, let us review what we have learned thus far. In Part 1, we studied combinational digital systems. We learned how to represent information in binary notation and how to operate on the information using Boolean algebra. Then, we learned how to design networks of Boolean operators to affect a desired input/output transformation, in other words, a truth table. Through the overview of logic families and integrated circuit technology in Part 2, we learned how the Boolean operators are physically realized.

An expanded view of the hierarchical classification of digital systems given in Figure 3.1 is shown in Figure 8.1, illustrating that digital systems are broadly subdivided into two categories: combinational and sequential. To complete our introduction to digital design, we will build on our understanding of combinational digital systems to examine *sequential* digital systems.

For combinational systems, the output is a function of only the applied input, regardless of any previous inputs. For sequential systems, the output is a function of the applied input *and* the *sequence* of previous inputs. To illustrate the difference between combinational and sequential behavior, consider the simple task of drawing a series of playing cards from a deck. If we wanted to detect when a card of the suite of clubs is drawn, we would build a combinational system. If, on the other hand, we wanted to detect a club card only when the previously drawn card was a heart, we would build a sequential system.

Figure 8.1 also shows that sequential digital systems are further subdivided into *synchronous* and *asynchronous* systems. For synchronous, also called *clocked*, sequential digital systems, the sequencing of new input values and the generation of new output values is strictly controlled to occur at regular intervals in time. For asynchronous, also called *self-timed*, sequential digital systems, the input excitation and output response occur at irregular intervals in time, dictated by internal circuit delays rather than an external agent. Presently, synchronous sequential digital design is more popular than asynchronous design because imposing a regular timing discipline simplifies the design process. In other words, synchronous design is generally easier than asynchronous design. Thus, we will restrict our discussion of sequential digital systems to synchronous analysis and design techniques. Except when noted, our use of the term *sequential* will imply synchronous sequential.

In this chapter, we will discuss synchronous sequential digital systems in more detail, explaining how such systems are constructed by combining combinational logic and *memory*. Various memory devices will be discussed. Then, we will learn how to use several analysis tools, such as *state tables*, *state diagrams*, and *timing diagrams*, to derive the input/output behavior of several example sequential systems. Finally, we will discuss several important timing aspects of synchronous sequential systems, including *setup* and *hold* times and *metastability*.

Figure 8.1

A taxonomy of electronic systems

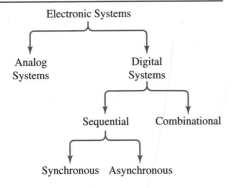

8.1 OVERVIEW OF SEQUENTIAL SYSTEMS

Figure 8.2 shows the general architecture of a synchronous sequential system. Basically, we add memory to combinational systems to form sequential systems. Memory provides the ability to store or "remember" what has occurred in the past, thus enabling the outputs to be a function of the previous inputs. By convention, the contents of memory collectively define the *system state*, also called the *present state*. The present state represents what has occurred to date. Each new set of inputs is recorded by appropriately updating memory; new memory values are collectively called the *next state*. Under control of the *clock* signal, combinational logic and memory work in concert to produce a desired sequence of outputs in response to a sequence of inputs in the following manner.

Step	Action
Step 1	Assume an initial set of values for memory that defines the initial or "power-up" state of the sequential system.
Step 2	Apply a set of inputs.
Step 3	Based on the new inputs and the present state, generate a new set of values for the outputs and memory (next state).
Step 4	Under the control of the clock signal, update the contents of memory; the next state values become the present state values.
Step 5	Return to Step 2.

Figure 8.3 (p. 218) is a more detailed look at the organization of the combinational logic block given in Figure 8.2. If the combinational logic driving the output is a function of the input *and* the present state, the synchronous sequential system is called a *Mealy system* or *Mealy machine*. If the combinational logic driving the output is a function of only the pre-

Figure 8.2

Synchronous sequential digital system architecture

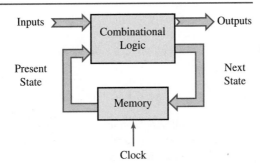

Figure 8.3

Moore and Mealy synchronous sequential digital system architectures

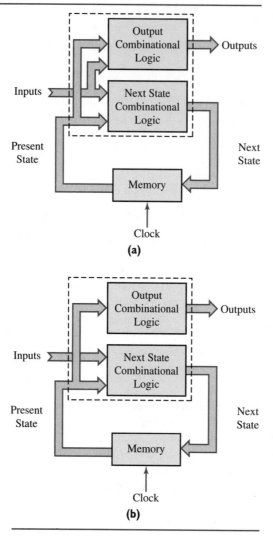

Figures (a) and (b) illustrate Mealy and Moore machines, respectively.

sent state, the synchronous sequential system is called a *Moore system* or *Moore machine*. The Moore machine can be further simplified by omitting the data inputs; in this case, the clocked sequence of only the present state values generates the output sequence. We will examine examples of this type of digital system when we study counters.

Mealy and Moore machines are not mutually exclusive. That is, a synchronous sequential system can be both a Mealy and a Moore machine, having some outputs dependent on the inputs and present state and some outputs dependent on only the present state.

8.2 MEMORY DEVICES

We need memory devices to build synchronous sequential systems. A memory device holds one bit of digital information, either a "1" or a "0." There are many ways to store a single bit of information. Figure 8.4 illustrates a taxonomy of several different types of memory devices.

Figure 8.4

Memory devices

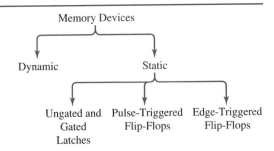

Similar to *static* and *dynamic* logic studied in Chapter 6, there is *static* and *dynamic* memory. A dynamic memory device must periodically refresh or restore its contents to prevent losing the bit of information. A static memory device, on the other hand, retains its value without having to be periodically refreshed. We will focus our discussions on static memory devices because these devices are generally more popular and easier to use. However, you should note that advanced digital design techniques involving dynamic logic and memory can be effective in satisfying challenging area/speed performance specifications.

Static storage devices can be further classified into one of three categories: *latches*, *pulse-triggered flip-flops*, and *edge-triggered flip-flops*. The pulse-triggered and edge-triggered flip-flops are *transition-sensitive* memory devices, which means that logic level transitions, $0 \Rightarrow 1$ or $1 \Rightarrow 0$, control device operation. In contrast, latches are *level-sensitive* memory devices, which means that logic levels, 0 or 1, control device operation. Regrettably, latches are sometimes called flip-flops, but we will reserve the term *flip-flop* for transition-sensitive memory devices.

Before we begin discussing various memory devices in detail, let us put in perspective their general comparative capabilities with respect to synchronous sequential digital systems. As bistable memory devices, all latches and flip-flops operate similar to the seesaw shown in Figure 8.5.

A seesaw can be considered a bistable device because it can be in one of the two states shown in Figure 8.5(a) and Figure 8.5(b). One state holds the rider on the right up (Figure 8.5(a)) and the other state holds the rider on the left up (Figure 8.5(b)). A latch or flip-flop can likewise be considered a bistable device because they can be in one of two states; one state holds a logic 1 and the other state holds a logic 0. Just as a little energy in the form of a rider kick transitions the seesaw from one stable state to another, a little energy in the form of a voltage potential transitions a latch or flip-flop from one stable state to another.

Figure 8.5

A seesaw analogy for latches and flip-flops

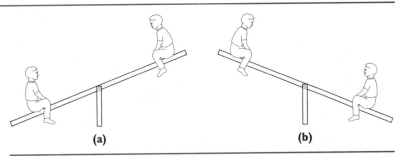

(a)   (b)

Figures (a) and (b) show two stable conditions or states of a seesaw.

Latches are generally simpler memory devices than pulse-triggered flip-flops, which are, in turn, simpler than edge-triggered flip-flops. Consequently, edge-triggered flip-flops offer more function and are more frequently used for synchronous sequential system design. Our tour of memory devices starts with the simplest devices, latches, to introduce storage concepts and popular control techniques, for example, SR, JK, T, and D. We will then use our understanding of latches to build progressively toward an understanding of flip-flops.

In the following sections, note that JK and T latches are not practical memory devices because they can oscillate under certain input conditions. You will never find a JK or T latch in a vendor data book, nor are you ever likely to find a JK or T latch in a conventional state machine. (JK and T flip-flops are, however, practical and useful memory devices.) We merely introduce JK and T latches here for instructional purposes to survey and understand popular set/reset control techniques before we tackle the more complex switching actions of flip-flops.

8.2.1 Ungated (Basic) Latches

As we noted, latches are the simplest of the three types of static memory devices shown in Figure 8.4. There are two types of latches: *ungated* and *gated*. Ungated latches, also called *basic* or *fundamental mode* latches, will be explained in this section; gated latches will be discussed in Section 8.2.2. We will examine four popular kinds of latches: *SR, JK, D,* and *T*.

The first latch we will examine, the *ungated* or *basic SR latch*, is shown in Figure 8.6. The figure shows that an ungated SR latch can be built by connecting two **nor** gates together in a *cross-coupled* configuration: The output of the top **nor** gate feeds back as an input to the bottom **nor** gate and vice versa. The regenerative or positive feedback is the basic mecha-

Figure 8.6

An SR latch using nor gates

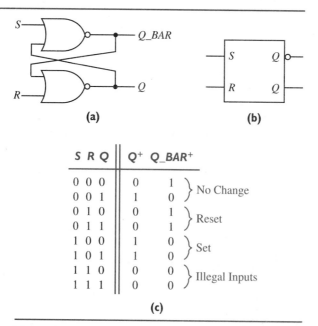

Figures (a) and (b) show the logic schematic and IEEE symbol for an SR latch using **nor** gates, respectively. Figure (c) gives the truth table.

nism that enables the cross-coupled **nor** gates to store information. The output Q denotes the value of the stored bit, either 0 or 1. The output Q_BAR is, by definition, the complement of Q. The signals S and R control the value of the stored bit.

The input/output behavior of the SR latch is described by the truth table, also called a *function table* or *characteristic table*, shown in Figure 8.6(c). The inputs to the truth table are S, R, and Q (remember that Q is the present value of the output fed back as an input). The outputs of the truth table are Q^+ and Q_BAR^+, which denote the next or new latch outputs. To understand how the signals S and R control the contents of the SR latch, we will examine in detail a few rows of the truth table.

Figure 8.7 shows a *literal* analysis of the *set* operation, where $S = 1$ and $R = 0$. Recall from Chapter 3 that literal analysis traces Boolean literals—1's and 0's—through a digital system to generate an output response to an input excitation. Figure 8.7(a) and Figure 8.7(b) show that the content of an SR latch is set to 1 ($Q^+ = 1$) when $S = 1$ and $R = 0$, regardless of the initial latch content.

The trick to analyzing a cross-coupled latch is to run around the feedback loop propagating new outputs to new inputs until the latch reaches a stable or quiescent state. As an example, consider Figure 8.7(a), which shows what happens when the SR latch is storing a 1 ($Q = 1$) and the inputs are set to $S = 1$ and $R = 0$. Start by selecting and evaluating one of the **nor** gates; the choice is arbitrary. Selecting, for instance, the top **nor** gate and propagating inputs to outputs shows that the new value of Q_BAR is the same as its old value, namely, 0.

$$Q\_BAR^+ = \overline{S + Q} = \overline{1 + 1} = 0$$

Propagating the new output value for Q_BAR back to the input and evaluating the bottom **nor** gate shows that the new value of Q is also the same as its old value, namely, 1.

$$Q^+ = \overline{R + Q\_BAR^+} = \overline{0 + 0} = 1$$

Since the new output values are the same as the old output values, the latch is in a stable state and the literal analysis is finished. You should check that the same logic analysis result

Figure 8.7

Sample literal analysis of an SR latch using nor gates

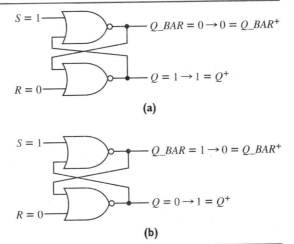

(a)

(b)

Figure (a) shows the output response when the initial content is 1, in other words, $Q = 1$. Figure (b) shows the output response when the initial content is 0, in other words, $Q = 0$.

is obtained by reversing the order in which the cross-coupled gates are evaluated, in other words, by evaluating the bottom **nor** gate followed by the top **nor** gate.

Figure 8.7(b) shows what happens when the SR latch is storing a 0 ($Q = 0$) and the inputs are set to $S = 1$ and $R = 0$. Again, evaluating the top **nor** gate first shows that Q_BAR changes from 1 to 0. Propagating the new output value for Q_BAR back to the input and evaluating the bottom **nor** gate shows that Q changes from 0 to 1. Propagating the new output value for Q back to the input and reevaluating the top **nor** gate shows that Q_BAR does not change value, so we are done running around the feedback loop, the latch is in a stable state, and the literal analysis is finished. In summary, the input conditions $S = 1$ and $R = 0$ set the contents of the SR latch to 1.

Figure 8.8 shows that when $S = 1$ and $R = 1$, $Q^+ = Q\_BAR^+ = 0$, which violates the requirement that Q_BAR always be the complement of Q. To get around this problem, we define the condition of $S = 1$ and $R = 1$ to be *illegal* or not allowed, for proper latch operation. In other words, we should not try to both set and reset an SR latch at the same time.

The *reduced truth table* shown below summarizes the operation of the ungated **nor** SR latch. An SR latch derives its name from the set/reset actions described by rows 1 and 2 of the reduced truth table.

Problem-Solving Skill
A solution does not always have to provide exactly the desired results. Sometimes the desired solution can be efficiently derived by judiciously constraining the behavior of a more general solution.

Row	S R	Q^+	Action
0	0 0	Q	No change
1	0 1	0	Reset
2	1 0	1	Set
3	1 1	–	Illegal inputs

Figure 8.9 shows how the input/output logic expression for the SR latch,

$$Q^+ = S + (\overline{R} \cdot Q) \tag{8.1}$$

can be derived from a Karnaugh map based on the reduced truth table entries. Equation 8.1 is also called a *characteristic equation* or *next state equation*.

An SR latch can also be built by connecting two **nand** gates together in a cross-coupled

Figure 8.8

Sample literal analysis of an SR latch using nor gates

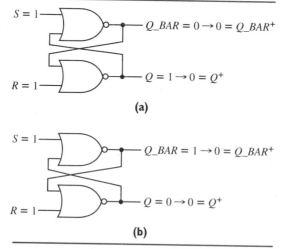

$S = 1$ — $Q\_BAR = 0 \rightarrow 0 = Q\_BAR^+$

$R = 1$ — $Q = 1 \rightarrow 0 = Q^+$

(a)

$S = 1$ — $Q\_BAR = 1 \rightarrow 0 = Q\_BAR^+$

$R = 1$ — $Q = 0 \rightarrow 0 = Q^+$

(b)

Figure (a) shows the output response when the initial content is 1, in other words, $Q = 1$. Figure (b) shows the output response when the initial content is 0, in other words, $Q = 0$.

Figure 8.9

Karnaugh map of an SR latch using nor gates

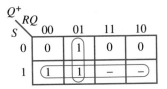

configuration, as shown in Figure 8.10(a). Notice from the reduced truth table in Figure 8.10(c) that the inputs are active-0, thus the signal names *S_BAR* and *R_BAR*. For instance, the **nand** SR latch is set when *S_BAR* is asserted (*S_BAR* = 0) and *R_BAR* is not asserted (*R_BAR* = 1). Similar to the **nor** SR latch, it is illegal to have both inputs (*S_BAR* = 0 and *R_BAR* = 0) asserted at the same time.

The second latch we will consider, the *JK latch*, is a variation of the SR latch that removes the restriction that the set and reset inputs cannot be asserted at the same time. A JK latch is shown in Figure 8.11.

The logic schematics of the ungated JK latch in Figure 8.11(a) emphasize that a **nor** SR latch is used to build a **nor** JK latch. The reduced truth table shown below summarizes the operation of the basic **nor** JK latch.

Row	J K	Q^+	Action
0	0 0	Q	No change
1	0 1	0	Reset
2	1 0	1	Set
3	1 1	$\overline{Q}$	Toggle

To understand the operation of a JK latch, analyze each truth table entry using the literal

Figure 8.10

An SR latch using nand gates

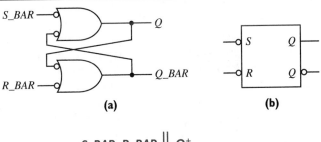

S_BAR	R_BAR	Q^+	
0	0	–	Illegal Input
0	1	1	Set
1	0	0	Reset
1	1	Q	No Change

(c)

Figures (a) and (b) show the logic schematic and IEEE symbol for an SR latch using **nand** gates, respectively. Figure (c) gives the reduced truth table.

Figure 8.11

A JK latch using nor gates

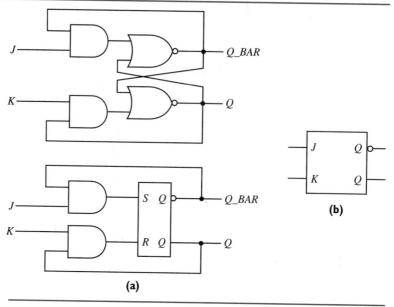

(a)

(b)

Figures (a) and (b) respectively show the logic schematic and IEEE symbol for a JK latch using **nor** gates.

analysis technique shown in Figures 8.7 and 8.8. When J and K inputs are both deasserted ($J = 0$) and ($K = 0$), the contents of a JK latch remains unchanged. Asserting J ($J = 1$) and deasserting K ($K = 0$) sets a JK latch; in other words, the new value of the output Q is 1. Asserting K ($K = 1$) and deasserting J ($J = 0$) resets a JK latch, in other words, the new value of the output Q is 0. If both J and K inputs are asserted at the same time ($J = 1$ and $K = 1$), a JK latch toggles, meaning that the new output value becomes the complement of the old output value. The content of a JK latch is changed to a 1 if it previously held a 0 or a 0 if it previously held a 1.

If the J and K inputs are both held 1 long enough for the JK latch to toggle state and for the new complemented outputs to "race" back as new inputs, the JK latch will toggle state again. This action repeats as long as J and K are 1 and, consequently, the JK latch oscillates. Oscillation can be great for generating periodic signals, but it is generally undesirable for sequential systems. When memory devices holding system state values oscillate, "life quickly becomes out of control" and it is difficult to reliably generate a desired sequence of output values in response to a sequence of input values. Hence, as discussed earlier, a JK latch is not a practical memory device for state machines. Following sections on flip-flops will discuss ways to improve latches to control multiple state changes or oscillations.

Figure 8.12 shows how the characteristic equation expression for the JK latch

$$Q^+ = (J \cdot \overline{Q}) + (\overline{K} \cdot Q) \tag{8.2}$$

can be derived from a Karnaugh map based on the reduced truth table entries.

Figure 8.13 shows that the third latch we will examine, the T *latch*, is formed by connecting together the inputs of a JK latch. By tying together these inputs, a T latch has only two of the four possible input conditions of a JK latch: Either $T = 0$, which implies $J = 0$ and $K = 0$; or $T = 1$, which implies $J = 1$ and $K = 1$. Thus, when $T = 0$, the contents of a T latch remain unchanged, $Q^+ = Q$. When $T = 1$, the contents of a T latch change to the complement of its previous value; a T latch derives its name from this "toggling" behavior.

Figure 8.12

Karnaugh map of a JK latch using nor gates

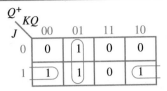

The reduced truth table and the characteristic equation are given below.

Row	T	Q^+	Action
0	0	Q	No change
1	1	$\overline{Q}$	Toggle

$$Q^+ = T \oplus Q \tag{8.3}$$

The problems concerning JK latch oscillation also apply to the T latch: If the input T is held 1 long enough, the outputs of the T latch will oscillate. Remember that at this point in our discussion, we are focusing on learning popular memory device control techniques and their associated truth tables and characteristic equations. In following sections we will discuss ways to improve basic latches to control multiple state changes or oscillations.

The last kind of latch we will study is the *D latch*. Similar to a JK latch, a D latch can also be viewed as a variation of an SR latch. Figure 8.14 shows a **nor** D latch, built by adding an inverter between the set and reset inputs of a **nor** SR latch.

Placing an inverter between the *S* and *R* inputs ensures that the set and reset inputs are never both asserted at the same time. The reduced truth table and the characteristic equation

Figure 8.13

A T latch using nor gates

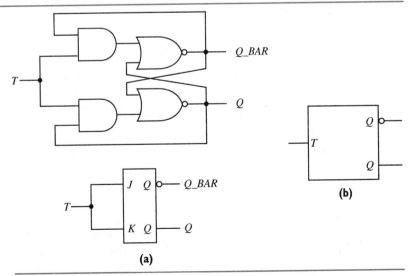

Figures (a) and (b) respectively show the logic schematic and IEEE symbol for a T latch using **nor** gates.

Figure 8.14

A D latch using nor gates

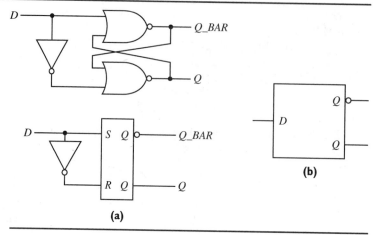

Figures (a) and (b) respectively show the logic schematic and IEEE symbol for a D latch using **nor** gates.

given below show the simple relationship between the input (D) and the contents of the latch (Q^+): The output equals the input.

Row	D	Q^+	Action
0	0	0	Reset
1	1	1	Set

$$Q^+ = D \tag{8.4}$$

The D latch is also called a *transparent* latch because of its property that the output equals or "follows" the input. D flip-flops (discussed in the following sections) are often the most commonly used memory devices in synchronous sequential design because of their straightforward input/output behavior.

8.2.2 Gated Latches

As previously noted, basic latches are not well suited for *sychronous* sequential digital systems because they operate in an *asynchronous* fashion: Whenever the inputs to a latch change value, the contents of the latch are updated according to the associated characteristic equation. This asynchronous behavior can result in erratic state changes, such as oscillation. In a synchronous sequential system, however, we want to control when memory devices change state. A simple way to do this is to add an extra signal, called an *enabling* or *clock* signal, that controls or "gates" when input signals can drive a memory device.

To understand the importance of a synchronizing clock signal, let us compare the asynchronous behavior of a sequential system using ungated latches with the synchronous sequential digital system model and the five-step operation given earlier. Like a synchronous system, a sequential system using ungated latches assumes an initial state, accepts a set of inputs, and computes new output and next state values. However, the timing of when the present state values are updated is not controlled by a clock signal, but rather by when the ungated latches change value. A desired input/output sequence can be generated, provided the inputs are applied and the outputs observed at just the right times. This timing can be difficult to achieve, especially when the latches do not all change value at the same time due to varying signal path delays in the combinational logic. To avoid these complications, we impose a synchronous timing discipline.

Gated versions of the **nor** SR and D latches are shown in Figure 8.15. The truth tables for the gated latches can be easily derived from the truth tables for the ungated latches. For example, the reduced truth table and characteristic equation for a gated D latch is given below.

Row	C D	Q⁺	Action
0	0 –	Q	No change
1	1 0	0	Reset
2	1 1	1	Set

$$Q^+ = (\overline{C} \cdot Q) + (C \cdot D)$$

As a notational convenience, the truth tables and characteristic equations described in the previous section for ungated latches are often used for gated latches, with the understanding that the clock signal is implied.

8.2.3 Master-Slave (Pulse-Triggered) Flip-Flops

Controlling when a latch changes value by gating its inputs with a clock signal has the disadvantage that the *clock pulse* (period of time the clock is asserted) must be kept short enough to ensure that the memory contents are updated only *once* per clock cycle or, equivalently, per set of inputs. Referring again to Figure 8.2, when the clock is asserted and the next state values are transferred to present state values, we do not want the new present state values to loop through the combinational logic and memory devices and again update present state values before the clock is deasserted. A clock pulse is also commonly called a *clock tick*.

Figure 8.15

Gated SR and D latches using nor gates

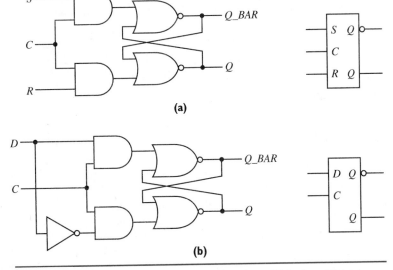

(a)

(b)

The logic schematic and IEEE symbol for a gated **nor** SR latch and D latch are respectively shown in Figures (a) and (b).

A better way to control when a memory device changes state is to arrange two gated latches in a *master-slave* configuration. Such a memory device is called a *master-slave* or *pulse-triggered* flip-flop. As an example, a master-slave SR flip-flop is shown in Figure 8.16.

Figure 8.16(b) shows that the IEEE denotes *pulse-triggered* behavior by the ¬ symbol. The term *pulse-triggered* is used because both the rising and falling edges of the clock signal are used to sample the inputs and generate the outputs. If the inputs are sampled on the falling edge of the clock signal and the outputs generated on the rising edge of the clock signal, the flip-flop is called a *positive* pulse-triggered or master-slave device. Alternatively, if the inputs are sampled on the rising edge of the clock signal and the outputs generated on the falling edge of the clock signal, the flip-flop is called a *negative* pulse-triggered or master-slave device.

The SR flip-flop shown in Figure 8.16(a) is a negative master-slave memory device. To understand the behavior of this master-slave configuration, examine the *timing diagram* given in Figure 8.16(c). A timing diagram is drawn so time starts at the left (initial state) and progresses to the right. For the timing diagram in Figure 8.16(c), the clock pulse labeled

Figure 8.16

A master-slave (pulse-triggered) SR flip-flop

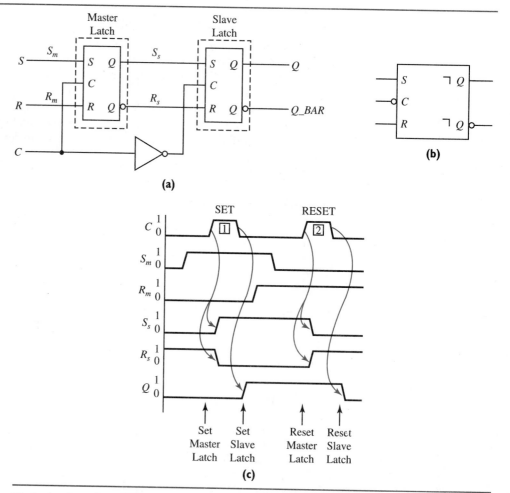

The logic schematic and IEEE symbol for a master-slave SR flip-flop are respectively shown in Figures (a) and (b). Figure (c) gives a timing diagram illustrating set and reset actions.

2 occurs after (later in time) the clock pulse labeled 1 . The arrows show cause-and-effect, in other words, which signal transitions cause other signal transitions.

Consider the set operation. When the clock signal is asserted ($C = 1$), the master SR latch is enabled and the slave SR latch is disabled. The stored bit in the master latch is updated according to the values of the input signals S_m and R_m. Since $S_m = 1$ and $R_m = 0$, the output of the master latch is set to 1; consequently, $S_s = 1$ and $R_s = 0$. When the clock is deasserted ($C = 0$), the master SR latch is disabled and the slave SR latch is enabled. The contents of the master latch are transferred to the slave latch and the flip-flop output is set ($Q = 1$).

Just as JK, T, and D latches are easily constructed from an SR latch, JK, T, and D master-slave flip-flops are easily constructed from an SR master-slave flip-flop. Figure 8.17 shows the logic schematics and IEEE symbols for JK, T, and D master-slave flip-flops.

Similar to a submarine hatchway, the master-slave configuration acts like a passageway having two doors. The double doors of a submarine hatchway are never opened at the same time to prevent water spilling from the ocean into the hull. Likewise, the master and slave latches of a pulse-triggered flip-flop are never enabled at the same time to prevent next state information from spilling into the present state information. Again, refer to Figure 8.2. The next state values are transferred to the present state values in two steps. First, the slave door is closed and the master door is opened, allowing the next state values to be transferred to an intermediate waiting place. Then, the master door is closed and the slave door is opened, transferring the next state values from the intermediate waiting place to the final destination—the present state values. The two doors are never open at the same time, which isolates the next state and present state values and ensures that the contents of the master-slave flip-flop will change only *once* per clock cycle.

Although the complementary actions of the master and slave latches address the problem of unwanted *multiple* state changes associated with latches, pulse-triggered flip-flops unfortunately are not completely free of timing problems; the flip-flops can mistakenly be set to the wrong *single* state. In other words, with pulse-triggered flip-flops, we are assured of one state change per clock cycle, but the change may not be the *desired* state change because during the time the master latch is enabled, transient signal transitions, called *glitches*, can incorrectly set/reset the master latch.

Figure 8.18 shows how the **nor** SR master-slave flip-flop shown in Figure 8.16 can be mistakenly set to 1, an action called *ones catching*. During the time the clock is 1, an unexpected and unwanted spike on S_m sets the master latch, which, in turn, sets the slave latch and the output of the flip-flop. Once set, a flip-flop remains set until explicitly reset, so there is no way to undo the errant effect of the glitch.

A similar timing problem called *zeros catching* can occur for a **nand** SR master-slave flip-flop. One way to avoid the ones and zeros catching problems is to synchronize the switching of memory devices on only the rising *or* falling edge of a clock signal; such memory devices are called *edge-triggered* flip-flops, and we discuss them next.

8.2.4 Edge-Triggered Flip-Flops

Like master-slave flip-flops, edge-triggered flip-flops are transition-sensitive memory devices that respond only to input changes, not input levels. However, as the name implies, edge-triggered flip-flops operate off only the rising *or* falling clock transition or edge, whereas master-slave flip-flops operate off both rising *and* falling clock transitions or edges.

Figure 8.19 shows a negative, edge-triggered SR flip-flop. In Figure 8.19(b), the arrowhead shape pointing to the right inside the rectangle on the clock (C) input designates an edge-triggered input, also more generally called a *dynamic input*. The bubble outside the rectangle on the clock (C) input designates a negative, or falling $1 \Rightarrow 0$, edge-triggered input;

Figure 8.17

Master-slave (pulse-triggered) JK, T, and D flip-flops

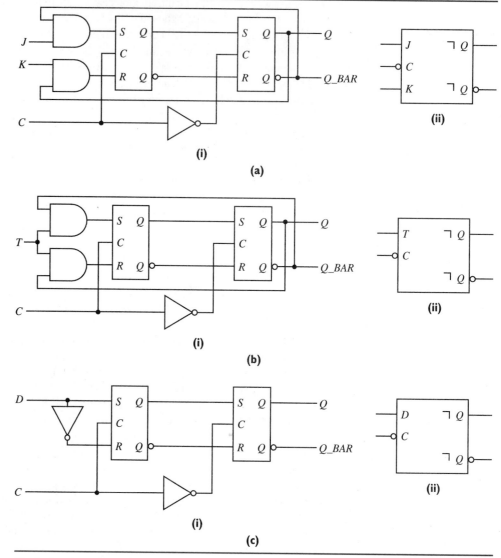

The logic schematics and IEEE symbols for master-slave JK, T, and D flip-flops are respectively shown in Figures (a)–(c).

the absence of a bubble designates a positive, or rising $0 \Rightarrow 1$, edge-triggered input. The truth table in Figure 8.19(c) explains the operation of the negative edge-triggered SR flip-flop. The down arrow ($\downarrow$) specifies that the outputs are generated on the negative-going edge of the clock signal C.

The first and second stages of logic respectively work in a master-slave configuration to operate the final stage of logic, which is an SR latch. When the clock signal is 1, the first (master) stage of logic is enabled, passing the S and R inputs on to the S_s and R_s inputs of the second (slave) stage of logic. Also, the second stage of logic is disabled, keeping the SR latch in a holding state. When the clock signal transitions from 1 to 0, the second stage of logic is enabled, operating the SR latch according to the S_l and R_l inputs. Then, after a period

Figure 8.18

Ones catching problem with master-slave flip-flops

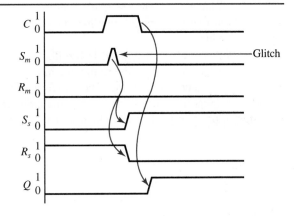

Figure 8.19

A negative edge-triggered SR flip-flop

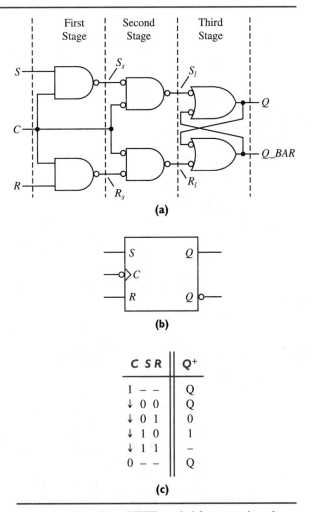

The logic schematic and IEEE symbol for a negative edge-triggered SR flip-flop are respectively shown in Figures (a) and (b). Figure (c) lists the truth table.

of time equal to the propagation delay of the first stage, the first stage again disables the second stage, placing the SR latch back into a holding state. Note that the propagation delay of the first (master) stage must be long enough to allow the second (slave) stage to operate the SR latch before being disabled.

Figure 8.20 shows a positive edge-triggered D flip-flop with asynchronous, active-0 set (*SET_BAR*) and clear (*CLR_BAR*) inputs. The truth table in Figure 8.20(c) explains the operation of the positive edge-triggered D flip-flop. The up arrow ($\uparrow$) in the truth table specifies that the D input sets or resets the flip-flop only when the clock transitions from $0 \Rightarrow 1$, in other words, the rising or positive-going edge. The truth table also shows that the contents of

Figure 8.20

A positive edge-triggered D flip-flop

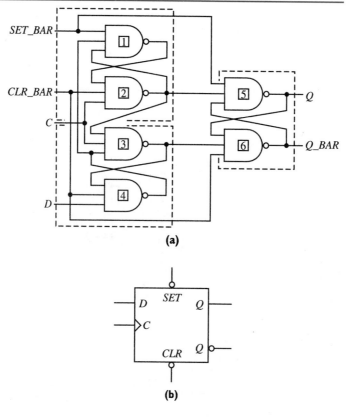

(a)

(b)

SET_BAR	CLR_BAR	C	D	Q	ACTION
0	0	–	–	–	Illegal Inputs
0	1	–	–	1	Asynchronous Set
1	0	–	–	0	Asynchronous Reset
1	1	0	–	Q	No Change
1	1	$\uparrow$	1	1	Synchronous Set
1	1	$\uparrow$	0	0	Synchronous Reset
1	1	1	–	Q	No Change

(c)

The logic schematic and IEEE symbol for a positive edge-triggered D flip-flop are respectively shown in Figures (a) and (b). Figure (c) lists the truth table.

the D flip-flop will be forced to 1 when *SET_BAR* is asserted (*SET_BAR* = 0) and *CLR_BAR* is deasserted (*CLR_BAR* = 1). Conversely, the contents will be forced to 0 when *CLR_BAR* is asserted (*CLR_BAR* = 0) and *SET_BAR* is deasserted (*SET_BAR* = 1). These actions occur regardless of the data (*D*) and clock (*C*) inputs. *SET_BAR* and *CLR_BAR* are called *asynchronous* inputs because they are not controlled by the clock signal; these signals can be applied at any time and are typically used to force a sequential system to a certain state to initialize the system or to troubleshoot a system failure. It is illegal to assert both the set and clear inputs at the same time because both flip-flop outputs become 1, violating their required complement behavior.

The dotted boxes in Figure 8.20(a) show that the positive edge-triggered D flip-flop is composed of three **nand** SR latches. When the clock *C* is 0, the outputs of the **nand** gates labeled $\boxed{2}$ and $\boxed{3}$ are both driven to 1, keeping the output SR latch (**nand** gates $\boxed{5}$ and $\boxed{6}$) in a holding state. When the clock *C* transitions from 0 to 1, the outputs of **nand** gates $\boxed{2}$ and $\boxed{3}$ will respectively follow $\overline{D}$ and *D*, driving the output of the SR latch to the appropriate set or reset state. When the clock *C* is 1, the outputs of the **nand** gates $\boxed{2}$ and $\boxed{3}$ keep the two input SR latches in holding states, blocking any changes on *D* from affecting the flip-flop. Thus, inputs affect outputs only on the rising clock edge.

8.3 LITERAL ANALYSIS

Now that we have learned about a variety of commonly used memory devices, we can proceed to discuss how these memory devices can be combined with combinational logic to build synchronous sequential digital systems. We will start by looking at several representative sequential systems and learning how to analyze their input/output behavior. Recall from Chapter 3 that *literal* analysis derives an output response to a specific input excitation; a set of 1's and 0's is applied to the input and propagated to the output. Literal analysis for synchronous sequential digital systems is summarized below.

Literal Analysis

• *Mealy Machine*
  1. *Initialize the system—set the contents of the memory devices.*
  2. *Apply a set of inputs.*
  3. *Propagate the set of inputs to the outputs and next state values.*
  4. *Observe the outputs.*
  5. *Apply a clock pulse and update the present state values.*
  6. *Return to Step 2.*
  Note: *The inputs and outputs are valid immediately before the clock pulse.*
• *Moore Machine*
  1. *Initialize the system—set the contents of the memory devices.*
  2. *Apply a set of inputs.*
  3. *Propagate the set of inputs to the next state values.*
  4. *Apply a clock pulse, update the present state values, and propagate the new present state values to the output values.*
  5. *Observe the outputs.*
  6. *Return to Step 2.*
  Note: *The inputs are valid during the clock pulse; the outputs are valid immediately following the clock pulse.*

Mealy and Moore machines' input/output timings have subtle, but important, differences. With a Mealy machine, the inputs are applied and the outputs are subsequently generated after a period of time that allows for combinational logic propagation delays. With a Moore machine, the inputs are applied, the sequential system is clocked, and *then* the outputs are generated after a period of time that allows for combinational logic propagation delays. Since the literal analyses of Mealy and Moore machines are slightly different, these systems will be addressed separately in Sections 8.3.1 and 8.3.2, respectively.

8.3.1 Mealy Machines

What does the sequential system given in Figure 8.21(a) do? Before we begin a detailed analysis, let us make a few general observations about the sequential system and its expected input/output behavior. First, the memory device is a negative edge-triggered D flip-flop,

Figure 8.21

A sequential digital system

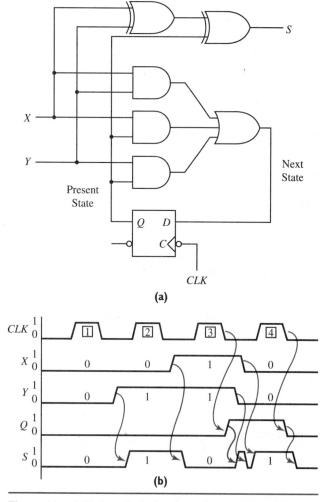

(a)

(b)

Figures (a) and (b) show the logic schematic and timing diagram of a synchronous sequential system, respectively.

which makes the sequential system a synchronous system. Second, the output (S) is a function of both the input and the present state, which makes the sequential system a Mealy synchronous system.

To generate the timing diagram given in Figure 8.21(b) using literal analysis, we start by assuming the initial state of the digital system, in other words, the initial content of the D flip-flop. The leftmost point or beginning of the timing diagram shows that the output of the D flip-flop is assumed to be initially set to 0 ($Q = 0$). Having initialized the sequential system, the output response for each input is generated by repeatedly following the literal analysis steps outlined earlier. A set of inputs is applied and the output and next state values are computed based on the inputs and present state values. The resulting outputs are observed. Next, a clock pulse is applied, the present state values are updated, and the system is then ready for a new set of inputs.

In conducting these steps, it is important to understand when the input and output values are considered to be *valid*, in other words, when they may be sampled. It is difficult to define the point during a cycle clock when Mealy outputs become valid because the response depends on both when stimuli were applied and various circuit propagation delays. However, the entire Mealy machine, including outputs, must reach a stable state just before activating and updating the memory devices to initiate a new clock cycle, otherwise we do not have a predictable sequential behavior. Thus, we will adopt the convention that Mealy inputs and outputs are valid, or may be sampled, immediately before the clock edge that activates and updates memory. Actually, a Mealy machine must reach a stable state for a period of time before activating and updating the memory devices, but we will postpone discussing detailed timing issues until Section 8.5.

We begin analyzing the Mealy machine in Figure 8.21(a) by considering the first set of inputs applied for the first clock pulse labeled $\boxed{1}$, $X = 0$, and $Y = 0$. With the D flip-flop initially set to 0 ($Q = 0$), the output

$$S = X \oplus Y \oplus Q = 0 \oplus 0 \oplus 0 = 0$$

and the next state

$$D = (X \cdot Y) + (X \cdot Q) + (Y \cdot Q) = (0 \cdot 0) + (0 \cdot 0) + (0 \cdot 0) = 0$$

Since the computed next state value (D) and the present state value (Q) are equal, the timing diagram shows that the contents of the flip-flop does not change value on the falling edge of clock pulse $\boxed{1}$.

Applying the same analysis procedure to the next set of inputs ($X = 0$ and $Y = 1$) shows that the change in input Y causes the output to change to 1.

$$S = X \oplus Y \oplus Q = 0 \oplus 1 \oplus 0 = 1$$

The next state, however, is again computed to be 0,

$$D = (X \cdot Y) + (X \cdot Q) + (Y \cdot Q) = (0 \cdot 1) + (0 \cdot 0) + (1 \cdot 0) = 0$$

so the flip-flop does not change value on the falling edge of clock pulse $\boxed{2}$.

The third set of inputs ($X = 1$ and $Y = 1$) shows that the change in X causes the output to return to 0.

$$S = X \oplus Y \oplus Q = 1 \oplus 1 \oplus 0 = 0$$

The next state also changes value from 0 to 1,

$$D = (X \cdot Y) + (X \cdot Q) + (Y \cdot Q) = (1 \cdot 1) + (1 \cdot 0) + (1 \cdot 0) = 1$$

so the flip-flop is set to a 1 ($Q = 1$) on the falling edge of clock pulse $\boxed{3}$.

Finally, X and Y return to 0 for the fourth set of inputs ($X = 0$ and $Y = 0$). The output S

changes from 0 to 1

$$S = X \oplus Y \oplus Q = 0 \oplus 0 \oplus 1 = 1$$

and the D flip-flop is reset to 0 when clock pulse ④ is applied.

$$Q = D = (X \cdot Y) + (X \cdot Q) + (Y \cdot Q) = (0 \cdot 0) + (0 \cdot 1) + (0 \cdot 1) = 0$$

The sequence of input/output values computed thus far are summarized below.

$$X = 0\ 0\ 1\ 0\ \ldots$$
$$Y = 0\ 1\ 1\ 0\ \ldots$$
$$S = 0\ 1\ 0\ 1\ \ldots$$
$$\longmapsto \text{time}$$

Interpreting X and Y as binary numbers (least significant bit on the left and most significant bit on the right) reveals that the output is the sum of X and Y and the Mealy machine is a binary adder. Starting with the least significant bits, corresponding operand bits are applied to the adder. The next state computes the carry and saves it in the D flip-flop to be applied to the next-higher-order column summation.

The last set of inputs applied for clock pulse ④ ($X = 0$ and $Y = 0$) illustrates the importance of knowing when to properly sample a valid output value. As the inputs change from $X = 1$ and $Y = 1$ (third set) to $X = 0$ and $Y = 0$ (fourth set), the output changes value several times. However, these signal changes are not of interest and are ignored because, by definition, the inputs and outputs are not sampled until just before the falling edge of the clock.

8.3.2 Moore Machines

Literal analysis of a Moore machine is similar to the literal analysis of a Mealy machine except for when the outputs are considered to be valid. For a Moore machine, the output is a function of only the present state, so the output is valid when the present state is valid. We will adopt the convention that Moore outputs are valid, in other words, may be sampled, immediately after the clock edge that activates and updates memory. Actually, a Moore machine does not reach a stable state for a period of time after activating and updating the memory devices due to the propagation delay of any combinational logic, but again, we will postpone discussing detailed timing issues until Section 8.5.

To understand Moore machine operation, consider the simple Moore machine given in Figure 8.22(a). Notice that the Moore machine uses a positive edge-triggered D flip-flop. The initial contents of the D flip-flop is assumed to be 0. The first input is applied, $A = 1$, and the next state is computed.

$$D = A \oplus Q = 0 \oplus 1 = 1$$

Then, clock pulse ① updates the present state, which also updates the output because the output X is simply the output of the D flip-flop, Q. Repeating this analysis procedure for clock pulses ② and ③ reveals that the sequence of inputs

$$A = 1\ 0\ 1 \ldots$$
$$\longmapsto \text{time}$$

generates the sequence of outputs

$$X = 1\ 1\ 0 \ldots$$
$$\longmapsto \text{time}$$

By running a few longer sequences through the Moore machine, we quickly realize that the

Figure 8.22

The Moore machine

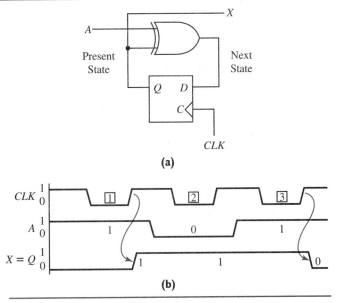

(a)

(b)

Figures (a) and (b) show the logic schematic and timing diagram of a synchronous Moore machine.

sequential machine checks for odd parity. That is, the output is a 1 if an odd number of 1's have been received, otherwise the output is a 0.

8.4 SYMBOLIC ANALYSIS

Analyzing a sequential system by tracing 1's and 0's works well for generating a few input/output transformations for simple designs. However, literal analysis quickly becomes tedious and error prone for a complex design involving many input patterns. For these more challenging sequential system analysis tasks, it is generally easier to use a *symbolic* approach that uses switching algebra expressions instead of individual binary values. Symbolic sequential system analysis is summarized below.

Symbolic Analysis

1. *Generate switching algebra expressions describing the outputs and flip-flop inputs.*
2. *Generate the next state equation for each flip-flop, using its input switching algebra expression(s) and characteristic equation.*
3. *Generate the next state Karnaugh map for each flip-flop, using its next state equation. Also, generate a Karnaugh map for each output.*
4. *Generate a state table.*
5. *Optionally, generate a state diagram and/or a timing diagram.*

Following the format used for literal analysis, symbolic analysis of Mealy and Moore machines will be discussed separately.

8.4.1 Mealy Machines

We will use symbolic analysis to derive the sequential input/output behavior of the Mealy machine given in Figure 8.23. The figure illustrates several conventions commonly used in drawing schematics of sequential machines. First, a memory device is typically identified by an uppercase letter, A, B, ... The uppercase letter is then used as a subscript to identify the inputs and outputs of the memory device. For instance, the J input and Q output of the A JK flip-flop are labeled J_A and Q_A, respectively. Second, to simplify the schematic, the feedback connections from the flip-flop outputs to the combinational logic generating the flip-flop inputs are not drawn. For example, K_A is driven by Q_A, which is the output of the A JK flip-flop; the connection is implied, but not shown.

In the following symbolic analysis, it is important to bear in mind that the clock signal CLK does not appear in any of the equations, maps, or tables we build to describe the sequential system. Rather, the presence of the clock signal is implied.

We begin by generating the switching algebra expressions for the system output and flip-flop inputs according to the combinational system symbolic analysis technique described in Chapter 3. The logic expressions for the output Z and the J and K inputs for the A and B flip-flops are given below.

$$Z = X \cdot Q_A \cdot \overline{Q_B}$$
$$J_A = X \cdot Q_B$$
$$K_A = Q_A$$
$$J_B = X \cdot \overline{Q_A} \cdot \overline{Q_B}$$
$$K_B = Q_B$$

Next, for each JK flip-flop, the switching algebra expressions for the J and K inputs are substituted into the flip-flop's characteristic equation (Equation 8.2) to yield a *next state equation*. The next state equation defines the new contents of a flip-flop as a function of the present state (flip-flop outputs) and system inputs. The next state equations Q_A^+ and Q_B^+ for flip-flops A and B are given below.

$$Q_A^+ = (J_A \cdot \overline{Q_A}) + (\overline{K_A} \cdot Q_A)$$
$$Q_A^+ = (X \cdot Q_B \cdot \overline{Q_A}) + (\overline{Q_A} \cdot Q_A)$$
$$Q_A^+ = X \cdot Q_B \cdot \overline{Q_A}$$

Figure 8.23

Mealy machine analysis example

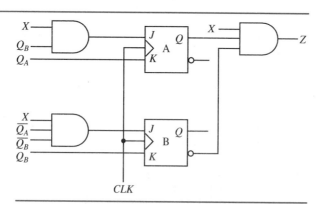

CLK

Note that to simplify the schematic, the connections of the flip-flop outputs to the combinational logic driving the flip-flop inputs are not shown.

$$Q_B^+ = (J_B \cdot \overline{Q_B}) + (\overline{K_B} \cdot Q_B)$$
$$Q_B^+ = (X \cdot \overline{Q_A} \cdot \overline{Q_B} \cdot \overline{Q_B}) + (\overline{Q_B} \cdot Q_B)$$
$$Q_B^+ = X \cdot \overline{Q_A} \cdot \overline{Q_B}$$

Again, note that the clock signal does not appear in the next state equations, although the switching actions of the flip-flops are assumed to be controlled by a clock signal.

The third step in symbolic analysis maps the output and next state equations onto Karnaugh maps, as shown in Figure 8.24. The Karnaugh maps for the next state equations in Figure 8.24(a) and Figure 8.24(b) are appropriately called *next state Karnaugh maps*. (Refer to Chapter 3 for a discussion of mapping switching algebra expressions onto Karnaugh maps.)

In generating Karnaugh maps, it is a good practice to try to group separately the input variables and the present state variables. For example, the Karnaugh maps for Q_A^+, Q_B^+, and Z have the input variable X listed vertically and the present state variables Q_A and Q_B grouped together horizontally. This practice makes the Karnaugh maps easier to read and facilitates generating a state table, which is the next step in the analysis procedure.

Table 8.1 shows the *state table*, also called a *next state table* or a *transition table*, for the Mealy machine in Figure 8.23. A state table has three major columns: Present state, Next state, and Output. The present state is the present contents of the memory devices. The next state is the new contents of the memory devices after the system is *clocked*; in other words, after a clock pulse is applied. The last column lists the output value(s). The next state and output columns are each further divided into additional columns, showing the dependency on the input as well as the present state.

The state table in Table 8.1 is generated directly from the Karnaugh maps in Figure 8.24. When $X = Q_A = Q_B = 0$, the upper leftmost squares of the next state Karnaugh maps show that $Q_A^+ = Q_B^+ = 0$ and the upper leftmost square of the output Karnaugh map shows that $Z = 0$. These values are recorded in the first row of Table 8.1; Q_A^+ and Q_B^+ are recorded in the second column and Z is recorded in the fourth column. When $X = 1$, $Q_A = 0$, and $Q_B = 0$, the lower leftmost squares of the next state Karnaugh maps show that $Q_A^+ = 0$ and $Q_B^+ = 1$ and the lower leftmost square of the output Karnaugh map shows that $Z = 0$. These values

Figure 8.24

Karnaugh maps for output and next state equations

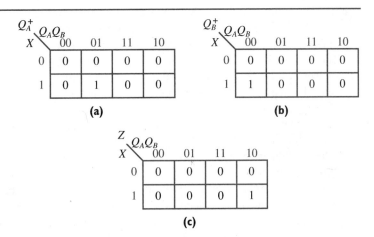

Table 8.1

Present state $Q_A Q_B$	Next state $Q_A^+ Q_B^+$		Output Z	
	$X = 0$	$X = 1$	$X = 0$	$X = 1$
00	00	01	0	0
01	00	10	0	0
10	00	00	0	1
11	00	00	0	0

State table for example Mealy machine

are also recorded in the first row of Table 8.1; Q_A^+ and Q_B^+ are recorded in the third column and Z is recorded in the last column.

Thus, we have filled in the entries for the first row of the state table, which reads as follows:

When the Mealy machine is in state 00 (contents of both flip-flops are 0), the output (Z) is 0 regardless of the input value. When a clock pulse is applied, the machine remains in state 00 if the input is 0, otherwise the machine transitions to the new state 01 ($Q_A = 0$ and $Q_B = 1$).

We fill in each row of the state table in a similar manner. For each combination of X, Q_A, and Q_B, we find the corresponding squares in the Karnaugh maps and record the values of Z, Q_A^+, and Q_B^+ in the next state table.

Since a Boolean equation and a Karnaugh map are two ways of representing the same information, a state table can be derived directly from the output and next state equations instead of the output and next state Karnaugh maps; the choice is largely a matter of personal preference.

Because a state table completely defines the behavior of a sequential machine, the symbolic analysis task is finished. Have you figured out what the Mealy machine does? No? Don't feel discouraged; sometimes it is difficult to abstract from a state table the behavior of a sequential machine. Thus, a *state diagram* and/or *timing diagram* are also often generated as alternative representations for the information contained in a state table. These alternative representations can sometimes offer a better way to view and understand a sequential machine. A state diagram for our Mealy machine is shown in Figure 8.25.

A state diagram uses circles to represent states and directed arcs to represent state-to-state transitions. States are often given abstract names, such as S_0 and S_1, as a labeling convenience because identifying each state by the associated contents of the memory devices, in other words, a bit string, can be tedious and error prone when describing large sequential systems. For a Mealy machine, each directed arc has at least one label of the form x/y, where x is an input value and y is an output value. This notation denotes that while the input is x and the sequential system is *in* the state at the tail of the arc, the sequential system generates the output y and transitions to the state at the head of the arc on next clock pulse.

The state diagram in Figure 8.25 is generated directly from the state table in Table 8.1. The four present states listed in the leftmost column of the state table are represented by four circles. For this simple analysis exercise, there is little difference between identifying a state by an abstract name or the associated 2-bit contents of the two JK flip-flops. We will use abstract names: S_0 represents the present state $Q_A = 0$ and $Q_B = 0$, S_1 represents the present

Figure 8.25

State diagram for example Mealy machine

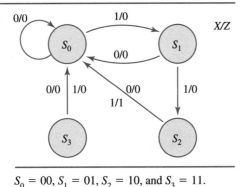

X/Z

$S_0 = 00, S_1 = 01, S_2 = 10,$ and $S_3 = 11.$

state $Q_A = 0$ and $Q_B = 1$, and so on. The next state and output information listed in the remaining columns of the state table are represented by the directed arcs.

For example, the $X = 0$ next state and output columns of the first row of the state table are described by the arc that starts at state S_0 and loops back to state S_0 having the label 0/0. If the input (X) is 0 and the sequential system is in state S_0, the output (Z) is 0 and the machine remains in state S_0 after a clock pulse is applied. As another example, the third row of the state table is described by the single arc from state S_2 to state S_0 having the labels 0/0 and 1/1. If the sequential system is in state S_2, the output (Z) is 0 if the input (X) is 0, otherwise the output is 1. When a clock pulse is applied, the machine transitions to state S_0 regardless of the input value.

Starting in state S_0 and running a few sample input sequences through the state diagram

$$X = 1\,1\,1\,1\,1\,1\,1\,0\,1\,1\,1 \ldots$$
$$Z = 0\,0\,1\,0\,0\,1\,0\,0\,0\,0\,1 \ldots$$
$$\longmapsto \text{time}$$

reveals that the Mealy machine detects when three successive 1's have been applied, in other words, the binary number 7.

As a final comment concerning this analysis example, notice that state S_3 should never occur under normal operation. Such states are called *don't care* or *disallowed* states; they are unused combinations of flip-flop values. Even though under normal operation state S_3 should never occur, it is important to account for the possibility that it *may* unexpectedly occur due to an unforeseen event, such as an operator error or a component failure. If state S_3 occurs in our sequence detector, the Mealy machine is designed so that the output will be zero and the next clock pulse will reset the system to the initial state S_0.

As a slightly more complex symbolic analysis exercise, let us derive the behavior of the Mealy machine given in Figure 8.26. Again, we begin by generating the switching algebra expressions for the system output $Z_2 Z_1 Z_0$ and flip-flop inputs $D_A, D_B,$ and D_C.

$$Z_2 = Q_A \cdot Q_B$$
$$Z_1 = Q_A \oplus Q_C$$
$$Z_0 = X$$
$$D_A = \overline{Q_B} \cdot \overline{Q_C} \cdot (Q_A + X)$$
$$D_B = (Q_A \cdot \overline{Q_B} \cdot \overline{Q_C}) + (\overline{Q_A} \cdot \overline{Q_B} \cdot Q_C)$$
$$D_C = (\overline{Q_B} \cdot \overline{Q_C} \cdot \overline{X}) + (\overline{Q_A} \cdot \overline{Q_B} \cdot Q_C \cdot X)$$

Problem-Solving Skill

A good solution always accounts for all possible eventualities.

Figure 8.26

Mealy machine analysis example

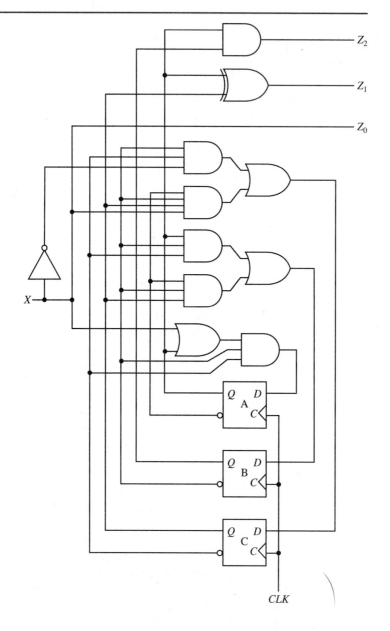

The next step, generating the next state equations, is easy because the characteristic equation for a D flip-flop is $Q^+ = D$. Thus, the D flip-flop input equations directly yield the next state D flip-flop equations.

$$Q_A^+ = D_A = \overline{Q_B} \cdot \overline{Q_C} \cdot (Q_A + X)$$
$$Q_B^+ = D_B = (Q_A \cdot \overline{Q_B} \cdot \overline{Q_C}) + (\overline{Q_A} \cdot \overline{Q_B} \cdot Q_C)$$
$$Q_C^+ = D_C = (\overline{Q_B} \cdot \overline{Q_C} \cdot \overline{X}) + (\overline{Q_A} \cdot \overline{Q_B} \cdot Q_C \cdot X)$$

Next, the next state and output switching algebra equations are mapped onto the Karnaugh maps shown in Figure 8.27. Then, the Karnaugh maps are mapped onto the state table shown in Table 8.2. Finally, a state diagram is generated, as shown in Figure 8.28.

Figure 8.27

Karnaugh maps for output and next state equations

Q_A^+

Q_AQ_B \ Q_CX	00	01	11	10
00	0	1	0	0
01	0	0	0	0
11	0	0	0	0
10	1	1	0	0

Q_B^+

Q_AQ_B \ Q_CX	00	01	11	10
00	0	0	1	1
01	0	0	0	0
11	0	0	0	0
10	1	1	0	0

Q_C^+

Q_AQ_B \ Q_CX	00	01	11	10
00	1	0	1	0
01	0	0	0	0
11	0	0	0	0
10	1	0	0	0

Z_2

Q_AQ_B \ Q_CX	00	01	11	10
00	0	0	0	0
01	0	0	0	0
11	1	1	1	1
10	0	0	0	0

Z_1

Q_AQ_B \ Q_CX	00	01	11	10
00	0	0	1	1
01	0	0	1	1
11	1	1	0	0
10	1	1	0	0

Z_0

Q_AQ_B \ Q_CX	00	01	11	10
00	0	1	1	0
01	0	1	1	0
11	0	1	1	0
10	0	1	1	0

Starting in state S_0, running a few sample input sequences through the state diagram and recording the results in the timing diagram in Figure 8.29 shows that the Mealy machine is a simple serial-to-parallel code converter. Three clock cycles enter three bits serially on the input X, most significant bit first. On the third clock cycle, the three bits appear in parallel on the output $Z_2Z_1Z_0$; intermediate outputs on the first and second clock cycles are ignored. (State S_7 is an unused state.)

8.4.2 Moore Machines

To complete our discussion of symbolic analysis, let us derive the behavior of the Moore machine given in Figure 8.30. The outputs $Z_2Z_1Z_0$ depend only on the present state, in other

Table 8.2

State table for example Mealy machine

Present state $Q_AQ_BQ_C$	Next state $Q_A^+Q_B^+Q_C^+$		Output Z	
	$X = 0$	$X = 1$	$X = 0$	$X = 1$
000	001	100	000	001
001	010	011	010	011
010	000	000	000	001
011	000	000	010	011
100	111	110	010	011
101	000	000	000	001
110	000	000	110	111
111	000	000	100	101

Figure 8.28

State diagram for example Mealy machine

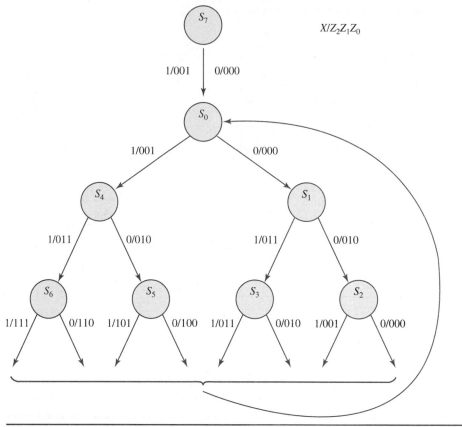

$X/Z_2Z_1Z_0$

$S_0 = 000, S_1 = 001, S_2 = 010, S_3 = 011, S_4 = 100, S_5 = 111, S_6 = 110,$ and $S_7 = 101$

Figure 8.29

Timing diagram for example Mealy machine

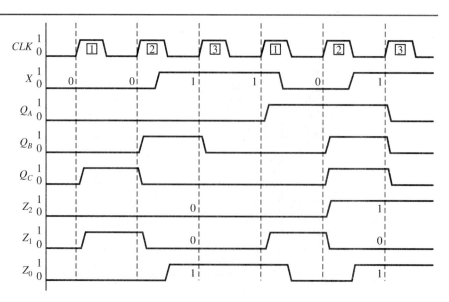

Figure 8.30

Moore machine analysis example

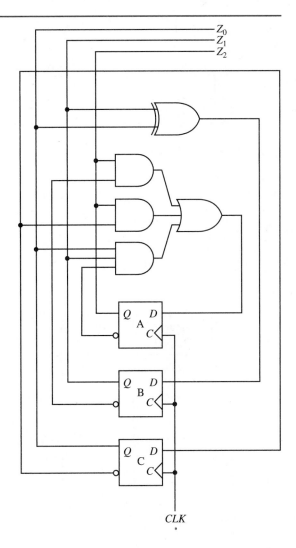

words, the flip-flop outputs, as is required of a Moore machine. Also, note that the Moore machine has no data inputs, only a clock input drives the output sequence.

The switching algebra equations for the output and flip-flop inputs are given below.

$$Z_2 Z_1 Z_0 = Q_A Q_B Q_C$$
$$D_A = (Q_A \cdot \overline{Q_B}) + (Q_A \cdot \overline{Q_C}) + (\overline{Q_A} \cdot Q_B \cdot Q_C)$$
$$D_B = \overline{Q_B} \oplus Q_C$$
$$D_C = \overline{Q_C}$$

Since the characteristic equation for a D flip-flop is

$$Q^+ = D$$

the flip-flop input equations directly yield the next state flip-flop equations.

$$Q_A^+ = D_A = (Q_A \cdot \overline{Q_B}) + (Q_A \cdot \overline{Q_C}) + (\overline{Q_A} \cdot Q_B \cdot Q_C)$$
$$Q_B^+ = D_B = \overline{Q_B} \oplus Q_C$$
$$Q_C^+ = D_C = \overline{Q_C}$$

Karnaugh maps of the next state expressions are given below.

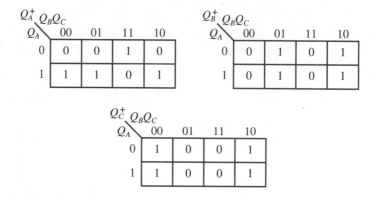

Finally, the state table and state diagram are obtained from the Karnaugh maps.

	Present state $Q_A Q_B Q_C$	Next state $Q_A^+ Q_B^+ Q_C^+$	Output $Z_2 Z_1 Z_0$
(S_0)	000	001	000
(S_1)	001	010	001
(S_2)	010	011	010
(S_3)	011	100	011
(S_4)	100	101	100
(S_5)	101	110	101
(S_6)	110	111	110
(S_7)	111	000	111

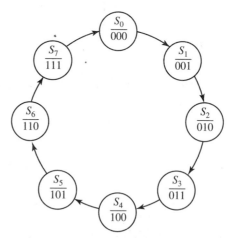

A state diagram for a Moore machine differs slightly from a state diagram for a Mealy machine. For a Moore machine, the output values are listed inside the state circles because the output depends only on the present state.

Like Mealy machine state diagrams, Moore machine state diagrams can have multiple arcs entering and leaving each state if state-to-state transitions are a function of input values;

each arc is labeled with the associated input value(s). Our example Moore machine state diagram has no arc labels because there are no data inputs; clock pulses trigger all state-to-state transitions.

The state diagram is generated directly from the state table. The states are given the symbolic labels S_0–S_7 and the present state to next state transitions are denoted by the directed arcs. Each state circle contains the associated 3-bit system output $Z_2 Z_1 Z_0$ that is also the contents of the three D flip-flops. For instance, when the machine is in state S_2, the output is $Z_2 Z_1 Z_0 = 010$.

Starting in state S_0 and clocking the sequential system several times generates the output sequence

$$Z_2 Z_1 Z_0 = 000\ 001\ 010\ \ldots$$
$$\longmapsto \text{time}$$

which reveals that the example Moore machine is a binary counter. When the count reaches 7_{10}, the next clock pulse resets the count to 0_{10} and the count sequence begins again.

8.5 TIMING ISSUES

In our earlier discussion of latches and flip-flops we saw that timing problems — oscillations and glitches — can occur in sequential systems. We will conclude this chapter by discussing in more detail timing issues required for proper operation of memory devices and sequential systems. These timing issues include flip-flop *setup*, *hold*, and *propagation* times. We will also discuss *clock skew*, *metastability*, and *maximum clocking frequency* for synchronous sequential systems.

8.5.1 Setup and Hold Times

To ensure proper flip-flop switching behavior, certain timing restrictions must be placed on when flip-flop input signals can change value; these timing restrictions are commonly called *setup* and *hold* times. Figure 8.31 shows the definition of setup (t_{su}) and hold (t_h) times for an edge-triggered flip-flop. For an edge-triggered flip-flop to work properly, the input signals that determine the next state must be stable for at least a period t_{su} before the clock transition and remain stable for a period t_h after the clock transition. The term *stable* means that the inputs must not change value. Setup and hold times need not be equal and are typically specified in associated product data books.

Setup and hold times are usually positive values, but can also be negative values. Negative timing constraints denote internal delays within a memory device or cell. Figure 8.32 shows an example of negative hold times.

Figure 8.31

Setup and hold times for an edge-triggered flip-flop

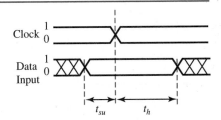

Figure 8.32(a) shows a D flip-flop memory macro or cell that includes a buffer on the data (D) input for signal refresh. The buffer is internal to the cell; the buffer and D flip-flop are packaged together as a single component. Figure 8.32(b) shows how the memory component timing constraints at the external ports are modified to account for the propagation delay t_p of the internal buffer. The top signal waveform is the reference clock signal, labeled *CLK*. The middle signal waveform is the external data signal, labeled D'. The bottom signal waveform is the internal data signal D that actually drives the D flip-flop, shifted in time by the propagation delay of the buffer.

The length of time the data input D' must remain stable is $t_{su} + t_h$, but the hold time t_h is negative. This convention implies the data/clock timing relationship shown for D' and *CLK*. Adhering to the negative timing constraints guarantees that the data D delayed by the buffer propagation delay t_p is correctly aligned with the clock to properly operate the internal D flip-flop.

Figure 8.33 shows an example of negative setup times. Figure 8.32(a) shows a D flip-flop memory macro or cell that includes a buffer on the clock (C) input for signal refresh. Again, the buffer and D flip-flop are packaged together as a single component. The top signal waveform in Figure 8.32(b) is the external clock signal, labeled *CLK*. The middle signal waveform is the internal, reference clock C that actually drives the D flip-flop, shifted in time by the propagation delay of the buffer. The bottom signal waveform is the data signal, labeled D.

The length of time the data input D must remain stable is again $t_{su} + t_h$, but the setup time t_h is negative. This convention implies the data/clock timing relationship shown for *CLK* and D. Adhering to the negative timing constraints guarantees that the data is correctly aligned with the internal clock C delayed by the buffer propagation delay t_p to properly operate the D flip-flop.

Figure 8.32

Negative hold time for an edge-triggered D flip-flop

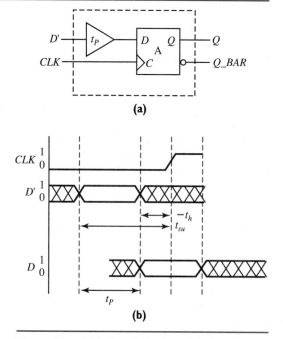

Figures (a) and (b) show the logic and timing diagrams, respectively.

Figure 8.33

Negative setup time for an edge-triggered D flip-flop

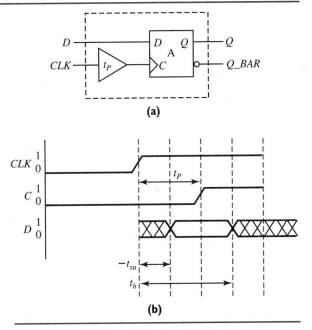

Figures (a) and (b) show the logic and timing diagrams, respectively.

8.5.2 Metastability

If setup and hold times are violated, the behavior of a memory device becomes nondeterministic, meaning that the memory device may or may not work properly. This nondeterministic behavior is called *metastability*. Metastability problems tend to occur when a synchronous sequential system is interfaced to the larger asynchronous environment. Data inputs supplied by the asynchronous environment may occasionally violate timing requirements.

Returning to our original seesaw metaphor of memory devices (shown in Figure 8.5), metastability can be viewed as a seesaw stuck in a balanced condition, as shown in Figure 8.34. In switching the seesaw states (shown at bottom) between low and high riders, the seesaw may get stuck in an interim state (shown at top) where both riders are balanced, in other words, are neither high nor low. The time the seesaw remains in the balanced interim state varies and can, in principle, be infinite. However, a little energy (jiggle) by either rider typically "kicks" the seesaw out of the balanced interim state back into one of the bottom states and the seesaw resumes normal operation.

Similar concepts apply to memory devices. Consider, for example, an edge-triggered JK flip-flop with both data inputs set equal to 1. Under normal operating conditions, the JK flip-flop should toggle state on the clock pulse. However, if the J and K inputs are not held stable long enough, the flip-flop may not toggle state. Alternatively, the flip-flop may start to toggle state, but be unable to finish the transition, "hanging" in an invalid state halfway between a logic 0 and 1. This latter condition is called a *metastable* state. A flip-flop never permanently remains in a metastable state. Eventually, new inputs or random electrical disturbances will force a flip-flop out of a metastable state, but the period of time a flip-flop remains in a metastable state is generally unpredictable.

Metastability can cause a memory device to fail to operate properly, which in turn can cause the larger sequential system to fail to operate properly. Metastability "gotchas" can

Figure 8.34

Bistable states and metastability

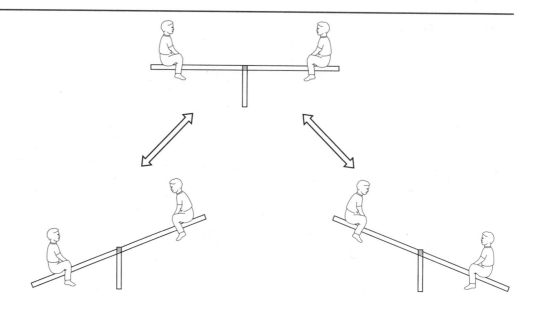

never be eliminated, but there are several common techniques for reducing the chances of their occurrence. The simplest technique is to use faster memory devices, which have shorter setup and hold times and thus provide smaller windows for unwanted signal changes that cause metastability to occur.

Another technique is to use double *synchronizer* stages, as shown below. Data from an

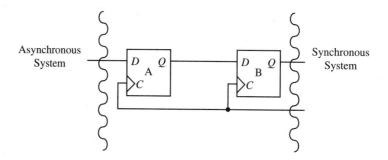

asynchronous source is clocked into the first D flip-flop (A) on the first clock cycle and then clocked into the second D flip-flop (B) on the next clock cycle. The second D flip-flop holds the data stable for the synchronous system. If asynchronous signal changes violate the setup and hold times of the first D flip-flop and, consequently, cause the flip-flop to go metastable, this flip-flop has an entire clock cycle to become stable before the metastability condition is propagated to the second flip-flop and, consequently, to the entire system. Thus, double synchronizer stages provide some reasonable insurance that the data inputs to the synchronous system adhere to setup and hold timing constraints to avoid metastability.

8.5.3 Propagation Delay

Another important timing parameter of memory devices is *propagation delay*. Propagation delay is the time between input excitation and output response, in other words, how fast a memory device can change state. As discussed in Chapter 6, propagation delay can be specified relative to rising (t_{PLH}) and falling (t_{PHL}) output transitions. Alternatively, a nominal propagation delay (t_p) can be specified, which represents the average of t_{PLH} and t_{PHL}.

Figure 8.35 shows that the flip-flop setup time, flip-flop propagation delay, and the delay of the combinational logic generating the next state signals collectively determine the speed of a synchronous sequential system or, equivalently, its maximum clocking frequency.

Assuming positive edge-triggered flip-flops, t_P represents the time required for flip-flops to change value in response to the rising edge of the clock signal. Then, t_{CL} represents the time required for the new present state values to propagate through the combinational logic to generate the new next state values (flip-flop inputs). Finally, these new flip-flop inputs must be stable for at least a period t_{su} before the next clock transition to comply with the setup timing requirement. Thus, the maximum clocking frequency (f_{max}) or, equivalently, the maximum rate at which system inputs can be applied is given by

$$f_{max} = \frac{1}{t_P + t_{CL} + t_{su}}$$

8.5.4 Clock Skew

The last synchronous sequential system timing issue we will discuss is *clock skew*. In theory, all memory devices in a synchronous sequential system change state at the same time under control of a centralized, governing clock signal. In practice, however, all memory devices in a synchronous sequential system may not change state at exactly the same time because clock signal transitions may arrive at different memory devices at different times due to varying circuit path delays. Clock circuit paths may have varying *logic* delays, due to different clock gating or buffering. Clock circuit paths may also have varying *wire* delays, due to different lengths or loads. If two memory devices receive the same clocking instruction at different times, the clock is *skewed* and the time differential is called *clock skew*.

Figure 8.36 illustrates clock skew and how it can cause incorrect memory device switching. Figure 8.36(a) shows the correct operation of two cascaded positive edge-triggered D flip-flops, assuming no clock skew. A single clock, *CLK*, activates flip-flops A and B at the same time; the new input value of *X* is loaded into flip-flop A while the old value of flip-flop A is loaded into flip-flop B. The propagation delay (t_p) of flip-flop A ensures that flip-flop B has time to load the old value of flip-flop A before flip-flop A assumes its new value.

Figure 8.35

Maximum clocking frequency of a synchronous sequential system

Clock

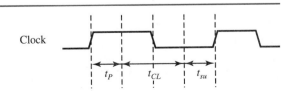

t_P t_{CL} t_{su}

Figure 8.36

Clock skew

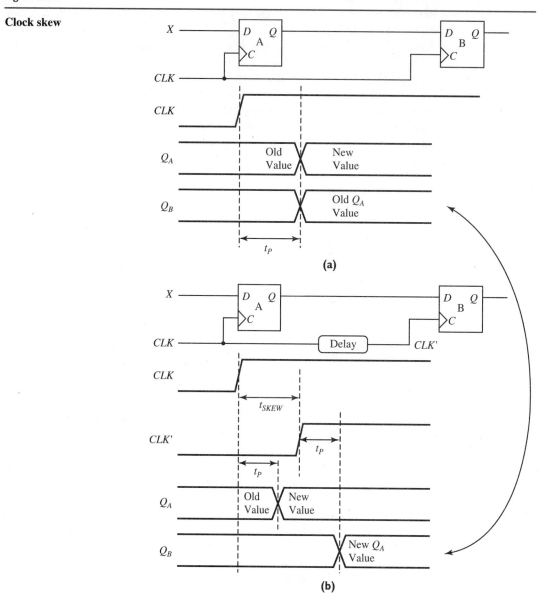

(a)

(b)

Figure 8.36(b) shows the incorrect operation of two cascaded positive edge-triggered D flip-flops, assuming clock skew. The clock signals that drive flip-flops A and B are skewed by t_{SKEW}, so the flip-flops do not switch at the same time. The rising edge of clock signal *CLK* activates flip-flop A and then t_{SKEW} units later in time activates flip-flop B. Since $t_{SKEW} > t_P$, flip-flop A has loaded the new input value of *X* by the time flip-flop B switches; thus, flip-flop B incorrectly loads the *new* value of flip-flop A instead of the *old* value of flip-flop A.

8.6 SUMMARY

In this chapter, we introduced sequential digital systems. A sequential digital system is composed of two principal elements: memory and combinational logic. Memory provides storage to remember past inputs. Combinational logic provides operators to transform inputs into outputs. Sequential digital system outputs are a function of present inputs and past inputs, generating a sequence of outputs in response to a sequence of inputs.

Sequential systems are further classified as either asynchronous or synchronous. We concentrated on synchronous sequential systems that impose a regular timing discipline to control when inputs are applied and outputs are generated. If the outputs are a function of the present state and inputs, the sequential system is called a Mealy machine. If the outputs are a function of only the contents of the memory devices (present state), the sequential system is called a Moore machine. With a Mealy machine in a given state, a set of inputs is applied and the output response is generated. The inputs and state also determine the next state, in other words, the flip-flop contents after the clock pulse. With a Moore machine in a given state, a set of inputs is applied, which determines the next state when clocked, which in turn determines the output response.

We discussed a variety of memory devices commonly used in sequential systems, including latches, pulse-triggered flip-flops, and edge-triggered flip-flops. Then, we analyzed several example sequential systems using literal and symbolic techniques to understand how combinational logic and memory components contribute to the overall sequential system's input/output behavior. Convenient notations were introduced for representing sequential behavior, including characteristic equations, state tables, state diagrams, and timing diagrams.

Correct synchronous sequential system operation requires adherence to several timing considerations. Signals controlling memory devices must be stable for a period of time before (setup) and after (hold) the devices are activated to ensure reliable state transitions. Failure to adhere to setup and hold times may cause metastability, where a memory device "hangs" for an indeterminate period of time in an illegal state that is neither a logic 0 nor 1. In addition to metastability, clock skew is another timing issue that can cause errant sequential behavior. Clock skew refers to the condition when circuit propagation delays generate multiple clock signals, delayed or skewed in time relative to each other, that cause memory devices to be activated at irregular intervals of time. Finally, the combination of memory device propagation delay, next state combinational gate propagation delay, and memory device setup times determines how fast a synchronous sequential system may operate or, in other words, the maximum clocking frequency.

In closing, sequential digital systems are amazingly powerful machines. The simple binary adder shown in Figure 8.21, composed of just six gates and one D flip-flop, can add two arbitrarily large numbers. The ability to control the timer on a microwave oven or send a spacecraft across our galaxy depends in part on the basic principles and concepts of sequential digital systems.

8.7 PROBLEMS

1. Consider the following sequential digital systems.

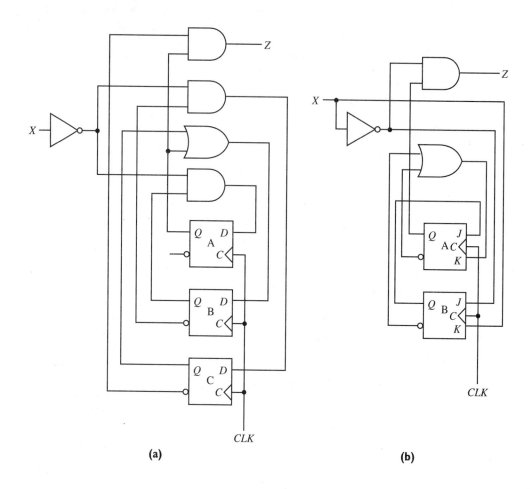

(a)   (b)

a. Classify each sequential digital system as a Mealy or a Moore machine.
b. Identify the next state logic, output logic, and memory.

2. It is generally undesirable to use simple mechanical switches as inputs to sequential digital systems because the switches suffer from a condition called *bounce*, illustrated below.

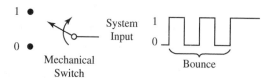

The mechanical parts and contact action cause the switch to bounce a couple of times before making contact. This bouncing action yields a series of intermediate values that can be misinterpreted by a sequential system as a series of false inputs.
a. Show how a simple SR latch can be used to *debounce* a mechanical switch.
b. Explain the operation of the switch debouncer.

3. Examine the following **nor** JK latch timing diagram.

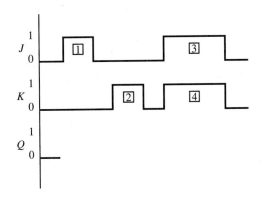

   a. Generate the timing diagram for the output Q. Use arrows to show cause-and-effect. Assume that the JK latch changes state only once during the input pulses ① and ② and twice during the input pulses ③ and ④ .
   b. What happens to a **nor** JK latch when the J and K inputs are both continually asserted, in other words, when $J = 1$ and $K = 1$?

4. Generate a truth table similar to Figure 8.6(c) for the circuit given below. Is there a stable, no change state?

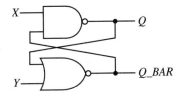

5. Data books often offer a $J\overline{K}$ flip-flop.
   a. Generate the truth table for a $J\overline{K}$ flip-flop.
   b. How might a $J\overline{K}$ flip-flop be used?

6. Any flip-flop, with some extra gates, can realize any other flip-flop. Use a D flip-flop to realize a JK flip-flop.

7. To satisfy a particular customer requirement, we have produced a new latch, called a "T_D latch" that can function as a T latch or a D latch. The new latch is shown below.

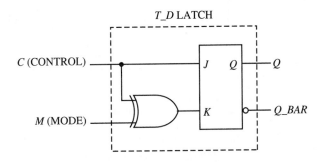

   a. Generate a truth table showing new outputs (Q^+ and Q_BAR^+) as a function of inputs and the present state.

b. Generate a Karnaugh map for Q^+.

c. Generate the characteristic equation for the T_D latch.

8. Consider the following timing diagram.

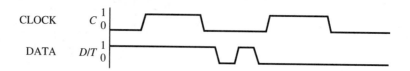

a. Assuming the D master-slave flip-flop given in Figure 8.17(c) and the initial master and slave latch contents to be 0, show the master and slave RS latch outputs in response to the clock and data inputs.

b. Assuming the T master-slave flip-flop given in Figure 8.17(b) and the initial master and slave latch contents to be 0, show the master and slave RS latch outputs in response to the clock and data inputs.

9. Examine the following positive edge-triggered D flip-flop timing diagram (see also Figure 8.20).

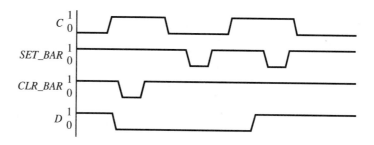

a. Generate the timing diagram showing output Q in response to the clock (C), set (SET_BAR), clear (CLR_BAR), and data (D) inputs. Assume that the initial flip-flop contents are 0.

10. Repeat Problem 9 assuming a negative edge-triggered D flip-flop.

11. Explain the operation of the following circuit.

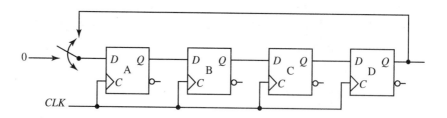

a. Assuming the initial state $Q_A = 1$, $Q_B = 0$, $Q_C = 0$, and $Q_D = 0$, generate a timing diagram showing the flip-flop contents after four clock pulses with the input switch set so that $D_A = 0$.

b. Assuming the initial state $Q_A = 1$, $Q_B = 0$, $Q_C = 0$, and $Q_D = 0$, generate a timing diagram showing the flip-flop contents after four clock pulses with the input switch set so that $D_A = Q_D$.

c. What are some possible applications of this circuit?

12. For the sequential machine in Problem 1(a),

a. Show the output response to the input sequence

$$0\ 1\ 0\ \ldots$$
$$\longmapsto \text{time}$$

using literal analysis. Assume the initial state $Q_A = 0$, $Q_B = 0$, and $Q_C = 0$.

b. Generate a timing diagram for X, Q_A, Q_B, Q_C, and Z, annotated by arrows showing cause-and-effect transition-to-transition relationships.

13. For each sequential machine in Problem 1, derive the input/output behavior using symbolic analysis.

a. Show system output and flip-flop input Boolean expressions, flip-flop next state equations, state table, and state diagram.

b. Show the output response to the input sequence

$$0\ 1\ 0\ \ldots$$
$$\longmapsto \text{time}$$

using the state diagram. Assume the initial state $Q_A = 0$, $Q_B = 0$, and $Q_C = 0$.

c. Compare the output responses. What is the relationship between the two sequential systems?

d. Make observations concerning relative complexities of sequential systems using Moore versus Mealy machines and D versus JK flip-flops.

14. Consider the following sequential digital system.

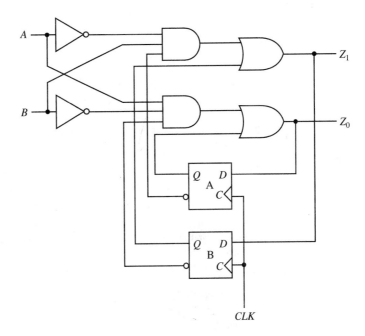

a. Is the sequential digital system a Mealy or a Moore machine?

b. Identify the next state logic, output logic, and memory. Comment on any unique aspects.

c. Derive the input/output behavior using symbolic analysis. Show the output and flip-flop input Boolean equations. Show the next state Karnaugh maps for each flip-flop. Generate a state table and a state diagram.

d. Comment on any unique aspects of the state diagram. Is this a good design?

e. What useful function does this sequential machine perform? (*Hint*: The digital system performs a popular logical operation on two bit vectors. Corresponding bits of the two operands are applied to the inputs A and B, starting with the *most significant* bits.)

15. The following digital system detects (active-1) a binary coded decimal (BCD) digit.

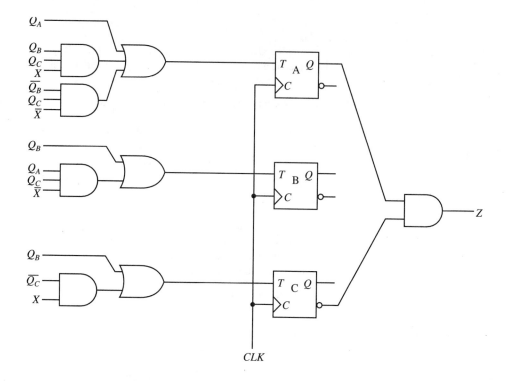

a. Derive the input/output behavior using symbolic analysis. Show the output and flip-flop input Boolean equations. Show the next state Karnaugh maps for each flip-flop. Generate a state table and a state diagram.

b. Using the state diagram, show which BCD digit is detected by this sequential machine. Assume the initial state $Q_A = 0$, $Q_B = 0$, and $Q_C = 0$.

REFERENCES AND READINGS

[1] Brookshear, G. *Theory of Computation: Formal Languages, Automata, and Complexity.* Redwood City, CA: Benjamin/Cummings, 1989.

[2] Carroll, J. and D. Long. *Theory of Finite Automata with an Introduction to Formal Languages.* Englewood Cliffs, NJ: Prentice-Hall, 1989.

[3] Chen, C. *Computer Engineering Handbook.* New York: McGraw-Hill, 1992.

[4] Davio, M., J. Deschamps, and A. Thayse. *Digital Systems with Algorithm Implementation.* New York: Wiley, 1983.

[5] Fabricius, E. *Modern Digital Design and Switching Theory.* Ann Arbor, MI: CRC Press, 1992.

[6] Garrod, S. and R. Borns. *Digital Logic: Analysis, Application, & Design.* Philadelphia, PA: Saunders, 1991.

[7] Lind, L. and J. Nelson. *Analysis and Design of Sequential Digital Systems.* New York: Wiley, 1977.

[8] Lkine, R. *Structured Digital Design Including MSI/LSI Components and Microprocessors.* Englewood Cliffs, NJ: Prentice-Hall, 1983.

[9] McCluskey, E. *Logic Design Principles with Emphasis on Testable Semicustom Circuits.* Englewood Cliffs, NJ: Prentice-Hall, 1986.

[10] Mealy, G. "A Method for Synthesizing Sequential Circuits," *Bell System Technical Journal,* vol. 34, 1045–1061, 1955.

[11] White, R. *How Computers Work.* Emeryville, CA: Ziff-Davis Press, 1993.

9
SEQUENTIAL DESIGN: SYNTHESIS

In Chapter 8, we discussed the basic operating concepts and principles of synchronous sequential digital systems. Such systems are constructed with combinational logic and memory; memory stores system state and combinational logic updates system state and generates outputs. Working together, combinational logic and memory transform a sequence of inputs into a desired sequence of outputs. We also learned various representations for sequential systems behavior, such as state tables and state diagrams, and how to use these representations to derive the input/output transformation of a sequential system.

With our understanding of sequential system *analysis*, we can now focus on the more challenging task of sequential system *design*. Instead of starting with a sequential system and generating its behavior, we will start with a desired behavior and generate a sequential system. Similar to combinational system design, sequential system design will be described in two stages: *synthesis* and *implementation*. Figure 9.1 shows that synthesis generates a mathematical description realizing the desired function, whereas implementation generates an actual, physical system realizing the mathematical description. The synthesis portion of sequential design is addressed in this chapter; the implementation portion of sequential design is addressed in Chapter 10.

Consistent with Chapter 8, we will investigate only *synchronous* sequential design; *asynchronous* sequential design will not be addressed. Thus, references to "sequential" will imply "synchronous sequential," unless otherwise noted.

Figure 9.1

Digital sequential system design

English Prose

Refining Functional Specification

State Diagram

⎫
⎬ **Synthesis**
⎭

Minimized Encoded State Table

Realizing Functional Specification

Software/ Hardware System

⎫
⎬ **Implementation**
⎭

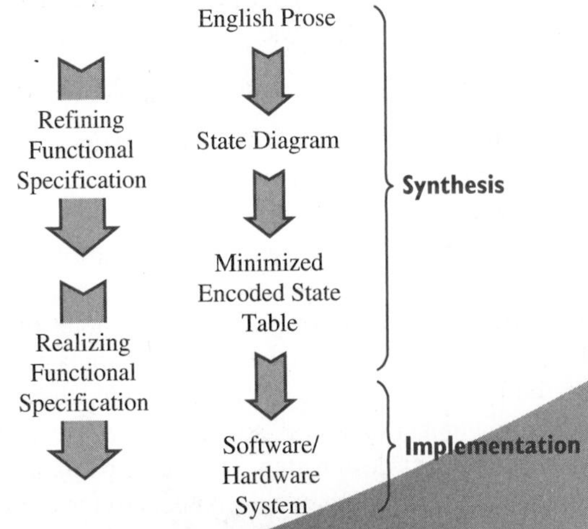

The general sequential digital system design process shown in Figure 9.1 is described in more detail below.

Synchronous Sequential System Design
1. *Generate a state diagram from the problem statement.*
2. *Minimize the number of states — state reduction.*
3. *Select a binary encoding for each state — state assignment.*
4. *Generate an encoded state table.*
5. *Generate next state Karnaugh maps for each memory device.*
6. *Select the type of memory device.*
7. *Generate a Karnaugh map for each memory device input using the associated next state Karnaugh map and appropriate excitation truth table. Also, generate a Karnaugh map for each output.*
8. *Generate minimized output and next state (memory device input) logic expressions.*
9. *Implement memory and combinational logic.*
10. *Document final sequential design.*

We are already familiar with many of the design steps because the design process is roughly the inverse of the analysis process. Also, sequential design includes elements of combinational design. Thus, we can exploit our understanding of sequential analysis and combinational design in learning sequential system design.

Before we investigate the details of each design step, the following section will illustrate the general synchronous sequential system design process with a simple example. The design example shows one of the most common applications of synchronous sequential systems: *counters.*

9.1 SIMPLE DESIGN EXAMPLE

A 3-bit counter that cyclically counts in the binary sequence 0_{10} to 7_{10}

$$000_2, 001_2, 010_2, \ldots 110_2, 111_2, 000_2, 001_2, 010_2, \ldots$$
$$\longmapsto \text{time}$$

was analyzed in Figure 8.30. Let us design a variation that counts in only even numbers. The desired behavior might be stated as follows:

Design a counter that counts in the cyclic pattern $0_{10}, 2_{10}, 4_{10}, 6_{10}, 0_{10}, 2_{10}, \ldots$.

Step 1 of the design process translates the English description of the desired behavior into more precise mathematical terms by generating a state diagram. The state diagram for the even counter is shown in Figure 9.2(a).

Choosing a Moore machine, a separate state is designated for each count in the desired counting sequence. Using the abstract state labeling convention introduced in Chapter 8, S_0 denotes the state in which the counter outputs 0_{10}, S_1 denotes the state in which the counter outputs 2_{10}, S_2 denotes the state in which the counter outputs 4_{10}, and S_3 denotes the state in which the counter outputs 6_{10}. For this simple example, there are no extra or redundant states (4 is the minimum number of states), so we can skip the state reduction step.

State assignment is the next design step. The objective is to select a set of memory devices and to choose a unique encoding of the contents of the memory devices for each state. Assuming bistable memory devices, only two memory devices are needed to represent four

Figure 9.2

Design of an even counter

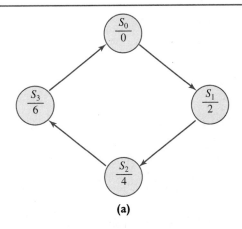

(a)

Present State $Q_A Q_B$	Next State $Q_A^+ Q_B^+$	Output $Z_2 Z_1 Z_0$
0 0	0 1	0 0 0
0 1	1 0	0 1 0
1 0	1 1	1 0 0
1 1	0 0	1 1 0

(b)

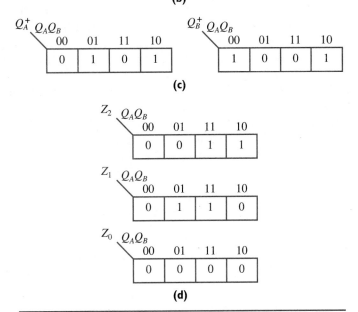

(c)

(d)

Figure (a) shows the state diagram. Figure (b) shows the encoded state table. Figure (c) shows the next state/flip-flop input Karnaugh maps. Figure (d) shows the output Karnaugh maps.

Table 9.1

State assignment for even counter	State	Binary encoding $(Q_A Q_B)$
	S_0	00
	S_1	01
	S_2	10
	S_3	11

states. The memory devices will be labeled A and B, and their contents (outputs) will be labeled Q_A and Q_B, respectively. A state assignment is given in Table 9.1. (For now, accept this suggested state assignment. We will discuss in more detail how to make state assignments in Section 9.4.)

Step 4 of the design process generates a *state table*, also called an *encoded state table*. The encoded state table for our even counter design is shown in Figure 9.2(b) and is derived from the state diagram and the state encodings. The "Present State" and "Next State" columns describe the state-to-state transitions. The last column describes the output per state. For example, the first row specifies that when the counter is in state 00 (S_0), the output $Z = 000$ (0_{10}) and the counter will transition to state 01 (S_1) on the next clock pulse.

Proceeding to Step 5, the next state Karnaugh maps for memory devices A (Q_A^+) and B (Q_B^+) are shown in Figure 9.2(c). Next state Karnaugh maps provide another way to describe state-to-state transitions; the present state forms the cell indices and the associated next state forms the cell contents. For example, the "Present State" to "Next State" transition given in the first row of the encoded state table is described by the leftmost cells of the next state Karnaugh maps. When memory devices A and B contain the present state $Q_A = 0$ and $Q_B = 0$, the next clock pulse will change the flip-flop contents to the entries given in the "00" cells, namely, $Q_A^+ = 0$ and $Q_B^+ = 1$.

Next, Step 6 selects the type of memory device; we will use positive edge-triggered D flip-flops for this example design. Then, for Step 7, we generate a Karnaugh map for each output and flip-flop input. The Karnaugh maps generated in Step 5 describing flip-flop *output* behavior can be converted to Karnaugh maps describing flip-flop *input* behavior by using the following D flip-flop *excitation table*.

QQ^+	D	Action
00	0	No change
01	1	Set
10	0	Reset
11	1	No change

Appropriately named, excitation tables describe the flip-flop input values (middle column) required to achieve desired flip-flop output transitions (left column). The D flip-flop excitation table shows that the next output value Q^+ is set by the input value D regardless of the present output value Q; in other words, $D = Q^+$. Thus, the next state Karnaugh maps for Q_A^+ and Q_B^+ in Figure 9.2(c) are also the flip-flop input Karnaugh maps for D_A and D_B, respectively. The three output Karnaugh maps for $Z = Z_2 Z_1 Z_0$ are shown in Figure 9.2(d). These output Karnaugh maps are derived directly from the first ("Present State") and third ("Output") columns of the encoded state table in Figure 9.2(b).

Figure 9.3

Logic specification of a
Moore machine design of
an even counter

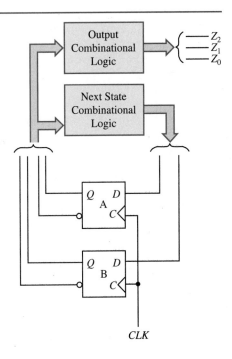

Step 8 uses the two flip-flop input Karnaugh maps and the three system output Karnaugh maps to generate minimized logic expressions for the next state and output combinational logic, respectively. Using the combinational design process explained in Chapter 4, the minimized forms of the even counter next state and output logic are given below.

$$D_A = Q_A \oplus Q_B$$
$$D_B = \overline{Q_B}$$
$$Z_2 = Q_A$$
$$Z_1 = Q_B$$
$$Z_0 = 0$$

At this point, the initial English specification has been successfully translated and refined into the logic specification for the Moore machine shown in Figure 9.3. The Boolean expressions for D_A and D_B form the input/output specification for the next state combinational logic and the Boolean expressions for the output $Z = Z_2 Z_1 Z_0$ form the input/output specification for the output combinational logic.

The last steps of the sequential design process implement the logic for D_A, D_B, and Z and document the final design. A schematic of the final even counter design is shown in Figure 9.4 (p. 264); the combinational logic is implemented using a two-level **nand-nand** network.

We encourage you to "reverse-engineer" the even counter design to ensure that the design actually realizes the desired function. That is, analyze the counter design and generate a timing diagram to verify that the output cyclically counts the sequence

$$000_2, 010_2, 100_2, 110_2, 000_2, 010_2, \ldots$$
$$\mapsto \text{time}$$

Figure 9.4

Even counter design

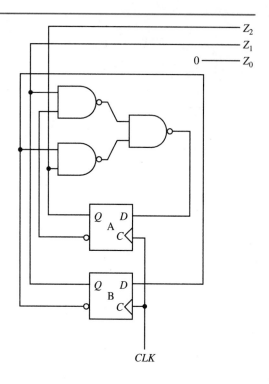

9.2 GENERATING STATE DIAGRAMS FROM PROSE

Having presented an overview of the synchronous sequential system design process, we will next examine the individual design steps in more detail. The first step in the sequential system design process transforms an English functional description into a state diagram.

Typically, the principal objective in generating a state diagram is to minimize the number of states, which tends to reduce hardware complexity. However, it is important to note that the caveats placed on the utility of logic minimization in Chapter 4 also apply to the utility of sequential minimization: Minimizing the number of states is not always necessary or desirable. Design time and/or labor budgets may not warrant spending the time to carefully generate a minimal state diagram just to save a few flip-flops and a few gates. Alternatively, nonminimal state diagrams are sometimes intentionally constructed to take advantage of standard parts, such as counters or programmable logic devices. Although it is important to understand how to generate a state diagram that minimizes the number of states, it is equally important to recognize that such techniques may or may not be required for a particular design.

There is no algorithm to systematically transform an English functional description into a state diagram. Generating a good state diagram may depend on opportunistically recognizing and exploiting unique aspects of a specification, such as regularity or commonality in various conditions or actions. Generating a good state diagram is further complicated when the English description is incomplete or imprecise so that the designer must either ask for clarification or make educated judgments on interpreting the specifications. In lieu of a formal procedure for generating state diagrams, we will use the following helpful set of guidelines.

Problem-Solving Skill
Practical limitations on schedules and budgets may not warrant investing resources to develop an optimal solution.

To illustrate these guidelines and gain experience in generating state diagrams from English specifications, several example exercises are presented in the following subsections.

9.2.1 Resetting Sequence Detector

Let us begin by reviewing a sequential machine studied in Chapter 8 that detects a sequence of three consecutive 1's. An English specification may be given as follows:

> When three consecutive 1's (7_{10}) have been received on the input, the output should be asserted active-1.

To help us understand the input/output specification, we will generate sample input/output sequences; the input is labeled X and the output is labeled Z. The following two sets of possible input/output sequences reveal an ambiguity in the specification.

$$X = 0101111110 \ldots$$
$$Z = 0000010010 \ldots$$

$$X = 0101111110 \ldots$$
$$Z = 0000011110 \ldots$$
$$\longmapsto \text{time}$$

When three consecutive 1's have been received, should the sequential machine start counting again from zero as illustrated in the first sample sequence, or should previously received 1's be reused in a new count as illustrated in the second sample sequence? The first case is an example of a *resetting* sequential behavior, and the second case is an example of a *nonresetting* sequential behavior. To proceed with our sequence detector exercise, we will assume the behavior given in the first set of test sequences: The machine will start counting again from zero when three consecutive 1's have been received.

The next step is to decide whether the design will be a Moore or a Mealy machine. Unfortunately, there is no simple way to make this decision; any sequential behavior can generally be realized as a Moore or a Mealy machine. A state diagram for a Mealy machine will sometimes yield fewer states than a comparable state diagram for a Moore machine. However, some sequential behaviors, such as counters, are typically implemented as Moore machines because the desired output sequences are closely related to well-defined state transition sequences. For our sequence detector, we will generate state diagrams for both Mealy and Moore machines and compare the results.

Now we can start drawing circles (states) and arcs (transitions). In specifying states, we try to give meaningful definitions by associating each state with a set of conditions that relate to the problem statement. Sometimes it helps to associate states with goals: we specify a series of goals that start at an *initial state* and build toward identifying desired input sequences or patterns. Then, we connect the goals to form a sequence of state transitions that describe successful paths, in other words, state diagram paths that result in asserting the out-

Figure 9.5

Mealy machine design of a sequence detector

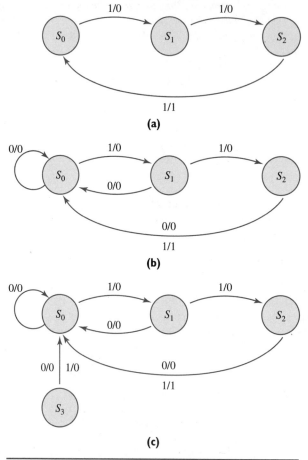

Figure (a) shows the successful path. Figures (b) and (c) complete the state diagram by showing the remaining transitions and unused state, respectively.

puts. For instance, Figure 9.5(a) shows the start of a Mealy state diagram for our sequence detector, identifying three states and a successful path. The meanings of the states are given in Table 9.2.

Starting in the initial state S_0, the three arcs define a successful path of detecting three consecutive 1's and then resetting or returning to the initial state. The remaining arcs for states S_0–S_2 (shown in Figure 9.5(b)) describe what happens when the design fails to traverse the successful path.

Table 9.2

State definitions

State	Definitions
S_0	Initial state.
S_1	One 1 has been received.
S_2	Two consecutive 1's have been received.

To determine whether the state diagram in Figure 9.5(b) is finished, we check to see if all arcs and states have been defined. A completely defined state diagram requires that each state have 2^N outputs for a system with N inputs. Since our sequence detector has only one input X, each state must have $2^1 = 2$ outputs: an output transition (arc) when $X = 0$ and an output transition (arc) when $X = 1$. Checking Figure 9.5(b), we see that all required arcs are defined; the single arc leaving state S_2 actually denotes two transitions. (Since state S_2 transitions to state S_0 when the input is 0 or 1, these two state-to-state transitions are represented by a single arc.)

A completely defined state diagram also requires the number of states to be a power of 2 (in other words, $2^1 = 2$, $2^2 = 4$, $2^3 = 8$, ...) because state information is realized as binary codes stored in bistable memory devices. Recall from Chapter 2 that a string of N bits has a total of 2^N possible unique values or bit combinations. Since a minimum of two memory devices are required to store three unique binary codes or states, our sequence detector will actually have four states. Three states will describe the desired sequential behavior and the remaining fourth state will be an *extra*, *unused*, or *don't care* state. We can ignore the extra state and, in this case, the state diagram in Figure 9.5(b) is finished.

Alternatively, we can include the extra state in the state diagram, as shown in Figure 9.5(c). Under normal operating conditions, the sequence detector will circulate between states S_0–S_2 and the extra fourth state S_3 will never be used. However, if the sequence detector inadvertently enters the extra state, the detector will reset (state S_0) and return to its normal operation on the next clock. Addressing unused states improves the robustness or integrity of a design by accounting for the unexpected.

As a final aspect of our resetting sequence detector exercise, Figure 9.6 shows an alternate state diagram realizing a Moore machine. Although the Moore machine state diagram uses one more state than the Mealy machine state diagram in Figure 9.5(b), the difference does not, in this case, affect hardware complexity; the Moore and the Mealy machines will both use two memory devices. The Moore state diagram also illustrates that the terms *resetting/nonresetting* can be somewhat misleading in that they describe sequential input/output behavior, but *not necessarily* associated sequential state transition behavior. When a sequence of three 1's has been received, the Mealy state machine always "resets" or returns to the initial state S_0, but the Moore machine need not return to the initial state S_0. Yet, both the Mealy and the Moore state machines describe the desired resetting or restarting sequential behavior. Perhaps restarting/nonrestarting would be better terms, but we will continue to use the conventionally accepted resetting/nonresetting.

Problem-Solving Skill

Be aware of the potential differences between the function specified for design and the function provided by implementation. The former reflects what is intended to be built; the latter reflects what actually is built.

Figure 9.6

Moore machine design of a sequence detector

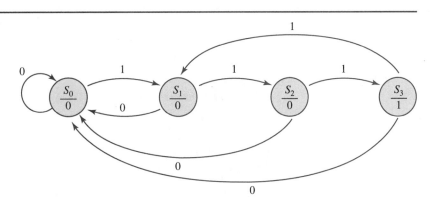

9.2.2 Nonresetting Sequence Detector

Once an input sequence has been detected, a resetting sequential machine "wipes the slate clean" and starts anew looking for the next input sequence. In contrast, a nonresetting sequential machine does not "wipe the slate clean," but instead may reuse portions of a previous input sequence in looking for the next input sequence. It is instructive to study an example of generating a state diagram for a nonresetting sequential machine because such diagrams illustrate the importance of *reusing states* and can be a little trickier to generate than state diagrams for resetting sequential machines.

Consider the following nonresetting sequential machine specification.

Design a digital system that detects either the input sequence 1100_2 or 1000_2. A sample input stimulus (X)/output response (Z) sequence is given below.

$$X = X_0 X_1 X_2 X_3 X_4 X_5 \ldots$$
$$X = 1 \; \; 1 \; 0 \; 0 \; \; 0 \; 1 \; 0 \; 0 \; 0 \; 0 \ldots$$

$$Z = Z_0 Z_1 Z_2 Z_3 \; Z_4 Z_5 \ldots$$
$$Z = 0 \; \; 0 \; 0 \; 1 \; \; 1 \; 0 \; 0 \; 0 \; 1 \; 1 \; 0 \ldots$$
$$\longmapsto \text{time}$$

Notice that input sequences can overlap. The input samples X_0–X_3 form the sequence 1100_2, whereas the input samples X_1–X_4 form the sequence 1000_2. Thus the input samples

$$
\begin{array}{ccccc}
X_0 & X_1 & X_2 & X_3 & X_4 \ldots \\
1 & 1 & 0 & 0 & 0 \ldots
\end{array}
$$

X_1–X_3 are "reused," which will be reflected in a reuse of states in the associated state diagram.

We will again choose a Mealy machine and begin constructing the state diagram by identifying the successful paths, in other words, the state transitions that result in asserting the output. The two successful paths for a Mealy machine state diagram that identify the desired sequences 1100_2 and 1000_2 are shown in Figure 9.7(a). The path $S_0 \Rightarrow S_1 \Rightarrow S_2 \Rightarrow S_3$ detects the input sequence 1100_2 and the path $S_0 \Rightarrow S_1 \Rightarrow S_4 \Rightarrow S_5$ detects the input sequence 1000_2. The meanings of the states in Figure 9.7 are given in Table 9.3.

The remaining state transitions are shown in Figure 9.7(b). The transition leaving S_3 goes to S_5 because this transition identifies the first three bits of the sequence $1\underline{000}$, as well as the end of the sequence $1\underline{100}$. The transition leaving state S_5 returns to the initial state S_0 to start looking for a new sequence because there are no subsequences of interest within the sequence 1000.

Since our Mealy machine has one input, X, each state must have $2^1 = 2$ output arcs.

Table 9.3

State definitions	State	Definitions
	S_0	Initial state.
	S_1	The pattern "1" has been detected.
	S_2	The pattern "11" has been detected.
	S_3	The pattern "110" has been detected.
	S_4	The pattern "10" has been detected.
	S_5	The pattern "100" has been detected.

Figure 9.7

Mealy machine design of a nonresetting sequence detector

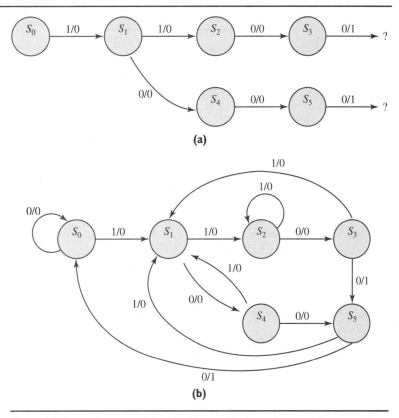

(a)

(b)

Figure (a) shows the successful paths. Figure (b) shows the remaining transitions.

Checking Figure 9.7(b) shows that all required arcs are defined. Since a minimum of three memory devices are required to store six unique binary codes or states, our sequence detector has two extra or unused states. In contrast to the previous resetting sequential state diagram example, we will elect not to address unused states at this point, so our state diagram is finished. Later sections in this chapter will show how unused states are addressed in the sequential system design process.

9.2.3 DRAM Controller Example

As our final state diagram exercise, we will use a *DRAM controller* to show an example of an *incompletely specified* state diagram, which has some state transitions and/or output values unspecified. As the name implies, a *DRAM controller* controls the interactions between a processor and dynamic random access memory (DRAM). To understand the role of a DRAM controller, we will digress briefly to discuss DRAMs.

Recall from Chapter 5 that *memory* stores digital information. The term *random access* means that any storage location can be directly accessed in a read or write operation. The term *dynamic* means that the data is stored only for a short period of time and must be periodically refreshed to prevent losing the stored data. A DRAM is typically organized as a grid of storage locations; a storage location can be a single bit (denoted by x1), 4 bits (denoted by x4), a byte (denoted by x8), and so on. A storage location in the grid is accessed by addressing a particular row and column; the intersection of the row and column identifies the storage location. The contents of the accessed storage location are either updated by a processor in a write operation or transferred to a processor in a read operation. To conserve integrated circuit package and printed circuit board space, DRAMs often multiplex the row and column addresses over a single set of input terminals or pins. With memory address multiplexing, a single DRAM location is accessed in two steps: First the row is accessed and then the column is accessed.

To mask the details of storage refreshing and memory address multiplexing from a processor, a DRAM controller is placed between a processor and dynamic random access memory. For example, Figure 9.8 illustrates how a DRAM controller intercedes on behalf of a processor to translate a processor's memory access request into the required sequence of DRAM-specific memory address multiplexing actions.

It is helpful to view processor memory as a book. Part of the processor memory address, processed via the decoder, selects a particular page in the book, which is typically a DRAM integrated circuit. Part of the processor memory address, processed via the multiplexer and DRAM controller, selects a particular location on a given page, which is a storage location within a DRAM integrated circuit. For simplicity, only one "page" or DRAM integrated circuit is shown in Figure 9.8.

The processor initiates a generic memory access request by simply specifying the location address and whether the operation is a read or write, *W_BAR*. Next, the decoder selects a DRAM controller by asserting the appropriate chip select (*CS_BAR*) signal. The selected DRAM controller then processes the memory request. More specifically, the function of the DRAM controller can be stated as follows:

When the chip select signal *CS_BAR* is asserted active-0, set up the row address by setting the multiplexer control input, *MUX* = 0, and then load the row address into DRAM by asserting the *RAS_BAR* (Row-Address-Strobe) signal active-0. Next, set up the column address by setting the multiplexer control input, *MUX* = 1, and then load the column address into the DRAM by asserting the *CAS_BAR* (Column-Address-Strobe)

Figure 9.8

A DRAM controller

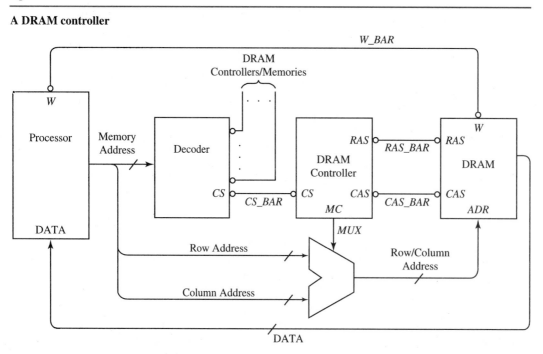

Figure 9.9

Moore machine design of a DRAM controller

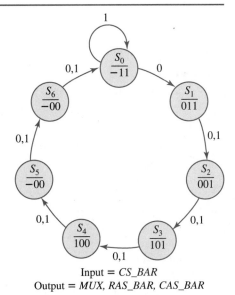

Input = CS_BAR
Output = MUX, RAS_BAR, CAS_BAR

signal active-0. Wait for two clock cycles to satisfy the minimum memory access time, deassert RAS_BAR and CAS_BAR, and return to the initial state to service the next DRAM access request.

Figure 9.9 shows the state diagram for the memory address multiplexing function of a DRAM controller, implemented as a Moore machine. States are defined in Table 9.4.

The DRAM controller starts out in the initial or reset state S_0 and waits for the input chip select line (CAS_BAR) to be asserted active-0. States S_1 and S_2 set up and load the row address. States S_3 and S_4 set up and load the column address. States S_5 and S_6 serve as *wait* states to ensure that another memory access is not serviced before the DRAM has completed its present access operation.

Notice that the value of the MUX output controlling the multiplexer is listed as a don't care (–) in states S_0, S_5, and S_6. The MUX output is not important in state S_0 because a DRAM access request has not yet been received from the processor. The MUX output is not important in states S_5 and S_6 because the row and column addresses have already been loaded into the DRAM. Since a don't care condition can be either a 0 or a 1 (see Chapter 4), incompletely specified state diagrams provide more flexibility in realizing a functional specification.

Table 9.4

State definitions

State	Definitions
S_0	Initial state.
S_1	Set up row address.
S_2	Load row address.
S_3	Set up column address.
S_4	Load column address.
S_5	Wait a clock cycle.
S_6	Wait a clock cycle.

9.3 STATE REDUCTION

The next step in the sequential system design process, *state reduction*, ensures that the state diagram has the minimum number of states by checking that there are no *equivalent* states, also called *redundant* or *duplicate* states. Equivalent states are defined as follows:

Term	Definition
State Equivalence	Two states S_x and S_y are equivalent if and only if they have the same or equivalent next states and generate the same outputs for every possible input.

If states S_x and S_y are equivalent, S_x can be used in place of S_y and S_y can be eliminated, thereby simplifying the state diagram by one state.

For simple sequential designs, state reduction is often unnecessary because the number of required states is obvious from the English functional description. However, for complex state diagrams, it is difficult to determine by simple inspection whether more states have been used than are really needed. In such cases, more formal state reduction techniques are required. We will use a two-step state reduction strategy: First we will try to eliminate equivalent states by state table *row comparison*, and then we will eliminate any remaining equivalent states by an *implication table*.

Row comparison is a simpler technique than an implication table, so we use row comparison as a coarse filter to quickly eliminate equivalent states. However, row comparison does not always yield a minimal state diagram, so we also use the more general technique of implication tables to complete state reduction.

9.3.1 Row Comparison

According to the definition of state equivalence, two present states are equivalent if their associated state table rows are identical. Consequently, row comparison eliminates redundant states by checking for state table rows having the same next state and output entries.

To illustrate row comparison state reduction, consider a simple state machine that might be used for computer communications. Figure 9.10 illustrates a popular computer communi-

Figure 9.10

Token ring local area network computer communications scheme

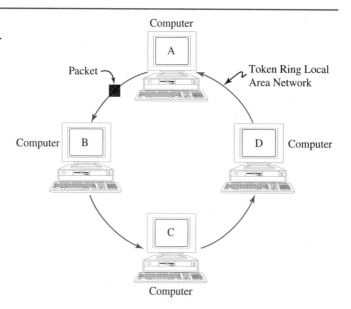

cation or networking scheme called a *token ring local area network* (LAN), which connects computers together in a ring configuration.

Computers communicate with each other by exchanging messages called *packets*. A packet is a series of 1's and 0's encoding the source address, destination address, and message content. As a packet travels around the ring from its source to its destination, each computer checks the packet's destination address. If the packet's destination address matches the computer's address, the computer receives the packet and sends a reply message back to the sender. If the packet's destination address does not match the computer's address, the computer passes the packet along to the next computer on the ring. For this communication scheme to work, each computer must have a device, called a *LAN adaptor*, that connects the computer to the token ring and handles the communication protocol functions, such as packet recognition and address decoding.

Figure 9.11 shows a Moore machine state diagram for a simple LAN adaptor. Assume that the start of a packet is denoted by two successive 0's. Also, assume that the next three bits following the packet start bits represent the packet destination address. States S_0–S_2 perform the packet recognition task, recognizing the start of a packet by detecting two successive 0's. States S_3–S_5 and S_6–S_8 perform the destination address decoding task. The LAN adaptor is "hardwired" to recognize either the address 100_2 or 010_2. The output signal *PACKET* is asserted if a packet has arrived and is addressed for either 100_2 or 010_2. The associated state table is shown in Table 9.5 (p. 274).

Although the identification of the successful paths and the construction of the LAN adaptor state diagram seem fairly straightforward, there are, unfortunately, redundant states. Comparing the rows for the present states S_5 and S_8 shows that these states have the same next state and output entries. Thus, S_5 and S_8 are equivalent; in other words, $S_5 \equiv S_8$. The reduced state table obtained by replacing S_8 with S_5 and deleting the present state row for S_8 is shown in Table 9.6.

Row reduction is not finished because Table 9.6 shows that the rows for the present states S_4 and S_7 are identical. Hence, S_4 and S_7 are equivalent; in other words, $S_4 \equiv S_7$. Table 9.6

Figure 9.11

State diagram of a LAN adaptor

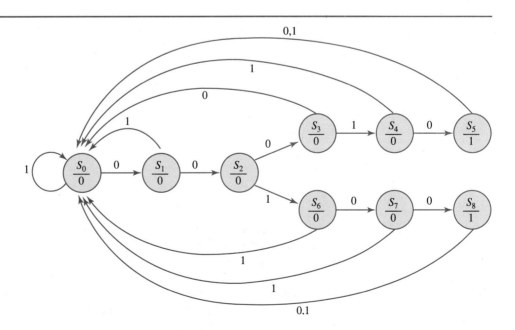

Table 9.5

LAN adaptor state table

Present state	Next state		Output (PACKET)
	X = 0	X = 1	
S_0	S_1	S_0	0
S_1	S_2	S_0	0
S_2	S_3	S_6	0
S_3	S_0	S_4	0
S_4	S_5	S_0	0
S_5	S_0	S_0	1
S_6	S_7	S_0	0
S_7	S_8	S_0	0
S_8	S_0	S_0	1

Table 9.6

Reduced LAN adaptor state table — pass #1

Present state	Next state		Output (PACKET)
	X = 0	X = 1	
S_0	S_1	S_0	0
S_1	S_2	S_0	0
S_2	S_3	S_6	0
S_3	S_0	S_4	0
S_4	S_5	S_0	0
S_5	S_0	S_0	1
S_6	S_7	S_0	0
S_7	S_5	S_0	0

Table 9.7

Reduced LAN adaptor state table — pass #2

Present state	Next state		Output (PACKET)
	X = 0	X = 1	
S_0	S_1	S_0	0
S_1	S_2	S_0	0
S_2	S_3	S_6	0
S_3	S_0	S_4	0
S_4	S_5	S_0	0
S_5	S_0	S_0	1
S_6	S_4	S_0	0

can be further simplified by replacing S_7 with S_4 and deleting the present state row for S_7. The resulting state table is shown in Table 9.7.

Now row comparison is completed because there are no more identical rows. For this particular LAN adaptor exercise, row comparison identifies all equivalent states; thus, state reduction is also completed. Row comparison, however, does not always identify all equivalent states, as will be shown in the following subsection.

9.3.2 Implication Tables

Although two present states are equivalent if their associated state table rows are identical, two present states are not necessarily inequivalent if their associated state table rows are different. This "loophole" is illustrated in the simple state diagram in Figure 9.12, showing why the row comparison state reduction technique is not always sufficient to guarantee a minimum number of states. There are no identical rows in the state table in Table 9.8, yet the state diagram contains redundant states: $S_0 \equiv S_1$ and $S_2 \equiv S_3$. Row comparison does not work for this state table because state equivalencies are interdependent: $S_0 \equiv S_1$ if $S_2 \equiv S_3$, and $S_2 \equiv S_3$ if $S_0 \equiv S_1$. These interdependencies are called *implications*, and an *implication table* is used to systematically check all implications to identify equivalent states.

To illustrate how to construct and use an implication table, we will use the Moore machine state diagram shown in Figure 9.13 and the associated state table shown in Table 9.9 (p. 276).

Figure 9.12

Nonminimal Mealy state diagram

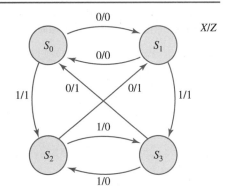

Table 9.8

Nonminimal Mealy machine state table

Present state	Next state		Output (Z)	
	X = 0	X = 1	X = 0	X = 1
S_0	S_1	S_2	0	1
S_1	S_0	S_3	0	1
S_2	S_1	S_3	1	0
S_3	S_0	S_2	1	0

Figure 9.13

Implication table example—Moore state diagram

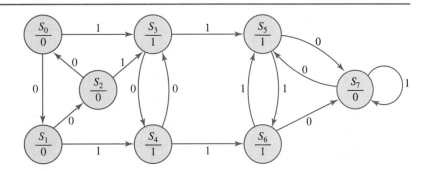

Table 9.9

Implication table example—Moore state table

Present state	Next state		Output (Z)
	X = 0	X = 1	
S_0	S_1	S_3	0
S_1	S_2	S_4	0
S_2	S_0	S_3	0
S_3	S_4	S_5	1
S_4	S_3	S_6	1
S_5	S_7	S_6	1
S_6	S_7	S_5	1
S_7	S_5	S_7	0

We construct an implication table by systematically checking all possible pairs of states for state equivalences. We start by checking S_0 with all other states: $S_0 \equiv S_1$, $S_0 \equiv S_2$, and so on. The results form the first column of the implication table shown below.

S_1	$1 \equiv 2$ $3 \equiv 4$
S_2	$0 \equiv 1$
S_3	X
S_4	X
S_5	X
S_6	X
S_7	$1 \equiv 5$ $3 \equiv 7$
	S_0

Comparing the first two rows, we see that S_0 and S_1 will have the same next state and output entries and thus be equivalent ($S_0 \equiv S_1$) if $S_1 \equiv S_2$ and $S_3 \equiv S_4$. We represent these implications by the shorthand notation $1 \equiv 2$ and $3 \equiv 4$ and enter that notation in the entry indexed by S_0 and S_1. Comparing the first and third rows, we see that $S_0 \equiv S_2$ if $S_0 \equiv S_1$. Thus, we enter the implication $0 \equiv 1$ in the entry indexed by S_0 and S_2. S_0 cannot possibly be equivalent to S_3 because the two states have different outputs. Thus, we enter an X in the entry indexed by S_0 and S_3. We also enter X's in the entries indexed by (S_0, S_4), (S_0, S_5), and (S_0, S_6) because the output of S_0 differs from states S_4–S_6. Finally, we compare the first row with the last row and find that $S_0 \equiv S_7$ if $S_1 \equiv S_5$ ($1 \equiv 5$) and $S_3 \equiv S_7$ ($3 \equiv 7$).

After checking all pairs of equivalent states involving S_0, we repeat the procedure to check all pairs of equivalent states involving S_1. We will not waste time, however, checking the second row with the first row because the pair (S_0, S_1) has already been checked in the preceding step. Thus, the second row need only be compared with the rows containing the present states S_2–S_7. The results form the second column in the implication table shown below.

	S_0	S_1	S_2	S_3	S_4	S_5	S_6
S_1	$1 \equiv 2$ $3 \equiv 4$						
S_2	$0 \equiv 1$	$0 \equiv 2$ $3 \equiv 4$					
S_3	X	X	X				
S_4	X	X	X	$5 \equiv 6$ $3 \equiv 4$			
S_5	X	X	X	$4 \equiv 7$ $5 \equiv 6$	$3 \equiv 7$		
S_6	X	X	X	$4 \equiv 7$	$3 \equiv 7$ $5 \equiv 6$	$5 \equiv 6$	
S_7	$1 \equiv 5$ $3 \equiv 7$	$4 \equiv 7$ $2 \equiv 5$	$0 \equiv 5$ $3 \equiv 7$	X	X	X	X

Each column of the implication table is filled in by comparing a row in Table 9.9 with all the rows beneath it. Remember, an X in an entry indexed by S_x and S_y denotes that S_x and S_y are *not* equivalent ($S_x \not\equiv S_y$).

The implications identified above may or may not be true. After constructing the implication table, we review the implications and eliminate any false implications; this could take several passes. For example, consider all squares containing the implication $3 \equiv 7$. Since there is an X in the entry indexed by S_3 and S_7, these states are not equivalent and the implication $3 \equiv 7$ is false. Likewise, all state equivalences depending on the implication $3 \equiv 7$ are false. Thus, $S_0 \not\equiv S_7$, $S_2 \not\equiv S_7$, $S_4 \not\equiv S_5$, and $S_4 \not\equiv S_6$, and X's are placed in these entries. The revised implication table is shown below.

	S_0	S_1	S_2	S_3	S_4	S_5	S_6
S_1	$1 \equiv 2$ $3 \equiv 4$						
S_2	$0 \equiv 1$	$0 \equiv 2$ $3 \equiv 4$					
S_3	X	X	X				
S_4	X	X	X	$5 \equiv 6$ $3 \equiv 4$			
S_5	X	X	X	$4 \equiv 7$ $5 \equiv 6$	X		
S_6	X	X	X	$4 \equiv 7$	X	$5 \equiv 6$	
S_7	X	$4 \equiv 7$ $2 \equiv 5$	X	X	X	X	X

We check all remaining entries containing X's to make sure that the associated state pairs do not appear as equivalences in the implication table. Such a check reveals that $S_1 \not\equiv S_7$, $S_3 \not\equiv S_5$, and $S_3 \not\equiv S_6$ because $S_4 \not\equiv S_7$. Thus, we add three additional X's to the implication table, and we have completed the first pass of reviewing the implications. The updated implication table is shown below.

	S_0	S_1	S_2	S_3	S_4	S_5	S_6
S_1	$1 \equiv 2$ $3 \equiv 4$						
S_2	$0 \equiv 1$	$0 \equiv 2$ $3 \equiv 4$					
S_3	X	X	X				
S_4	X	X	X	$5 \equiv 6$ $3 \equiv 4$			
S_5	X	X	X	X	X		
S_6	X	X	X	X	X	$5 \equiv 6$	
S_7	X	X	X	X	X	X	X

Problem-Solving Skill
For every problem, there is an inverse problem. Sometimes it is easier to solve the inverse problem than the problem.

Based on the new set of X's, we check the implications again. If an implication is found to be false, we place X's in all the squares containing the implication. We continue rechecking the implications and updating the implication table until all false implications have been found. For our example, another pass through the implications in the table above shows that no further implications can be eliminated. When all false implications have been found, entries containing an X denote states that are *not* equivalent; the remaining entries not containing an X denote state equivalences.

For our example, the five entries not containing an X identify the following state equivalences.

$$S_0 \equiv S_1, \quad S_1 \equiv S_2, \quad S_0 \equiv S_2, \quad \therefore \quad S_0 \equiv S_1 \equiv S_2$$
$$S_3 \equiv S_4$$
$$S_5 \equiv S_6$$

Notice that the first set of three state equivalences identifies a set of three equivalent states $(S_0 \equiv S_1 \equiv S_2)$.

The implication table state reduction technique is summarized below.

Implication Table State Reduction

1. *Compare each row of the state table with all the rows underneath it; the results form the columns in the implication table.*
    a. *If the two rows having present states S_x and S_y have different outputs, the states are not equivalent. Place an X in the entry in the implication table indexed by S_x and S_y.*
    b. *If the two rows having present states S_x and S_y have the same outputs, the states may be equivalent. Place the implication(s) required to make the next states equal in the entry in the implication table indexed by S_x and S_y.*
2. *Eliminate all false implications. Place X's in the entries containing implications involving states that have been shown to be not equivalent.*
3. *Entries not containing an X denote state equivalences.*

9.4 STATE ASSIGNMENT AND ENCODED STATE TABLES

Thus far, we have discussed how to translate an English functional description into a state diagram having a minimal number of states. This section will address the next two steps in the design process: deriving a *state assignment* and generating an *encoded state table*.

State assignment determines how a set of states will be represented by a set of memory devices. As explained earlier (Section 9.2), a state is represented by a binary code stored in a set of memory devices. Making a good state assignment is important because the complexity of the output and next state combinational logic is often strongly influenced by the relationship between system state transitions and corresponding memory device transitions. A good state assignment can simplify system combinational logic and, consequently, the associated design effort.

For instance, the state assignment given in Table 9.1 for the even counter design discussed in Section 9.1 resulted in the following next state Boolean expressions.

$$D_A = \underline{Q_A} \oplus Q_B$$
$$D_B = \overline{Q_B}$$

Switching the binary representations for S_0 and S_1, such that $S_0 = 01$ and $S_1 = 00$, yields the following simpler set of next state combinational logic equations.

$$D_A = \overline{Q_B}$$
$$D_B = Q_A$$

More substantial savings can be obtained with more complex designs.

State assignment involves the two tasks listed below:
1. Select the number of memory devices.
2. Assign a unique binary code (contents of memory devices) to each state.

Consider the first task of selecting the number of memory devices. Using the fewest number of memory devices to minimize hardware costs is probably the most obvious strategy, but there are alternative strategies for selecting the number of memory devices. For example, a state assignment technique called *one-hot encoding* uses one memory device per state; only one memory device is "hot," or asserted, in any given state. Using a few more memory devices can sometimes simplify the next state logic, thereby reducing hardware costs. Regrettably, alternative state assignment techniques are beyond the scope of this text and will not be addressed.

In the following discussions, we will use the minimum number of memory devices required to represent N_S states. Since N_D bistable memory devices can store a total of 2^{N_D} unique binary codes, the minimum number of memory devices required to represent N_S states is given by the smallest integer greater than or equal to $\log_2 N_S$.[1]

The second task assigns a unique binary code to each state. Unfortunately, the difficulty of this task increases as the number of states increases because there are many ways to choose a set of N_S codes out of 2^{N_D} possible codes. The laws of permutations and combinations of large numbers make an exhaustive, trial-and-error state assignment approach infeasible for most designs. Therefore, we will use the set of guidelines, sometimes called the *Armstrong-Humphrey Rules* (in the left margin), for state assignment.

State Assignment Guidelines (Armstrong-Humphrey Rules)
1. *States having the same next state for a given input should have "adjacent state assignments."*
2. *The next states of a given state should have "adjacent state assignments."*
3. *States having the same output for a given input should have "adjacent state assignments."*

---

[1] Since most calculators and logarithm tables support only base 10 and natural (base e) logarithms, we reformulate

$$N_D = \text{ceil}(\log_2 N_S)$$

as

$$N_D = \text{ceil}(\log_2 10 \times \log_{10} N_S)$$

or

$$N_D = \text{ceil}(3.322 \times \log_{10} N_S)$$

where $\text{ceil}(x)$ yields the smallest integer not less than x.

These rules help to obtain a *good* state assignment, but there is no guarantee of obtaining an *optimal* state assignment. Additionally, sometimes none of the rules apply, in which case we simply try to make a reasonable state assignment. This situation often occurs with counters; a reasonable state assignment matches output values with state values. In other words, the output sequence defines the state transition sequence.

The Armstrong-Humphrey Rules suggest state assignments that will yield groups of 1's and 0's in the output and memory device input Karnaugh maps, which, in turn, respectively minimize the output and next state combinational logic. Specifically, Rules 1 and 2 attempt to minimize the next state combinational logic and Rule 1 is generally more effective than Rule 2. Rule 3 attempts to minimize the output combinational logic.

State assignments are specified in terms of *state adjacencies*. To understand the meaning of an "adjacent state assignment," it is helpful to think of a state as a minterm. Recall from Chapter 3 that a minterm is the conjunction of a combination of Boolean variables, representing a truth table entry or, more simply, a binary code. A state assignment is also a binary code. Like minterms, states having adjacent binary code assignments can be grouped together on a Karnaugh map to affect logic minimization. This concept will be illustrated by an example.

Consider generating a state assignment for the nonresetting sequence detector example discussed in Section 9.2.2, which detects the input sequences 1100_2 or 1000_2. The state table for the nonresetting sequence detector is shown in Table 9.10.

According to Rule 1, S_0 and S_5 should have adjacent state assignments because they both transition to state S_0 when the input $X = 0$. Also, states S_0, S_3, S_4, and S_5 should have adjacent state assignments because they all transition to state S_1 when the input $X = 1$. Continuing in a similar fashion, the following set of adjacent state assignments can be identified using Rule 1.

$$S_0, S_5 \quad S_3, S_4 \quad S_0, S_3, S_4, S_5 \quad S_1, S_2$$

According to Rule 2, S_0 and S_1 should have adjacent state assignments because they are the next states of state S_0. Likewise, S_4 and S_2 should have adjacent state assignments because they are the next states of state S_1. Continuing in a similar fashion, the following set of adjacent state assignments can be identified using Rule 2.

$$S_0, S_1 \quad S_4, S_2 \quad S_3, S_2 \quad S_5, S_1$$

Finally, the following set of adjacent state assignments can be identified using Rule 3.

$$S_0, S_1, S_2, S_4 \quad S_3, S_5 \quad S_0, S_1, S_2, S_3, S_4, S_5$$

States S_0, S_1, S_2, and S_4 form an adjacency set because they all have the output 0 when the input $X = 0$. States S_3 and S_5 form an adjacency set because they have the output 1 when the

Table 9.10

Nonresetting sequence detector state table

Present state	Next state		Output (Z)	
	$X = 0$	$X = 1$	$X = 0$	$X = 1$
S_0	S_0	S_1	0	0
S_1	S_4	S_2	0	0
S_2	S_3	S_2	0	0
S_3	S_5	S_1	1	0
S_4	S_5	S_1	0	0
S_5	S_0	S_1	1	0

input $X = 0$. The last adjacency set includes all states because all states have an output 0 when the input $X = 1$.

Having generated suggested state adjacencies, we next show a state assignment in Figure 9.14 and Table 9.11 that tries to satisfy the suggested state adjacencies. The state assignment in Figure 9.14 satisfies all the Rule 1 adjacencies. For example, the states S_0 and S_5 have adjacent assignments because their Karnaugh map entries can be grouped together. The state assignment also satisfies the first two Rule 2 adjacencies and the first two Rule 3 adjacencies. For example, the states S_0, S_1, S_2, and S_4 have adjacent assignments because their Karnaugh map entries can be grouped together.

The state assignment in Table 9.11 does not satisfy all suggested state adjacencies. When a state assignment cannot be obtained that satisfies all suggested state adjacencies, we try to satisfy as many as possible, giving the state adjacencies associated with Rule 1 the highest priority and the state adjacencies associated with Rule 3 the lowest priority.

Now that we have generated a state assignment, we update the state table by replacing symbolic state labels with their binary codes. Table 9.12 shows the encoded state table for the nonresetting sequence detector.

Figure 9.14

Karnaugh map for state assignment

State Assignment

Q_A \ Q_BQ_C	00	01	11	10
0	S_5	S_0	S_4	S_3
1		S_1	S_2	

Table 9.11

State assignment for nonresetting sequence detector

State	Binary encoding ($Q_AQ_BQ_C$)
S_0	001
S_1	101
S_2	111
S_3	010
S_4	011
S_5	000

Table 9.12

Nonresetting sequence detector encoded state table

Present state $Q_AQ_BQ_C$	Next state ($Q_A^+Q_B^+Q_C^+$) $X = 0$	Next state $X = 1$	Output (Z) $X = 0$	Output (Z) $X = 1$
001	001	101	0	0
101	011	111	0	0
111	010	111	0	0
010	000	101	1	0
011	000	101	0	0
000	001	101	1	0

9.5 KARNAUGH MAPS AND BOOLEAN EXPRESSIONS

To keep this discussion in perspective, let us review where we are in the sequential design process. Thus far, we generated a state diagram, checked for redundant states, and obtained state assignments. The resulting minimized encoded state table described the input/output behavior of the desired sequential machine. Now we need to derive Boolean expressions defining the combinational logic to generate the output and next state signals.

Output combinational logic is generated by transcribing the values for each output variable from the encoded state table onto Karnaugh maps or Quine-McCluskey tables and obtaining minimized Boolean expressions. Continuing with the nonresetting sequence detector example, we show the output Karnaugh map in Figure 9.15.

For each combination of the present state, Q_A, Q_B, and Q_C, and the input X, we look up the output entry in Table 9.12. The present state indexes the row, and the input indexes the column. We copy the output entry into the respective output Karnaugh map, filling in the square indexed by the same combination of the present state, Q_A, Q_B, and Q_C, and the input X.

The output Karnaugh map contains don't cares for the unused states, 100_2 and 110_2. Recall we elected not to include unused states in generating the state diagram for the nonresetting sequence detector. For the present, we will assume that we can ignore the output Z in the two unused states and use the don't cares to assign logic minimization. Thus, the unused states in the incompletely specified state diagram yield don't care output values.

Examining the output Karnaugh map in Figure 9.15 shows that our earlier efforts to obtain a good state assignment have paid off; the 1's are conveniently positioned to be grouped together. The minimized Boolean expression for the output Z is given by

$$Z = \overline{Q_C} \cdot \overline{X} \tag{9.1}$$

Next, we seek the minimized Boolean expressions for the memory devices. We start by copying the values for each next state variable from the encoded state table onto *next state* Karnaugh maps; these Karnaugh maps describe output behavior of the memory devices. The next state Karnaugh maps for the three next state variables Q_A^+, Q_B^+, and Q_C^+ are shown in Figure 9.16.

For each combination of the present state, Q_A, Q_B, and Q_C, and the input X, we look up the next state entry in Table 9.12. The present state indexes the row, and the input indexes the column. We copy each bit of the next state entry into the respective next state Karnaugh map, filling in the square indexed by the same combination of the present state, Q_A, Q_B, and Q_C, and the input X.

For example, let us locate the entry in the encoded state table and the cells of the next state Karnaugh maps indexed by the present state and input values

$$Q_A = 1, \quad Q_B = 0, \quad Q_C = 1, \quad \text{and} \quad X = 0$$

Figure 9.15

Output Karnaugh map for nonresetting sequence detector

Figure 9.16

Next state Karnaugh maps for nonresetting sequence detector

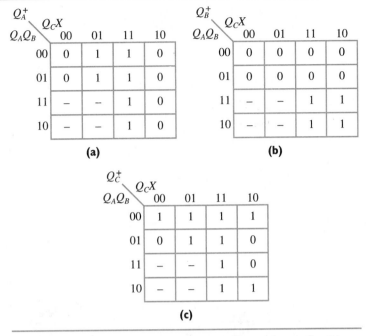

(a)

(b)

(c)

Figures (a), (b), and (c) illustrate the next state Karnaugh maps for Q_A^+, Q_B^+, and Q_C^+, respectively.

The entry is in the second row and second column of the encoded state table in Table 9.12; the cells are at lower right corners of the next state Karnaugh maps in Figure 9.16. Now, we simply transcribe the next state values.

$$Q_A^+ = 0, \quad Q_B^+ = 1, \quad \text{and} \quad Q_C^+ = 1$$

found in the encoded state table entry into the respective next state Karnaugh map cells. The value of $Q_A^+ = 0$ goes in the Karnaugh map for Q_A^+; the value for $Q_B^+ = 1$ goes in the Karnaugh map for Q_B^+; and the value for $Q_C^+ = 1$ goes in the Karnaugh map for Q_C^+.

Like the output Karnaugh map, the next state Karnaugh maps contain don't cares, denoting state transitions that have not yet been defined for the two unused states. These state transitions will be defined in the process of defining the next state logic.

The next state Karnaugh maps in Figure 9.16 describe the present state/next state behavior of the memory devices, but this is unfortunately not the information we need to define the next state logic. The present state/next state behavior describes the desired memory device *output* response; we need to find the memory device *input* stimuli or excitation that will cause this response. In other words, based on the present state, Q_A, Q_B, and Q_C, and the input X, the next state combinational logic needs to set up the memory device inputs to cause the desired transitions to Q_A^+, Q_B^+, and Q_C^+. The next state Karnaugh maps describing the desired output response can be transformed into another set of Karnaugh maps describing the required input excitation via *excitation tables*. The truth tables and associated excitation tables for SR, JK, T, and D memory devices are listed on p. 284. Think of an excitation table as the "inverse" of a truth table. A truth table describes output response as a function of input excitation, whereas an excitation table describes input

Truth Tables

SR	Q^+	Action
0 0	Q	No change
0 1	0	Reset
1 0	1	Set
1 1	–	Illegal inputs

JK	Q^+	Action
0 0	Q	No change
0 1	0	Reset
1 0	1	Set
1 1	$\overline{Q}$	Toggle

T	Q^+	Action
0	Q	No change
1	$\overline{Q}$	Toggle

D	Q^+	Action
0	0	Reset
1	1	Set

Excitation Tables

QQ^+	SR	Action
0 0	0 –	No change
0 1	1 0	Set
1 0	0 1	Reset
1 1	– 0	No change

QQ^+	JK	Action
0 0	0 –	No change
0 1	1 –	Set
1 0	– 1	Reset
1 1	– 0	No change

QQ^+	T	Action
0 0	0	No change
0 1	1	Set
1 0	1	Reset
1 1	0	No change

QQ^+	D	Action
0 0	0	No change
0 1	1	Set
1 0	0	Reset
1 1	1	No change

excitation as a function of output response. (Kind of a funny twist in thinking about digital systems.)

Excitation tables can be easily derived from truth tables. For example, consider the first row in the JK excitation table shown above. What values of the inputs J and K will cause a JK memory device that is holding a 0 value ($Q = 0$) to continue to hold a 0 value ($Q^+ = 0$) (in other words, not change state)? Referring to the truth table, if $J = 0$, $K = 0$, and $Q = 0$, there is no change in the output and Q stays at a 0 value. Alternatively, if $J = 0$, $K = 1$, and $Q = 0$, the output is forced to 0. Combining these two possibilities fills in the first row of the JK excitation table. If the desired output response is a "transition" from $Q = 0$ to $Q^+ = 0$, then the inputs can be either $J = 0$ and $K = 0$ or $J = 0$ and $K = 1$, in other words, $J = 0$ and $K = –$. We encourage you to derive the remaining rows of the excitation tables in a similar fashion.

Returning to our sequence detector design example, let us use the excitation tables and the next state Karnaugh maps to derive the memory device input Karnaugh maps. First, we select a memory device; let us opt for positive edge-triggered D flip-flops. The D excitation table shows that $D = Q^+$; thus, the next state Karnaugh maps for Q_A^+, Q_B^+, and Q_C^+ in Figure 9.16 are also respectively the flip-flop input Karnaugh maps for D_A, D_B, and D_C. No further work is required. The next state minimized Boolean expressions are listed below, taken from the groupings shown in Figure 9.17.

$$D_A = X$$
$$D_B = Q_A$$
$$D_C = \overline{Q_B} + X$$

(9.2)

Figure 9.17

D flip-flop input Karnaugh maps for non-resetting sequence detector

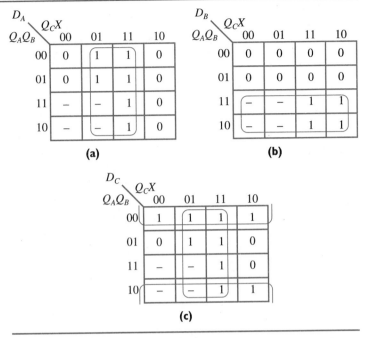

(a)

(b)

(c)

Figures (a), (b), and (c) illustrate the flip-flop input Karnaugh maps for D_A, D_B, and D_C, respectively.

It is instructive to select a different memory device and repeat the process of defining the next state combinational logic for the nonresetting sequence detector. For instance, let us use JK flip-flops instead of D flip-flops. Figure 9.18 (p. 286) shows the resulting Karnaugh maps for the J and K inputs of each flip-flop.

For each flip-flop, we use the next state Karnaugh maps in Figure 9.16 and the JK excitation table to generate the input Karnaugh maps in Figure 9.18. For example, look at the top left square of the next state Karnaugh map for Q_A^+ (indexed by $Q_A = 0$, $Q_B = 0$, $Q_C = 0$, and $X = 0$) and note the values for $Q_A = 0$ and $Q_A^+ = 0$. Referring to the JK excitation table, we see that this flip-flop output response can be caused by setting the inputs $J_A = 0$ and $K_A = -$. We enter these values into the top left squares (indexed by $Q_A = 0$, $Q_B = 0$, $Q_C = 0$, and $X = 0$) of the input Karnaugh maps for J_A and K_A, respectively. We then fill in the remaining squares in a similar fashion. If a next state Karnaugh map entry contains a don't care (–), we enter don't cares in the corresponding squares in the flip-flop input Karnaugh maps. (If the flip-flop outputs are not important (don't cares), neither are the flip-flop inputs.)

Having completed the JK flip-flop input Karnaugh maps, we generate the next state minimized Boolean expressions for the JK flip-flops listed below.

$$J_A = X$$
$$K_A = \overline{X}$$
$$J_B = Q_A$$
$$K_B = \overline{Q_A}$$
$$J_C = \overline{Q_B} + X$$
$$K_C = Q_B \bullet \overline{X}$$

(9.3)

Figure 9.18

JK flip-flop input Karnaugh maps for nonresetting sequence detector

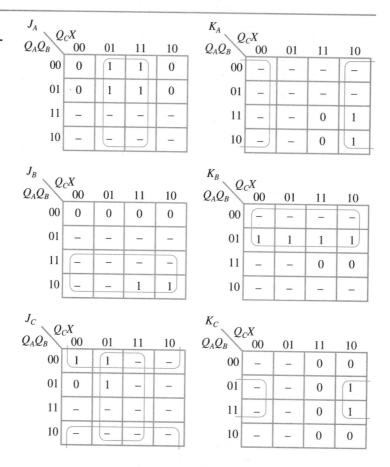

Using JK flip-flops can sometimes result in less complex next state logic because the associated input Karnaugh maps typically contain many don't care entries that can be exploited for logic minimization. The sequence detector next state logic for JK flip-flops, however, is slightly more complex than the next state logic for D flip-flops.

To gain proficiency in generating flip-flop input Karnaugh maps, try working with Karnaugh map *regions* instead of individual squares. Consider, for instance, JK memory devices. When $Q_x = 0$, copy the contents of the next state Karnaugh map to the J_x Karnaugh map and fill in the K_x Karnaugh map with don't cares. When $Q_x = 1$, copy the complement of the contents of the next state Karnaugh map to the K_x Karnaugh map and fill in the J_x Karnaugh map with don't cares. For our sequence detector example, we would generate the J_A and K_A Karnaugh maps by filling in the first and second rows ($Q_A = 0$ region) and then the third and fourth rows ($Q_A = 1$ region). We would generate the J_B and K_B Karnaugh maps by filling in the first and last rows ($Q_B = 0$ region) and then the second and third rows ($Q_B = 1$ region). Finally, we would generate the J_C and K_C Karnaugh maps by filling in the first and second columns ($Q_C = 0$ region) and then the third and fourth columns ($Q_C = 1$ region).

To conclude this section, let us examine the implications of our design actions with respect to unused states. In generating the output and next state Boolean expressions, the don't care values were selected to be either 0 or 1 to facilitate logic minimization, in other words,

Figure 9.19

Nonresetting sequence detector state diagram with unused states

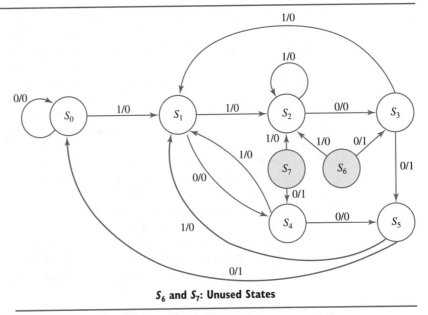

S_6 and S_7: Unused States

D flip-flops and JK flip-flops yield the same state diagram.

to obtain the largest groupings. Figure 9.19 shows the effect of these don't care assignments on the transitions associated with the unused states $S_6 = 110_2$ and $S_7 = 100_2$. The D flip-flop and JK flip-flop designs yield identical state diagrams, but this need not always be the case.

The outputs for the unused states are computed by updating the don't care entries in the output Karnaugh map (Figure 9.15). For example, for the unused states $Q_A = 1$, $Q_B = 0$, and $Q_C = 0$ and the input condition $X = 0$, the logic expression in Equation 9.1 yields an output value of $Z = 1$. Note that this output is a false assertion of a detected sequence, but is acceptable assuming the unused state will never occur.

The transitions for the unused states are computed by updating the don't care entries in the next state Karnaugh maps (Figure 9.16), based on the values assigned to the don't care entries in the flip-flop input Karnaugh maps (Figure 9.17 and Figure 9.18). Again, consider the unused state $Q_A = 1$, $Q_B = 0$, and $Q_C = 0$ and the input condition $X = 0$. The next state logic expressions given in Equation (9.2) for the D flip-flops yield

$$D_A = 0, \quad D_B = 1, \quad \text{and} \quad D_C = 1$$

Similarly, the next state logic expressions given in Equation (9.3) for the JK flip-flops yield

$$J_A K_A = 01, \quad J_B K_B = 10, \quad \text{and} \quad J_C K_C = 10$$

Both sets of inputs imply that the A flip-flop will reset to 0, the B flip-flop will set to 1, and the C flip-flop will set to 1 at the next clock pulse, making the next state $Q_A^+ = 0$, $Q_B^+ = 1$, and $Q_C^+ = 1$. This state transition is denoted by the arc from S_7 to S_4 in Figure 9.19. Remaining transitions for the unused states are generated in a similar fashion.

If the outputs or transitions associated with the unused states are unacceptable, don't care entries must be replaced by desired logic values. Then, the output and next logic combinations must be regenerated.

9.6 EXERCISE: MARKOV SPEECH PROCESSOR

To conclude this chapter on synchronous sequential system synthesis and review the associated steps, we will execute a finite state machine synthesis example, starting with an English description and ending with specifications for memory and combinational logic. This exercise will bring together the procedures involved in generating optimized state machines.

The behavior of many physical systems cannot be mathematically modeled by deterministic, closed-form algorithms or equations because they are either too complex to understand and/or they act in an erratic, unpredictable manner. Such systems can sometimes be modeled as *stochastic* processes, where behavior is described as a timed sequence of random transitions between a fixed number of states. The modeling practice is called *Hidden Markov Modeling* (HMM), and the sequence of states is called a *Markov Chain* [7, 8, 10].

As an example, human speech is often modeled using Hidden Markov Modeling. Configurations of a vocal system that produce distinct sounds are modeled by states in a Markov Chain. Speech is modeled as a series of sounds generated by statistically controlled state transitions. Since there is no precise correspondence between voice and states, a Markov Chain is only an approximation of human speech. The exact definition of the states is unknown or "hidden," hence the term Hidden Markov Modeling [1, 5].

Our task is to design a sequential system for speech processing based on Hidden Markov Modeling. The problem statement is given below.

Design a speech processor based on a unidirectional six-stage Markov Chain; each stage generates a unique binary code denoting a sound or utterance. Two successive sounds may be the same sound, sounds associated with adjacent stages, or sounds separated by one stage. Assume stage-to-stage transitions are controlled by a probabilistic function.

We begin by generating the state diagram shown in Figure 9.20. Movement around the Markov Chain is unidirectional in a clockwise direction. Allowable state-to-state transitions are (1) no transition, (2) move one stage, or (3) move two stages. In other words, at each state, say S_2, the speech processor can either stay in that state (S_2), transition to the next ad-

Figure 9.20

State diagram for a six-stage Markov Chain

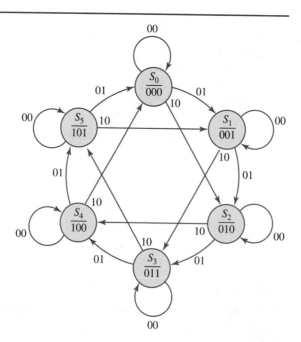

jacent state (S_3), or skip the next adjacent state and transition to the following state (S_4). The input that determines state transition is assumed to be generated by a suitable probabilistic distribution that models human speech patterns. The outputs represent phonemes. A sequence of phonemes constitutes an utterance, in other words, a word.

Next we check for redundant states. Inspecting the state table in Table 9.13 immediately shows that there are no redundant states because each state has a unique output.

Next, we generate state binary encodings using the Armstrong-Humphrey Rules. Rule 1 cannot be used because there are no states having the same next state for a given input. Rule 2, concerning grouping the next states of a given state, suggests the following adjacent state assignments:

$$S_0, S_1, S_2 \quad S_1, S_2, S_3 \quad S_2, S_3, S_4 \quad S_3, S_4, S_5 \quad S_0, S_4, S_5 \quad S_0, S_1, S_5$$

Finally, Rule 3 cannot be used because there are no states having the same output for a given input. The binary state encoding shown in Figure 9.21 and listed in Table 9.14 satisfies four of the six suggested state adjacencies. Using the state assignment in Table 9.14, the encoded state table for our speech processor is given in Table 9.15.

Table 9.13

Speech processor state table

Present state	Next state			Output (SPEECH)
	X = 00	X = 01	X = 10	
S_0	S_0	S_1	S_2	000
S_1	S_1	S_2	S_3	001
S_2	S_2	S_3	S_4	010
S_3	S_3	S_4	S_5	011
S_4	S_4	S_5	S_0	100
S_5	S_5	S_0	S_1	101

Figure 9.21

Karnaugh map for state adjacencies and assignments

State Assignment

Q_A \ $Q_B Q_C$	00	01	11	10
0	S_0	S_1	S_3	S_2
1	S_4	S_5		

Table 9.14

State assignment for speech processor

State	Binary encoding $(Q_A Q_B Q_C)$
S_0	000
S_1	001
S_2	010
S_3	011
S_4	100
S_5	101

Table 9.15

Speech processor encoded state table

Present state $Q_A Q_B Q_C$	Next state $(Q_A^+ Q_B^+ Q_C^+)$			Output (SPEECH)
	X = 00	X = 01	X = 10	
000	000	001	010	000
001	001	010	011	001
010	010	011	100	010
011	011	100	101	011
100	100	101	000	100
101	101	000	001	101

We complete the synthesis process by generating the minimized output and next state logic. The minimized output logic is easily generated by inspection because the output *SPEECH* equals the present state $Q_A Q_B Q_C$. Minimized Boolean expressions are given below.

$$SPEECH_2 = Q_A$$
$$SPEECH_1 = Q_B$$
$$SPEECH_0 = Q_C$$

We generate the minimized next state logic by first generating the next state Karnaugh maps shown in Figure 9.22 and then transforming the next state Karnaugh maps into the flip-flop input Karnaugh maps shown in Figure 9.23. We will assume T flip-flops. The minimized

Figure 9.22

Next state Karnaugh maps for speech processor

Q_A^+

$Q_B Q_C$ \ $X_1 X_0$	00	01	11	10
00	0	0	–	0
01	0	0	–	0
11	0	1	–	1
10	0	0	–	1

$Q_A = 0$

Q_A^+

$Q_B Q_C$ \ $X_1 X_0$	00	01	11	10
00	1	1	–	0
01	1	0	–	0
11	–	–	–	–
10	–	–	–	–

$Q_A = 1$

Q_B^+

$Q_B Q_C$ \ $X_1 X_0$	00	01	11	10
00	0	0	–	1
01	0	1	–	1
11	1	0	–	0
10	1	1	–	0

$Q_A = 0$

Q_B^+

$Q_B Q_C$ \ $X_1 X_0$	00	01	11	10
00	0	0	–	0
01	0	0	–	0
11	–	–	–	–
10	–	–	–	–

$Q_A = 1$

Q_C^+

$Q_B Q_C$ \ $X_1 X_0$	00	01	11	10
00	0	1	–	0
01	1	0	–	1
11	1	0	–	1
10	0	1	–	0

$Q_A = 0$

Q_C^+

$Q_B Q_C$ \ $X_1 X_0$	00	01	11	10
00	0	1	–	0
01	1	0	–	1
11	–	–	–	–
10	–	–	–	–

$Q_A = 1$

Boolean expressions are

$$T_A = (Q_A \cdot X_1) + (Q_A \cdot Q_C \cdot X_0) + (Q_B \cdot X_1) + (Q_B \cdot Q_C \cdot X_0)$$
$$T_B = (\overline{Q_A} \cdot X_1) + (\overline{Q_A} \cdot Q_C \cdot X_0)$$
$$T_C = X_0$$

Generating the T_A, T_B, and T_C Karnaugh maps is easy because the T flip-flop excitation table defines a simple input/output relationship: T is 0 if there is no state change, otherwise T is 1. Thus, we copy regions of the next state Karnaugh maps where Q_x is 0 directly to the associated regions of the T flip-flop input Karnaugh maps. Then, we copy the complement of the regions of the next state Karnaugh maps where Q_x is 1 to the associated regions of the T flip-flop input Karnaugh maps.

Figure 9.23

T flip-flop input Karnaugh maps for speech processor

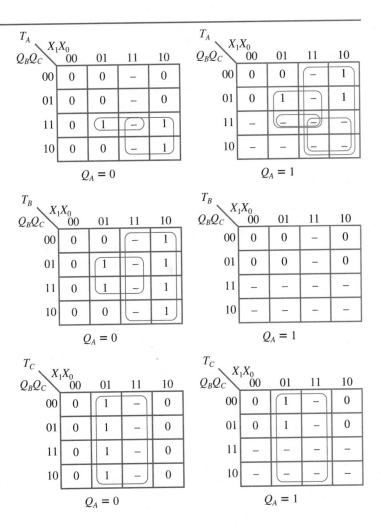

We have finished the sequential system synthesis process! The remaining design task of implementing the finite state machine with actual hardware is the topic of Chapter 10.

9.7 SUMMARY

In this chapter we discussed the general process of designing synchronous sequential machines, examining how to transform an English description of a desired function into the logical definition of a sequential machine. We studied several examples of developing from prose a variety of state diagrams describing Moore/Mealy and resetting/nonresetting behavior. We saw that defining a state diagram involves identifying a set of states and specifying the state-to-state transitions to affect the desired input/output behavior. We also discussed how to use row reduction and implication tables to check for equivalent states. If two states are equivalent, one of the states can be discarded, thereby simplifying the state diagram without changing the function of the overall state machine.

Continuing to the next step in the sequential design process, we saw that state assignment assigns a unique binary number or code to each state in the state diagram so that system states are represented by the contents of a set of memory devices. Making a good state assignment is important because the complexity of the next state and output combinational logic can be strongly influenced by the state assignment.

In the final section of this chapter, we explained how to generate minimized Boolean expressions defining the next state and output combinational logic. The next state logic implements the desired state transitions. The output logic implements the desired system response. We discussed constructing the required Karnaugh maps using excitation tables and addressed don't cares. We also studied the trade-offs in using different kinds of memory devices.

Through representations such as state diagrams, state tables, and Karnaugh maps and operations such as state reduction, state assignment, and logic minimization, we are able to translate and refine an initial English functional specification into a more detailed and precise logical specification. This logical specification defines the memory and combinational logic for a sequential machine that realizes the desired function.

9.8 PROBLEMS

1. Generate a minimal resetting state diagram that detects an active-1 signal pulse, . . . 010 . . . or an active-0 signal pulse, . . . 101 . . . . The output is active-1. Do not include any unused states. Show the implication table demonstrating that there are no redundant states.
   a. For a Moore Machine
   b. For a Mealy Machine
   c. Compare the Moore and the Mealy machines relative to used states, unused states, and the required number of bistable memory devices.

2. Generate a minimal nonresetting state diagram that detects an active-1 signal pulse, . . . 010 . . . or an active-0 signal pulse, . . . 101 . . . . The output is active-1. Do not include any unused states. Show the implication table demonstrating that there are no redundant states.
   a. For a Moore Machine
   b. For a Mealy Machine
   c. Compare the Moore and the Mealy machines relative to used states, unused states, and the required number of bistable memory devices.

3. Using positive edge-triggered D flip-flops, design two versions of the Moore state machine generated in Problem 1. Show state assignment and logic minimization.

a. Denote unused states as don't care states.
b. Define unused states such that they do not falsely assert the output and always transition to the state that initiates a new input signal pulse sequence detection.

4. Using positive edge-triggered D flip-flops, design two versions of the Mealy state machine generated in Problem 1. Show state assignment and logic minimization.
   a. Denote unused states as don't care states.
   b. Define unused states such that they do not falsely assert the output and always transition to the state that initiates a new input signal pulse sequence detection.

5. Does the following state diagram have any equivalent states?

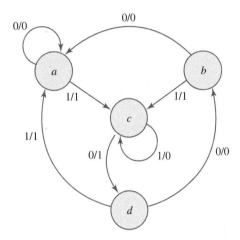

6. Does the following state diagram have any equivalent states?

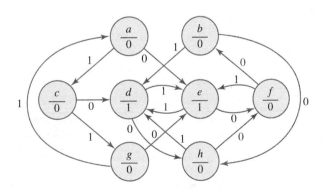

7. Design a variation of the common binary counter that counts in only odd numbers. The desired behavior might be stated as follows:

   Design a counter that counts in the cyclic pattern $1_{10}, 3_{10}, 5_{10}, 7_{10}, 1_{10}, 3_{10}, \cdots$.

   Show the state reduction, state assignment, and logic minimization.
   a. Use positive edge-triggered D flip-flops.
   b. Use positive edge-triggered JK flip-flops.

8. Design a 2-bit up/down binary counter with enable having the following function table.

EN_BAR	UP_DWN	Z^+	Action
0	0	$Z + 1$	Up count
0	1	$Z - 1$	Down count
1	–	Z	Hold count

Show the state reduction, state assignment, and logic minimization.
   a. Use positive edge-triggered T flip-flops.
   b. Use positive edge-triggered SR flip-flops.

9. Counters can also perform clock frequency division. Draw the output timing diagram for the binary counter designed in Problem 8 for eight clock cycles, assuming $EN\_BAR = 0$ and $UP\_DWN = 0$.
   a. What is the relationship between the frequency of the least significant output bit (LSB) and the frequency of the clock?
   b. What is the relationship between the frequency of the most significant output bit (MSB) and the frequency of the clock?

10. Design a sequential system that generates two outputs. One output divides the clock frequency by 10 and the other output divides the clock frequency by 6, as shown below.

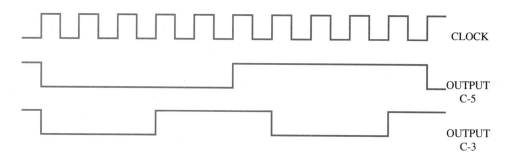

11. Analyze the sequential system given in the figure shown on p. 295 by generating a state diagram.
   a. What function does this system perform? (*Hint:* Starting in state "0000," list the output for ten clock cycles and assume states 1010_2 to 1111_2 are unused states.)
   b. What happens if the sequential system inadvertently enters one of the unused states?

12. Redesign the sequential system in Problem 11 to be *self-correcting*, whereby the sequential system always recovers from inadvertently entering an unused state. If an unused state becomes active, the sequential system will eventually transition to the desired state sequencing behavior.

13. Design a decade counter that directly drives the seven-segment light emitting diode (LED) display shown below. Each LED display segment is turned on, in other words, lit up, by asserting the associated pin active-1. Each LED display segment is turned off by deasserting the associated pin. A 1 is displayed by asserting segments b and c.

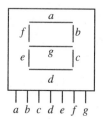

Figure for Problem 11

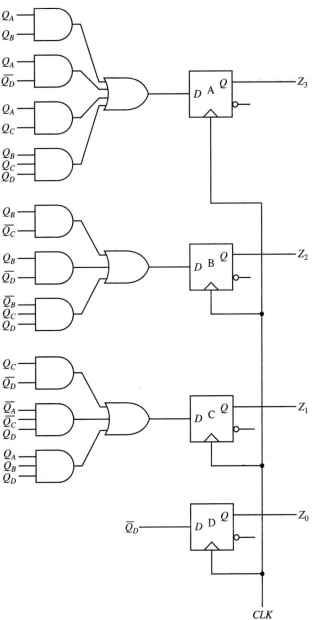

14. Show how the decade counter designed in Problem 13 can be cascaded to form higher-order counters.

15. Design a sequential binary subtracter that accepts two operands and yields the difference. Corresponding bits of the operands are entered least significant bit (LSB) to most significant bit (MSB). The truth table for a *full subtracter* is given in Table 9.16 (p. 296). A full subtracter performs the arithmetic operation

$$\text{Difference} = \text{Minuend} - \text{Subtrahend} - \text{Borrow-In}$$

Table 9.16

Rules for subtracting two binary numbers. These rules describe how to generate the difference and borrow bits on a column-by-column basis when subtracting two binary numbers.	Borrow-in bit from previous column	Minuend bit	Subtrahend bit	Difference bit	Borrow-out bit to next column
	0	0	0	0	0
	0	0	1	1	1
	0	1	0	1	0
	0	1	1	0	0
	1	0	0	1	1
	1	0	1	0	1
	1	1	0	0	0
	1	1	1	1	1

and generates a borrow-out bit if

$$\text{Subtrahend} + \text{Borrow-In} > \text{Minuend}$$

Show the state diagram, state assignment, and logic minimization.

a. Use positive edge-triggered D flip-flops.

b. Use positive edge-triggered SR flip-flops.

16. Design a resetting sequential majority function that asserts (active-1) the output if the past three inputs contain two or more 1's. Sample input/output sequences are given below.

$$X = 0101110110 \ldots$$
$$Z = 0000010010 \ldots$$
$$\mapsto \text{time}$$

Assume a Mealy machine, using positive edge-triggered D flip-flops. Show the state diagram, state reduction, state assignment, and logic minimization.

17. Design a 2-bit counter that counts in the cyclic Gray codes 00_2, 01_2, 11_2, 10_2, 00_2, 01_2, $\ldots$. The counter accepts two control signals to set and clear the count, as described in the function table below.

SET	CLR	Z^+	Action
0	0	$Z + 1$	Up count
0	1	00	Clear count
1	0	11	Set count
1	1	–	Illegal input

Assume a Moore machine and use positive edge-triggered D flip-flops.

18. Redesign the 2-bit Gray code counter described in Problem 17, using positive edge-triggered D flip-flops with synchronous reset and clear inputs.

REFERENCES AND READINGS

[1] Waibel, A., and K. Lee, eds. *Readings in Speech Recognition.* San Mateo, CA: Morgan Kaufman, 1990.

[2] Comer, D., ed. *Digital Logic and State Machine Design.* Philadelphia, PA: Saunders, 1990.

[3] Buchanan, J., ed. *CMOS/TTL Digital Systems Design.* New York: McGraw-Hill, 1990.

[4] Holdsworth, B., and G. Martin, eds. *Digital Systems Reference Book.* Boston, MA: Butterworth-Heinemann, 1990.

[5] Furui, S., and M. Sondhi, eds. *Advances in Speech Signal Processing.* New York: Dekker, 1992.

[6] Herbst, L., ed. *High Speed Digital Electronics.* Englewood Cliffs, NJ: Prentice-Hall, 1992.

[7] Anderson, W. *Continuous-Time Markov Chains.* New York: Springer-Verlag, 1991.

[8] Clymer, J. *Systems Analysis Using Simulation & Markov Models.* Englewood Cliffs, NJ: Prentice-Hall, 1990.

[9] Garrod, S., and R. Borns. *Digital Logic: Analysis, Application, & Design.* Philadelphia, PA: Saunders, 1991.

[10] Gillespie, D. *Markov Processes: An Introduction for Physical Scientists.* Boston, MA: Academic Press, 1992.

[11] Haznedar, H. *Digital Microelectronics.* Benjamin Cummings, 1991.

[12] Kleitz, W. *Digital and Microprocessor Fundamentals: Theory and Applications.* Englewood Cliffs, NJ: Prentice-Hall, 1990.

[13] McDonald, J. *Fundamentals of Digital Switching.* New York: Plenum Press, 1990.

[14] Pucknell, D. *Fundamentals of Digital Logic Design with VLSI Circuit Applications.* Englewood Cliffs, NJ: Prentice-Hall, 1990.

[15] Ryan, R. *Basic Digital Electronics.* Blue Ridge Summit, PA: TAB Books, 1990.

10
SEQUENTIAL DESIGN: IMPLEMENTATION

In Chapter 9, we discussed the first stage of sequential digital system design: synthesis. Starting from an English statement of a desired input/output function, we learned how to develop a state diagram. From the state diagram, we then developed expressions describing the memory and combinational logic elements of the sequential machine architecture using state reduction, state assignment, and logic minimization techniques.

In this chapter, we will discuss the second stage of sequential digital system design: *implementation*. As shown in Figure 10.1, implementation involves translating or mapping a minimized state machine into hardware. We will examine several approaches for implementing sequential machines. First, we will review combinational logic implementation strategies and study their application in realizing output and next stage logic. Then, we will introduce regular collections of flip-flops, called *registers*, and study how they can realize memory and certain kinds of state machines. We will also discuss counters and their application in realizing certain kinds of state machines. We will conclude with a look at sequential programmable logic devices, such as registered programmable read-only memories and programmable logic arrays.

10.1 COMBINATIONAL LOGIC

Once we have synthesized Boolean expressions for the next state and output combinational logic, the next step is to implement the logic. To that end, we will briefly review some of the following combinational logic implementation techniques studied in Chapter 5.

- Two-level networks
- Multi-level networks
- Multiplexers
- Decoders
- Memory
- Programmable logic devices

We will also examine the interdependencies and trade-offs between combinational versus sequential and synthesis versus implementation aspects of design. Aspects of a sequential design may affect the choice of a combinational implementation technique and, conversely, a combinational implementation technique may affect aspects of a sequential design.

As an example, consider designing the combinational logic for the local area network (LAN) adaptor discussed in Chapter 9 using a two-level gate network. The Moore machine minimal state diagram is given in Figure 10.2. The output signal *PACKET* is asserted active-1 if a packet is detected (00_2) and a valid destination address is decoded (100_2 or 010_2). A state assignment is given in Table 10.1, using the state assignment guidelines discussed in Section 9.4. Table 10.2 (p. 300) gives the encoded state table. The output and D flip-flop input Karnaugh maps are given in Figure 10.3 (p. 300). Again, our state assignment efforts have paid off, in that the 1's in Figure 10.3 are generally grouped together. The minimized sum-of-products (SOP) Boolean expressions

$$PACKET = (Q_A \cdot \overline{Q_B} \cdot Q_C)$$
$$D_A = (\overline{Q_A} \cdot Q_B \cdot \overline{X}) + (Q_A \cdot Q_B \cdot X) + (Q_A \cdot \overline{Q_C} \cdot \overline{X})$$
$$D_B = (Q_B \cdot \overline{Q_C}) + (\overline{Q_A} \cdot \overline{Q_B} \cdot Q_C \cdot \overline{X})$$
$$D_C = (Q_B \cdot \overline{Q_C}) + (\overline{Q_C} \cdot \overline{X})$$

Figure 10.1

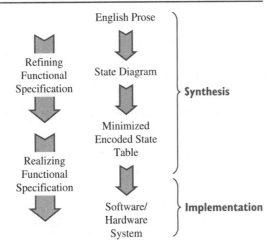

Digital sequential system implementation

English Prose

Refining
Functional
Specification

State Diagram

Minimized
Encoded State
Table

Realizing
Functional
Specification

Software/
Hardware
System

Synthesis

Implementation

Figure 10.2

A Moore machine state diagram of a LAN adaptor

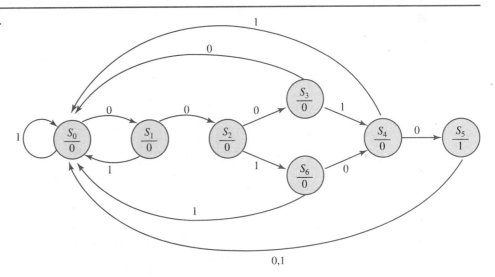

Table 10.1

State assignment for a LAN adaptor	State	**Binary encoding** $(Q_A Q_B Q_C)$
	S_0	000
	S_1	001
	S_2	010
	S_3	111
	S_4	100
	S_5	101
	S_6	011

Table 10.2

LAN adaptor encoded
state table

Present state $Q_A Q_B Q_C$	Next state ($Q_A^+ Q_B^+ Q_C^+$)		Output (*PACKET*)
	$X = 0$	$X = 1$	
000	001	000	0
001	010	000	0
010	111	011	0
111	000	100	0
100	101	000	0
101	000	000	1
011	100	000	0

Figure 10.3

LAN adaptor Karnaugh
maps

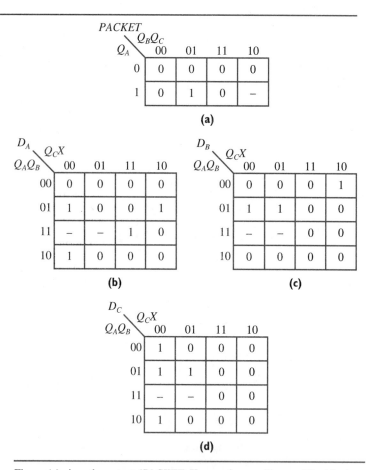

(a)

(b)

(c)

(d)

Figure (a) gives the output (*PACKET*) Karnaugh map. Figures (b)–(d) give
the flip-flop input Karnaugh maps.

are implemented in Figure 10.4(a) using a two-level **nand-nand** network. Alternatively, the minimized product-of-sums (POS) Boolean expressions

$$PACKET = \overline{(\overline{Q_A} + Q_B + \overline{Q_C})}$$
$$D_A = (Q_A + Q_B) \cdot (Q_A + \overline{X}) \cdot (Q_B + \overline{X}) \cdot (\overline{Q_A} + \overline{Q_C} + X)$$
$$D_B = (Q_B + Q_C) \cdot (\overline{Q_C} + \overline{X}) \cdot (\overline{Q_B} + \overline{Q_C}) \cdot \overline{Q_A}$$
$$D_C = \overline{Q_C} \cdot (Q_B + \overline{X})$$

are implemented in Figure 10.4(b) using a two-level **nor-nor** network.

Figure 10.4

**LAN adaptor output and
next state logic**

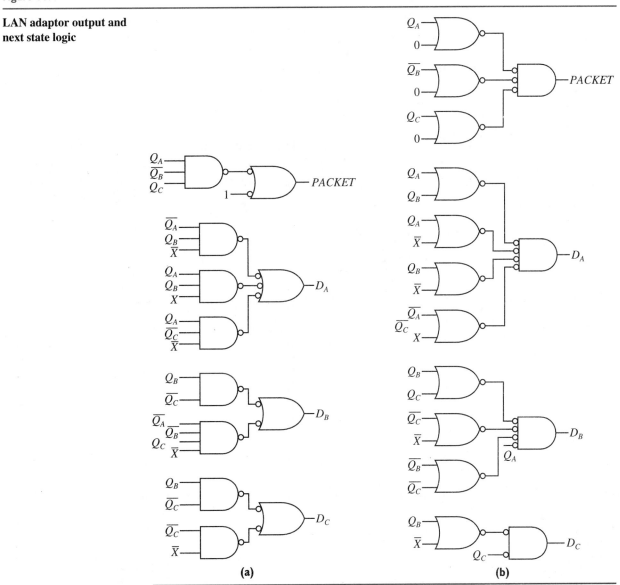

(a) **(b)**

Figure (a) implements the combinational logic using a two-level **nand-nand** network. Figure (b) implements the combinational logic using a two-level **nor-nor** network.

It is instructive to make a few observations about the **nand-nand** and **nor-nor** implementations. First, we have assumed that the flip-flops provide their outputs in uncomplemented and complemented forms and have taken advantage of this feature in the implementations. Second, note that **nand-nand** and **nor-nor** logic implementations do *not* realize the same sequential machine. Both implementations realize the desired LAN adaptor function, but address differently the unused states. For example, the **nand-nand** implementation of D_B realizes the don't cares as 1's, whereas the **nor-nor** implementation of D_B realizes the don't cares as 0's. Thus, a particular unused state assignment may favor a particular combinational logic implementation strategy and vice versa.

As another example, consider designing the combinational logic for the nonresetting sequence detector discussed in Chapter 9, using a *medium-scale-integration* (MSI) combinational logic implementation option. Specifically, memory will be used because memory supports multiple logic functions, which are characteristic of sequential machines. Recall that the output (Z) is asserted active-1 whenever the input (X) sequences 1100_2 or 1000_2 are detected.

$$X = 11000110000 \ldots$$
$$Z = 00011000110 \ldots$$
$$\longmapsto \text{time}$$

The encoded state table developed in Chapter 9 is repeated in Table 10.3.

Recall that memory implements combinational logic by storing the entire truth table. The input variables collectively constitute memory addresses. For each memory address, the associated storage location holds the output value(s). The truth table for the nonresetting sequence detector next state and output logic is given below. Assuming D flip-flops, truth table entries for the next state (flip-flop input) and output logic are taken directly from the encoded state table.

$Q_A Q_B Q_C X$	$D_A D_B D_C Z$	MINTERM
0000	0011	m_0
0001	1010	m_1
0010	0010	m_2
0011	1010	m_3
0100	0001	m_4
0101	1010	m_5
0110	0000	m_6
0111	1010	m_7
1000	----	m_8
1001	----	m_9
1010	0110	m_{10}
1011	1110	m_{11}
1100	----	m_{12}
1101	----	m_{13}
1110	0100	m_{14}
1111	1110	m_{15}

Figure 10.5 shows how a 16×4 memory realizes the nonresetting sequence detector next state and output logic. The memory stores the entire truth table; don't care values are arbitrarily set to 0. The four inputs Q_A, Q_B, Q_C, and X address one of sixteen rows. The four bits comprising the addressed row contain the values for the logic functions D_A, D_B, D_C, and Z.

Table 10.3

Nonresetting sequence detector encoded state table

Present state $Q_A Q_B Q_C$	Next state ($Q_A^+ Q_B^+ Q_C^+$)		Output (Z)	
	X = 0	X = 1	X = 0	X = 1
001	001	101	0	0
101	011	111	0	0
111	010	111	0	0
010	000	101	1	0
011	000	101	0	0
000	001	101	1	0

Figure 10.5

Nonresetting sequence detector logic using a 16 × 4 memory

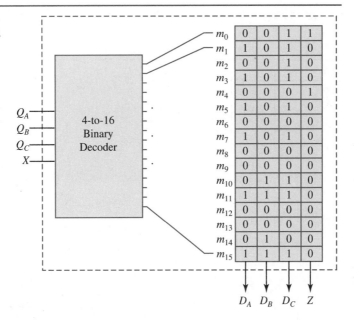

Note that memory uses the *canonical* instead of the *minimized* form of the logic expressions. Thus, any steps taken during the sequential design process to obtain minimized combinational logic expressions are irrelevant (in other words, not required) if memory is used to realize combinational logic. For example, using JK flip-flops to interject don't cares into logic Karnaugh maps to facilitate logic minimization is pointless because logic minimization itself is not necessary. Also, state assignment may be omitted. We use the word *may* because state assignment may be required to address other design aspects, such as minimizing the simultaneous switching of physically close devices or wires. These observations again emphasize the interdependencies between combinational and sequential system design.

10.2 REGISTERS

Having discussed strategies for implementing combinational logic, let us move on and discuss strategies for implementing memory. Individual latches or flip-flops can be obtained from a variety of semiconductor manufacturers as either application-specific integrated circuit (ASIC) library cells or small scale integrated (SSI) circuits. (Refer to Chapter 7 for a discussion of integrated circuit technology.) However, latches and flip-flops are more commonly packaged as medium scale integrated (MSI) circuits in sets of four (quad), six (hex), eight (octal), and so on. These sets of binary storage devices are called *registers*. A sample quad register (which stores four bits) is shown in Figure 10.6(a); Figure 10.6(b) shows the IEEE symbol.

The quad register contains four positive edge-triggered D flip-flops. On the positive edge of the clock signal *CLK*, the inputs, D_A–D_D, are transferred to the outputs, Q_A–Q_D. *CLR_BAR* is an asynchronous, active-0, clear signal. When *CLR_BAR* = 0 (asserted active-0), the quad register clears or resets to 0000_2, independent of the clock signal.

The notation used in the IEEE symbol requires some explanation. The top section is a *control block* that defines the control signals *CLK* and *CLR_BAR*. The bottom section defines an *array* of data signals. A *control dependency notation* defines which control sig-

Figure 10.6

A quad register

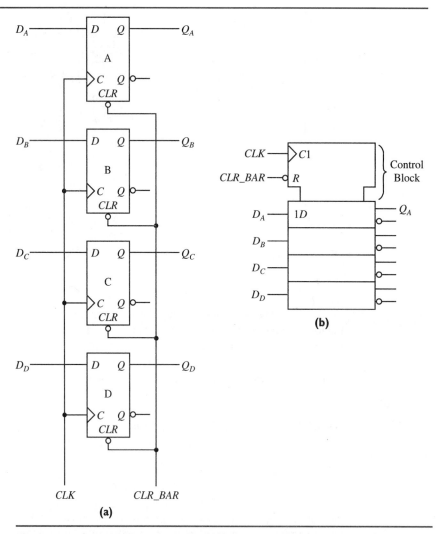

(a)

(b)

Figures (a) and (b) show the schematic and IEEE symbol for a quad register, respectively.

nals affect which data signals; a control signal labeled Cn affects data inputs labeled nD. Thus, the positive edge-triggered clock $C1$ controls the data input labeled $1D$ (D_A). The clock signal also controls the data inputs D_B-D_D. If all the elements of the data array section are identical, it is customary to define the properties of only the first (top) element and to assume that the properties apply to all elements (to save a little writing). The reset control action is denoted by the letter R and applies to all flip-flops; the bubble means that CLR_BAR is active-0.

Manufacturer data books list several types of registers providing a wide spectrum of function and performance. For example, Figure 10.7 shows a *quad 2-port* register. The *SEL* signal controls whether the register loads the inputs A_0-D_0 or A_1-D_1 on the rising edge of the clock signal *CLK*.

As another example, Figure 10.8 shows a register called a *register file* or a *register bank*. A register file is a random access memory (RAM) configured as multiple registers. A 128-bit RAM configured as a 16-word-by-8-bit array provides 16 octal registers. The read and write

Figure 10.7

A quad 2-port register

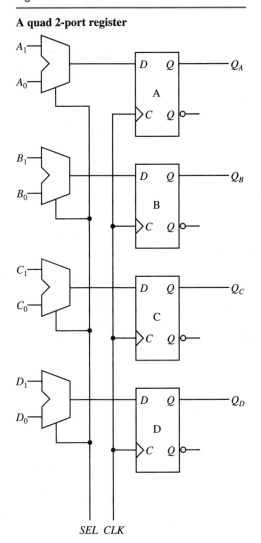

SEL CLK

Figure 10.8

A 16 × 8 register file

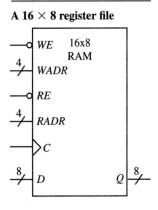

address inputs, *RADR* and *WADR*, provide simultaneous reading and writing. The contents of an octal register addressed by *RADR* can be accessed on the output Q by asserting active-0 the read enable *RE*. At the same time, the contents of an octal register addressed by *WADR* can be updated with the value on the input D by asserting active-0 the write enable *WE*.

10.3 SHIFT REGISTERS

Another popular class of registers are *shift registers*. The quad register in Figure 10.6 can be easily changed into a shift register by connecting flip-flop inputs and outputs, as shown in Figure 10.9. On the rising edge of the clock signal *CLK*, the contents of the shift register "shifts" one position to the right; the value on the serial input D_A is loaded into the leftmost flip-flop FF_A, the contents of FF_A are loaded into FF_B, the contents of FF_B are loaded into FF_C, and the contents of FF_C are loaded into the rightmost flip-flop FF_D. FF_D drives the serial output Q_D.

The shift register shown in Figure 10.9 is more accurately called a *serial in/serial out* shift register because the input data is applied and the output data is generated one bit at a time. Data can be input and output in a serial or parallel format. With a parallel format, the

Figure 10.9

A quad shift register

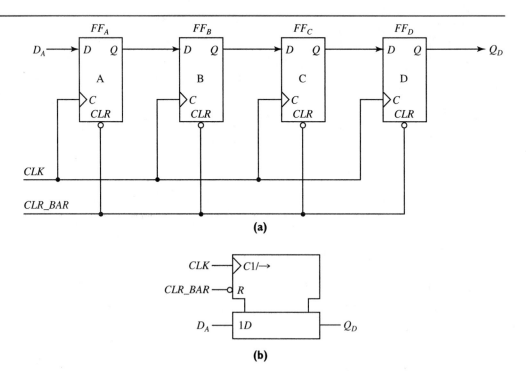

Figures (a) and (b) show the schematic and IEEE symbol for a quad shift register, respectively. The "→" symbol designates that the register contents shift one position from left to right on each rising clock edge.

input/output of each flip-flop in the shift register is accessible, enabling all flip-flops to be written simultaneously with a parallel input action or read simultaneously with a parallel output action. Combining the serial and parallel formats for input and output yields the following four common types of shift registers:

1. serial input/serial output
2. serial input/parallel output
3. parallel input/serial output
4. parallel input/parallel output

A *universal shift register* is a versatile device that can be programmed to be any of the four shift registers listed above. In addition, a universal shift register can be programmed to shift left, shift right, or hold (no shift). It is instructive to examine the design of a universal shift register because it illustrates an application of *design practices*. To that end, the inputs/outputs and operating modes of a universal shift register are respectively given in Figure 10.10 and Table 10.4 (p. 308).

A universal shift register could be designed using the formal sequential design techniques discussed in Chapter 9, but this approach would be difficult. The state diagram contains only 16 states for the 16 possible values of the four registers, but 8 inputs (excluding the clock input) means there are many possible state-to-state transitions. Enumerating these transitions and keeping track of them during the synthesis process would be timing consuming and error prone.

We can quickly design the universal shift register less formally by directly implementing the desired sequential behavior with multiplexers, as shown in Figure 10.11. When $OP_1 OP_0 = 00$, the output of each flip-flop is circulated back to its input through the I_0 multiplexer input, thereby leaving the contents of the flip-flops unchanged. When $OP_1 OP_0 = 01$, the output of each flip-flop is routed to the input of its left neighbor through the I_1 multiplexer input, thereby shifting the contents of the flip-flops one position to the left. When $OP_1 OP_0 = 10$, the output of each flip-flop is routed to the input of its right neighbor through the I_2 multiplexer input, thereby shifting the contents of the flip-flops one position to the right. Finally, when $OP_1 OP_0 = 11$, the inputs D_A–D_D are routed to the flip-flop inputs through the I_3 multiplexer inputs, thereby loading the flip-flops.

There are several common applications for shift registers. One is signal delay: A bit entering on the serial right (*SR*) input of the universal shift register does not appear on the serial output Q_D until four clock cycles later in time. Another application is binary multiplication and division. Shifting a binary number one position to the right divides the number by 2, while shifting one position to the left multiplies the number by 2. We will study

Problem-Solving Skill

Always try to estimate the resources required to solve a problem before beginning the work.

Figure 10.10

A universal quad shift register

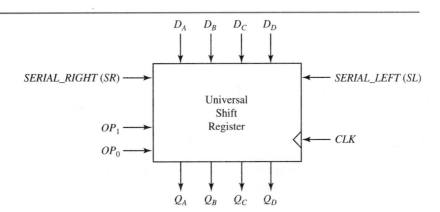

Table 10.4

Universal shift register modes of operation	Operating modes OP_1OP_0	Description
	00: hold	Contents of the four flip-flops remain unchanged; in other words, $Q_A^+Q_B^+Q_C^+Q_D^+ = Q_AQ_BQ_CQ_D.$
	01: shift left	Contents of the flip-flops shift one position to the left. The rightmost flip-flop (Q_D) takes on the value of the input signal *SERIAL_LEFT* (*SL*); in other words, $Q_A^+Q_B^+Q_C^+Q_D^+ = Q_BQ_CQ_DSL.$
	10: shift right	Contents of the flip-flops shift one position to the right. The leftmost flip-flop (Q_A) takes on the value of the input signal *SERIAL_RIGHT* (*SR*); in other words, $Q_A^+Q_B^+Q_C^+Q_D^+ = SRQ_AQ_BQ_C.$
	11: parallel load	Contents of the four flip-flops are loaded in parallel; in other words, $Q_A^+Q_B^+Q_C^+Q_D^+ = D_AD_BD_CD_D.$

Figure 10.11

A universal shift register

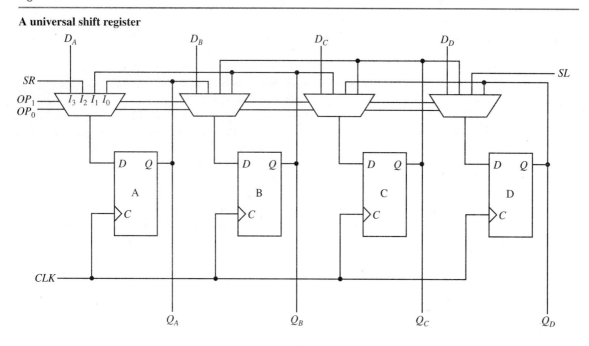

additional uses of shift registers in implementing sequential systems in the following sections.

10.3.1 Ring Counters

A shift register can be combined with combinational logic to realize sequential counting functions often used for control applications. For example, a shift register can easily realize a *ring counter*, also called a *walking ring counter*. A "ring count" is generated by configuring a linear shift register into a "ring" by connecting the output of the rightmost flip-flop back to the input of the leftmost flip-flop and circulating a single 1 around the ring. For instance, a four-stage shift register configured as a ring counter generates the cyclic four-state sequence ($Q_A Q_B Q_C Q_D$)

$$1000, 0100, 0010, 0001, 1000, \ldots$$
$$\longmapsto \text{time}$$

This pattern is not particularly useful as a numeric counting sequence, but the output is useful as a sequential 1-of-4 decoder output to control the sequencing of certain actions. Figure 10.12 shows that on each clock cycle, the asserted output enables a set of actions.

Figure 10.13 shows how to configure a quad universal shift register as a ring counter. Initially, $OP_1 OP_0 = 11$ to load the universal shift register with $Q_A Q_B Q_C Q_D = 1000$. Then, $OP_1 OP_0 = 10$ to configure the device as a circular right shift register.

The feedback logic driving the serial right SR input may be a little confusing, as you might expect the output Q_D to be directly connected to SR to configure the ring. But the **nor** gate logically performs the desired ring connectivity and also addresses unused states.

Figure 10.12

Ring count

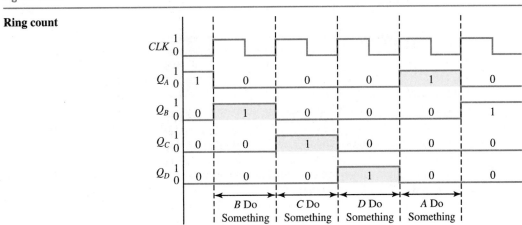

Figure 10.13

A ring counter

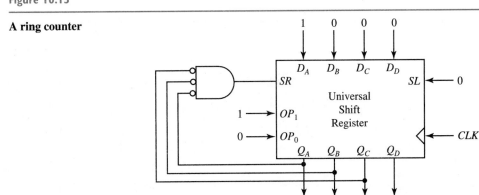

The advantage of using an existing MSI device to quickly realize the ring counting function is gained at the expense of wasting available states. In general, an N-bit shift register configured as a ring counter generates an N-count sequence, leaving $2^N - N$ unused states. The quad (4-bit) universal shift register ring counter, for example, uses only 4 of the $2^4 = 16$ available states. With so many unused states, it is a good idea to make sure that the ring counter returns to one of the allowed counting states if it mistakenly enters an unused state. In other words, the ring counter should be *self-correcting*. The **nor** gate ensures that the ring counter in Figure 10.13 is self-correcting. If the ring counter mistakenly enters an unused state, it will return to the desired counting sequence within three clock cycles, at most. For example, if the ring counter mistakenly enters the state $Q_A Q_B Q_C Q_D = 1111$, it will sequence through the states 0111, 0011, and 0001 to reenter the desired counting sequence. Problem 9 generates the state diagram for the ring counter in Figure 10.13.

The number of unused states can be reduced by modifying the feedback combinational logic. For example, an N-bit shift register configured as a *twisted ring counter* generates a $2N$-count sequence, leaving only $2^N - 2N$ unused states. A twisted ring counter is also called a *switch tail* or *Johnson* counter. A "twisted ring count" is generated by circulating the *complement* of the output of the rightmost flip-flop back to the input of the leftmost flip-flop. For instance, a four-stage shift register configured as a twisted ring counter generates the cyclic 8-count sequence $(Q_A Q_B Q_C Q_D)$

$$1000, 1100, 1110, 1111, 0111, 0011, 0001, 0000, 1000, \ldots$$
$$\mapsto \text{time}$$

Figure 10.14 shows how to configure a quad universal shift register as a twisted ring counter. Initially, $OP_1 OP_0 = 11$ to load the universal shift register with $Q_A Q_B Q_C Q_D = 1000$. Then, $OP_1 OP_0 = 10$ to configure the device as a circular right shift register. The combinational logic, $SR = \overline{Q_D} + (Q_A \cdot \overline{Q_B}) + (Q_A \cdot \overline{Q_C})$ ensures that the twisted ring counter is self-correcting. For example, if the twisted ring counter mistakenly enters the state $Q_A Q_B Q_C Q_D = 0101$, it will sequence through the states 0010, 1001, and 1100 to reenter the desired counting sequence. Problem 11 generates the state diagram for the twisted ring counter in Figure 10.14.

Unlike the ring counter, the twisted ring counter unfortunately does not directly provide the sequential 1-of-N decoder output for control applications. However, Figure 10.15 shows how a twisted ring 8-count sequence can be easily transformed into a 1-of-8 decoder output using only **and** gates. O_1 is asserted active-1 when the twisted ring count is $Q_A Q_B Q_C Q_D = 1000$, O_2 is asserted active-1 when the twisted ring count is $Q_A Q_B Q_C Q_D = 1100$, and so on.

Figure 10.14

A twisted ring counter

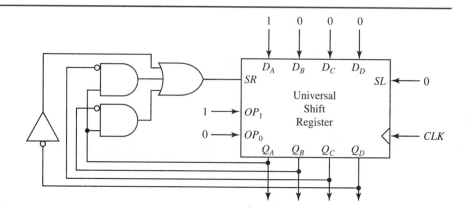

Figure 10.15

**A twisted ring counter
with 1-of-8 output**

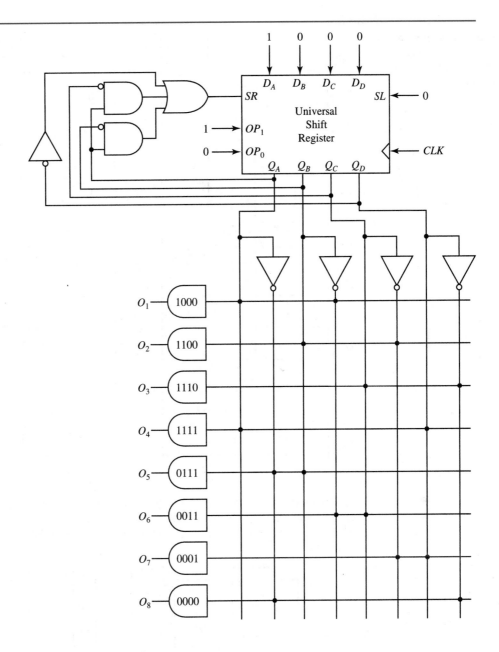

The timing diagram in Figure 10.16 (p. 312) shows an 8-count twisted ring sequence with associated 1-of-8 decoder output.

10.3.2 Sequential Machines

Shift registers are not limited to realizing just ring counters. Shift registers can be used to realize any sequential system having an output dependent on a *fixed* number of previous inputs and/or outputs. The fixed number of previous inputs and/or outputs are saved in a shift register and combinational logic detects the desired bit patterns.

Figure 10.16

The timing diagram of an 8-count twisted ring sequence with 1-of-8 decoder output

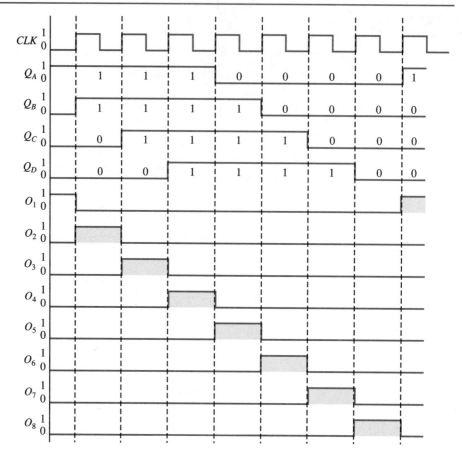

As a simple example, the serial-to-parallel code converter Mealy machine shown in Figures 8.28 and 8.29 can be easily implemented using a three-stage serial input/parallel output shift register. The input X drives D_A. Three clock cycles serially shift three inputs into the

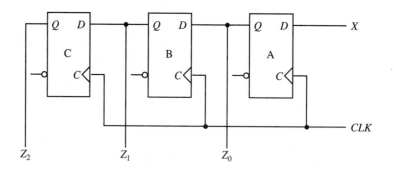

shift register. On the third clock cycle, the three inputs appear in parallel on the output $Z_2 Z_1 Z_0$.

As another example, let us consider implementing the nonresetting sequence detector discussed in Section 10.1 with a universal shift register. The output of the sequence detector depends on the previous four inputs: the output (Z) is asserted active-1 whenever the input

Figure 10.17

Implementing a nonresetting sequence detector with a shift register

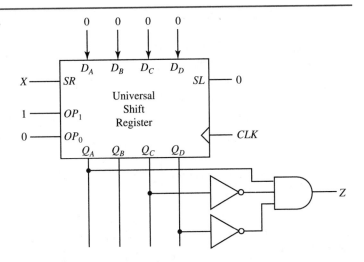

(X) sequences 1100_2 or 1000_2 are detected. This nonresetting sequence detector design is shown in Figure 10.17. The input X is tied to the serial right (SR) input, and the universal shift register is configured to shift right, $OP_1OP_0 = 10$. The **and** gate and two inverters decode the input sequences 1100_2 or 1000_2.

10.4 COUNTERS

Using a shift register to realize a ring counter is one example of exploiting a medium scale integrated (MSI) device to build a counter instead of custom designing the counter "from scratch." There are, however, many MSI devices available that support counting applications and offer a wide range of function and performance. The following set of characteristics are helpful in sorting through and understanding MSI counters.

Characteristic	Description
Modulus	Length of count sequence
Coding	Count sequence
Direction	Count up (add) or down (subtract)
Resetable	Ability to reset counter to all zeros
Loadable	Ability to load or set counter to a specific value

To illustrate counter characteristics, we will look at a few representative MSI counters. Figure 10.18 (p. 314) shows a 4-bit binary counter. The 4-bit binary counter generates the cyclic 16-count sequence

$$0_{10}, 1_{10}, 2_{10}, 3_{10}, \ldots, 12_{10}, 13_{10}, 14_{10}, 15_{10}, 0_{10}, 1_{10}, 2_{10}, 3_{10}, \ldots$$
$$\mapsto \text{time}$$

illustrated in the timing diagram in Figure 10.18(b). A 4-bit binary counter is called a *hex counter* or a *modulo-16 counter* because the count sequence has 16 unique states before repeating. A 4-bit binary counter is also called a *divide-by-16 ($\div 16$) counter* because the most significant output bit Q_A changes value at 1/16 the rate of the input clock frequency. A 4-bit counter is also a $\div 8$ counter (Q_B output), $\div 4$ counter (Q_C output), and $\div 2$ counter (Q_D output). The ripple carry output (RCO), sometimes called terminal count (TC), is asserted active-1 when $Q_AQ_BQ_CQ_D = 1111_2$.

Figure 10.18

A binary counter

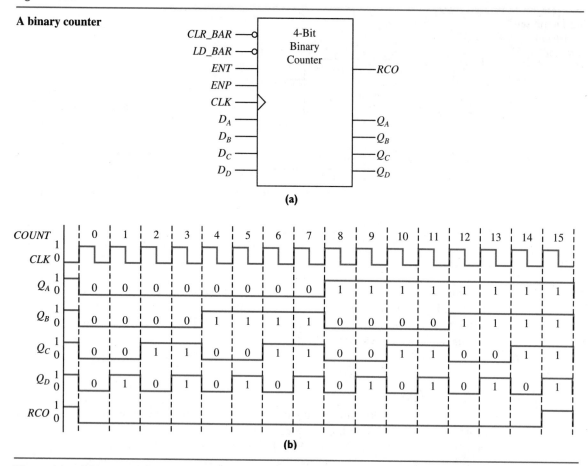

(a)

(b)

Figures (a) and (b) respectively show the inputs/outputs and timing diagram for a 4-bit binary counter.

Counter reset is controlled by the active-0 synchronous clear signal (*CLR_BAR*), also sometimes called master reset (*MR_BAR*). The counter resets to 0000_2 when *CLR_BAR* is asserted active-0 and the clock signal transitions from 0 to 1. Counter loading is controlled by the active-0 synchronous load signal (*LD_BAR*), also sometimes called parallel enable (*PE_BAR*). The counter loads the value on the data inputs $D_A D_B D_C D_D$ when *LD_BAR* is asserted active-0 and the clock signal transitions from 0 to 1. The count enable trickle (*ENT*) and count enable parallel (*ENP*) signals provide a count enable/disable capability and are also used with the $\overline{RCO}$ signal to extend the counting modulus. These capabilities will be explained in more detail in the next section.

As another example of an MSI counter, Figure 10.19 shows an up/down binary coded decimal (BCD) counter. (See Chapter 2 for a discussion of binary coded decimal.) The 4-bit BCD counter generates the cyclic 10-count sequence

$$0_{10}, 1_{10}, 2_{10}, 3_{10}, \ldots, 6_{10}, 7_{10}, 8_{10}, 9_{10}, 0_{10}, 1_{10}, 2_{10}, 3_{10}, \ldots$$
$$\mapsto \text{time}$$

illustrated in the state diagram in Figure 10.19(b). A 4-bit BCD counter is called a *decade*

Figure 10.19

An up/down BCD counter

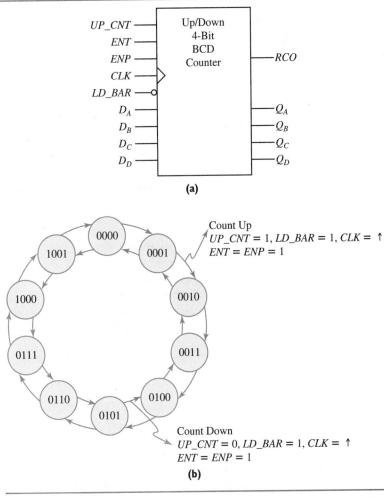

(a)

Count Up
$UP\_CNT = 1$, $LD\_BAR = 1$, $CLK = \uparrow$
$ENT = ENP = 1$

Count Down
$UP\_CNT = 0$, $LD\_BAR = 1$, $CLK = \uparrow$
$ENT = ENP = 1$

(b)

Figures (a) and (b) respectively show the inputs/outputs and state diagram for a 4-bit up/down BCD counter.

counter or a *modulo-10 counter* because the count sequence has 10 unique states before repeating. The *RCO* signal is asserted active-1 when $Q_A Q_B Q_C Q_D = 1001_2 = 9_{10}$.

Counting direction is controlled by the *UP_CNT* signal. On the rising edge of the clock input *CLK*, the BCD counter counts up (clockwise state transitions) if *UP_CNT* = 1 or counts down (counterclockwise state transitions) if *UP_CNT* = 0. The *LD_BAR*, *ENP*, and *ENT* signals serve the same functions described earlier for the binary counter.

10.4.1 Variable Modulus Counters

If a particular MSI counter does not offer the counting modulus required for a design, it is usually straightforward to modify the MSI counter to accommodate larger or smaller counting sequences. For a larger modulus, most MSI counters are designed to be *cascaded*; connecting several small counters together to form a larger counter. For example, Figure 10.20 (p. 316) shows three 4-bit (1 decimal digit) BCD counters with the range $0_{10}-9_{10}$ cascaded to form a 12-bit (3 decimal digits) BCD counter with the range $0_{10}-999_{10}$. The counters are

Figure 10.20

Extending the counting modulus by ripple carry cascading counters

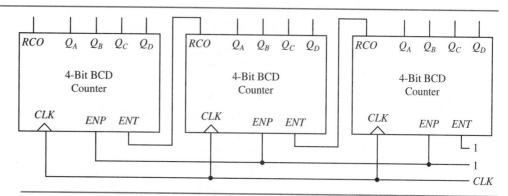

For clarity, not all signals and connections are shown.

connected in a ripple carry configuration with the RCO of each 4-bit counter connected to the next-higher-order 4-bit counter. Each counter is controlled by the enable parallel ENP and enable trickle ENT inputs. Both ENP and ENT must be asserted active-1 to enable a counter. Deasserting ENP disables a counter; deasserting ENT disables a counter *and* the ripple carry output. ENP typically serves as a master control, either enabling or disabling the entire cascade of counters. ENT typically serves as a stage control, enabling or disabling selective counters to propagate interstage carries.

To understand the operation of the counters in Figure 10.20, examine Table 10.5. Table columns list counter outputs. Table rows list clock cycles. $COUNT_1$ and RCO_1 are respectively the count and ripple carry outputs of the first, or rightmost, counter. $COUNT_2$ and RCO_2 are respectively the count and ripple carry outputs of the second, or middle, counter. Finally, $COUNT_3$ and RCO_3 are respectively the count and ripple carry outputs of the third, or leftmost, counter.

The entire cascade of counters is globally enabled by setting $ENP = 1$. However, only the first counter is actually enabled during the first nine clock cycles because the ripple carry outputs of the first (RCO_1) and second counters (RCO_2) respectively set $ENT = 0$ for the second and third counters. On clock cycle 9, the first counter reaches its terminal count 1001_2 and asserts its RCO output, $RCO_1 = 1$. Now $ENT = 1$ for the first and second coun-

Table 10.5

A three-stage ripple carry BCD counter

RCO_3	$COUNT_3$	RCO_2	$COUNT_2$	RCO_1	$COUNT_1$	Clock cycle
0	0000	0	0000	0	0000	0
↓	↓	↓	↓	↓	↓	↓
0	0000	0	0000	1	1001	9
0	0000	0	0001	0	0000	10
↓	↓	↓	↓	↓	↓	↓
0	0000	0	1000	1	1001	89
0	0000	0	1001	0	0000	90
↓	↓	↓	↓	↓	↓	↓
0	0000	1	1001	1	1001	99
0	0001	0	0000	0	0000	100
↓	↓	↓	↓	↓	↓	↓

ters, and these stages increment their counts to $0001\ 0000_2$ on clock cycle 10. In this manner, lower-order counters record their tallies and pass along the results to higher-order counters.

A disadvantage of the ripple carry configuration is that the carry propagation limits the speed of the multistage (chip) counter. To understand the accumulation of carry propagation delays, consider in Table 10.5 the sequence of actions that occur during clock cycle 99. On the rising edge of the clock signal for cycle 99, the first counter increments to 1001_2, then RCO_1 sets to 1, and then RCO_2 sets to 1. These three events must complete to properly set and enable counters before initiating the next count on clock cycle 100.

We can build a faster counter by cascading the BCD counters in the *lookahead carry* configuration illustrated in Figure 10.21. The ripple carry configuration propagates carries through the higher-order stages *before* clocking the low-order stage through its count cycle, whereas the lookahead carry configuration propagates carries through the higher-order stages *while* clocking the low-order stage through its count cycle. To understand this difference, compare the sample counting sequences for the ripple carry and lookahead carry configurations given in Table 10.5 and Table 10.6, respectively.

Specifically, note the different RCO_2 values for clock cycle 90. When the first counter in the lookahead carry configuration transitions from 1001_2 to 0000_2, RCO_1 transitions from 1 to 0 and appropriately disables the second and third counters by deasserting ENP. However,

Problem-Solving Skill

Closely coupling the control of all elements of a design can result in the slowest element unnecessarily gating the performance of the entire design. In other words, all actions do not have to "march to the same drummer."

Figure 10.21

Extending the counting modulus by lookahead carry cascading counters

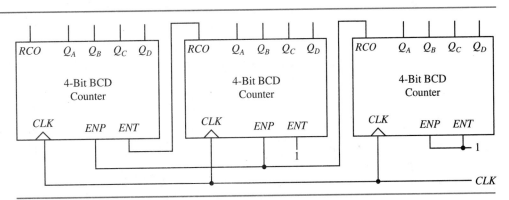

For clarity, not all signals and connections are shown.

Table 10.6

A three-stage lookahead carry BCD counter

RCO_3	$COUNT_3$	RCO_2	$COUNT_2$	RCO_1	$COUNT_1$	Clock cycle
0	0000	0	0000	0	0000	0
↓	↓	↓	↓	↓	↓	↓
0	0000	0	0000	1	1001	9
0	0000	0	0001	0	0000	10
↓	↓	↓	↓	↓	↓	↓
0	0000	0	1000	1	1001	89
0	0000	1	1001	0	0000	90
↓	↓	↓	↓	↓	↓	↓
0	0000	1	1001	1	1001	99
0	0001	0	0000	0	0000	100
↓	↓	↓	↓	↓	↓	↓

carry propagation remains enabled. While the first counter repeats its count sequence during cycles 91 to 99, RCO_2 is set to 1 because $COUNT_2 = 1001_2$ and $ENT = 1$. On the rising edge of the clock signal for cycle 99, all higher-order carry signals are set. Only two events must complete before initiating the next count on clock cycle 100: The first counter increments to 1001_2 and then RCO_1 sets to 1.

The modulus of a counter can be reduced by adding combinational logic to control reset and load actions to force the counter to cycle through only a subset of its entire count sequence. For example, Figure 10.22(a) shows how to change a binary counter having modulus 16 into a BCD counter having modulus 10.

When the binary counter reaches $9_{10} = 1001_2$, the single **nand** gate asserts the CLR_BAR input and resets the counter to 0000_2 on the next clock pulse. The binary counter also resets on counts $11_{10} = 1011_2$, $13_{10} = 1101_2$, and $15_{10} = 1111_2$, but this has no effect other than to minimize the reset logic. Note the importance of the *scynchronous* reset. What would happen if the **nand** gate drove an *asynchronous* reset? When the output $Q_A Q_B Q_C Q_D$ changes to 1001_2 and the **nand** gate asserts an asynchronous reset, the asynchronous reset would immediately attempt to change the output to 0000_2. If this reset action is accomplished before the next clock pulse, the 1001_2 count is skipped over, generating an incorrect count sequence. With a synchronous reset, asserting CLR_BAR does not reset the counter until the next clock pulse.

Figure 10.22(b) shows a simple example of using load logic to change a modulo-10 counter into a modulo-5 counter having the count sequence ($Q_A Q_B Q_C Q_D$)

$$0010_2, 0011_2, 0100_2, 0101_2, 0110_2, 0010_2, \ldots$$
$$\longmapsto \text{time}$$

When the modulo-10 counter reaches $0110_2 = 6_{10}$, the single **nand** gate asserts active-0 the synchronous load LD_BAR input and sets the counter to

$$Q_A Q_B Q_C Q_D = D_A D_B D_C D_D = 0010_2$$

on the next clock pulse. The modulo-10 counter also loads on counts $0111_2 = 7_{10}$, $1110_2 = 14_{10}$, and $1111_2 = 15_{10}$, but this has no effect other than to minimize the load logic.

Figure 10.22

Reducing the counting modulus by reset/load logic

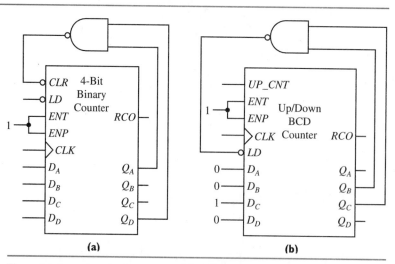

(a)   (b)

Figure (a) shows how to change a modulo-16 counter into a modulo-10 counter.
Figure (b) shows how to change a modulo-10 counter into a modulo-5 counter.

10.4.2 Cyclic Sequential Machines

MSI counters are not limited to realizing just counting sequences. Counters can be used to realize general sequential systems having state diagrams that are cyclic or nearly cyclic. A *cyclic state diagram* contains a loop of states. A near-cyclic state diagram contains a loop of states with a few "off-loop" transitions.

Figure 10.23 shows an example of a near-cyclic state diagram composed of a main loop of states S_0–S_4 with three off-loop transitions. Off-loop transitions from one state to a different state can be addressed with load logic. Off-loop transitions to and from the same state can be addressed with load logic or count enable logic.

As an example, reconsider the DRAM controller design discussed in Chapter 9. The near-cyclic state diagram given in Figure 9.9 is repeated in Figure 10.24. Figure 10.25 (p. 320) shows the implementation of the DRAM controller using a 4-bit BCD counter.

Figure 10.23

A near-cyclic state diagram

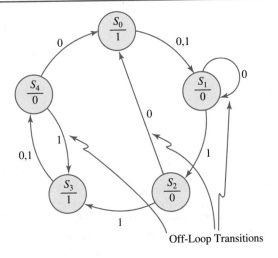

Off-Loop Transitions

Figure 10.24

A DRAM controller state diagram

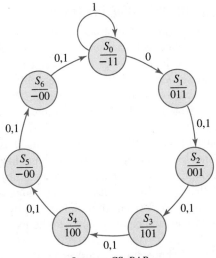

Input = CS_BAR
Output = MUX, RAS_BAR, CAS_BAR

Figure 10.25

A DRAM controller implementation using a 4-bit BCD counter

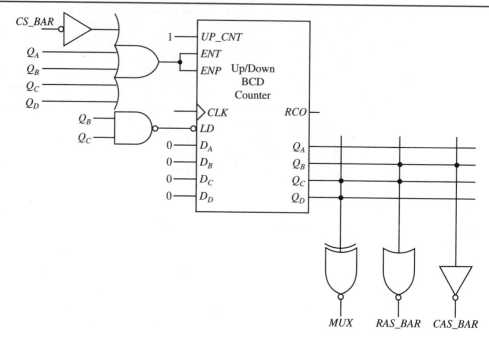

States S_0-S_6 are respectively assigned the binary codes 0000_2-0110_2. The transition from S_6 to S_0 is addressed by load logic. When the BCD counter reaches $0110_2 = S_6$, the load input LD_BAR is asserted active-0 and the counter is set to the next state $0000_2 = S_0$. The transition from and to S_0 when the input $CS\_BAR = 1$ is addressed by count disable/enable logic. While the BCD counter is in state S_0 and the input $CS\_BAR = 1$, the count enable parallel/trickle inputs are deasserted, which inhibits counting.

10.5 SEQUENTIAL PROGRAMMABLE LOGIC DEVICES (PLDs)

The programmable logic devices (PLDs) introduced in Chapter 5 for implementing combinational digital systems are collectively categorized as *combinational* PLDs. There is also a wide variety of programmable logic devices for implementing sequential digital systems, collectively categorized as *sequential* PLDs. Many sequential PLDs are simply combinational PLDs with registers added to store state information. Thus, this section will serve as a review of various programmable logic device architectures and an introduction to typical sequential system applications.

Recall that in Section 10.1 and Section 10.2 we examined using memory to realize output/next state combinational logic and registers to store state. Instead of using separate memory and register integrated circuits, manufacturers offer both memory and registers on a single integrated circuit called a *registered memory*. The logical structure of a registered memory is shown in Figure 10.26, derived from Figure 5.34 by simply adding flip-flops and feeding some of the flip-flop outputs back as inputs to the fixed **and** plane. These flip-flops hold system state.

For example, Figure 10.27 shows how to implement the nonresetting sequence detector discussed in Section 10.1 using a 512×8 registered programmable read-only memory

Figure 10.26

Logical view of a registered memory using shorthand notation

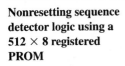

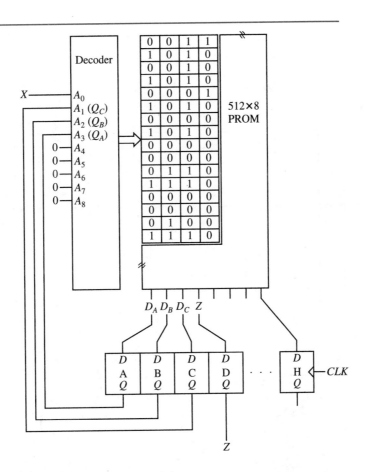

Figure 10.27

Nonresetting sequence detector logic using a 512×8 registered PROM

(PROM). Notice that the output Z is a registered output, being driven by the D flip-flop output Q_D. Driving Z by a clocked flip-flop has the advantage that Z is *synchronized* and any undesired glitches that may appear on the PROM output do not appear on the flip-flop output Q_D. However, implementing Z as a registered output has the disadvantage that a latency of one clock cycle is introduced between the inputs and outputs; that is, the output will lag the input by one clock cycle.

Registers can be added to a combinational programmable logic array (PLA) to yield a sequential PLA. Figure 10.28 shows the schematic for a sequential PLA, sometimes called a sequential *field-programmable logic array* (FPLA) or a *programmable logic sequencer* (PLS).

Figure 10.29 (p. 324) shows a sequential PLA implementation of the Markov speech processor Moore state machine discussed in Chapter 9. The **and** and **or** planes realize the next state logic equations for D_A, D_B, and D_C. These flip-flop input expressions are generated directly from the next state Karnaugh maps shown in Figure 9.22.

$$D_A = (Q_A \bullet \overline{X_1} \bullet \overline{X_0}) + (Q_A \bullet \overline{Q_C} \bullet \overline{X_1}) + (Q_B \bullet Q_C \bullet X_0) + (Q_B \bullet X_1)$$
$$D_B = (\overline{Q_A} \bullet \overline{Q_B} \bullet X_1) + (Q_B \bullet \overline{X_1} \bullet \overline{X_0}) + (Q_B \bullet \overline{Q_C} \bullet \overline{X_1}) + (\overline{Q_A} \bullet \overline{Q_B} \bullet Q_C \bullet X_0)$$
$$D_C = (\overline{Q_C} \bullet X_0) + (Q_C \bullet \overline{X_0})$$

The output *SPEECH* equals $Q_A Q_B Q_C$.

Figure 10.28

Logical view of a sequential PLA using shorthand notation

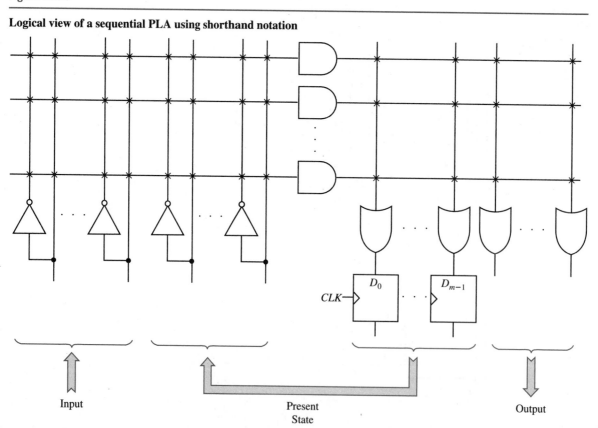

Figure 10.29

Sequential PLA implementation of a Markov speech processor

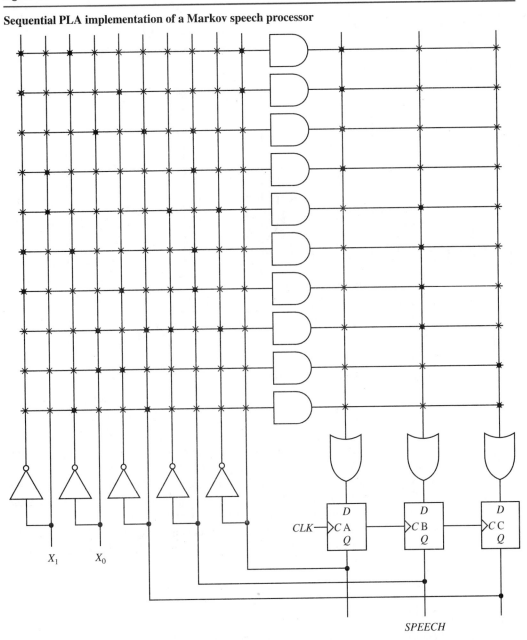

Registers can also be added to a combinational programmable array logic (PAL[1]) to yield a sequential PAL. Figure 10.30 (p. 324) shows the schematic for a sequential PAL, sometimes called a *registered PAL*.

A wide variety of sequential PALs are available, varying in function and complexity. Some sequential PALs offer **xor** gates for implementing counting and comparing functions. Others offer *programmable macrocells* for defining device pin direction (input or output),

---

[1] PAL® is a registered trademark of Advanced Micro Devices.

Figure 10.30

Logical view of a sequential PAL using shorthand notation

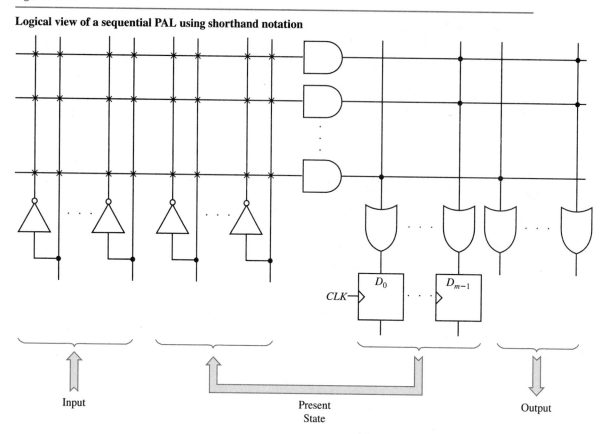

| Input | Present State | Output |

active state (active-0 or active-1), and logic (combinational or registered). A sequential PAL with programmable macrocells is illustrated in Figure 10.31.

A simple programmable macrocell is shown in Figure 10.32 [1]. Depending on the fuses F_1 and F_2 and the product term IO_CTL controlling the tri-state inverter, the programmable macrocell can be set to any one of the six configurations given in Table 10.7.

Figure 10.33 (p. 326) shows how the nonresetting sequence detector can be easily implemented using a sequential PAL with programmable macrocells. The rightmost programma-

Table 10.7

Programmable macrocell configurations	Configuration	Description
	Registered output active-0 logic state	Route Q output of D flip-flop through MUX_1 multiplexer and I_2 tri-state inverter to IO pin.
	Registered output active-1 logic state	Route Q_BAR output of D flip-flop through MUX_1 multiplexer and I_2 tri-state inverter to IO pin.
	Combinational output active-0 logic state	Route sum term OUT through MUX_1 multiplexer and I_2 tri-state inverter to IO pin.
	Combinational output active-1 logic state	Route sum term OUT through I_1 inverter, MUX_1 multiplexer, and I_2 tri-state inverter to IO pin.
	Primary input	Route IO signal through MUX_2 multiplexer to **and** plane.
	Present state input	Route Q_BAR output of D flip-flop through MUX_2 multiplexer to **and** plane.

Figure 10.31

A sequential PAL with programmable macrocells

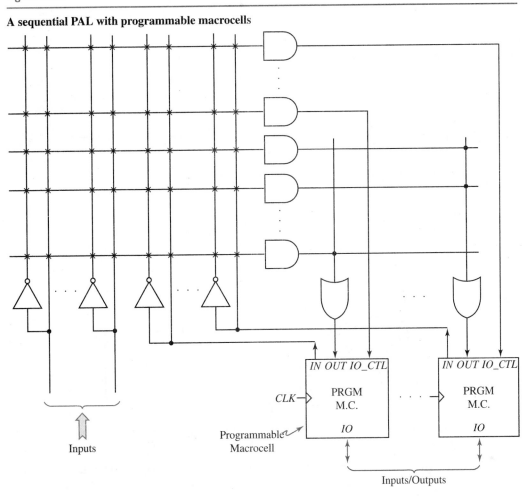

Figure 10.32

A simple programmable macrocell

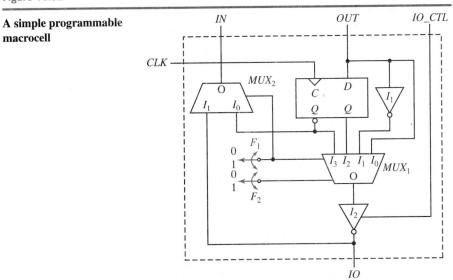

Figure 10.33

Implementing a nonresetting sequence detector with a sequential PAL

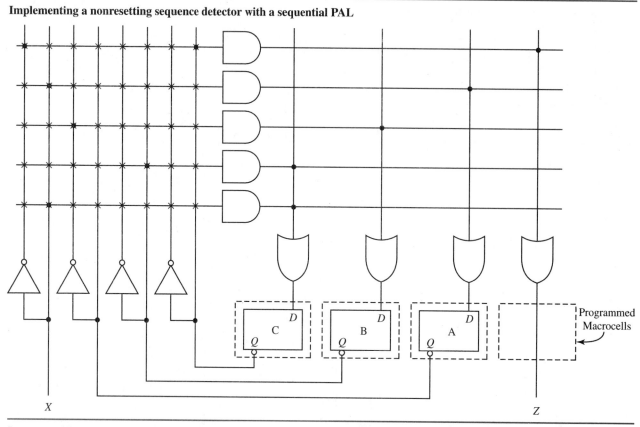

Programmable macrocells are shown configured; configuration and clocking details are omitted for clarity.

ble macrocell is configured as a combinational output with an active-1 logic state to provide the system output Z. The remaining programmable macrocells are configured as present state inputs to provide $Q_A/\overline{Q_A}$, $Q_B/\overline{Q_B}$, and $Q_C/\overline{Q_C}$.

The system output (Z) and flip-flop input (D_A, D_B, and D_C) logic expressions for the nonresetting sequence detector derived in Chapter 9 (Equations 9.1 and 9.2) are repeated below.

$$Z = \overline{Q_C} \bullet \overline{X}$$
$$D_A = X$$
$$D_B = Q_A$$
$$D_C = \overline{Q_B} + X$$

The output logic expression for Z, in other words, product term $\overline{Q_C} \bullet \overline{X}$, is generated by the two programmed connections denoted by "**X**" driving the top **and** gate. The next state logic expressions are generated in a similar manner. For instance, the flip-flop input logic expression for D_B, in other words, the product term Q_A, is generated by the single programmed connection driving the third **and** gate from the top.

10.6 PUTTING IT ALL TOGETHER

We conclude this chapter by executing some digital system design examples that bring together and summarize all that we have learned about combinational and sequential logic synthesis and implementation. We will look at a *sequence generation* example in the design

Figure 10.34

A simplified illustration of a stepper motor

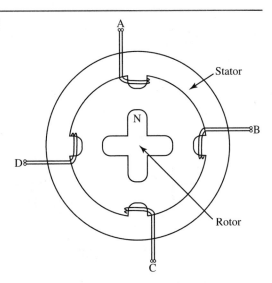

of a stepper motor controller and a *sequence detection* example in the design of a sequential majority monitor.

10.6.1 Stepper Motor Controller

Stepper motors convert electricity into mechanical motion using the basic force of magnetic attraction and repulsion. Electricity is used to establish a rotating electromagnetic field, which creates a torque to rotate a shaft.

Figure 10.34 shows a simplified construction of a stepper motor. The outside ring is called a *stator* and has four coils of wire, labeled A through D. Passing electrical current through any one of the coils creates a magnet, appropriately called an *electromagnet*.[2] In the center of the stator is a permanent magnet, called a *rotor*; the "N" marks its north pole.

Figure 10.35 (p. 328) shows that selectively turning on and off the four coils creates a clockwise rotating magnetic field. This rotating magnetic field forces the rotor to turn, thereby creating mechanical motion. Each incremental movement of the rotor of 45° is called a *half-step*. Controlling excitation of the stator coils determines the position of the rotor and its rotational speed.

A sample design specification for a stepper motor controller is given as follows:

Design a controller for a four-phase stepper motor having half-step resolution.

Refering to Figure 10.35, the design specification implies that we need to generate the following 8-count sequence, assuming active-1 "on" conditions.

$$ABCD = 1000, 1100, 0100, 0110, 0010, 0011, 0001, 1001, 1000, \dots$$
$$\longmapsto \text{time}$$

The eight codes refer to the eight stepper motor configurations in Figure 10.35. Starting at the top of the figure and moving clockwise, the first configuration is denoted by the first code $ABCD = 1000$; coil A is on and coils B, C, and D are off. The next configuration (upper right) is denoted by the second code $ABCD = 1100$; coils A and B are on and coils C and D are off. The remaining configurations are denoted in a similar fashion.

---

[2] The polarity of an electromagnet is determined by the right hand rule: Wrap the fingers of your right hand around the coil in the direction of current flow and your thumb points to the north pole.

Figure 10.35

Creating a rotating magnetic field and torque

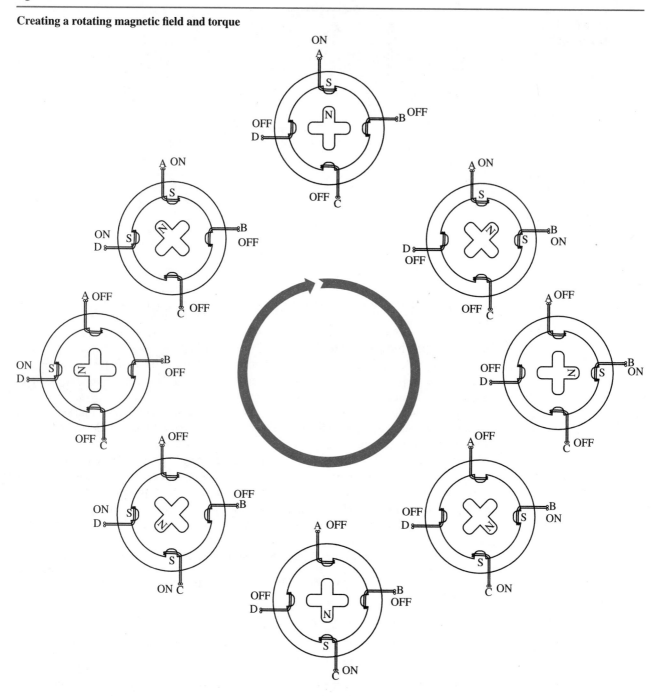

Associating count sequencing with state transitions suggests the Moore machine state diagram shown in Figure 10.36. There are no redundant states because each state generates a unique output.

Since the Armstrong-Humphrey Rules for state assignment do not yield any state adjacencies, we will assume the straight binary encoding given in Table 10.8. The encoded state table for the stepper motor controller is given in Table 10.9.

Figure 10.36

A Moore machine state
diagram for a stepper
motor controller

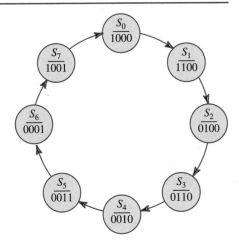

Table 10.8

State assignment for a stepper motor controller	State	Binary encoding $(Q_A Q_B Q_C)$
	S_0	000
	S_1	001
	S_2	010
	S_3	011
	S_4	100
	S_5	101
	S_6	110
	S_7	111

Table 10.9

Stepper motor controller encoded state table	Present state $(Q_A Q_B Q_C)$	Next state $(Q_A^+ Q_B^+ Q_C^+)$	Output $(C_A C_B C_C C_D)$
	000	001	1000
	001	010	1100
	010	011	0100
	011	100	0110
	100	101	0010
	101	110	0011
	110	111	0001
	111	000	1001

We generate the output Karnaugh maps given in Figure 10.37 directly from the encoded state table. We then derive the minimized sum-of-products Boolean expressions given below:

$$C_A = (\overline{Q_A} \cdot \overline{Q_B}) + (Q_A \cdot Q_B \cdot Q_C)$$
$$C_B = (\overline{Q_A} \cdot Q_C) + (\overline{Q_A} \cdot Q_B)$$
$$C_C = (Q_A \cdot \overline{Q_B}) + (\overline{Q_A} \cdot Q_B \cdot Q_C)$$

$$C_D = (Q_A \cdot Q_C) + (Q_A \cdot Q_B)$$

The flip-flop input Karnaugh maps are given in Figure 10.38. We have selected T flip-flops, which are often used for binary counters. We generate the T flip-flop input maps directly from the encoded state table using the simple excitation rule: $T_x = 0$ when $Q_x = Q_x^+$, otherwise $T_x = 1$. Again, we derive the minimized sum-of-products Boolean expressions.

$$T_A = Q_B \cdot Q_C$$
$$T_B = Q_C$$
$$T_C = 1$$

The implementation of our stepper motor controller is given in Figure 10.39. Next state and output logic is realized using **nand** gates.

10.6.2 Sequential 4-Input Majority Monitor

Having employed the 4-input majority function throughout our discussions of digital systems, it is fitting that we build a sequential 4-input majority monitor as our final design exercise. The design specification might be given as follows:

Design a sequential 4-input majority monitor that accepts four inputs and asserts the output if at least three of the four inputs are asserted. Groups of four inputs do not overlap.

Figure 10.37

Output Karnaugh maps for a stepper motor controller

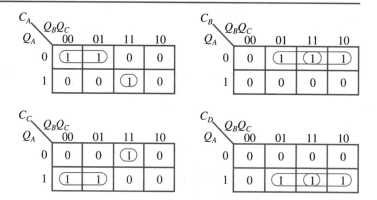

Figure 10.38

T flip-flop Karnaugh maps for a stepper motor controller

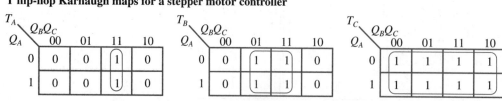

Figure 10.39

**Stepper motor controller
implementation**

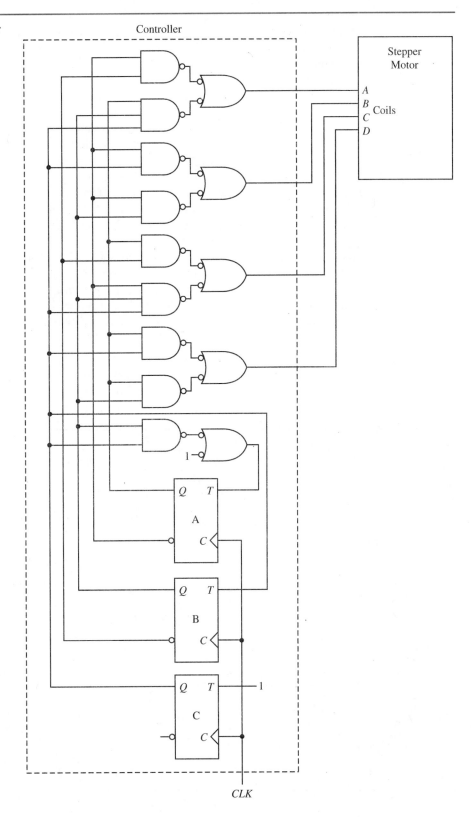

First, we construct the state diagram given in Figure 10.40. Since the stepper motor controller design exercise in the previous section is a Moore machine, we opted for a Mealy machine this time. It is convenient to think of the organization of the state diagram as being similar to the layout of a pinball machine. Depending on the values of four inputs, one of the paths down through the state diagram is traversed and the appropriate output is generated.

Next, we check for redundant states. Inspecting the state table in Table 10.10 shows that the rows for the present states S_7, S_8, S_9, and S_{11} are identical; likewise the rows for the present states S_{10}, S_{12}, and S_{13} are identical. Thus, $S_7 \equiv S_8 \equiv S_9 \equiv S_{11}$ and $S_{10} \equiv S_{12} \equiv S_{13}$. Ooops, looks like we were a bit too hasty in constructing the original state diagram!

A reduced state table is given in Table 10.11, derived from Table 10.10 by deleting the rows for the present states S_8, S_9, and S_{11} and replacing all remaining references to these states with S_7. Similarly, the rows for the present states S_{12} and S_{13} are also deleted, and all remaining references to these states are replaced with S_{10}.

This reduced state table shows that there is yet another redundant state; the rows for the present states S_4 and S_5 are identical, so $S_4 \equiv S_5$. Deleting the row for the present state S_5 and replacing all other references to S_5 with S_4 yields the new reduced state table in Table 10.12. The state table in Table 10.12 cannot be reduced any further (we encourage you to form an implication table to verify this result).

Now we can generate binary encodings for the nine states comprising our sequential 4-input majority monitor. Adjacencies for the Armstrong-Humphrey Rules are given below.

- States having the same next state for a given input should have adjacent state assignments.

$$S_3, S_4 \quad S_7, S_{10}, S_{14}$$

- The next states of a given state should have adjacent state assignments.

$$S_1, S_2 \quad S_3, S_4 \quad S_4, S_6 \quad S_7, S_{10} \quad S_{10}, S_{14}$$

Figure 10.40

A state diagram for a sequential 4-input majority monitor

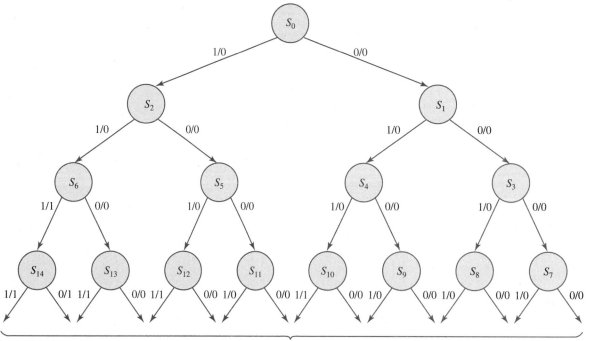

Table 10.10

State table for a sequential 4-input majority monitor	Present state	Next state		Output (MAJORITY)	
		$X = 0$	$X = 1$	$X = 0$	$X = 1$
	S_0	S_1	S_2	0	0
	S_1	S_3	S_4	0	0
	S_2	S_5	S_6	0	0
	S_3	S_7	S_8	0	0
	S_4	S_9	S_{10}	0	0
	S_5	S_{11}	S_{12}	0	0
	S_6	S_{13}	S_{14}	0	1
	S_7	S_0	S_0	0	0
	S_8	S_0	S_0	0	0
	S_9	S_0	S_0	0	0
	S_{10}	S_0	S_0	0	1
	S_{11}	S_0	S_0	0	0
	S_{12}	S_0	S_0	0	1
	S_{13}	S_0	S_0	0	1
	S_{14}	S_0	S_0	1	1

Table 10.11

Reduced state table for a sequential 4-input majority monitor	Present state	Next state		Output (MAJORITY)	
		$X = 0$	$X = 1$	$X = 0$	$X = 1$
	S_0	S_1	S_2	0	0
	S_1	S_3	S_4	0	0
	S_2	S_5	S_6	0	0
	S_3	S_7	S_7	0	0
	S_4	S_7	S_{10}	0	0
	S_5	S_7	S_{10}	0	0
	S_6	S_{10}	S_{14}	0	1
	S_7	S_0	S_0	0	0
	S_{10}	S_0	S_0	0	1
	S_{14}	S_0	S_0	1	1

Table 10.12

Reduced state table for a sequential 4-input majority monitor	Present state	Next state		Output (MAJORITY)	
		$X = 0$	$X = 1$	$X = 0$	$X = 1$
	S_0	S_1	S_2	0	0
	S_1	S_3	S_4	0	0
	S_2	S_4	S_6	0	0
	S_3	S_7	S_7	0	0
	S_4	S_7	S_{10}	0	0
	S_6	S_{10}	S_{14}	0	1
	S_7	S_0	S_0	0	0
	S_{10}	S_0	S_0	0	1
	S_{14}	S_0	S_0	1	1

- States having the same output for a given input should have adjacent state assignments.

$$S_0, S_1, S_2, S_3, S_4, S_7, S_{10} \quad S_0, S_1, S_2, S_3, S_4, S_6, S_7 \quad S_{10}, S_{14} \quad S_6, S_{14}$$

A state assignment satisfying the suggested state adjacencies is shown in Figure 10.41 and Table 10.13.

Assuming D flip-flops, Karnaugh maps for D_A, D_B, D_C, and D_D are given in Figure 10.42. We now derive the minimized sum-of-products Boolean expressions.

$$D_A = (X \cdot Q_B \cdot Q_C \cdot Q_D)$$
$$D_B = Q_C + (X \cdot \overline{Q_B} \cdot Q_D)$$
$$D_C = (\overline{Q_B} \cdot Q_D) + (X \cdot \overline{Q_B} \cdot \overline{Q_C})$$
$$D_D = (\overline{Q_B} \cdot \overline{Q_C} \cdot \overline{Q_D}) + (\overline{Q_A} \cdot Q_B \cdot Q_C \cdot Q_D) + (X \cdot Q_C \cdot Q_D) + (X \cdot Q_B \cdot Q_C)$$

The following Karnaugh map for the output *MAJORITY* yields

$$MAJORITY = Q_A + (X \cdot Q_B \cdot Q_D)$$

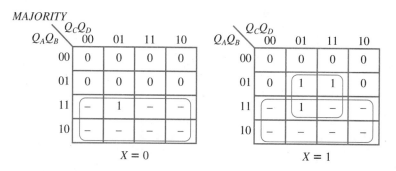

Figure 10.41

State assignment adjacencies for a sequential 4-input majority monitor

Q_AQ_B \ Q_CQ_D	00	01	11	10
00	S_0	S_1	S_2	S_3
01	S_7	S_{10}	S_6	S_4
11		S_{14}		
10				

Table 10.13

State assignment for a sequential 4-input majority monitor

State	Binary encoding $(Q_AQ_BQ_CQ_D)$
S_0	0000
S_1	0001
S_2	0011
S_3	0010
S_4	0110
S_6	0111
S_7	0100
S_{10}	0101
S_{14}	1101

Figure 10.42

Karnaugh maps for D flip-flop inputs

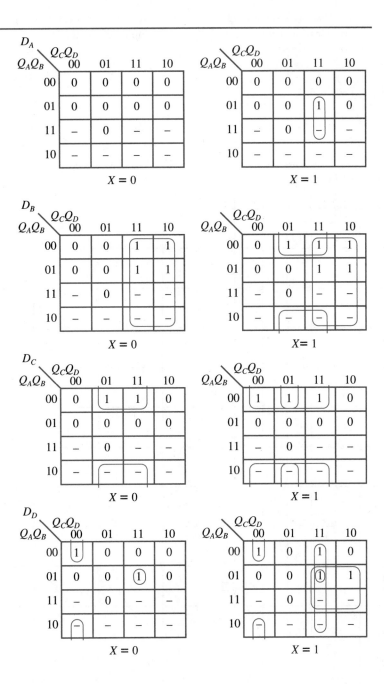

The final sequential 4-input majority monitor design is shown in Figure 10.43; the combinational logic and memory are implemented using a sequential programmable logic array.

Figure 10.43

Final design for the sequential 4-input majority monitor

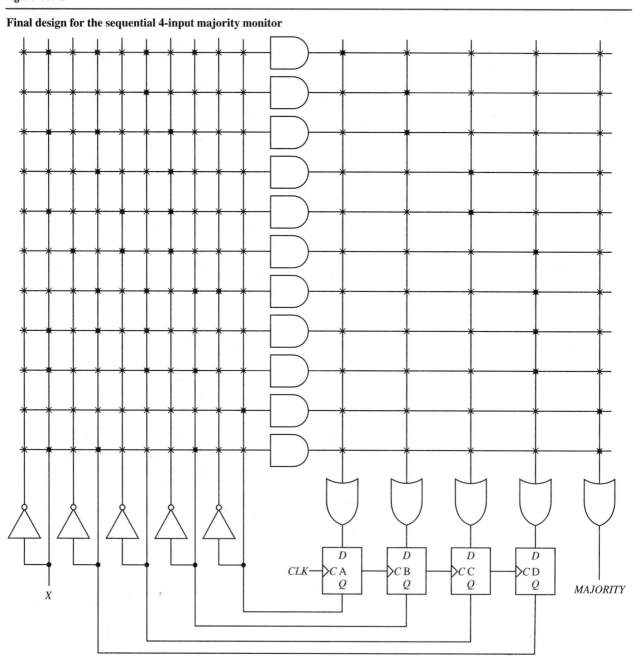

10.7 SUMMARY

With this chapter we have finished our study of sequential system design. Chapter 8 discussed the definition and analysis of sequential systems. Chapter 9 discussed the synthesis of sequential systems, generating logic expressions describing a desired state machine from an initial functional specification. Finally, this chapter discussed the implementation of sequential systems, realizing the logic expressions in hardware.

We saw that the combinational logic implementation strategies studied in Chapter 5 can be used to realize the next state and system output logic portions of sequential systems. Many factors can influence the choice of a particular combinational logic implementation strategy, such as don't care states, logic complexity, and design time. Memory and decoders that are well suited to realize multiple-output logic functions are often associated with sequential systems, but these options unfortunately do not take advantage of logic minimization.

We also learned that registers can implement state storage or, with a little additional logic, entire sequential systems, such as ring counters or finite sequence detectors. Sequential systems can often be easily and quickly implemented using counters. A wide selection of counters are available for realizing a variety of counting functions and near-cyclic state machines.

Finally, we discussed using registered memory, registered programmable logic array, and registered programmable array logic to realize sequential systems. Take care to understand these sequential programmable logic devices and their applications; they are rapidly emerging as popular techniques for realizing low-to-moderately complex digital systems.

10.8 PROBLEMS

1. Implement the nonresetting sequence detector described in Table 10.3 using a 4-to-16 decoder, **or** gates, and positive edge-triggered D flip-flops.

2. Design a resetting Moore machine that detects the number 9_{10} encoded in binary, 1001_2 and excess-3 1100_2. Sample input (X) and output (Z) sequences are given below.

$$X = 100110001100 \ldots$$
$$Z = 000100000001 \ldots$$
$$\longmapsto \text{time}$$

Show the minimal state diagram, state assignment, and logic minimization. Use the quad register given in Figure 10.6 to implement the flip-flops.
   a. Use two-level **nand-nand** networks to implement the combinational logic.
   b. Use two-level **nor-nor** networks to implement the combinational logic.
   c. Generate a complete state diagram showing unused states for the implementations in (a) and (b), and compare the designs.

3. For the resetting Moore machine described in Problem 2, use the programmable logic array shown below to implement the combinational logic.

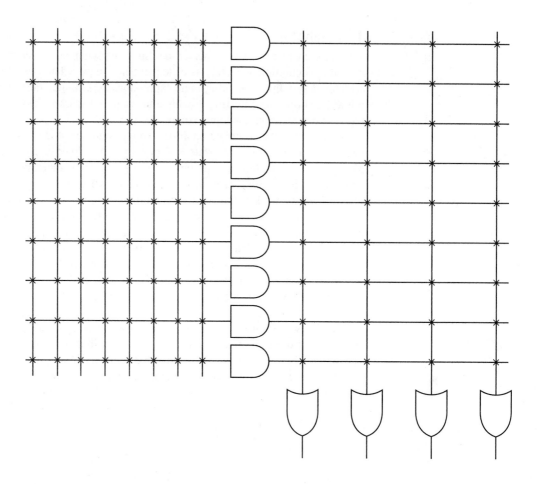

4. For the resetting Moore machine described in Problem 2,
   a. Use 8-input and 4-input multiplexers to implement the combinational logic.
   b. Use a 4-to-16 decoder and some **or** gates to implement the combinational logic.
   c. Comment on the advantages/disadvantages and similarities/differences of the implementations in (a) and (b).

5. Design a Mealy machine to control an elevator. The controller accepts a 2-bit input that represents the floor request, *FR*, as shown below.

$FR_1 FR_0$	Floor request
00	No request
01	First floor
10	Second floor
11	Third floor

The controller also generates two output signals to control the up (*UP*) and down (*DOWN*) movement of the elevator.

UP DOWN	Elevator movement
00	No movement
01	Down movement
10	Up movement
11	Undefined

Once a passenger enters a floor request, the input persists until the elevator reaches the desired floor and then the input resets to "no request." Show the minimal state diagram, state assignment, and logic minimization. Use the quad register given in Figure 10.6 to implement the flip-flops.

  a. Use two-level **nand-nand** networks to implement the combinational logic.
  b. Use two-level **nor-nor** networks to implement the combinational logic.
  c. Generate a complete state diagram showing unused states for the implementations in (a) and (b), and compare the designs.

6. As a safety measure, design the elevator controller implementations in Problem 5 so that unused states never assert elevator movement (in other words, always output 00).

7. For the elevator controller described in Problem 5, use the programmable logic array shown in Problem 3 to implement the combinational logic.

8. For the elevator controller described in Problem 5,
  a. Use 8-input multiplexers to implement the combinational logic.
  b. Use a 4-to-16 decoder and some **or** gates to implement the combinational logic.
  c. Comment on the advantages/disadvantages and similarities/differences of the implementations in (a) and (b).

9. Draw the state diagram of the ring counter shown in Figure 10.13 with and without the self-correcting feedback logic.

10. Generate a timing diagram showing the outputs of the ring counter given in Figure 10.13 for 10 clock cycles. Assume an initial state of $Q_A Q_B Q_C Q_D = 1000$.

11. Draw the state diagram of the twisted ring counter shown in Figure 10.14 with and without the self-correcting feedback logic.

12. Generate a timing diagram showing the outputs of the twisted ring counter given in Figure 10.14 for 13 clock cycles. Assume an initial state of $Q_A Q_B Q_C Q_D = 1111$. What is the relationship between the Q_A, Q_B, Q_C, and Q_D waveforms?

13. Long twisted ring counters are often used in digital waveform generation applications. Use the four waveforms generated in Problem 12 to generate another waveform. For each clock cycle, the magnitude of the new waveform is the sum of the number of 1's present on the outputs $Q_A Q_B Q_C Q_D$. What is the general shape of this new digital waveform?

14. A *linear feedback shift register* can be built by driving the input D_A of the four-stage shift register shown in Figure 10.9 with $Q_C \oplus Q_D$.
  a. Draw the linear feedback shift register and show that it generates a $2^4 - 1$ count sequence.
  b. What happens when $Q_A Q_B Q_C Q_D = 0000$?

15. Show how the resetting Moore machine described in Problem 2 can be implemented using the universal shift register shown in Figure 10.10.

16. For the 4-bit shift register shown in Figure 10.9,
  a. Configure the shift register to divide the clock frequency by 2.
  b. Configure the shift register to divide the clock frequency by 3.
  c. Configure the shift register to divide the clock frequency by 4.
  d. Configure the shift register to divide the clock frequency by 5.

17. Design a modulo-100 counter using the 4-bit binary counter shown in Figure 10.18.

18. Modify the 4-bit binary counter shown in Figure 10.18 to count in only even numbers, that is,

$$0_{10}, 2_{10}, 4_{10}, \ldots \ 12_{10}, 14_{10}, 0_{10}, 2_{10}, \ldots$$

19. Design a modulo-50 counter using the BCD counter shown in Figure 10.19.

20. Implement the elevator controller designed in Problem 5 using the BCD counter in Figure 10.19. Explain how the states and state transitions are realized.

21. An *asynchronous* or *ripple* counter can be designed by cascading a series of flip-flops; the flip-flop output of each stage is connected to the *clock* input of the next stage. For the following ripple counter, generate a timing diagram showing the outputs of the T flip-flops for 16 clock cycles. Assume an initial state of $Q_A Q_B Q_C Q_D = 0000$.

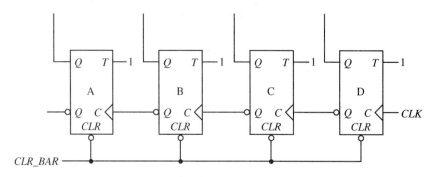

22. Design a 4-bit ripple up-counter using positive edge-triggered D flip-flops.

23. Two disadvantages of a ripple counter are slow counting speed and false intermediate count outputs. Assuming that the propagation delay of each T flip-flop in Problem 21 is T_{FF}, generate a timing diagram showing $Q_A Q_B Q_C Q_D$ transitioning from 1111_2 to 0000_2. Identify false intermediate counts.

24. Design a 4-bit ripple down counter using positive edge-triggered T flip-flops. Use the 4-bit ripple down-counter to build an 8-bit ripple down counter.

REFERENCES AND READINGS

[1] *Advanced Micro Devices.* Sunnyvale, CA: PAL Device Data Book, 1992.
[2] Barna, A. *High Speed Pulse & Digital Techniques.* New York: Wiley, 1980.
[3] Boyce, J. *Digital Logic.* Englewood Cliffs, NJ: Prentice-Hall, 1982.
[4] Castellucis, R. *Digital Circuits and Systems.* Reston, VA: Reston Publishing, 1981.
[5] Deem, B., and K. Muchow. *Digital Computer Circuits and Concepts.* Reston, VA: Reston Publishing, 1980.
[6] Frenzel, L. *Digital Computer Handbook.* Indianapolis, IN: H. W. Sams Publishing, 1981.
[7] Heiserman, D. *The Handbook of Digital IC Applications.* Englewood Cliffs, NJ: Prentice-Hall, 1980.
[8] Lenk, J. *Handbook of Digital Electronics.* Englewood Cliffs, NJ: Prentice-Hall, 1981.
[9] Levine, M. *Digital Theory and Experimentation Using Integrated Circuits.* Englewood Cliffs, NJ: Prentice-Hall, 1982.
[10] Mead, C. and L. Conway. *Introduction to VLSI Systems.* Reading, MA: Addison-Wesley, 1980.
[11] Middleton, R. *Digital Logic Circuits.* Indianapolis, IN: H. W. Sams Publishing, 1982.
[12] Strangio, C. *Digital Electronics: Fundamental Concepts and Applications.* Englewood Cliffs, NJ: Prentice-Hall, 1980.
[13] Taub, H. *Digital Circuits and Microprocessors.* New York: McGraw-Hill, 1982.
[14] Triebel, W. *Integrated Digital Electronics.* Englewood Cliffs, NJ: Prentice-Hall, 1979.
[15] Winkel, D. *The Art of Digital Design: An Introduction to Top-Down Design.* Englewood Cliffs, NJ: Prentice-Hall, 1980.

Part 4

VHDL:

COMBINATIONAL

SYSTEMS

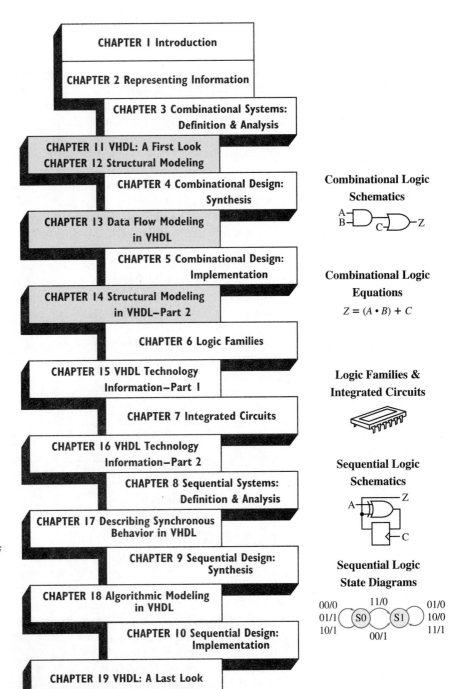

**Structural
Modeling**

```
I1: NOT_OP
  port map (A, B);
```

**Asynchronous
Data Flow
Modeling**

```
B <= not A;
```

**Technology
Modeling**

```
B <= not A after 5 ns;
```

**Synchronous
Data Flow
Modeling**

```
B1: block (C = '1')
begin
  B <= guarded not A after 5 ns;
end block B1;
```

**Algorithmic
Modeling**

```
P1: process (PS)
begin
  NS <= STATE_TABLE(PS);
end process P1;
```

CHAPTER 1 Introduction

CHAPTER 2 Representing Information

CHAPTER 3 Combinational Systems:
Definition & Analysis

CHAPTER 11 VHDL: A First Look
CHAPTER 12 Structural Modeling

CHAPTER 4 Combinational Design:
Synthesis

**CHAPTER 13 Data Flow Modeling
in VHDL**

CHAPTER 5 Combinational Design:
Implementation

**CHAPTER 14 Structural Modeling
in VHDL–Part 2**

CHAPTER 6 Logic Families

**CHAPTER 15 VHDL Technology
Information–Part 1**

CHAPTER 7 Integrated Circuits

**CHAPTER 16 VHDL Technology
Information–Part 2**

CHAPTER 8 Sequential Systems:
Definition & Analysis

**CHAPTER 17 Describing Synchronous
Behavior in VHDL**

CHAPTER 9 Sequential Design:
Synthesis

**CHAPTER 18 Algorithmic Modeling
in VHDL**

CHAPTER 10 Sequential Design:
Implementation

CHAPTER 19 VHDL: A Last Look

**Combinational Logic
Schematics**

$Z = (A \cdot B) + C$

**Combinational Logic
Equations**

$Z = (A \cdot B) + C$

**Logic Families &
Integrated Circuits**

**Sequential Logic
Schematics**

**Sequential Logic
State Diagrams**

00/0 11/0 01/0
01/1 (S0) (S1) 10/0
10/1 00/1 11/1

11
VHDL: A FIRST LOOK

VHDL is a language for describing digital systems. There are two objectives behind introducing VHDL as part of digital system design. First, VHDL embodies the basic concepts and fundamentals of digital systems. Thus, learning VHDL reenforces digital design theory. Second, VHDL is rapidly emerging as an international interoperability standard for design automation—in other words, a formal mechanism to "converse" with design automation tools. Thus, learning VHDL enables engineers to "speak the language" of design automation systems in common use in industry and to contribute more readily to digital design projects.

It is important to understand the relationship illustrated in Figure 11.1 between a VHDL description, also called a *model*, and a piece of digital hardware. VHDL describes what a digital system does and how the system does it. Once written, a VHDL model can be executed by a software program called a *simulator*. A simulator runs a VHDL description, computing the outputs of the modeled digital system in response to a series of inputs applied over time. In other words, a simulator exercises the dynamic behavior of a VHDL model. If a VHDL model accurately reflects the function of the associated actual hardware, then the input/output behavior obtained from the simulator in Figure 11.1(a) should match the input/output behavior obtained from the actual hardware in Figure 11.1(b).

Thus, VHDL provides a "high-tech blackboard" for designing digital systems. An initial design described as a VHDL model is progressively expanded and refined by repeatedly simulating the model, examining the results, and modifying the model. The final model is sufficiently detailed to serve as both a specification for actual hardware implementation and as a documentation aide to communicate the design to other groups or organizations.

We begin our study of VHDL by learning how to describe combinational digital systems defined in Chapter 3. This chapter discusses VHDL models of basic switching algebra operators; Chapter 12 discusses VHDL models of interconnected switching algebra operators forming logic schematics. As a precursor to these discussions, the following section explains and motivates the manner in which VHDL models will be presented.

11.1 VHDL PRESENTATION AND EXAMPLES

VHDL is presented from a *design perspective* versus a language theoretic perspective. That is, VHDL is explained by example. First we propose a digital system, then we present a VHDL model of the digital system, and then we explain the salient aspects of the VHDL model. We use this instructional approach so that you will understand the motivation for using a particular VHDL construct before learning the details of the construct. However, with this approach, constructs closely related from a language theoretic perspective may sometimes appear in different sections or even different chapters. To help maintain continuity of language issues, we will often keep a "running tab" of similar constructs discussed in different places (and, an extensive index is included at the end of the text).

Figure 11.1

Relationship between a VHDL model and hardware

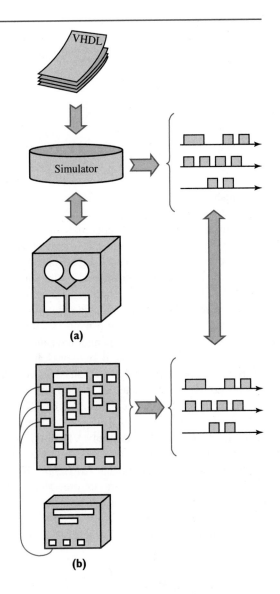

(a)

(b)

In our discussion of VHDL we also emphasize modeling *rules* and *practices*. Rules strictly define what constructs are allowed in the language and how to write the constructs. Rules are typically given in a language reference manual (LRM) [7]. Practices are guidelines on how to write constructs to reduce modeling errors and improve readability. Unlike modeling rules, which must be adhered to, modeling practices are suggestions that a designer may or may not choose to adopt. Practices are typically given in a user's guide (UG).

As an example, consider the portion of a typical VHDL model shown in Figure 11.2 (p. 344). It is not important at this time to understand this VHDL code segment; just note that a VHDL model is composed of *statements* and the code segment contains a **port** statement embedded within an **entity** statement.

Figure 11.2 illustrates modeling rules, such as that an **entity** statement starts with the word "**entity**," followed by the name of the entity "EXAMPLE_DESIGN," followed by the word "**is**," and so on. Figure 11.2 also illustrates modeling practices, such as that certain

Figure 11.2

VHDL modeling rules
versus practices

```
entity EXAMPLE_DESIGN is      -- Entity statement
  port                        -- Port statement
    (A, B : in  BIT;
     X, Y : out BIT);

    ⋮

  end EXAMPLE_DESIGN;
```

words are written in uppercase, certain parts of statements are written on separate lines, and certain lines are indented.

We will adopt the modeling practice of indenting statements that are embedded within other statements. Also, we will show words that the designer is free to choose, called *identifiers*, in uppercase letters in standard weight (such as the name of the entity, EXAMPLE_DESIGN). For words that are mandated by the language that the designer must use, called *keywords*, we will use lowercase letters in bold weight.

VHDL keywords are *reserved* keywords, meaning that they cannot be used for any other purpose. Anywhere we are allowed to choose a name, such as an entity name, we cannot choose a reserved keyword. For example, we cannot have an entity named "PORT" because **port** is a reserved keyword. (VHDL reserved keywords are listed in Appendix B.) Additional useful VHDL modeling practices will be explained as we discuss the language.

As a final point concerning the style and organization of VHDL presentation and examples, in this text we present two versions of VHDL: VHDL-87 and VHDL-93. As explained in Chapter 1, VHDL was first adopted as an IEEE standard in 1987, and the associated version is designated *IEEE Std 1076-1987* or simply VHDL-87. VHDL was recertified as an IEEE standard in 1993, and the recertification process introduced several modifications and enhancements yielding a new version designated *IEEE Std 1076-1993* or simply VHDL-93. Presently, both VHDL-87 and VHDL-93 are in use. Hence, we emphasize VHDL-87, but introduce, where appropriate, new language and hardware modeling capabilities of VHDL-93. Minor VHDL-87/VHDL-93 differences are discussed in footnotes; major VHDL-87/VHDL-93 differences are discussed in separate sections.

11.2 BASIC LANGUAGE ORGANIZATION

Let us introduce the general organization of VHDL by looking at example descriptions of basic switching algebra operations studied in Chapter 3. For instance, Figure 11.3 shows the graphical symbol and VHDL model of a 2-input **xor** operation. The **xor** logic operator is defined in Figure 3.10.

The VHDL model shown in Figure 11.3(b) comprises a *design entity*, which is the basic construct in VHDL for modeling a digital system. Figure 11.4 shows that a design entity can represent an arbitrarily complex digital system, ranging from a simple combinational system to an entire network of computers. The digital system can be a physical piece of hardware that has been designed or a conceptual piece of hardware that is being designed.

Figures 11.3 and 11.4 show that a VHDL design entity is composed of two parts: an *interface* and a *body*. The interface is denoted by the keyword **entity** and the body is denoted by the keyword **architecture**. A convenient way to view the roles of the interface and body is to imagine a digital system enclosed inside a casing or black box. The *interface*

Figure 11.3

A VHDL model of 2-input exclusive disjunction operation

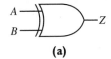

(a)

```
-- Interface
entity XOR2_OP is
  -- Input/output ports
  port
   (A, B : in  BIT;
    Z    : out BIT);
end XOR2_OP;


-- Body
architecture EX_DISJUNCTION of XOR2_OP is
begin
 Z <= A xor B;
end EX_DISJUNCTION;
```

(b)

Figures (a) and (b) show the logic symbol and VHDL description of a 2-input **xor** operator, respectively.

Figure 11.4

A VHDL design entity

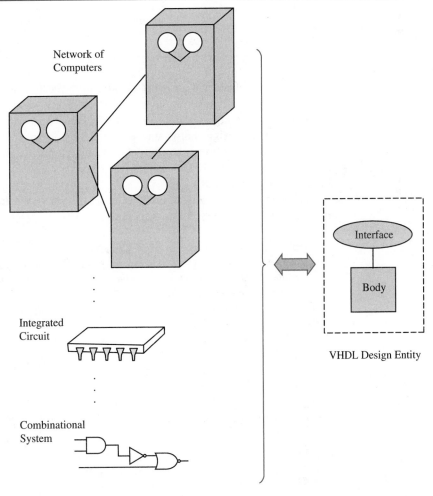

Network of Computers

Integrated Circuit

Combinational System

Digital System

Interface

Body

VHDL Design Entity

describes aspects of the digital system visible outside the black box that define the boundary between the system and its environment, such as signals that flow into and out of the box. The *body* describes aspects of the digital system inside the black box that define how the outputs respond to the inputs.

The division of the design entity into an interface and body is an instance of an important design principle of VHDL: the separation of *specification* from *implementation*. It is often helpful to separate *what* a design does (specification) from *how* a design operates (implementation) because there are usually many different ways to implement or realize a desired function. Each possible implementation possesses a unique set of performance/cost trade-offs.

11.3 INTERFACE

Let us investigate in more detail the interface of the 2-input **xor** logic operator model, repeated below.

```
entity XOR2_OP is
  -- Input/output ports
  port
   (A, B : in  BIT;
    Z    : out BIT);
end XOR2_OP;
```

The interface is defined via an *entity declaration*. The first line of the entity declaration, called the *header*, starts with the keyword **entity**, followed by the name of the design entity, followed by the keyword **is**. An entity declaration closes with the keyword **end**, followed again by the design entity name. Repeating the design entity name at the closing of an entity declaration is optional; however, this modeling practice is recommended to improve readability. The entity declaration is terminated with a semicolon. In general, semicolons terminate statements in VHDL.[1]

The design entity name is an example of an *identifier*. We are free to choose the name, provided we adhere to the following rules.

Identifiers
- *An identifier can be any length, in other words, as many characters as desired.*
- *An identifier is case insensitive, meaning that there is no difference between uppercase and lowercase letters. (For example XOR2_OP and XOr2_Op denote the same identifier.)*
- *The allowed characters are a–z (lowercase letters), A–Z (uppercase letters), 0–9 (numerals), and _ (underscore).*
- *The first character must be a letter and the last character must not be an underscore.*
- *No adjacent underscores are allowed.*

---

[1] <u>VHDL-93</u>: In VHDL-93, an entity declaration may close with the keywords **end entity**.

```
entity XOR2_OP is
      ⋮
end entity XOR2_OP;
```

We discuss a more general set of rules for identifiers provided in VHDL-93, *extended identifiers*, in Subsection 11.3.2.

The lines beginning with two dashes, "--," are *comments*. *All text* after the leading two dashes to the end of the line is considered part of the comment. Since the end of a line terminates a comment, comments that span multiple lines must have two leading dashes on each line. Comments can be placed anywhere within a VHDL model, even in the middle of a statement.

```
port
  (A,                    -- Input from machine A
   B : in  BIT;          -- Input from machine B
   Z : out BIT);
```

Comments provide a free form or unstructured way to annotate a VHDL model. Comments help document a VHDL model and should be used liberally to make a model easy to read and understand.

The keyword **port** introduces *port* declarations. In VHDL, ports are signals that flow into and/or out of a design entity. Ports are declared by specifying (1) the name, (2) the direction of information flow, and (3) the type of information the signal carries. The name of a port can be any legal identifier; we will follow the signal naming conventions discussed in Chapter 3.

The direction of information flow is called the *mode* of a port. VHDL provides the following five port modes.

1. **in**   (port information flows only into the design entity)
2. **out**   (port information flows only out of the design entity)
3. **inout**   (port information flows into and out of the design entity)
4. **linkage**
5. **buffer**

The mode may be omitted in a port declaration, in which case port mode **in** is assumed. The **linkage** and **buffer** port modes are used in special-purpose hardware modeling situations and will be discussed in later chapters.

The type of information declares the set of legal values for the port. VHDL provides a set of *predefined* information types that are commonly used in hardware modeling. For instance, the ports of the XOR2_OP design entity are of the predefined type BIT that has two values '0' and '1' representing the binary values "0" and "1," respectively. The values '0' and '1' are called BIT *literals*. The enclosing single quotes are mandatory to differentiate logic and integer values. Ports carrying the same type of information with the same direction of information flow can be collectively defined in a single declaration or individually defined in separate declarations.

```
entity XOR2_OP is
  -- Input/output ports
  port
    (A : in  BIT;
     B : in  BIT;
     Z : out BIT);
end XOR2_OP;
```

11.3.1 Types and Declarations

The type of information of a port introduced in the previous section illustrates an important characteristic of VHDL: the language is *strongly typed*. Every object that can represent a value, such as a signal, has a name and a *type*. The name allows referencing the value to perform useful computations or transformations. The type strictly defines the set of values the object may take on. The name and type of an object must be declared before the object can be used. For the XOR2_OP model, the ports A, B, and Z must be declared in the interface before they can be used in the architecture body to describe the exclusive disjunction logic operation.

Data typing provides a means to organize and structure information and, as such, facilitates creating models that more precisely describe hardware. Data typing also helps reduce modeling errors. If a designer mistakenly uses a signal improperly by trying to assign an illegal value to the signal, the VHDL compiler and/or simulator will catch the error. Without strong typing, such errors may be difficult to find in a large model.

11.3.2 VHDL-93: Extended Identifiers

The identifiers discussed in the previous section are called *basic identifiers* and their construction rules can, at times, be overly restrictive. For example, digital system data books often designate a part name with leading numerals, such as 74HC00. Digital system data books also often designate an active-0 signal state with a leading slash, such as /ALARM, or a trailing dash, such as ALARM-. These sample names are not legal VHDL names. The inability to express these names in VHDL impedes describing existing designs in VHDL and often leads to cumbersome and awkward work-arounds. Hence, VHDL-93 provides an enhanced set of rules for constructing identifiers that include both the basic identifiers and more general identifiers, called *extended identifiers*.

Rules for constructing *extended identifiers* in VHDL-93 are given below.

Extended Identifiers

- *An extended identifier can be any length, in other words, as many characters as desired.*
- *An extended identifier must be delimited by leading and trailing backslashes, \ (for example, \ 2XOR_OP \ ).*
- *The allowed characters are any graphic character. Graphic characters include all the characters allowed for VHDL-87 basic identifiers plus special characters such as dash "-," asterisk "*," A circumflex "Â," and e umlaut "ë." (The complete set of graphic characters is defined in the Language Reference Manual (LRM), Chapter 13 [8].)*
- *Within the enclosing backslashes, graphic characters can appear in any order, except that a backslash used as part of an extended identifier must be denoted by two adjacent backslashes. (For example, XOR \ 2 is denoted by \ XOR \\ 2 \ ).*
- *An extended identifier is case sensitive, meaning that there is a difference between uppercase and lowercase letters. (For example, \ XOR2_OP \ and \ XOr2_Op \ denote <u>different</u> identifiers.)*
- *An extended identifier is different from any keyword or basic identifier. (Thus, \entity\ is a legal extended identifier because it is different from the keyword **entity**. Also, the extended identifier \ XOR2_OP \ and the basic identifier XOR2_OP do not denote the same name.)*

11.4 ARCHITECTURE BODY

Having examined the interface of the XOR2_OP design entity, we can now consider the body, repeated below. Recall that the body describes how the design operates and is separate from the design entity interface to allow for different implementations.

```
-- Body
architecture EX_DISJUNCTION of XOR2_OP is
begin
 Z <= A xor B;
end EX_DISJUNCTION;
```

A body is also often called an *architecture* because the header of the definition of a body begins with the reserved keyword **architecture**. The header also declares the name of the architecture and the name of the associated interface. Thus, the line

```
architecture EX_DISJUNCTION of XOR2_OP is
```

declares an architecture named EX_DISJUNCTION and associates the architecture with the XOR2_OP design entity interface. An architecture is always associated with an entity interface. An architecture name can be any legal identifier and can even be the same as the associated design entity interface name. For example,

```
architecture XOR2_OP of XOR2_OP is
```

is legal and sometimes used as a convenient, though unimaginative, naming convention. Similar to a design entity interface declaration, an architecture body definition closes with the reserved keyword **end**, followed optionally by the name of the architecture. We will adopt the modeling practice of always repeating the architecture name at the closing.[2]

Following the header, an architecture is organized into two parts: *declarative part* and *statement part*.

```
-- Body
architecture EX_DISJUNCTION of XOR2_OP is
 -- Declarative part

   ⋮

begin
 -- Statement part

   ⋮

end EX_DISJUNCTION;
```

---

[2] <u>VHDL-93</u>: In VHDL-93, an architecture definition may close with the keywords **end architecture**.

```
architecture EX_DISJUNCTION of XOR2_OP is

   ⋮

end architecture EX_DISJUNCTION;
```

The declarative part extends from the header of the architecture body to the keyword **begin**, while the statement part extends from the keyword **begin** to the end of the architecture body. The division of an architecture into declarative and statement parts reflects the general strongly typed characteristic of VHDL: Objects must be declared before they are used. The EX_DISJUNCTION architecture does not contain any declarations; all objects (in other words, signals) used in the statement part of the architecture are declared in the interface. VHDL models using the architecture declaration section are shown in later sections.

11.4.1 Signal Transformations

The EX_DISJUNCTION architecture contains a single statement that describes the input/output behavior of the design entity XOR2_OP; the single statement is called a *signal assignment statement*.

```
Z <= A xor B;
```

Any time the input signals A and/or B change value, the signal assignment statement executes and computes a new value for the output signal Z as $A \oplus B$.

The exclusive disjunction logic operation is described by the **xor** *logic operator*. (VHDL formally defines **xor** as a *logical* operator, but we will use the term *logic* operator to be consistent with switching algebra.) The **xor** is a predefined logic operator having the conventional definition given in the truth table below.

X	Y	X **xor** Y
'0'	'0'	'0'
'0'	'1'	'1'
'1'	'0'	'1'
'1'	'1'	'0'

The symbol <= is an assignment operator; the value generated by evaluating the **xor** logic operator on the right-hand side is assigned to the target signal on the left-hand side.

11.5 LOGIC OPERATORS

The XOR2_OP design entity explained in the previous sections introduces logic operators in VHDL. Like types, VHDL provides a set of predefined logic operators that are commonly used in hardware modeling and are given below. (Refer to Chapter 3 for operator definitions.)

Operator	Definition
and	conjunction
or	disjunction
xor	exclusive disjunction
xnor[3]	complement exclusive disjunction
nand	complement conjunction
nor	complement disjunction
not	complement

---

[3] <u>VHDL-93</u>: The predefined logic operator **xnor** is provided only in VHDL-93.

The **not** logic operator is a *unary* operator taking one operand; all other logic operators are *binary* operators taking two operands.

The VHDL predefined logic operators can be combined to describe complex switching algebra operations. For example, Figure 11.5 shows how multi-input logic operations are modeled using a series of binary logic operators. A 3-input **and** operation forming the conjunction of three signals A, B, and C (in other words, $A \cdot B \cdot C$) is described using two **and** logic operators. As another example, Figure 11.6 shows an alternate description of the **xor** operation using **and**, **or**, and **not** VHDL logic operators.

There are a few rules for evaluating VHDL logic operators that unfortunately do not align exactly with switching algebra theory, so it is important to note them. The **and**, **or**, **nand**, and **nor** logic operators are called *short-circuit* operators because the right operand is *not* evaluated if the value of the left operand determines the result of the logic operator. For example, consider the following signal assignment statement.

```
TEST_SIG <= OPR_A and OPR_B;
```

Figure 11.5

A VHDL model of a 3-input conjunction switching algebra operation

```
-- Interface
entity AND3_OP is
  -- Input/output ports
  port
    (A, B, C : in  BIT;
     Z       : out BIT);
end AND3_OP;


-- Body
architecture CONJUNCTION of AND3_OP is
begin
  Z <= A and B and C;
end CONJUNCTION;
```

Figure 11.6

A VHDL model of a 2-input exclusive disjunction operation

```
-- Interface
entity XOR2_OP is
  -- Input/output ports
  port
    (A, B : in  BIT;
     Z    : out BIT);
end XOR2_OP;


-- Body
architecture AND_OR of XOR2_OP is
begin
  Z <= (not A and B) or (A and not B);
end AND_OR;
```

If the left operand, signal OPR_A, evaluates to '0', there is no need to evaluate the value of the right operand, signal OPR_B, because $0 \cdot OPR\_B = 0$, regardless of the value of OPR_B (Theorem 2b, Chapter 4). Although switching algebra operations do not have the short-circuit property, it is associated with VHDL logic operations.

VHDL logic operators are evaluated in an order determined by operator precedence: Operators of a higher precedence are evaluated before operators of a lower precedence. The unary **not** logic operator has a higher precedence than the binary logic operators, which all have an *equal*, lesser precedence. Although switching algebra defines the **and** operation to have a higher precedence than the **or** operation (see Chapter 3), the VHDL **and** and **or** logic operators have the *same* precedence.

Logic operators of the same precedence are evaluated from left to right in the textual order they appear. Thus, the signal assignment statement given in Figure 11.5,

```
Z <= A and B and C;
```

executes any time the signals A, B, and/or C change value and $A \cdot B \cdot C$ is generated by first **and**ing A and B and then **and**ing the result with C.

Parentheses further control the order of operator evaluation. Logic operators within parentheses are evaluated first, starting with the innermost, (most deeply nested) set of parentheses and working outward. Thus, the signal assignment statement given in Figure 11.6,

```
Z <= (not A and B) or (A and not B);
```

executes any time the signals A and/or B change value and $A \oplus B$ is generated by first evaluating the logic operators within the parentheses. By precedent rules, (**not** A **and** B) is evaluated by first complementing A and then **and**ing the result with B. Similarly, (A **and** **not** B) is evaluated by first complementing B and then **and**ing the result with A. Finally, the results of evaluating the two parenthetical expressions are **or**ed together and assigned to the output signal Z. Notice that the parentheses are *required* to correctly control the order in which the logic operators are evaluated. Since VHDL **and** and **or** operators have equal precedence, the parentheses force the **and** operators to be evaluated before the **or** operators.

11.6 CONCURRENCY

The signal assignment statements shown in Figures 11.3, 11.5, and 11.6 are more precisely called *concurrent signal assignment statements*. The term *concurrent* refers to how the signal assignment statements execute. Most statements found in popular software programming languages execute in a familiar *sequential* fashion. That is, statements execute only when encountered by the master, procedural flow-of-control and the textual order in which statements appear determines the order in which they execute.

In contrast, as the name implies, VHDL concurrent statements execute in a *concurrent* fashion. That is, statements execute only when associated signals change value. There is no master, procedural flow of control; each concurrent statement executes in a nonprocedural stimulus/response. Again, it is worth repeating that, for example, the single concurrent signal assignment statement describing the input/output behavior of the XOR2_OP entity executes any time the signals appearing on the right-hand side, A and/or B, change value. The semantics of concurrent signal assignment statements are designed to mimic the nonprocedural nature of hardware. All hardware elements of a design run in parallel or concurrently. Each hardware element responds to input changes and generates associated new outputs; the new outputs in turn cause new input changes, and the stimulus/response or activation/suspension cycle repeats.

To illustrate the execution semantics of concurrent signal assignment statements, consider yet another alternate description of the **xor** operation, given in Figure 11.7. The declarative part of the AND_OR_CONCURRENT architecture contains one declaration, namely, a *signal declaration*. The signal declaration

```
signal INT1, INT2 : BIT;
```

defines two signals named INT1 and INT2 of type BIT. Again, type BIT is a predefined type representing the binary values '0' and '1'. INT1 and INT2 are *internal* signals used within the architecture AND_OR_CONCURRENT to compute $A \oplus B$.

Figure 11.7

A VHDL model of a 2-input exclusive disjunction operation

```
-- Interface
entity XOR2_OP is
  -- Input/output ports
  port
    (A, B : in  BIT;
     Z    : out BIT);
end XOR2_OP;


-- Body
architecture AND_OR_CONCURRENT of XOR2_OP is
  -- Signal declarations
  signal INT1, INT2 : BIT;
begin
  -- Concurrent signal assignment statements
  INT1 <= A and not B;
  INT2 <= not A and B;
  Z <= INT1 or INT2;
end AND_OR_CONCURRENT;
```

A signal declaration starts with the reserved keyword **signal**, followed by the signal name(s) and type. Unlike ports, signals internal to a design entity do not have a mode or a direction of information flow. Similar to port declarations, signals of the same type can be collectively defined in a single declaration or individually defined in separate declarations.

```
signal INT1 : BIT;
signal INT2 : BIT;
```

The statement part of the architecture AND_OR_CONCURRENT contains three concurrent signal assignment statements.

```
INT1 <= A and not B;
INT2 <= not A and B;
Z    <= INT1 or INT2;
```

Each concurrent signal assignment statement executes only when associated signals appearing on the right-hand side change value. When the input signals A and/or B change value, the first two concurrent signal assignment statements execute and update the internal signals INT1 and INT2. Then, if INT1 and/or INT2 change value, the last concurrent signal assignment statement executes and updates the output signal Z.

The textual order in which concurrent statements appear has no effect on the order in which they execute. The signal assignment statements could have been written as

```
INT1 <= A and not B;
Z    <= INT1 or INT2;
INT2 <= not A and B;
```

and the input/output signal transform would remain unchanged.

11.7 DESIGN UNITS AND LIBRARIES

Although we have emphasized VHDL from a design perspective, we conclude this introductory chapter by examining VHDL from a language perspective to understand how various language elements that will be discussed in future chapters fit together and to introduce some computing environment issues associated with compiling and simulating VHDL models. To that end, Figure 11.8 shows a way of viewing the language organization of VHDL [9].

Starting from the base of the pyramid, predefined and user-defined *types* define data "templates" or sets. *Objects*, such as signals, hold values of the defined data types. *Expressions* combine operations with objects to yield new values, which are used by *statements* to

Figure 11.8

Language organization of VHDL

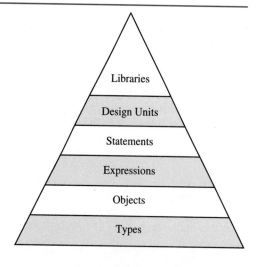

describe aspects of digital hardware. Statements are contained within *design units* and design units are, in turn, contained within *libraries*.

Design units and libraries pertain to how the analyzer (in other words, the compiler) and the underlying computing environment perceive VHDL. VHDL should ideally be independent of any particular computing environment. However, VHDL is actually realized by software that executes within a computing environment. Design units and libraries are links between the implementation-independent hardware description aspects of VHDL and the implementation-dependent aspects of the underlying computing environment.

Any portion of a VHDL model that can by itself by analyzed constitutes a design unit. The term *analyze* means checking the syntax and *static* semantics, in other words, checking for any errors that can be identified before simulation. The term *analyze* is used instead of the more familiar term *compile* because compiling typically infers the generation of executable or object code. With VHDL, code generation has historically been associated with simulation, which checks *dynamic* semantics or time-dependent behavior. The design units in VHDL are the following:

- Primary Design Units
  1. Entity Declaration
  2. Package Declaration
  3. Configuration Declaration
- Secondary Design Units
  1. Architectural Body
  2. Package Body

Configurations and packages will be discussed in later chapters. The division of *primary* and *secondary* design units conveys a restriction on the order in which design units may be analyzed. A secondary design unit may not be analyzed before its associated primary design unit. For instance, an architecture may not be analyzed before its associated design entity declaration is analyzed.

Design units are *conceptually* stored in design libraries. There may, for example, be a design library containing experimental design models. The creation of design libraries and their *physical* implementation via an operating system storage management system or a database management system is accomplished "outside" of VHDL. VHDL provides only the basic ability to specify a design library, thereby granting access to a certain set of design units.

There are two predefined libraries in VHDL: STD and WORK. The library STD provides declarations for predefined constructs in VHDL, such as predefined types. We have been able to use the predefined type BIT because its declaration is provided in the library STD and assumed to be automatically available. As the name implies, the library WORK is a synonym for the "working library," the library into which design units are presently being analyzed. For convenience, we will usually use the library WORK to store design entities, but will show in later chapters how to use user-defined libraries as well.

11.8 SUMMARY

We have taken our first look at VHDL by examining how to model the input/output signal transformations of basic switching algebra operations, such as **and** and **xor**. A VHDL model of a digital system is called a design entity, which is composed of an interface and a body. The division of a design entity into the interface and body represents the separation of specification from implementation. The interface defines external information, such as the name of the design entity and the input/output signals. The body or architecture defines internal information, namely, the input/output transformation.

The input/output signal transformation is described using concurrent signal assignment statements. Input signals appearing on the right-hand side of the assignment operator <= are combined with logic operators to yield a value that is assigned to the target signal on the left-hand side of the assignment operator. VHDL provides a standard set of logic operators to describe arbitrarily complex switching algebra expressions. Operator precedences, short-circuit operator properties, and parentheses determine how logic operators are evaluated.

Concurrent signal assignment statements execute in a nonprocedural fashion, meaning that there is no master, procedural flow-of-control that executes statements based on their textual order. Rather, each concurrent signal assignment statement controls its own activation. Whenever one or more signals appearing on the right-hand side of a concurrent signal assignment statement change value, the concurrent signal assignment statement activates, recomputes the desired logic transform, and updates the target signal. Understanding concurrent execution semantics can sometimes be challenging because it differs substantially from sequential execution semantics commonly found in most popular software programming language, and keeping track of the aggregate behavior of parallel actions can be difficult. However, over the course of this text, we will examine many example VHDL models using concurrent statements to gain proficiency.

We closed the chapter with an overview of the basic language structure of VHDL, involving a hierarchy of types, objects, expressions, statements, design units, and libraries. Types define data sets for objects. Objects hold values and these values are combined with operators within expressions to yield new values. Statements use expressions to describe hardware. Design units are independently analyzable pieces of VHDL. Libraries are conceptual groupings of design units. This structure serves as a conceptual scaffolding for understanding and relating various parts of VHDL that will be discussed in ensuing chapters. This structure also served as an introduction to issues defining the interface between the language and the underlying computing environment.

11.9 PROBLEMS

1. Describe the general roles of the following VHDL constructs in modeling switching algebra operations.
   a. Design entity   **b.** Interface   **c.** Architecture
   d. Signal declaration   **e.** Concurrent signal assignment statement

2. Data types are important to the structure and use of VHDL.
   a. What is a data type?
   b. What is the motivation for having data types in VHDL?
   c. What values does the predefined type `BIT` represent?

3. Determine whether each of the following identifiers are legal or illegal. If an identifier is illegal, explain why.
   a. `PORT`   **b.** `Status_of_Processor_X10672`   **c.** `3_IN_AND`
   d. `ASCII__DECODE`   **e.** `ALARM_SET2`   **f.** `MEMORY#ACCESS`
   g. `DATA_`

4. <u>VHDL-93</u>: Determine whether each of the following identifiers are legal or illegal. If an identifier is illegal, explain why.
   a. `signal`   **b.** `\signal\`   **c.** `/ALARM`   **d.** `\34μm\`
   e. `74HC00_102`   **f.** `\Address\\Enable`   **g.** `ON*`

5. For the following switching algebra operations, generate VHDL models and show by simulation that the models yield the correct outputs for all possible inputs.
   a. not   **b.** 2-input **and**   **c.** 2-input **or**   **d.** 2-input **xnor**

6. For the following switching algebra operations, generate VHDL models and show by simulation that the models yield the correct outputs for all possible inputs.
   a. buffer   **b.** 3-input **nand**   **c.** 4-input **and-or-invert**   **d.** 4-input **nor**

7.

```
entity LOGIC_OPERATION is
  port
    (A, B, C  : in  BIT;
     Z        : out BIT);
end LOGIC_OPERATION;

architecture MODEL of LOGIC_OPERATION is
begin
    Z <= (not A and not B and C) or
         (not A and B and not C) or
         (A and not B and not C) or
         (A and B and C);
end MODEL;
```

For the above VHDL model,
a. Generate the logic expression for LOGIC_OPERATION.
b. Generate a truth table, listing the inputs A, B, and C and the corresponding output Z.
c. What does LOGIC_OPERATION do? Write another version of LOGIC_OPERATION simplifying the concurrent signal assignment statement.
d. Verify part (c) with simulation, showing that the new version of LOGIC_OPERATION yields the truth table you generated in part (b).

8.

```
entity LOGIC_OPERATION is
  port
    (A, B  : in  BIT;
     Z     : out BIT);
end LOGIC_OPERATION;

architecture MODEL of LOGIC_OPERATION is
begin
    Z <= (A or B) and (A or not B);
end MODEL;
```

For the above VHDL model,
a. Generate the logic expression for LOGIC_OPERATION.
b. Generate a truth table, listing the inputs A and B and the corresponding output Z.
c. What does LOGIC_OPERATION do? Write another version of LOGIC_OPERATION simplifying the concurrent signal assignment statement.
d. Verify part (c) with simulation, showing that the new version of LOGIC_OPERATION yields the truth table you generated in part (b).

9. For each of the switching algebra operations below, generate a truth table and a VHDL model. Also, verify by simulation that the VHDL model exhibits the truth table behavior.

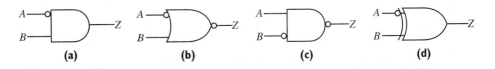

(a)　　　　(b)　　　　(c)　　　　(d)

10. A VHDL model that is *syntactically* correct is not necessarily *semantically* correct. It is possible to correctly describe the wrong logic. Explain why the logic expression

$$Z = A + B \cdot C$$

is not described by the following design entity and show how the problem can be fixed.

```
entity NICE_TRY is
  port
    (A, B, C : in  BIT;
     Z         : out BIT);
end NICE_TRY;

architecture MODEL of NICE_TRY is
begin
  Z <= A or B and C;
end MODEL;
```

11.

```
entity LOGIC_OPERATION is
  port
    (A, B, : in  BIT;
     Z      : out BIT);
end LOGIC_OPERATION;

architecture MODEL of LOGIC_OPERATION is
begin
  Z <= not (not A or not B);
end MODEL;
```

For the above VHDL model,
   a. Explain when the concurrent signal assignment statement executes.
   b. Explain how the right-hand side of the concurrent signal assignment statement is evaluated, in other words, the order in which the logic operators execute.
   c. Generate the logic expression for LOGIC_OPERATION.
   d. Generate a truth table, listing the inputs A and B and the corresponding output Z.
   e. What does LOGIC_OPERATION do? Write another version of LOGIC_OPERATION simplifying the concurrent signal assignment statement.
   f. Verify part (e) with simulation, showing that the new version of LOGIC_OPERATION yields the truth table you generated in part (d).

12.

```
entity LOGIC_OPERATION is
  port
    (A, B  : in  BIT;
     Z      : out BIT);
end LOGIC_OPERATION;
```

```
architecture MODEL of LOGIC_OPERATION is
   signal INT1, INT2, INT3 : BIT;
begin
   INT1 <= not A;
   INT2 <= not B;
   INT3 <= INT1 or INT2;
   Z <= not INT3;
end MODEL;
```

Assuming the inputs change from $A = 0$ and $B = 0$ to $A = 0$ and $B = 1$ for the above VHDL model,

a. Which concurrent signal assignment statements execute and in what order?

b. What are the old and new values for INT1, INT2, INT3, and Z?

13. Repeat Problem 12 assuming the inputs change from $A = 0$ and $B = 0$ to $A = 1$ and $B = 1$.

REFERENCES AND READINGS

[1] Mermet, J., ed. *VHDL for Simulation, Synthesis, and Formal Proofs of Hardware.* Boston, MA: Kluwer Academic Publishers, 1992.

[2] Aylor, J., R. Waxman and C. Scarratt. "VHDL—Feature Description and Analysis," *IEEE Design & Test of Computers,* vol. 3, no. 2, pp. 17–27, April 1986.

[3] Dewey, A. "VHDL: Implications of a Modern Hardware Description Language," *Proceedings of the IEEE International Conference on Computer Design,* pp. 78–81, October 1984.

[4] Dewey, A. "Candidate Standard Development Activities within the VHSIC Program," *Proceedings of the IEEE Computer Standards Conference,* May 1986.

[5] Donlin, M. "ASIC Complexity Fuels Drive to HDL Design," *Computer Design,* vol. 30, no. 8, pp. 77–84, May 1991.

[6] Egan, B. "ASIC Designers Turn to VHDL Tools Despite Obstacles," *Computer Design,* vol. 31, no. 2, pp. 55–64, February 1992.

[7] IEEE, *IEEE Standard VHDL Language Reference Manual,* 1076, 1987.

[8] IEEE, *IEEE Standard VHDL Language Reference Manual,* 1076, 1993.

[9] Kumar, K. *Tutorial: An Introduction to VHDL.* Cadence Design Systems, May 1992.

[10] Leung, S. and M. Shanblatt. *ASIC System Design with VHDL: A Paradigm.* Boston, MA: Kluwer Academic Publishers, 1989.

[11] Lipsett, R., E. Marschner, and M. Shahdad. "VHDL—The Language," *IEEE Design and Test of Computers,* vol. 3, no. 2, pp. 28–43, April 1986.

[12] Perry, D. *VHDL.* New York: McGraw-Hill, 1991.

[13] Shahdad, M. "An Overview of VHDL Language and Technology," *23rd ACM/IEEE Design Automation Conference,* pp. 320–326, June 1986.

[14] Zimmerman, A. "Designing with VHDL," *Design Automation,* vol. 2, no. 1, pp. 26–31, February 1991.

12

STRUCTURAL MODELING IN VHDL: PART I

In Chapter 11, we learned how to describe basic switching algebra operations in VHDL. In this chapter, we will learn how to describe the interconnection of switching algebra operations in VHDL. In other words, we will develop VHDL models of logic schematics.

Modeling logic schematics or netlists introduces a descriptive style called *structural modeling*. Structural modeling defines the behavior of a design entity by defining the *components* that comprise the design entity and their interconnection. A structural model *implicitly* or indirectly defines function because the design entity input/output transform can be derived knowing the constituent components and their behaviors.

We will discuss how to describe in VHDL hierarchical structures having multiple levels, in which design entities are composed of design entities, which are, in turn, composed of design entities, and so on. We will also discuss how to encapsulate and efficiently reuse design entities via VHDL *packages*. Finally, we will examine the extent of declarations, called *scope*, and structural modeling enhancements found in VHDL-93.

12.1 EXAMPLE SCHEMATIC

Let us begin by looking at an example VHDL description of the combinational logic schematic for the majority function discussed in Chapter 3 and repeated in Figure 12.1(a) (p. 362). The interface of the design entity is defined by the entity declaration, which gives the name of the design entity, MAJORITY, and the input/output signals. The ports A_IN, B_IN, and C_IN are input signals; the port Z_OUT is an output signal. All ports are of type BIT, carrying only the values '0' and '1'.

The body of the MAJORITY design entity illustrates *structural modeling*, defining the interconnection of logic operators. The MAJORITY structural model *implicitly*, or indirectly, defines the design entity input/output transform. The input/output transform of the MAJORITY design entity can be derived from a knowledge of the logic operations of the AND2_OP and OR3_OP *components* and the combinational system analysis techniques discussed in Chapter 3.

Before we examine the details of the various VHDL statements, let us first discuss the general aspects of the structural modeling style. Since VHDL descriptions are called *design entities* and structural models define interconnections of VHDL descriptions, you might well be wondering why we are interconnecting *components* instead of design entities. The roles of *components* and *design entities* in VHDL structural models can be a little confusing. It is convenient to think of components as placeholders for design entities, sort of "virtual" design entities. Design entities are interconnected by first interconnecting components and then associating the components with design entities; we will discuss associating components with design entities in more detail in the following section.

These structural concepts are illustrated in Figure 12.2. The figure shows that the STRUCTURE architecture describes the schematic for the majority logic by defining the in-

Figure 12.1

A VHDL model of combinational logic for the majority function

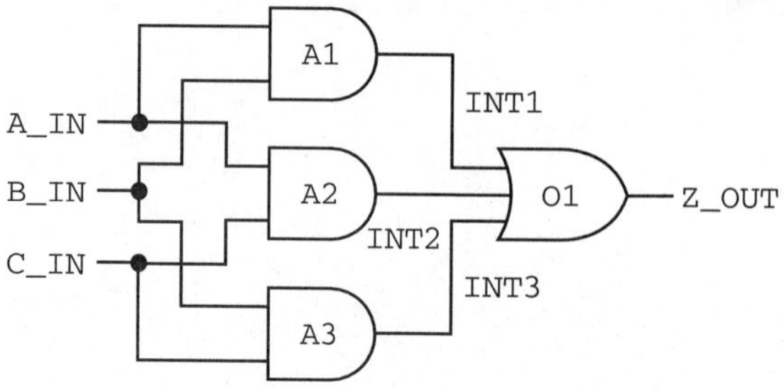

(a)

```
-- Interface
entity MAJORITY is
  -- Input/output ports
  port
    (A_IN, B_IN, C_IN : in BIT;
     Z_OUT            : out BIT);
end MAJORITY;


-- Body
architecture STRUCTURE of MAJORITY is
 --Declare logic operators
 component AND2_OP
  port (A, B : in BIT; Z : out BIT);
 end component;
 component OR3_OP
  port (A, B, C : in BIT; Z : out BIT):
 end component;


-- Declare signals to interconnect logic operators
 signal INT1, INT2, INT3 : BIT;
begin
 -- Connect logic operators to describe schematic
 A1: AND2_OP port map (A_IN, B_IN, INT1);
 A2: AND2_OP port map (A_IN, C_IN, INT2);
 A3: AND2_OP port map (B_IN, C_IN, INT3);
 O1: OR3_OP  port map (INT1, INT2, INT3, Z_OUT);

end STRUCTURE;
```

(b)

Figures (a) and (b) show the logic schematic and VHDL description of the majority function, respectively.

terconnection of components, which are, in turn, associated with design entities describing the **and** and **or** switching algebra operations. (Such design entities were developed in Chapter 11.)

Figure 12.2

VHDL structural descriptions

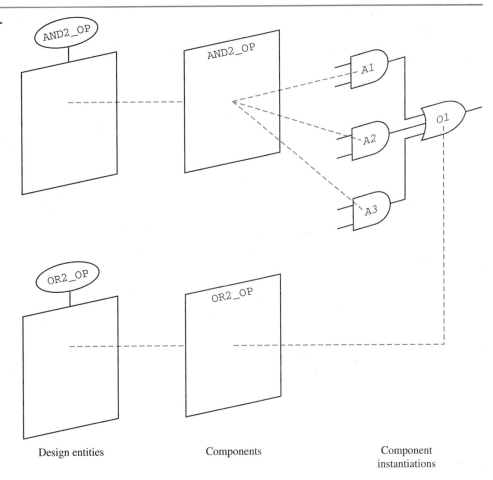

Design entities   Components   Component
                                              instantiations

Another way of viewing the roles of components and design entities in VHDL structural models is illustrated in Figure 12.3 (p. 364) [7]. Here, a component represents a chip (integrated circuit) socket. A design entity represents a chip that plugs into a chip socket. A VHDL structural model then defines the interconnection of instances of chip sockets holding chips.

For simple structural descriptions, the process of interconnecting components and then associating components with design entities admittedly seems a little cumbersome. It would be easier to bypass components and directly interconnect design entities. (This option is available in VHDL-93 and is discussed in Section 12.7.) However, we will continue to use components in example structural descriptions because components are *required* in VHDL-87 structural descriptions, and components provide powerful modeling capabilities that will be exploited in future chapters.

In the following discussions, keep in mind that for VHDL structural models of logic schematics, components are associated with design entities, which, in turn, describe logic operators. Hence, we will often use the terms *component instance*, *design entity instance*, and *logic operator instance* interchangeably.

Figure 12.3

Component instances, components, and design entities

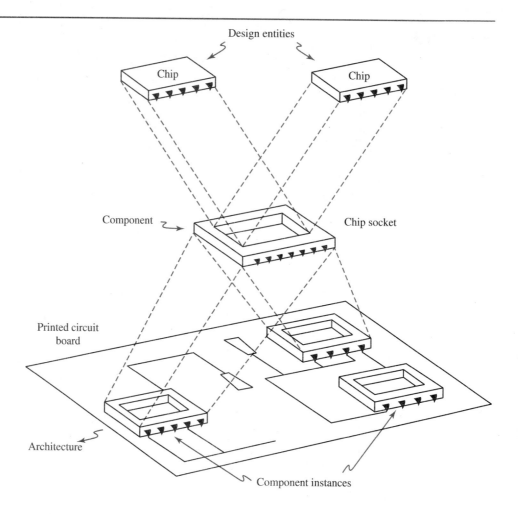

12.2 COMPONENT AND SIGNAL DECLARATIONS

The declarative part of the architecture STRUCTURE contains three declarations: two *component declarations* and one *signal declaration*. The signal declaration declares three signals.

The AND2_OP and OR3_OP component declarations respectively declare the **and** and **or** logic operators used in the majority combinational logic design.

```
-- Declare logic operators
component AND2_OP
 port (A, B : in BIT; Z : out BIT);
end component;
component OR3_OP
 port (A, B, C : in BIT; Z : out BIT);
end component;
```

Component declarations start with the reserved keyword **component**, followed by the name of the component. Like design entity and architecture names, component names can be any legal identifiers. Component ports are declared in the same manner in which design entity ports are declared. The AND2_OP component has two input ports, A and B, and one output port, Z, all of type BIT. The OR3_OP component has three input ports, A, B, and C, and one output port Z, all of type BIT. Finally, the keywords **end component** complete a component declaration.[1]

The similarities between component and design entity declarations are not coincidental; as explained earlier, components are virtual design entities. Component-to-entity association is provided by a *configuration*. For now, we will rely on *default configurations* that automatically associate components and design entities having the *same interface*, in other words, names and ports. Thus, returning to the structural description of the majority logic shown in Figure 12.2, we see that VHDL models for the design entities shown at the far left that can be automatically associated or "configured" with the AND2_OP and OR3_OP components shown in the middle can be easily written as follows:

```
entity AND2_OP is                  entity OR3_OP is
  port                               port
    (A, B : in BIT;                    (A, B, C : in BIT;
     Z    : out BIT);                   Z        : out BIT);
end AND2_OP;                        end OR3_OP;


architecture MODEL of AND2_OP is   architecture MODEL of OR3_OP is
begin                              begin
  Z <= A and B;                      Z <= A or B or C;
end MODEL;                         end MODEL;
```

Notice that the entity declarations and component declarations match; in other words, they have the same names and ports.

We also need signals to interconnect the component instances. The input/output signals for the majority function, A_IN, B_IN, C_IN, and Z_OUT, are declared in the definition of the design entity interface. The internal signals connecting the outputs of the **and** operators to the input of the **or** operator are declared via the following signal declaration statement.

```
-- Declare signals to interconnect logic operators
signal INT1, INT2, INT3 : BIT;
```

Signal declarations, which were introduced in Chapter 11, create one or more signals by specifying the name(s) and type. The name defines a "handle" used to refer to the signal; a signal name can be any legal identifier. The type defines the legal values the signal(s) may hold.

---

[1] In VHDL-93, a component declaration may contain the keyword **is** in the header and may close with the name of the component.

```
component AND2_OP is
  .
  .
  .
end component AND2_OP;
```

12.3 COMPONENT INSTANTIATION STATEMENTS

The statement part of the architecture STRUCTURE of the MAJORITY design entity contains four *component instantiation* statements that describe the logic schematic of the majority function.

```
-- Connect logic operators to describe schematic
A1: AND2_OP port map (A_IN, B_IN, INT1);
A2: AND2_OP port map (A_IN, C_IN, INT2);
A3: AND2_OP port map (B_IN, C_IN, INT3);
O1: OR3_OP  port map (INT1, INT2, INT3, Z_OUT);
```

Figure 12.4 shows that each component instantiation statement creates an *instance* of one of the declared components AND2_OP or OR3_OP. Each instance of a declared component represents a logic operator in the majority function schematic. The first three component instantiation statements define the three **and** operators. The last component instantiation statement defines the output **or** operator.

A component instantiation statement starts with a *label*, followed by a colon and a component name. A label can be any legal identifier the designer chooses to denote an instance of the named component. For the MAJORITY model, the labels A1, A2, and A3 denote three instances of the AND2_OP component. Similarly, the label O1 denotes an instance of the OR3_OP component. Following the name of the declared component is a **port map** clause, which describes how each component instance is connected to the rest of the system.[2]

Figure 12.4

Component instantiations

```
component AND2_OP                          component OR3_OP
  port (A, B: in BIT; Z: out BIT);           port (A, B, C: in BIT; Z: out BIT);
end component;                              end component;
```

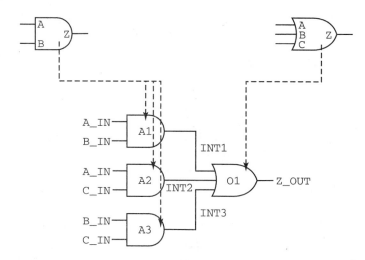

---

[2] VHDL-93: In VHDL-93, a component instantiation statement may include the keyword **component** before the name of the component.

```
A1: component AND2_OP port map (···);
```

Figure 12.5

Positional association connectivity

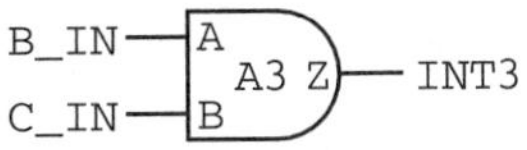


```
component AND_OP                                A3: AND_OP
  port (A, B: in BIT; Z: out BIT);                port map (B_IN, C_IN, INT3);
end component;
```

Aptly named, the **port map** clause defines the mapping between which signals are connected to which component ports, in other words, how a component is "wired-up." Figure 12.5 shows how connectivity is denoted by *positional association* for the A3 **and** logic operator. With positional association, the signals listed in the **port map** clause of the component instantiation are paired up or associated with the ports listed in the component declaration based on respective textual orderings. Moving from left to right, the first signal B_IN is connected to the first port A, the second signal C_IN is connected to the second port B, and so on. We encourage you to study the connectivity of the other components to understand how the logic schematic is described in VHDL.

Connectivity in a **port map** clause can also be denoted by *named association.* Figure 12.6 shows how the structure of the logic schematic of the majority function can be described using named association.

The only difference between positional and named association is the **port map** clause for the component instantiation statement. With named association, the name of the signal is explicitly associated with the name of the component port to which the signal is connected. The symbol => denotes the association or connectivity. For example, Z => INT1 specifies that the INT1 signal is connected to the Z port of the A1 instance of the AND2_OP component, in other words, the **and** operator. Remember the association is "*port name => signal name.*" Unlike positional associations, named associations may appear in any textual order in the **port map** clause. Positional and named association may be mixed in a single **port map**.

Figure 12.6

Named association connectivity

```
-- Body
architecture STRUCTURE of MAJORITY is
 -- Declare logic operators and signals
     .
     .
     .
begin
 -- Connect logic operators with named association
 A1: AND2_OP port map (A => A_IN, Z => INT1, B => B_IN);
 A2: AND2_OP port map (A => A_IN, B => C_IN, Z => INT2);
 A3: AND2_OP port map (Z => INT3, A => B_IN, B => C_IN);
 O1: OR3_OP  port map (A => INT1, B => INT2, C => INT3, Z => Z_OUT);
end STRUCTURE;
```

```
A2: AND2_OP port map (A_IN, B => C_IN, Z => INT2);
```

However, positional associations must come before named associations. Once a named association is given, all remaining associations must be named.

Note that component instantiation statements are considered concurrent statements. Although component instantiation statements do not actually execute during simulation, the textual order in which the statements are written has no bearing on the definition of the structure and its associated aggregate behavior. The following four component instantiation statements

```
-- Connect logic operators to describe schematic
A1: AND2_OP port map (A_IN, B_IN, INT1);
A2: AND2_OP port map (A_IN, C_IN, INT2);
A3: AND2_OP port map (B_IN, C_IN, INT3);
O1: OR3_OP  port map (INT1, INT2, INT3, Z_OUT);
```

could have been written as

```
-- Connect logic operators to describe schematic
A1: AND2_OP port map (A_IN, B_IN, INT1);
O1: OR3_OP  port map (INT1, INT2, INT3, Z_OUT);
A3: AND2_OP port map (B_IN, C_IN, INT3);
A2: AND2_OP port map (A_IN, C_IN, INT2);
```

and the logic schematic remains unchanged. As we study VHDL and introduce statements that may appear within an architecture, we will learn that all such statements are considered concurrent statements.

Figure 12.7 shows another example VHDL structural description. Figure 12.7(b) shows a VHDL model of the combinational schematic for the temperature controller analyzed in Chapter 3 and repeated in Figure 12.7(a). The documentation conventions discussed in Chapter 3 have been adopted in writing the VHDL structural model. Namely, the output signal of the inverter is active-0, hence _BAR is appended to the associated signal name.

12.4 HIERARCHICAL STRUCTURES

VHDL places no restrictions on the depth or breadth of structural models. Design entities can use components associated with design entities which, in turn, use components, yielding a multilevel structure. For example, Figure 12.8 (p. 370) shows how to combine two 3-way majority circuits and an **and** operator to build a combinational system that asserts the output active-1 if two groups of three voters both yield majority results.

Figure 12.9 shows a VHDL structural description of the two 3-way majority functions. Again, assuming an association between the MAJORITY *component* in Figure 12.9 and the MAJORITY *design entity* in Figure 12.1(b) via default configuration rules, the MAJORITY_2X3 VHDL model describes a two-level, hierarchical structure. The MAJORITY_2X3 design entity uses instances of the MAJORITY component, which, in turn,

Figure 12.7

VHDL model of combinational logic for temperature controller

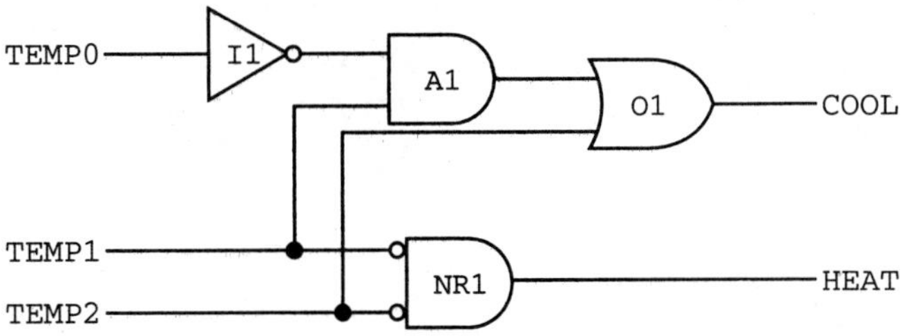

(a)

```
-- Interface
entity TEMP_CNTRL is
  port
    (TEMP2, TEMP1, TEMP0 : in  BIT;
     COOL, HEAT          : out BIT);
end TEMP_CNTRL;


-- Body
architecture STRUCTURE of TEMP_CNTRL is
  -- Declare logic operators
  component AND2_OP
   port (A, B : in BIT; Z : out BIT);
  end component;
  component OR2_OP
   port (A, B : in BIT; Z : out BIT);
  end component;
  component NOR2_OP
   port (A, B : in BIT; Z : out BIT);
  end component;
  component NOT_OP
   port (A : in BIT; A_BAR : out BIT);
  end component;

  -- Declare signals to interconnect logic operators
  signal TEMP0_BAR, INT1 : BIT;
begin
  -- Logic for COOL
  I1: NOT_OP  port map (TEMP0, TEMP0_BAR);
  A1: AND2_OP port map (TEMP0_BAR, TEMP1, INT1);
  O1: OR2_OP  port map (INT1, TEMP2, COOL);

  -- Logic for HEAT
  NR1: NOR2_OP port map (TEMP1, TEMP2, HEAT);
  end STRUCTURE;
```

(b)

Figures (a) and (b) show the logic diagram and VHDL description of the temperature controller function, respectively.

Figure 12.8

A two-level structural model of two 3-way majority functions

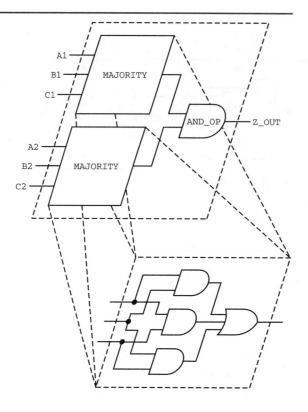

Figure 12.9

Hierarchical structural VHDL model

```
entity MAJORITY_2X3 is
  -- Input/output ports
  port
    (A1, B1, C1 : in BIT;
     A2, B2, C2 : in BIT;
     Z_OUT      : out BIT);
end MAJORITY_2X3;


architecture STRUCTURE of MAJORITY_2X3 is
 component AND2_OP
  port (A, B : in BIT; Z : out BIT);
 end component;
 component MAJORITY
  port (A_IN, B_IN, C_IN : in  BIT;
        Z_OUT            : out BIT);
 end component;

 -- Declare internal signals
 signal INT1, INT2 : BIT;
begin
 -- Connect component instances to describe schematic
 M1: MAJORITY port map (A1, B1, C1, INT1);
 M2: MAJORITY port map (A2, B2, C2, INT2);
 A1: AND2_OP  port map (INT1, INT2, Z_OUT);

end STRUCTURE;
```

Figure 12.10

VHDL hierarchy of design entities

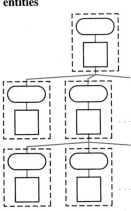

uses instances of the AND2_OP and OR3_OP components. We are also assuming that the AND2_OP and OR3_OP components are appropriately associated via default configurations with design entities respectively describing the **and** and **or** logic operations.

A hierarchical structural description is a powerful modeling construct in VHDL because it provides a mechanism to decompose the description of a large, complex digital system into smaller pieces. It is a good modeling practice to have VHDL structural hierarchies reflect convenient functional/physical digital system decompositions. In other words, if there is a natural way to partition and decompose a digital system, use this information to partition and decompose the associated VHDL structural model. Figure 12.10 modifies the simple notion of a design entity introduced in Figure 11.4 to include the ability to construct hierarchical structures of design entities.

12.5 PACKAGES

VHDL structural descriptions of logic schematics often use the same set of components and rewriting component declarations for every logic schematic can be tedious; this problem, however, can be addressed by *packages*. A package serves as a central repository for frequently used utilities, such as component declarations. Instead of repeating commonly used declarations every time they are needed in VHDL models, the declarations may be encoded once and grouped together in a package. The declarations may then be reused by any VHDL model by simply accessing the package.

Figure 12.11 (p.372) shows a revised VHDL structural model of the single 3-way majority function given in Figure 12.1 using a package. The only difference between the majority function models given in Figure 12.1(b) and Figure 12.11 is that the component declarations are taken out of the architecture and placed in the package LOGIC_OPS. The package LOGIC_OPS also includes the declaration of an inverter, NOT_OP. Including the inverter component makes LOGIC_OPS a more general-purpose package because any logic function can be implemented using only **and**, **or**, and **not** logic operators; in other words, the **and**, **or**, and **not** logic operators form a *minimal set* (see Chapter 3). The architecture accesses the component declarations in the package LOGIC_OPS via the **use** statement.

The **use** statement, more commonly called a *use clause*,

```
use WORK.LOGIC_OPS.all;
```

placed just before the architecture gives the architecture access to *all* declarations in the package LOGIC_OPS located in the *library* WORK, namely, the components AND2_OP, OR3_OP, and NOT_OP. Put another way, the use clause allows the package LOGIC_OPS to export its declarations. Thus, the identifiers AND2_OP and OR3_OP can be used in the STRUCTURE architecture as legal references to components. Assembling a name by concatenating a series of names together separated by periods, like WORK.LOGIC_OPS.**all**, is called *selected name* notation. We will discuss use clauses and libraries in more detail in the following sections.

Figure 12.11 shows that a *package declaration* begins with the reserved keyword **package**, followed by the name of the package. The name of a package can be any legal identifier. A package declaration closes with the keyword **end**, followed optionally by the name of the package. We will adopt the modeling practice of always repeating the package

Figure 12.11

A VHDL model of the majority function using a package

```vhdl
-- Package
package LOGIC_OPS is
  -- Declare logic operators
  component AND2_OP
   port (A, B : in BIT; Z : out BIT);
  end component;
  component OR3_OP
   port (A, B, C : in BIT; Z : out BIT);
  end component
  component NOT_OP
   port (A : in BIT; A_BAR : out BIT);
  end component;
end LOGIC_OPS;

-- Interface
entity MAJORITY is
  -- Input/output ports
  port
    (A_IN, B_IN, C_IN : in BIT;
     Z_OUT            : out BIT);
end MAJORITY;

-- Body
-- Use components in package LOGIC_OPS in library WORK
use WORK.LOGIC_OPS.all;
architecture STRUCTURE of MAJORITY is
  -- Declare signals to interconnect logic operators
  signal INT1, INT2, INT3 : BIT;
begin
  -- Connect logic operators
  A1: AND2_OP port map (A_IN, B_IN, INT1);
  A2: AND2_OP port map (A_IN, C_IN, INT2);
  A3: AND2_OP port map (B_IN, C_IN, INT3);
  O1: OR3_OP  port map (INT1, INT2, INT3, Z_OUT);

end STRUCTURE;
```

name in the closing.[3] Utilities to be exported appear within the package, such as component declarations and signal declarations. As we build an understanding of VHDL, we will note other statements that may appear within a package declaration. Recall from Chapter 11 that package declarations are primary design units; thus, a package declaration can be analyzed by itself.

---

[3] VHDL-93: In VHDL-93, a package declaration may close with the keywords **end package.**

```vhdl
package LOGIC_OPS is
  .
  .
  .
end package LOGIC_OPS;
```

Figure 12.12

The VHDL structural model of combinational logic for the *EVEN* function

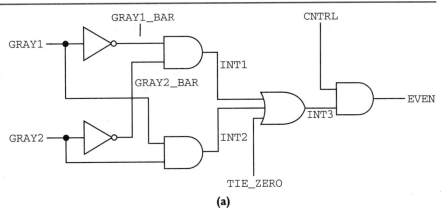

(a)

```
-- Interface
entity EVEN_GRAY is
  -- Input/output ports
  port
    (GRAY1, GRAY2 : in BIT;
     CNTRL         : in BIT;
     EVEN          : out BIT);
end EVEN_GRAY


-- Body
-- Use components in package LOGIC_OPS in library WORK
use WORK.LOGIC_OPS.all;
architecture STRUCTURE of EVEN_GRAY is
  -- Declare signals to interconnect logic operators
  signal GRAY1_BAR, GRAY2_BAR, INT1, INT2, INT3 : BIT;
  -- Signal TIE_ZERO will always have the value '0'
  signal TIE_ZERO : BIT := '0';
begin
  -- Connect logic operators
  N1: NOT_OP  port map (GRAY1, GRAY1_BAR);
  N2: NOT_OP  port map (GRAY2, GRAY2-BAR);
  A1: AND2_OP port map (GRAY1_BAR, GRAY2_BAR, INT1);
  A2: AND2_OP port map (GRAY1, GRAY2, INT2);
  O1: OR3_OP  port map (INT1, INT2, TIE_ZERO, INT3);
  A3: AND2_OP port map (INT3, CNTRL, EVEN);

end STRUCTURE;
```

(b)

Figures (a) and (b) show the logic schematic and VHDL description of the *EVEN* function, respectively.

To reenforce the use of packages and structural modeling, Figure 12.12 presents a VHDL model of the combinational digital system that detects the even Gray codes analyzed in Chapter 3. Again, the documentation conventions discussed in Chapter 3 have been adopted in writing the VHDL structural model. The data inputs, GRAY1 and GRAY2, are declared separately from the control input, CNTRL. The output signals of the inverters are active-0, hence _BAR is appended to the associated signal names.

The NOT_OP, AND2_OP, and OR3_OP components declared in the package LOGIC_OPS given in Figure 12.11 are used in the architecture to describe the logic schematic. The unused port of the 3-input **or** operator is "tied-off" to the constant BIT literal '0' with the signal TIE_ZERO. The TIE_ZERO signal is initialization to '0' by adding the expression := '0'; to the signal declaration. We will give a more detailed discussion of unused ports in structural models in Chapter 14.

Packages are a powerful feature of VHDL that have been introduced early in our study of the language to emphasize their importance and encourage their use in developing good modeling practices. In addition to facilitating code reuse, packages also serve to help coordinate the decentralized development of large VHDL models. When several engineers contribute segments of code to a single VHDL model, it is important to have standard naming conventions for components, types, and signal names. Standard naming conventions can be enforced by declaring the commonly used names within a package and then establishing the modeling practice that the engineers use the package. Another benefit of packages is that they can hide language details from engineers who do not need to have a detailed understanding of VHDL. For example, if the package LOGIC_OPS is supplied to an engineer, the engineer need only know how to use VHDL components and not how the components are declared.

12.6 NAME SPACES AND SCOPE

The concept of *scope* is closely associated with the theory of strongly typed languages and packages. Recall that the term *strongly typed* refers to the philosophy that everything must be declared before it can be used and that a declaration defines what a name represents. However, once properly declared, where is it legal to use the declared name? Scope defines the extent of a declaration, in other words, where a declared name can be used. The extent of a declaration is also called a *name space*. Let us review some of the scoping rules for VHDL constructs we have encountered thus far and, in the process, discuss *use clauses* and *library clauses* in detail.

Anything declared within the declarative part of an architecture may be used only within the enclosing architecture body; in other words, its scope is from the point of declaration to the end of the enclosing architecture. For example, the components AND2_OP and OR3_OP and signals INT1, INT2, and INT3 declared within the declarative part of the architecture STRUCTURE in Figure 12.1(b) may be used only within the enclosing architecture and are not known to any other VHDL model.

Anything declared within a design entity declaration may be used only within the enclosing entity declaration *and* associated architectures. The latter scope rule allows, for example, the input/output signals (A_IN, B_IN, C_IN, and Z_OUT) declared in the MAJORITY design entity interface in Figure 12.1(b) to be accessible to the structural description in the associated architecture STRUCTURE.

Anything declared within a package declaration may be used within the enclosing package. Furthermore, as explained earlier, the scope of declarations within a package may be extended by the use clause to other parts of a VHDL model.

12.6.1 Use Clauses

A use clause can be placed before an entity declaration, giving the design entity and associated architectures access to the package contents. A use clause can also be placed before a package declaration, giving a package access to another package. Thus, a hierarchy of pack-

Figure 12.13

Building a hierarchy of packages

```
package PACKAGE_A is
    declaration
    declaration
        .
        .
        .
end PACKAGE_A;


use WORK.PACKAGE_A.all;
package PACKAGE_B is
    declaration
    declaration
        .
        .
        .
end PACKAGE_B;


use WORK.PACKAGE_B.all;
architecture STRUCTURE of EXAMPLE is
begin
        .
        .
        .
end STRUCTURE;
```

ages may be constructed in which declarations in a package may build on declarations in another package, which, in turn, may build on declarations in yet another package, and so on. Figure 12.13 shows an example of a simple two-level package hierarchy. The declarations in PACKAGE_B use the declarations in PACKAGE_A. The declarations in PACKAGE_B are, in turn, used by the STRUCTURE architecture. Note that the use clause grants access to *only* the named package. For example, the architecture body STRUCTURE has access to only the declarations in PACKAGE_B, not the declarations in PACKAGE_A.

A use clause can provide varying degrees of export control and extended scoping. A use clause can allow a package to export, or make *directly visible*, (1) just its name, (2) a subset of its declarations, or (3) all its declarations.

The following example makes just a package name directly visible by including only the package name in the use clause.

```
use WORK.LOGIC_OPS;
architecture STRUCTURE of MAJORITY is
  ...

begin
  A1: LOGIC_OPS.AND2_OP port map (...);
  A2: LOGIC_OPS.AND2_OP port map (...);
  A3: LOGIC_OPS.AND2_OP port map (...);
  O1: LOGIC_OPS.OR3_OP  port map (...);

end STRUCTURE;
```

The architecture has direct access to the package name, LOGIC_OPS, but none of the declarations contained within LOGIC_OPS. Individual component declarations are indirectly accessed using a selected name notation, prepending the package name LOGIC_OPS as a qualifier to the component names.

The next example makes a subset of declarations directly visible by listing the declarations individually, separated by commas, in the use clause.

```
use WORK.LOGIC_OPS.AND2_OP, WORK.LOGIC_OPS.OR3_OP;
architecture STRUCTURE of MAJORITY is
  ...

begin
  A1: AND2_OP port map ( . . .);
  A2: AND2_OP port map ( . . .);
  A3: AND2_OP port map ( . . .);
  O1: OR3_OP  port map ( . . .);

end STRUCTURE;
```

The MAJORITY model does not require the inverter component, NOT_OP, so the use clause grants limited access to *only* the AND2_OP and OR3_OP components. The selected name notation is used to list the exported component declarations, prepending the library name WORK as a qualifier to the package name LOGIC_OPS, which, in turn, is prepended as a qualifier to the component names AND2_OP and OR3_OP.

Finally, appending the keyword **all** to a package name in a use clause makes all declarations within the package directly visible or accessible. Figures 12.11 and 12.12 show examples of this practice.

Limiting the export control and extended scoping of a package is often a matter of personal preference. However, if a modeling application uses only a few declarations from a fairly large set of package declarations, then the selected access techniques should be used to limit the declarations visible in a name space. In other words, it is generally a good modeling practice to use only what you need. Declarations from multiple packages can be imported by writing multiple use clauses. If two or more declarations having the same name are mistakenly imported from different packages, all of the conflicting declarations are discarded because it is impossible for the VHDL software to determine which declaration the designer actually intends to use and which declarations are imposters.

12.6.2 Nested Scopes

Scopes can be *nested*, meaning that a name space may be contained within a larger name space. VHDL follows the traditional scoping rules that outer name spaces extend into the embedded, inner name spaces. However, an inner name space has priority over enclosing, outer name spaces, meaning that a declaration in an inner name space can override or "hide" a declaration having the same name in an outer name space.

Consider, for example, the nested scopes shown in Figure 12.14. The scope of the TEST_SIG signal declaration in the package extends into the entity TEST and associated architecture STRUCTURE. However, another signal also named TEST_SIG is declared within the architecture STRUCTURE. So what is the value of the TEST_SIG signal used in the component instantiation statement? The answer is '1' because the declaration within the

Figure 12.14

Nested scopes

```
package TEST_PKG is
   signal TEST_SIG : BIT :='0';
end TEST_PKG;

use WORK.TEST_PKG.all;
entity TEST is
  port
    (. . . .);

end TEST;

architecture STRUCTURE of TEST is
 signal TEST_SIG : BIT := '1';

 component UNIT
  port (. . . .);
 end component;
begin

 U1 : UNIT port map(TEST_SIG, . . . .); -- Which TEST_SIG?
 end STRUCTURE;
```

inner architecture name space has priority over the declaration within the outer package name space.

12.6.3 Library Clauses

A *library clause* declares a name that denotes a library. A library name can be any legal identifier and a library clause can declare more than one library by listing the library names separated by commas. We have been able to use the library WORK in our previous examples because every design unit assumes the following library clause.

```
library WORK;
```

The library clause simply declares WORK to be a library.

Using user-defined libraries instead of the predefined library WORK involves adding appropriate library and use clauses. For instance, assume the LOGIC_OPS package in Figure 12.11 is located in the LOGIC_LIB library instead of the WORK library. The MAJORITY model is easily updated by adding a library clause for LOGIC_LIB and replacing the old library WORK with the new library LOGIC_LIB in the use clause.

```
-- Use components in package LOGIC_OPS in library WORK
library LOGIC_LIB;
use LOGIC_LIB.LOGIC_OPS.all;
architecture STRUCTURE of MAJORITY is
   ...
```

Every design unit also assumes the following two statements:

```
library STD;
use STD.STANDARD.all;
```

The first line is a library clause that declares STD to be a library. The second line is a use clause that grants direct visibility to all declarations contained within the package STANDARD, such as BIT, located within the library STD. The *IEEE Standard VHDL Language Reference Manual* gives the complete definition of the package STANDARD [5].

12.7 VHDL-93: DIRECT DESIGN ENTITY INSTANTIATION

Components serve as intermediaries between design entities and structural models. Structural models "talk to" components and components "talk to" design entities. Although this level of indirection provides powerful modeling capabilities that will be exploited in later chapters, many simple structural descriptions do not need such advanced modeling capabilities. For these situations, components tend to be "excess baggage," and it would be convenient to be able to bypass the component level of indirection and to instantiate directly design entities within structural models. VHDL-93 addresses this problem by relaxing the rules for structural modeling and allowing architectures to instantiate design entities as well as components.

Figure 12.15 modifies Figure 12.2 to show the structural model for the MAJORITY design entity using direct design entity instantiation. Figure 12.16 modifies Figure 12.1 to instantiate directly the AND2_OP and OR3_OP design entities given in Section 12.2.

For consistency, the four architecture statements in Figure 12.16 are still called component instantiation statements, although they actually instantiate design entities rather than components. To instantiate design entities, the references to components in Figure 12.1 are replaced by references to design entities. Such references begin with the keyword **entity**, followed by the selected name of the design entity. The selected name prepends the library name to the design entity name and the associated architecture name enclosed in parentheses. The **port map** clauses connect the ports of the instantiated design entities.

Direct design entity instantiation is useful for "bottom-up" design, where higher-level design entities are built out of known, existing, lower-level design entities. In this situation, the overhead of components, or chip sockets, is not needed as design entities, or chips, are directly interconnected.

12.8 SUMMARY

In this chapter, we have studied the basic mechanics for describing logic schematics in VHDL. The VHDL models describe netlists of logic operators using a structural modeling style. Component declarations define the logic operators and component instantiations define how the logic operators are interconnected. Connectivity is denoted by positional or named association. Positional association is a shorthand notation where signal/port connectivity is

Figure 12.15

VHDL structural descriptions using direct design entity instantiation

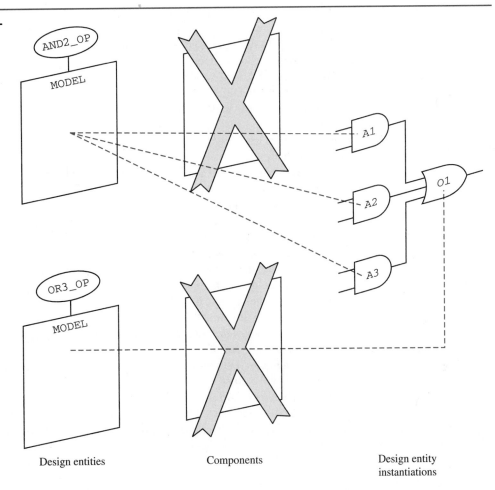

Design entities   Components   Design entity instantiations

Figure 12.16

A VHDL model of majority combinational logic using direct instantiation

```
-- Interface
entity MAJORITY is
  -- Input/output ports
  port
    (A_IN, B_IN, C_IN : in  BIT;
     Z_OUT            : out BIT);
end MAJORITY;

-- Body
architecture STRUCTURE of MAJORITY is
  -- Declare signals to interconnect logic operators
  signal INT1, INT2, INT3 : BIT;
begin
  -- Connect logic operators to describe schematic
  A1: entity WORK.AND2_OP(MODEL) port map (A_IN, B_IN, INT1);
  A2: entity WORK.AND2_OP(MODEL) port map (A_IN, C_IN, INT2);
  A3: entity WORK.AND2_OP(MODEL) port map (B_IN, C_IN, INT3);
  O1: entity WORK.OR3_OP(MODEL)  port map (INT1, INT2, INT3, Z_OUT);

end STRUCTURE;
```

implicitly defined via the textual order in which both are listed. Named association is a long-hand notation where signal/port association is explicitly defined.

We saw that structural modeling is not limited to one level of decomposition. A design entity can use design entities, which can, in turn, use design entities, and so on. Thus, the hierarchical structure of the description of a large design can contain several levels.

Also, we learned that packages serve as convenient central repositories for commonly used declarations, such as components. Packages reside in libraries; libraries are declared using library clauses. The scope of declarations within a package can be extended via the use clause. Scope defines where a declaration may be used.

Finally, we discussed a recent enhancement to VHDL that allows architectures to directly instantiate design entities as well as components. Writing a structural description by directly instantiating design entities has the advantage of avoiding component declarations and component-to-entity configurations, making VHDL structural models more succinct and readable. However, directly instantiating design entities has the disadvantage that the structure is hardwired, making the VHDL structural models more difficult to update.

We need to point out that neither component instantiation nor design entity instantiation provide predefined elements, in other words, a set of "commonly" used building blocks. This aspect illustrates an important design principle of VHDL: *VHDL tends to provide the general mechanism for generating solutions rather than providing the solutions themselves.* For instance, providing a set of building blocks for describing logic schematics may simplify some modeling tasks, but there is the problem of selecting a "commonly" used set of building blocks. Should we use 2-input **nand** operators, or maybe **not** and 3-input **and** operators? The logic operators used for a particular digital system design may depend on how the system is designed and/or manufactured. Hence, specifying a set of predefined building blocks may unfairly bias the hardware description language toward a particular design methodology or implementation technology—not a good characteristic of a standard communication medium. Even if we were able to agree on a set of predefined elements, there is the additional problem that this set would likely become quickly outdated as digital systems technology constantly changes. So each time we wanted to use a new set of logic operators, the language definition would have to change—again, not a good characteristic of a standard communication medium. Hence, to support design methodology independence and language extensibility, VHDL does not provide any predefined elements for structural descriptions.

12.9 PROBLEMS

1.

```
entity LOGIC_DESIGN is
  port
    (A, B, C, D : in BIT;
     Z          : out BIT);
end LOGIC_DESIGN;

architecture STRUCTURE of LOGIC_DESIGN is
 component AND2_OP
  port (A, B : in BIT; Z : out BIT);
 end component;
 component OR2_OP
  port (A, B : in BIT; Z : out BIT);
 end component;
```

```
component XNOR2_OP
 port (A, B : in BIT; Z : out BIT);
end component;
component NOT_OP
 port (A : in BIT; A_BAR : out BIT);
end component;

 signal A_BAR, B_BAR, C_BAR, I1, I2, I3, I4 : BIT;
begin
N1: NOT_OP port map (A, A_BAR);
N2: NOT_OP port map (B, B_BAR);
N3: NOT_OP port map (C, C_BAR);

A1:  AND2_OP  port map (A_BAR, C_BAR, I1);
NX1: XNOR2_OP port map (B, D, I2);
A2:  AND2_OP  port map (B_BAR, C, I3);

A3: AND2_OP port map (I1, I2, I4);
O1: OR2_OP  port map (I3, I4, Z);

end STRUCTURE;
```

For the above VHDL structural model,
  a. Show the combinational logic schematic.
  b. Obtain the output logic expression using the symbolic analysis technique discussed in Chapter 3.
  c. What is the output (Z) of LOGIC_DESIGN for the inputs ABCD = 0000, 0101, 1010, 1111?
  d. Write a design entity to be associated by default configuration rules with the AND2_OP, OR2_OP, NOT_OP, and XNOR2_OP components and verify part (c) with simulation.

2.

```
entity LOGIC_DESIGN is
  port
    (A, B, C, D : in BIT;
     Z          : out BIT);
end LOGIC_DESIGN;

architecture STRUCTURE of LOGIC_DESIGN is
 component AND2_OP
  port (A, B : in BIT; Z : out BIT);
 end component;
 component NOR2_OP
  port (A, B : in BIT; Z : out BIT);
 end component;

 signal I1, I2 : BIT;
begin
 A1: AND2_OP port map (A, B, I1);
 A2: AND2_OP port map (C, D, I2);
 N1: NOR2_OP port map (I1, I2, Z);

end STRUCTURE;
```

For the preceding VHDL structural model,

a. Show the combinational logic schematic.

b. Obtain the output logic expression using the symbolic analysis technique discussed in Chapter 3.

c. What is the output (Z) of LOGIC_DESIGN for all possible inputs?

d. Write design entities to be associated by default configuration rules with the AND2_OP and NOR2_OP components and verify part (c) with simulation.

3.

```vhdl
entity AND_DESIGN is
  port
    (A : in BIT;
     Z : out BIT);
end AND_DESIGN

architecture STRUCTURE of AND_DESIGN is
 component AND2_OP
  port (A, B : in BIT; Z : out BIT);
  end component;

  signal I1, I2, I3 : BIT;
  signal HIGH : BIT := '1';
begin
A1: AND2_OP port map (A,  HIGH, I1);
A2: AND2_OP port map (I1, HIGH, I2);
A3: AND2_OP port map (I2, HIGH, I3);
A4: AND2_OP port map (I3, HIGH, Z);

end STRUCTURE;
```

For the above VHDL structural model,

a. Show the combinational logic schematic.

b. Obtain the output logic expression using the symbolic analysis technique discussed in Chapter 3.

c. What is the output (Z) of AND_DESIGN for the inputs A = 0 and A = 1?

d. Write a design entity to be associated by default configuration rules with the AND2_OP component and verify part (c) with simulation.

4. Explain packages and list three benefits of using packages.

5.

```vhdl
package PKG1 is
 component AND2_OP
  port (A, B : in BIT; Z : out BIT);
  end component;
 component OR2_OP
  port (A, B : in BIT; Z : out BIT);
  end component;
end PKG1;
```

```
package PKG2 is
 component AND2_OP
  port (A, B : in BIT; Z : out BIT);
 end component;
end PKG2;

use PKG1.all; use PKG2.all;
entity SCOPE is
  port
    (. . . .);
end SCOPE;

architecture STRUCTURE of SCOPE is
 component NAND2_OP
  port (A, B : in BIT; Z : out BIT);
 end component;

begin

 . . . .
end STRUCTURE;
```

For the above VHDL structural model, what is the scope of the following component declarations? List design units and associated regions.

a. AND2_OP

b. OR2_OP

c. NAND2_OP

6. Describe the following logic expression

$$(\overline{A} \bullet \overline{B} \bullet D) + (A \bullet B \bullet C) + (\overline{B} \bullet \overline{C})$$

with a VHDL structural model using the following package located in library WORK.

```
package LOGIC_PKG is
 component AND3_OP
  port (A, B, C : in BIT; Z : out BIT);
 end component;
 component NAND2_OP
  port (A, B : in BIT; Z : out BIT);
 end component;
 component OR4_OP
  port (A, B, C, D : in BIT; Z : out BIT);
 end component;
end LOGIC_PKG;
```

a. Use positional association in the port map lists.

b. Use named association in the port map lists.

c. Draw the logic schematic.

d. What is the output for the inputs ABCD = 0101, 0110, 1001, 1110?

e. Write design entities to be associated by default configuration rules with the AND3_OP, NAND2_OP, and OR4_OP components and verify part (d) with simulation.

7. Describe the following logic expression

$$(A + B + \overline{C}) \bullet (\overline{A} + C) \bullet (\overline{A} + \overline{B})$$

with a VHDL structural model using the LOGIC_PKG package in Problem 6.

a. Use positional association in the port map lists.

b. Use named association in the port map lists.

c. Draw the logic schematic.

d. What is the output for the inputs ABC = 000, 100, 110, 111?

e. Write design entities to be associated by default configuration rules with the AND3_OP, NAND2_OP, and OR4_OP components and verify part (d) with simulation.

8. Repeat Problem 7 using only five component instantiation statements.

9. A VHDL model that is *syntactically* correct is not necessarily *semantically* correct. Does the following design entity LOGIC correctly describe the given logic expression? Explain your answer.

$$Z = (A \cdot C) + (B \cdot D)$$

```
entity LOGIC is
  port
    (A, B, C, D : in BIT;
     Z          : out BIT);
end LOGIC;

architecture STRUCTURE of LOGIC is
 component AND2_OP
  port (A, B : in BIT; Z : out BIT);
 end component;
 component OR2_OP
  port (A, B : in BIT; Z : out BIT);
 end component;

 signal I1, I2 : BIT;
begin
 A1: AND2_OP port map (A, C, I1);
 A2: AND2_OP port map (B, D, I2);
 O1: OR2_OP  port map (I2, I2, Z);

 end STRUCTURE;
```

REFERENCES AND READINGS

[1] Berge, J., A. Fonkoua, S. Maginot, and J. Rouillard. *VHDL'92*. Boston MA: Kluwer Academic Publishers, 1993.

[2] Fazakerly, B. "Mixed Level Simulation Technology—Part I," *Electronic Product Design*, October 1994, pp. 26–34.

[3] Maliniak, L. "A Beginners Guide to VHDL," *Electronic Design*, vol. 42, no. 21, October 1994, pp. 75–82.

[4] Graf, A. "Harness the Power of VHDL for PLD Design," *Electronic Design*, vol. 42, no. 19, September 1994, pp. 108–115.

[5] IEEE, *IEEE Standard VHDL Language Reference Manual*, 1076, 1987.

[6] Karnik, T. and S. Kang. "Hierarchical Partitioning of High-Level VHDL Structures," *Proceedings of VHDL International Users Forum Conference*, Spring 1994, pp. 36–45.

[7] Perry, D. *VHDL*. New York: McGraw-Hill, 1991.

[8] Randon, E. "Basic Concepts of HDLs: VHDL," *IEEE International Symposium on Industrial Electronics*, May 1994, pp. 42–46.

[9] Toivanen, J., J. Honola, J. Nurmi, and J. Tuominan. "A VHDL-Based Bus Model for Multi PCB System Design," *IEEE Proceedings of EURODAC with EUROVHDL*, September 1994, pp. 492–497.

[10] VanAlmsick, W. and W. Daehn. "Distributed Simulation for Structural VHDL Netlists," *IEEE Proceedings of EURODAC with EUROVHDL*, September 1994, pp. 598–603.

[11] Wilson, J. "Creating Efficient Memory Models in VHDL," *Microelectronics Journal*, vol. 25, no. 6, September 1994, pp. XIII–XVIII.

[12] Zhang, T. and S. Yongchao. "VHDL Design Automation and Data Modeling," *IEEE Proc. Computer, Communications, Control, and Power Eng.*, September 1993, pp. 585–588.

13
DATA FLOW MODELING IN VHDL

The previous two chapters introduced VHDL constructs aligned with the *definition* and *analysis* of combinational digital systems. Chapter 11 discussed how to describe in VHDL basic switching algebra operations. Chapter 12 discussed how to describe in VHDL the interconnection of basic switching algebra operations forming logic schematics. This chapter will introduce VHDL constructs aligned with the *synthesis* or *design* of combinational digital systems. We will learn how to describe general logic expressions that serve as specifications for combinational systems.

We will take a more detailed look at concurrent signal assignment statements and will define *conditional* and *selected* concurrent signal assignment statements. We will also examine the relationship between the parallelism of concurrent signal assignment statements and the parallelism of hardware. This modeling aspect leads naturally into a discussion of using *alternative* architecture bodies to encode different possible descriptions and/or implementations of a given design entity. Multiple alternative architectures will expand our understanding of the composition of design entities.

We will also expand our understanding of VHDL operators and data types. Logic operators will be reviewed and *relational*, *arithmetic*, and *shift* operators will be introduced. Two new predefined data types will be introduced: BOOLEAN and BIT_VECTOR. The type BOOLEAN represents the truth values TRUE and FALSE. The type BIT_VECTOR represents a contiguous set of objects of type BIT. An object, such as a signal, of type BIT_VECTOR models a group of conceptually related bits, such as a memory address or an ASCII character code.

13.1 MODELING STYLES

As we build our knowledge of VHDL constructs, we will also build an understanding of how to use the VHDL constructs, in other words, *modeling styles*. Thus, it is instructive to review the modeling styles introduced in Chapter 11 and Chapter 12 and to examine how these modeling styles relate to a general taxonomy of VHDL modeling styles.

A way of categorizing and relating hardware representations or abstractions is shown in Figure 13.1, based on the Gajski-Khun Y diagram [6]. Figure 13.1 shows that hardware representations can be thought of as constituting a three-dimensional space: structural, behavioral, and physical. For the present, we will concentrate on the structural and behavioral dimensions; aspects of the physical representation will be discussed in Chapters 15 and 16.

A *behavioral* representation or modeling style *explicity* defines the input/output function by specifying some sort of mathematical transfer function. Chapter 11 introduced behavioral modeling in VHDL. For example, the VHDL description given in Figure 11.7 specifies a logic expression. Such a description lies at the "switching algebra" point along the "behavioral representation" axis. Moving away from the center origin along the behavioral axis, one finds more abstract descriptions that are typically used for describing larger, more complex hardware systems. A behavioral description defines *what* a digital design does, but not

Figure 13.1

Gajski-Khun Y diagram

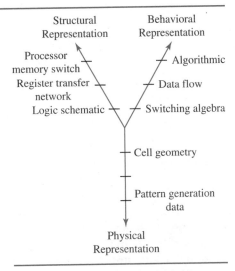

Hardware modeling space is divided into three dimensions: structural, behavioral, and physical.

necessarily *how* the design is implemented. There is no visibility into how the input/output function is partitioned and decomposed into smaller, more elementary functions.

In contrast, a *structural* representation or modeling style describes a digital system by specifying the interconnection of components that comprise the system. Chapter 12 introduced structural modeling in VHDL. For example, the VHDL description given in Figure 12.1 specifies the interconnection of logic operators comprising the combinational system implementing the majority function. Such a description lies at the "logic schematic" point along the "structural representation" axis in Figure 13.1. Moving away from the center origin along the structural axis, one again finds more abstract descriptions. For example, a component used in a register transfer network description might be an entire arithmetic logic unit (ALU).[1] The input/output function of structural descriptions is *implicitly* defined, in that the input/output function can be derived knowing the input/output functions of the individual components and their interconnection.

The difference between behavioral and structural representations can perhaps best be reviewed by an example. As an alternative to the structural model in Figure 12.1, Figure 13.2 (p.388) presents a behavioral model of the majority function using the concurrent signal assignment statement introduced in Chapter 11. Compare the VHDL models given in Figure 13.2 and Figure 12.1. The interfaces, or entity declarations, for the two VHDL models are identical; only the architecture statements differ.

The structural modeling style shown in Figure 12.1 defines the input/output function by component instantiations describing a netlist, whereas the behavioral modeling style shown in Figure 13.2 defines the input/output function by concurrent assignments describing signal transformations. The signal transformations define what the majority design does, but not how. Although the logic expression defining the majority input/output transform shown in

---

[1] An ALU is a subsystem within a processor that performs arithmetic and logic operations.

Figure 13.2

Behavioral description of the majority function

```
-- Interface
entity MAJORITY is
  -- Input/output ports
  port
    (A_IN, B_IN, C_IN : in BIT;
     Z_OUT            : out BIT);
end MAJORITY;

-- Body
architecture BEHAVIOR of MAJORITY is
begin
  -- Use concurrent signal assignment to describe
  -- input/output function.
  Z_OUT <= (not A_IN and B_IN and C_IN) or
           (A_IN and not B_IN and C_IN) or
           (A_IN and B_IN and not C_IN) or
           (A_IN and B_IN and C_IN);

end BEHAVIOR;
```

Figure 13.2 is given in canonical sum-of-products (SOP) form, the design may actually be implemented in the minimized form shown in Figure 12.1.

13.1.1 Data Flow and Algorithmic Modeling

Using concurrent assignments to describe signal transformations is one kind of behavioral modeling, called *data flow* modeling. Figure 13.3 shows that there are two general kinds of behavioral modeling: *data flow* and *algorithmic*. *Data flow* modeling styles use *concurrent statements*, whereas algorithmic modeling styles use *sequential statements*. The difference between a concurrent statement and a sequential statement is the manner in which the statements execute. The rules for executing sequential statements are based on the familiar execution semantics of most popular software programming languages. That is, such statements are principally executed from top to bottom in the sequence in which they are written, recog-

Figure 13.3

VHDL modeling style taxonomy

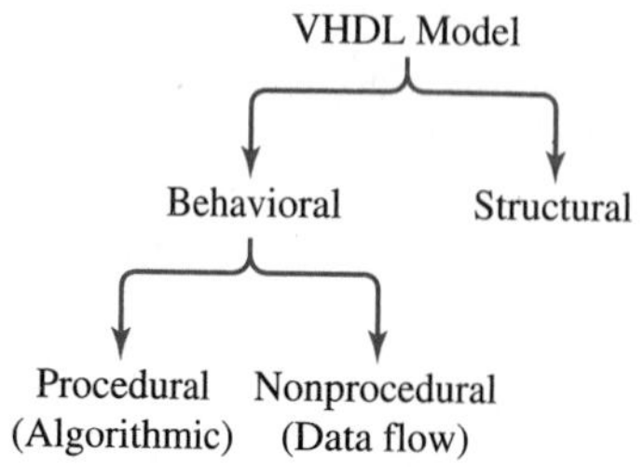

nizing that conditional and iteration constructs can alter this flow of control. In contrast, recall from Chapter 11 that the sequence in which concurrent statements are written has no bearing on the order in which the statements execute. Rather, the execution of concurrent statements is dependent on certain data conditions, hence the term *data flow*. The rules for executing concurrent statements are more aligned with the execution semantics of parallel software programming languages.

To understand the differences between concurrent and sequential statements, consider, for instance, signal assignment statements in VHDL. VHDL provides both a *concurrent* signal assignment statement and a *sequential* signal assignment statement. The two signal assignment statements can have the same syntax (in other words, they look the same), however, the signal assignment statements differ substantially in the manner in which they execute.

Three sample signal assignment statements are listed in the following segment of VHDL code.

```
            .
            .
            .
SIG_A <= IN_A and IN_B;
SIG_B <= IN_A nor IN_C;
SIG_C <= not IN_D;
            .
            .
            .
```

These signal assignment statements could be either concurrent or sequential signal assignment statements, depending on the context in which they appear within a VHDL model. If the statements are sequential signal assignment statements, they execute from top to bottom in the textual order listed; in other words, the signal assignment for SIG_A executes, then the signal assignment for SIG_B executes, and finally the signal assignment for SIG_C executes.

If, on the other hand, the statements are concurrent signal assignment statements, the textual listing of the statements has no bearing on the order in which they execute. The assignment statement for SIG_A executes only when either signal IN_A or IN_B change value. Likewise, the assignment statement for SIG_B executes only when either signal IN_A or IN_C change value. Finally, the assignment statement for SIG_C executes any time the signal IN_D changes value.

In summary, it is instructive to develop a perspective of how VHDL is used; such a perspective involves an understanding of common hardware representations and associated modeling styles. Modeling styles can be broadly partitioned into structural and behavioral descriptions; behavioral descriptions can be further partitioned into data flow and algorithmic descriptions. As we study VHDL, use these modeling styles as conceptual "buckets" for collecting various VHDL statements and understanding how these statements work together to generate models of digital systems.

13.2 CONDITIONAL CONCURRENT SIGNAL ASSIGNMENT STATEMENT

Let us examine the data flow VHDL model of the majority function in Figure 13.2 in more detail. The entity declaration for MAJORITY declares three input signals (A_IN, B_IN, and C_IN) and one output signal (Z_OUT), all of type BIT. Recall, type BIT is a predefined type representing the values of '0' and '1'.

The architecture body for the VHDL model contains the following statement defining the relationship between the inputs and the output. Introduced in Chapter 11, the statement is called a *conditional concurrent signal assignment statement.*

```
Z_OUT <=  (not A_IN and B_IN and C_IN) or
          (A_IN and not B_IN and C_IN) or
          (A_IN and B_IN and not C_IN) or
          (A_IN and B_IN and C_IN);
```

The signal assignment statement for Z_OUT is a concurrent statement because it appears within the context of an architecture. As will be discussed in later chapters, a sequential statement can only appear within the context of a *process statement* or a *subprogram.*

The conditional concurrent signal assignment has the format

```
target_signal <= value;
```

and executes whenever A_IN, B_IN, or C_IN change value. When the assignment statement executes, the value computed on the right-hand side is assigned to the target signal on the left-hand side.

The right-hand side of the conditional concurrent signal assignment statement for Z_OUT is an *expression* composed of the input signals and several logic operators, **not**, **and**, and **or**. VHDL logic operators were introduced in Chapter 11 and have the conventional meanings explained in Chapter 3. In general, an arbitrarily complex expression can be constructed to describe the right-hand/left-hand, stimulus/response behavior. However, VHDL's data typing rules mandate that the right-hand side expression yield a legal value for the type of the left-hand side target signal.

As another example of the conditional concurrent signal assignment statement, consider describing the printer controller discussed in Chapter 4; the truth table is repeated below.

Row	ENABLE	SELECT	USER_1	USER_2	PRINT_1	PRINT_2
0	0	0	0	0	0	0
1	0	0	0	1	0	0
2	0	0	1	0	0	0
3	0	0	1	1	0	0
4	0	1	0	0	0	0
5	0	1	0	1	0	0
6	0	1	1	0	0	0
7	0	1	1	1	0	0
8	1	0	0	0	0	0
9	1	0	0	1	0	1
10	1	0	1	0	1	0
11	1	0	1	1	1	1
12	1	1	0	0	0	0
13	1	1	0	1	1	0
14	1	1	1	0	0	1
15	1	1	1	1	1	1

The data flow VHDL model of the printer controller shown in Figure 13.4 illustrates a more general format for the conditional concurrent signal assignment statement. The conditional concurrent signal assignment statements describing the input/output function of the printer controller have the format:[2]

```
target_signal <= value1 when condition1 else
                 value2 when condition2 else
                         .

                 valueN-1 when conditionN-1 else
                 valueN;
```

Figure 13.4

Data flow description of a printer controller

```
-- Interface
entity PRNT_CNTRL is
  -- Input/output ports
  port
    (USER_1, USER_2 : in BIT;
     ENABLE, SEL    : in BIT;
     PRINT_1, PRINT_2 : out BIT);
end PRNT_CNTRL;


-- Body
architecture DATA_FLOW of PRNT_CNTRL is
begin
  -- Use conditional concurrent signal assignment statement
  -- to describe input/output function.
  PRINT_1 <= USER_1 when (ENABLE = '1' and SEL = '0') else
             USER_2 when (ENABLE = '1' and SEL = '1') else
             '0'; -- ENABLE = '0'


  PRINT_2 <= USER_2 when (ENABLE = '1' and SEL = '0') else
             USER_1 when (ENABLE = '1' and SEL = '1') else
             '0'; -- ENABLE = '0'
end DATA_FLOW;
```

---

[2] VHDL-93: In VHDL-93, conditional concurrent signal assignment statements may also have the formats:

```
target-signal <= value1 when condition1 else
                 value2 when condition2 else
                         .

                 valueN-1 when conditionN-1 else
                 valueN when conditionN;
```

or
```
    target_signal <= value when condition;
```

If no condition evaluates to TRUE, then no assignment is made to the target signal.

The **else** clauses are optional, allowing the preceding conditional concurrent signal assignment statement format to collapse neatly down to the simpler format

```
target_signal <= value;
```

used earlier in the `MAJORITY` model.

As the name implies, the *conditional* concurrent signal assignment statement is modeled after the conventional semantics of the *conditional* "if statement" in software programming languages. When one or more of the signals appearing on the right-hand side change value, the conditional concurrent signal assignment statement executes, evaluating the condition clauses in the textual order listed from top to bottom, in other words, `condition1`, `condition 2`, and so on. If a condition is found to be true, the corresponding expression is evaluated and the resulting value is assigned to the target signal. If none of the conditions are found to be true, the expression following the last condition clause is evaluated, assigning the value `valueN` to the target signal.

The conditions must be Boolean expressions; in other words, they must evaluate to a result of type `BOOLEAN`. We have not discussed the type `BOOLEAN`. `BOOLEAN` is another example of a predefined type, representing the values `TRUE` and `FALSE`. The equals operator, =, in the condition (`ENABLE` = ′1′ **and** `SEL` = ′0′) is an example of a *relational* operator, which will be discussed in the next section.

The conditional concurrent signal assignment statement for `PRINT_1` thus reads as follows:

If `ENABLE` is ′1′ and `SEL` is ′0′, then the value of `USER_1` is assigned to `PRINT_1`. Else, if `ENABLE` is ′1′ and `SEL` is ′1′, then the value of `USER_2` is assigned to `PRINT_1`. Otherwise, `PRINT_1` is assigned the value ′0′.

You should compare the `PRNT_CNTRL` VHDL design entity with the truth table to understand how the conditional concurrent signal assignment statements encode the logic function.

As a third example of data flow modeling, Figure 13.5 gives two models of the weather vane combinational system discussed in Chapter 4 showing two forms of the conditional concurrent signal assignment statement. The inputs N, S, E, and W respectively denote the wind directions north, south, east, and west. The output N_EW is asserted whenever the wind direction is northeast (inputs N and E asserted) or northwest (inputs N and W asserted).

The ability to transform the general form of the conditional concurrent signal assignment statement into alternate forms illustrates again the VHDL design principle of providing general mechanisms for generating solutions rather than the solutions themselves. A core set of basic, general-purpose statements are provided that can be used in a variety of special-purpose ways.

Problem-Solving Skill
Always try to distill the aspects of a solution down to a few basic, general-purpose mechanisms. Such an approach enables many specific problem situations to be readily addressed as customized applications of a core set of solution mechanisms.

13.3 RELATIONAL OPERATORS

Recall from Chapter 11 that VHDL provides a set of *logic* operators, as shown in Table 13.1. Logic operators can accept operand(s) of the predefined types `BIT` and `BOOLEAN`; however, operands of type `BIT` and `BOOLEAN` cannot be mixed in a given operator. The `BOOLEAN` value `TRUE` corresponds to the `BIT` value ′1′; the `BOOLEAN` value `FALSE` corresponds to the `BIT` value ′0′ Although similar, types `BIT` and `BOOLEAN` typically represent different information and, as such, are used in different ways. Type `BOOLEAN` represents truth values and is used to express conditions. Type `BIT` represents the binary values of a digital signal and is used to express switching algebra operations.

Figure 13.5

Behavioral description of the weather vane function

```
entity WEATHER_VANE is
  port
    (N, S, E, W : in BIT;
     N_EW        : out BIT);
end WEATHER_VANE;

architecture DATA_FLOW1 of WEATHER_VANE is
begin
 N_EW <= N and (E or W);

end DATA_FLOW1;
```

(a)

```
entity WEATHER_VANE is
  port
    (N, S, E, W : in BIT;
     N_EW        : out BIT);
end WEATHER_VANE;

architecture DATA_FLOW2 of WEATHER_VANE is
begin
 N_EW <= '1' when (N='1' and E=;1;) else
         '1' when (N='1' and W='1') else
         '0';

end DATA_FLOW2;
```

(b)

Figures (a) and (b) use two different forms of the conditional concurrent signal assignment statement to describe the same logic function.

VHDL also provides a set of *relational* operators, as shown in Table 13.2 (p. 394). Relational operators test for equality, inequality, and ordering, having the traditional meanings listed below.

Operator	Definition
=	equals
/=	not equals
<	less than
<=	less than or equal to
>	greater than
>=	greater than or equal to

Table 13.1

VHDL logic operators

Class	Operators						
Logic	and	or	nand	nor	xor	not	xnor*

Note: (*) The logic operator **xnor** is provided only in VHDL-93.

Table 13.2

VHDL relational operators	Class	Operators					
	Relational	=	/=	<	<=	>	>=

Relational operators are binary operators and yield a result of type BOOLEAN, either TRUE or FALSE. For instance, for the predefined type BIT, the expression ('0' < '1') yields the value TRUE. Be careful not to confuse the signal assignment operator "<=" with the less-than-or-equals relational operator "<=." The operators use the same symbols, but they have different meanings and are used in different contexts.

13.3.1 Operator Precedence and Evaluation

VHDL defines a set of rules that determine the order in which VHDL operators comprising an expression are evaluated. The rules are based on a ranking or *precedence* assigned to each operator. Logic and relational operator *precedences* are listed in Table 13.3. The table shows that for the logic and relational operators, the **not** logic operator has the highest precedence, followed by the relational operators, followed by the remaining logic operators.

Using operator precedences, the rules for evaluating VHDL expressions are given below.
1. Operators of higher precedence are evaluated first. Operators with the next highest precedence are evaluated second, and so on.
2. Operators of equal precedence are evaluated from left to right in the textual order they appear.
3. Parenthetical expressions, if present, are evaluated first according to the above rules, starting with the most deeply nested set of parentheses and working outward.

Notice that the parentheses in the logic expression for the conditional concurrent signal assignment statement driving Z_OUT in Figure 13.2 are required to force the **and** operators to be evaluated before the **or** operators and correctly yield a sum-of-products signal transformation. The parentheses in the conditions in the conditional concurrent signal assignment statements driving PRINT_1 and PRINT_2 are not required; they are a recommended modeling practice to improve readability.

Table 13.3

Logic and relational operator precedence	Precedence	Operators					
	Highest				not		
	↓	=	/=	<	<=	>	>=
	Lowest	**and**	**or**	**nand**	**nor**	**xor**	**xnor***

Note: (*) The logic operator **xnor** is provided only in VHDL-93.

13.4 SELECTED CONCURRENT SIGNAL ASSIGNMENT STATEMENT

We will now discuss another kind of concurrent signal assignment statement, which is illustrated in the VHDL model given in Figure 13.6. The VHDL model presents a data flow description of a combinational system that detects even Gray codes analyzed in Chapter 3 and structurally modeled in Chapter 12. If the input control signal CNTRL is asserted and the input data signal GRAY encodes an even number, then the output signal EVEN is asserted.

Figure 13.6

Data flow description of the even Gray code detector

CNTRL	GRAY(1)	GRAY(0)	EVEN
0	0	0	0
0	0	1	0
0	1	0	0
0	1	1	0
1	0	0	1
1	0	1	0
1	1	0	0
1	1	1	1

(a)

```vhdl
-- Interface
entity EVEN_GRAY is
  -- Input/output ports
  port
    (CNTRL : in BIT;
     GRAY  : in BIT_VECTOR(1 downto 0);
     EVEN  : out BIT);
end EVEN_GRAY;

-- Body
architecture DATA_FLOW of EVEN_GRAY is
begin
  -- Use selected concurrent signal assignment statement
  -- to describe input/output function.
  with CNTRL select
  EVEN <= not(GRAY(1) xor GRAY(0)) when '1',
          '0'                       when '0';
end DATA_FLOW;
```

(b)

Figures (a) and (b) respectively give a truth table and VHDL model for a combinational digital system that detects even 2-bit Gray codes.

The interface declares two input signals and one output signal. The input signal CNTRL controls if the even Gray code detector is active and is of the predefined type BIT. The input signal GRAY encodes numbers as 2-bit Gray codes and is of the predefined type BIT_VECTOR. BIT_VECTOR is the first example we have seen of an *array type* in VHDL; each element of the array type is of type BIT. The type BIT_VECTOR provides an easy way to declare a signal to be a group or string of conceptually related bits, more commonly called a *bit vector*. The declaration of the port GRAY in the EVEN_GRAY example

```vhdl
GRAY : in BIT_VECTOR(1 downto 0);
```

sets the length of the BIT_VECTOR type to 2 with a *descending* index range 1 **downto** 0. Individual elements of the bit vector GRAY are accessed by the *indexed name* notations,

GRAY(1) and GRAY(0). We will tend to use descending index ranges to be consistent with our earlier discussions of the binary number system in Chapter 2; array indices will match bit string subscripts. Also, we will usually adopt the convention that the leftmost array position represents the most significant bit. Thus, the first element of GRAY, accessed by GRAY(1), is the most significant bit. The next, and in this case, last, element of GRAY, accessed by GRAY(0), is the least significant bit. The type BIT_VECTOR will be discussed in more detail in Section 13.7.

The architecture uses a single *selected concurrent signal assignment statement* to encode the desired input/output function having the following format.

```
with discriminant select
target_signal <= value1 when choice1,
                 value2 when choice2,
                         .

                 valueN when choiceN;
```

The conditional concurrent signal assignment statement is modeled after the "if statement," whereas the selected concurrent signal assignment statement is modeled after the "case statement" in software programming languages. The selected concurrent signal assignment statement executes when any signal appearing in the discriminant, value, or choice expressions changes value. When a selected concurrent signal assignment statement executes, the choice clauses are evaluated, in other words, choice1, choice2, and so on. For the choice that equals the value of the discriminant, the corresponding value (value1, value2, and so on) is assigned to the target signal.

There are a few rules concerning the proper construction of a selected concurrent signal assignment statement. First, the discriminant following the reserved keyword **with** that controls which value is assigned to the target signal must be of a type having *discrete* values, in other words, a finite set of values that can be enumerated. This requirement is satisfied by the signal CNTRL because it is of type BIT, having only two values '0' and '1'. Second, all possible values of the discriminant must be accounted for by the set of choices. In other words, one of the choices has to match the value of the discriminant. Moreover, only *one* of the choices must match the value of the discriminant. These rules prevent indeterminate behavior, ensuring that there is a complete and consistent correspondence between discriminant values and target signal values.

The selected concurrent signal assignment statement for EVEN thus reads as follows:
- When CNTRL is '1', then $EVEN = \overline{(GRAY(1) \oplus GRAY(0))}$; in other words, $EVEN = (GRAY(1) \odot GRAY(0))$.
- When CNTRL is '0', then $EVEN = 0$.

Compare the VHDL model of the EVEN_GRAY design entity with the truth table to understand how the selected concurrent signal assignment statement encodes the logic function.

13.4.1 Choices

In cases where the controlling discriminant has many possible values, VHDL provides several mechanisms to simplify specifying the choices. For example, consider the following selected concurrent signal assignment statement that is controlled by a 3-bit signal, COMMAND.

```
signal COMMAND : BIT_VECTOR(2 downto 0);
                              .
                              .
with COMMAND select
target_signal <= value1 when B"000",
                 value2 when B"010" to B"100",
                 value3 when B"001" | B"101",
                 value4 when others;
```

The four choices account for the eight possible values of the signal COMMAND.

The first choice specifies a single value, B"000", which is an example of a *bit string literal*. Literals specify value in VHDL. A bit string literal is a shorthand notation for specifying the value of an object of type BIT_VECTOR. The characters enclosed in double quotes define the individual bit values; the leading B defines the base to be binary. The second choice specifies a *range* of values, covering the three values B"010", B"011", and B"100". A range specifies a contiguous set of values of the discriminant and can be ascending or descending. Individual values and ranges can be combined with the vertical bar, |, as illustrated with the third choice. The third choice specifies the two values B"001" or B"101". Finally, the last choice uses the keyword **others** to account for all other values of COMMAND not specified by the first three choices, in other words, B"110" and B"111". If the **others** choice is used, it must be the *last* choice.

As another example of the selected concurrent signal assignment statement, Figure 13.7 shows a VHDL model of the sine wave limit detector combinational system discussed in Chapter 4. The input SINE_WAVE encodes sine wave amplitude as a 4-bit two's complement number in the symmetrical range -7_{10} to 7_{10}. The output LIMIT is asserted whenever the sine wave amplitude equals or exceeds an upper limit of 6_{10} or a lower limit of -6_{10}.

Figure 13.7

A VHDL model of a sine wave amplitude limit detector

```
entity LIMIT_DETECTOR is
  port
    (SINE_WAVE : in BIT_VECTOR(3 downto 0);
     LIMIT     : out BIT);
end LIMIT_DETECTOR;

architecture DATA_FLOW of LIMIT_DETECTOR is
begin
  -- Use selected concurrent signal assignment statement
  -- to describe input/output function.
  with SINE_WAVE select
  LIMIT <= '1' when B"0110" to B"1010",
           '0' when others;
end DATA_FLOW;
```

13.5 DATA FLOW AND HARDWARE PARALLELISM

To gain a better understanding of concurrent signal assignment statements, let us explore how to exploit the concurrency in data flow behavioral descriptions to indicate the parallelism of the actual hardware, thereby *suggesting* a possible implementation. To illustrate this concept, we will develop a data flow VHDL model of an *even parity generator*.[3]

13.5.1 Parity Encoding

Parity encoding is used to ensure the integrity of a digital signal. Sometimes a digital signal can get corrupted during transmission by unwanted noise or random disturbances that change a 1 to a 0 or vice versa. One simple way of checking for such transmission errors is to add an extra bit to the signal, called a *parity bit*. For an even parity error checking scheme, the parity bit is set so the total number of 1's in the digital signal including the parity bit is an even number. The signal is then transmitted and the total number of 1's are again counted at the receiving end of the communications link. If the total number of 1's is not an even number, then the receiver knows that the digital signal has been corrupted during transmission and requests a retransmission.

The truth table for a 4-input, or 4-bit, even parity generator is shown below.

ABCD	PARITY
0000	0
0001	1
0010	1
0011	0
0100	1
0101	0
0110	0
0111	1
1000	1
1001	0
1010	0
1011	1
1100	0
1101	1
1110	1
1111	0

The total number of 1's contained in the four input signals, A, B, C, and D, and the 1-bit *PARITY* signal is always an even number.

Using the switching algebra minimization techniques discussed in Chapter 4, the *PARITY* output can be shown to be the exclusive disjunction of all four inputs.

$$
\begin{aligned}
PARITY &= (\overline{A} \bullet \overline{B} \bullet \overline{C} \bullet D) + (\overline{A} \bullet \overline{B} \bullet C \bullet \overline{D}) + (\overline{A} \bullet B \bullet \overline{C} \bullet \overline{D}) + (\overline{A} \bullet B \bullet C \bullet D) \\
&\quad + (A \bullet B \bullet \overline{C} \bullet D) + (A \bullet B \bullet C \bullet \overline{D}) + (A \bullet \overline{B} \bullet \overline{C} \bullet \overline{D}) + (A \bullet \overline{B} \bullet C \bullet D) \\
&= \overline{A} \bullet [(\overline{B} \bullet \overline{C} \bullet D) + (\overline{B} \bullet C \bullet \overline{D}) + (B \bullet \overline{C} \bullet \overline{D}) + (B \bullet C \bullet D)] \\
&\quad + A \bullet [(B \bullet \overline{C} \bullet D) + (B \bullet C \bullet \overline{D}) + (\overline{B} \bullet \overline{C} \bullet \overline{D}) + (\overline{B} \bullet C \bullet D)] \qquad \text{P4b} \\
&= \overline{A} \bullet [\overline{B} \bullet [(\overline{C} \bullet D) + (C \bullet \overline{D})] + B \bullet [(\overline{C} \bullet \overline{D}) + (C \bullet D)]] \\
&\quad + A \bullet [B \bullet [(\overline{C} \bullet D) + (C \bullet \overline{D})] + \overline{B} \bullet [(\overline{C} \bullet \overline{D}) + (C \bullet D)]] \qquad \text{P4b} \\
&= \overline{A} \bullet [\overline{B} \bullet (C \oplus D) + B \bullet (\overline{C \oplus D})] + A \bullet [B \bullet (C \oplus D) + \overline{B} \bullet (\overline{C \oplus D})] \qquad \textbf{xor} \\
&= \overline{A} \bullet [B \oplus C \oplus D] + A \bullet [\overline{B \oplus C \oplus D}] \qquad \textbf{xor} \\
&= A \oplus B \oplus C \oplus D \qquad \textbf{xor}
\end{aligned}
$$

---

[3] **parity**: The comparative odd-even relationship between two integers. *The American Heritage Dictionary*, Houghton Mifflin Company. Reprinted by permission.

Extending the above analysis, the even parity bit for an *N*-bit digital signal can be computed as an *N*-bit **xor**.

13.5.2 VHDL Parity Generator Model

Given the above explanation of even parity encoding, a VHDL data flow description of an 8-bit even parity generator is given in Figure 13.8. The interface declares an 8-bit input signal BVEC and a 1-bit output signal PARITY. The architecture contains a conditional concurrent signal assignment statement that computes the value for the output signal PARITY by taking the exclusive-or of all the input signals.

As was explained earlier, the data flow description of EVEN_PARITY is a behavioral model defining function but not necessarily implementation. The reader of the description may "see" a logic implementation in a data flow description, but this is only "in the eye of the beholder." There are many ways to realize a logic expression. For example, EVEN_PARITY could be implemented in the tree configuration shown in Figure 13.9(a) (p. 400) or the cascade configuration shown in Figure 13.9(b). The tree configuration uses a parallel-oriented scheme to compute the parity, whereas the cascade configuration uses a serial-oriented scheme to compute the parity.

Concurrency in data flow descriptions can be exploited to indicate hardware parallelism, thereby *suggesting* an implementation, in other words, helping the reader to "see" a particular implementation. The "reader" may be a designer or a computer program such as a synthesis tool. For instance, the VHDL data flow description of EVEN_PARITY given in Figure 13.10(a) more closely reflects the parallel nature of the tree-configured even parity logic, whereas the VHDL data flow description of EVEN_PARITY given in Figure 13.10(b) more closely reflects the serial nature of the cascade-configured even parity logic.

In summary, all three VHDL models of EVEN_PARITY given in Figure 13.8 and Figure 13.10 describe exactly the same input/output function. However, the manner in which the concurrent signal assignment statements are used to describe the input/output function can be exploited to indicate varying degrees of parallelism of the associated hardware, thereby suggesting an implementation.

Problem-Solving Skill

Algorithms can often be cast into two equivalent forms: serial and parallel. In a serial form, an algorithm is decomposed into several tasks that are performed sequentially. In a parallel form, an algorithm is decomposed into a different set of tasks that are performed concurrently.

Figure 13.8

Data flow description of an 8-bit even parity generator

```
-- Interface
entity EVEN_PARITY is
  -- Input/output ports
  port
    (BVEC   : in BIT_VECTOR(7 downto 0);
     PARITY : out BIT);
end EVEN_PARITY;

-- Body
architecture DATA_FLOW of EVEN_PARITY is
begin
  -- If BVEC holds an odd number of 1's, the xor of all
  -- the bits yields a 1.
  PARITY <= BVEC(0) xor BVEC(1) xor BVEC(2) xor BVEC(3) xor
            BVEC(4) xor BVEC(5) xor BVEC(6) xor BVEC(7);

end DATA_FLOW;
```

Figure 13.9

Alternative logic diagrams for the even parity function

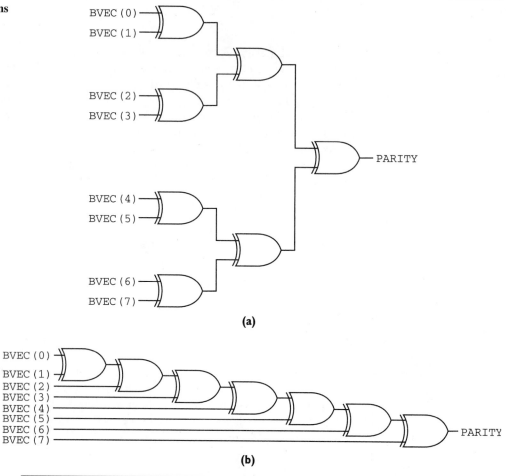

Figure (a) shows a tree configuration of the **xor** operators. Figure (b) shows a cascade configuration of the **xor** operators.

Figure 13.10

Alternative data flow descriptions of the even parity function

```
-- Interface
entity EVEN_PARITY is
  -- Input/output ports
  port
    (BVEC    : in BIT_VECTOR(7 downto 0);
     PARITY : out BIT);
end EVEN_PARITY;
```

Figure 13.10 continued

```
-- Body
architecture TREE of EVEN_PARITY is
  signal INT1, INT2, INT3, INT4, INT5, INT6 : BIT;
begin
  -- First row of tree
  INT1 <= BVEC(0) xor BVEC(1); INT2 <= BVEC(2) xor BVEC(3);
  INT3 <= BVEC(4) xor BVEC(5); INT4 <= BVEC(6) xor BVEC(7);
  -- Second row of tree
  INT5 <= INT1 xor INT2; INT6 <= INT3 xor INT4;
  -- Last row of tree
  PARITY <= INT5 xor INT6;

end TREE;
```

(a)

```
-- Interface
entity EVEN_PARITY is
  -- Input/output ports
  port
    (BVEC   : in BIT_VECTOR(7 downto 0);
     PARITY : out BIT);
end EVEN_PARITY;

-- Body
architecture CASCADE of EVEN_PARITY is
  signal INT1, INT2, INT3, INT4, INT5, INT6 : BIT;
begin
  INT1 <= BVEC(0) xor BVEC(1); INT2 <= INT1 xor BVEC(2);
  INT3 <= INT2 xor BVEC(3); INT4 <= INT3 xor BVEC(4);
  INT5 <= INT4 xor BVEC(5); INT6 <= INT5 xor BVEC(6);
  PARITY <= INT6 xor BVEC(7);

end CASCADE;
```

(b)

Figure (a) shows a data flow model reflecting the parallel nature of the tree configuration of the **xor** operators. Figure (b) shows a data flow model reflecting the serial nature of the cascade configuration of the **xor** operators.

13.6 ALTERNATIVE ARCHITECTURAL BODIES

Since the three VHDL descriptions of the even parity generator shown in Figure 13.8, Figure 13.10(a), and Figure 13.10(b) have the same interface, the three descriptions do not actually define three different parity generators. Rather, the descriptions define three alternative implementations, or architectures, of a single parity generator, EVEN_PARITY. The interface definition of EVEN_PARITY was repeated for each description only for clarity. In practice, the entity declaration is given only once. Then, associated alternative architecture bodies are described, giving a unique name to each architecture, such as DATA_FLOW, TREE, and CASCADE.

```
architecture DATA_FLOW of EVEN_PARITY is
       .
       .

architecture TREE of EVEN_PARITY is
       .
       .

architecture CASCADE of EVEN_PARITY is
       .
       .
```

Thus, an entity declaration or interface can have one or more associated architectures. Figure 13.11 modifies Figure 12.10 to update our understanding of design entities. Each design entity is composed of an interface and one or more associated architecture bodies. Like building blocks, design entities can be connected together to form larger design entities, which, in turn, can be connected together to form even larger design entities, and so on.

Alternative architectures can be used effectively in a variety of situations during design. For example, alternative architectures can represent multiple versions of a design entity generated by a single designer or competing design teams. Alternative architectures can also represent multiple views of a design entity, supporting different design activities. An abstract, behavioral view might support system-level simulation or synthesis design activities, whereas a more detailed structural view might support test or physical design activities (see Chapter 1). Finally, alternative architectures can judiciously control design information exchanged between two organizations. A high-level, behavioral description can be used to transmit the input/output functions without revealing proprietary information concerning detailed implementation.

Figure 13.11

VHDL design entity composition and structure

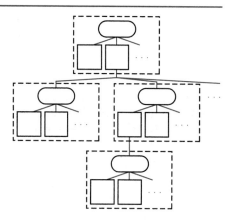

13.7 BIT VECTORS

Our earlier example VHDL models introduced bit vectors and the predefined type BIT_VECTOR; we will now examine them in more detail. Recall that the predefined BIT_VECTOR is an array type, where each array element is of type BIT. Moreover, BIT_VECTOR is called a *composite type* because it represents a group or collection of values, which are the individual array elements or bits. Remember that BIT_VECTOR, by itself, is *not* an array; rather, BIT_VECTOR is a template for an array. In the following port declaration, for instance, taken from Figure 13.8,

```
BVEC: in BIT_VECTOR(7 downto 0);
```

the type BIT_VECTOR defines the form of the array and the port BVEC defines the actual array or bit vector. The signal BVEC is a particular "instantiation," more accurately called an *elaboration*, of the type BIT_VECTOR.

An array in VHDL has the following three defining characteristics:
1. the type of the elements in the array,
2. the length of the array, and
3. the indices of the array.

All the elements of the array denoted by type BIT_VECTOR are of type BIT, representing the values '0' and '1'.

The length of the array denoted by the type BIT_VECTOR is *not* specified. Figure 13.12(a) shows that BIT_VECTOR represents a special kind of array in which the number of bits or the array length is unconstrained or arbitrary; such an array is aptly called an *unconstrained array*. The length of an array of type BIT_VECTOR is set at object declaration, which allows us to use BIT_VECTOR to declare many different bit vectors of different lengths. Figure 13.12(b) shows a diagram of the input/output signal BVEC; the port declaration given above sets the length of the array (bit vector) to 8.

The indices of the array denoted by the type BIT_VECTOR are the natural numbers, in other words, 0, 1, 2, ... The index range can be ascending or descending. Several possible alternative declarations for the port BVEC are given below, illustrating a variety of indexing schemes.

```
-- Ascending indexing scheme, starting at 0
BVEC : in BIT_VECTOR(0 to 7);
-- Ascending indexing scheme, starting at 8
BVEC : in BIT_VECTOR(8 to 15);
-- Descending indexing scheme, starting at 15
BVEC : in BIT_VECTOR(15 downto 8);
```

Figure 13.12

Type BIT_VECTOR

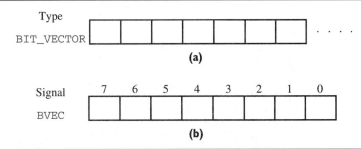

Figure (a) shows that the type BIT_VECTOR represents an array of bits of an unconstrained or arbitrary length. Figure (b) shows an array of type BIT_VECTOR that is constrained to be of length 8.

Again, we will tend to use a descending range for bit vectors. The most significant bit is associated with the leftmost array position, and the least significant bit is associated with the rightmost array position.

13.7.1 Specifying Values

A *bit string literal* is a shorthand notation for specifying the value of a bit vector. Figure 13.13(a) declares signal SIG_A to be an 8-bit vector and uses the bit string literal B"1110_0011" to assign the value 227_{10} to SIG_A. Several additional examples of specifying values for bit vectors are given below.

B"11111010"   B"1111_1010"

(a)   **(b)**

X"FA" equivalent to B"11111010"

(c)

O"372" equivalent to B"011111010"

(d)

A bit string literal starts with the single character "B," "O," or "X," which denotes the base.

Designation	Base
B	Binary
O	Octal
X	Hexadecimal

The octal and hexadecimal bases provide shorthand notations for specifying long, cumbersome bit strings. (See Chapter 2 for an explanation of based numbering systems.)

The value of the bit string literal is enclosed in *double* quotes. If the base is binary, only the numerals 0 and 1 can be used. If the base is octal, only the numerals 0–7 can be used. If the base is hexadecimal, only the numerals 0–9 and the letters A–F or a–f can be used. An underscore character may also be used. The only purpose of the underscore character is to improve readability; it has no effect on the numeric value. That is, examples (a) and (b) specify the same bit string literal, however, example (b) is more readable.

A bit string literal in octal is converted to an equivalent bit string literal in binary by replacing each octal digit by its 3-bit binary representation given in Table 2.4. Similarly, a bit string literal in hexadecimal is converted to an equivalent bit string literal in binary by replacing each hexadecimal digit by its 4-bit binary representation given in Table 2.5. Notice

Figure 13.13

Bit string literals

```
signal SIG_A : BIT_VECTOR(7 downto 0)

SIG_A <= B"1110_0011";
```

(a)

(b)

Figure (a) illustrates a segment of VHDL using a bit string literal to assign a value to a bit vector. Figure (b) illustrates the individual elements (bits) and their values.

that although the four bit string literal examples all specify the same numeric value, 250_{10}, they do *not* all specify the same VHDL bit string. Examples (a), (b), and (c) specify an 8-bit vector containing the value 11111010_2. However, example (d) specifies a 9-bit vector containing the value 011111010_2.

In Chapter 16, we will discuss enhancements to bit string literals provided by VHDL-93. The enhancements enable bit string literal notation to be used to specify values of types other than BIT_VECTOR.

13.7.2 Slices

A *slice* provides a way of referring to a part of a one-dimensional array, in general, and a bit vector, in particular. For example, if we want to refer to just the lower four bits of the signal BVEC in the EVEN_PARITY design entity, then we could simply write the following:

```
BVEC(3 downto 0)
```

Slices of bit vectors are useful because bit vectors are often viewed as collections of fields, where each field carries a piece of information. For example, a bit vector representing a computer instruction may contain a field representing the kind of instruction. To execute the instruction, we could use a slice to extract the field containing the instruction for decoding.

Slices are associated only with one-dimensional arrays. The directionality of the index range of the slice must be the same as the directionality of the index range of the referenced one-dimensional array. For example,

```
BVEC(0 to 4)
```

is illegal because the slice has an ascending index range and the referenced bit vector, BVEC, has a descending index range.[4] In addition, the bounds of the index range of the slice must be legal indices for the referenced one-dimensional array. For example,

```
BVEC(8 downto 5)
```

is illegal because the slice uses the index 8 and the referenced bit vector, BVEC, uses only indices $7, 6, \ldots, 1, 0$.

13.7.3 Aliases

An *alias* is an alternative name. VHDL aliases can be used to rename objects, in general, and bit vectors, in particular.

---

[4] To be precise, a mismatch between a slice index direction and the referenced one-dimensional array direction is not an error, it simply yields a *null slice*, in other words, a slice denoting *no* elements.

For example, consider the following design entity declaration.

```
-- Interface
entity SOME_UNIT is
  -- Input/output ports
  port
    (A_VERY_LONG_SIGNAL_NAME : in BIT;
     AN_EVEN_LONGER_SIGNAL_NAME : out BIT_VECTOR(7 downto 0);
end SOME_UNIT;
```

The input/output signals have long, cumbersome names. Recall that there are no restrictions on the length of an identifier in VHDL. If we want to use shorter signal names, one approach is to change the design entity declaration. Unfortunately, we may not have the freedom to change the entity declaration because it may have been defined to support systemwide configuration or function information.

Another approach is to declare *aliases* for the input/output signals, thereby giving the signals a new set of names according to personal preference. In architecture EXAMPLE, SIG_IN is an alias for the input signal A_VERY_LONG_SIGNAL_NAME. Similarly, SIG_OUT is an alias for the output signal AN_EVEN_LONGER_SIGNAL_NAME.

```
-- Body
architecture EXAMPLE of SOME_UNIT is
  -- Alias declaration
  alias SIG_IN : BIT is A_VERY_LONG_SIGNAL_NAME;
  alias SIG_OUT : BIT_VECTOR(7 downto 0) is
                  AN_EVEN_LONGER_SIGNAL_NAME;

begin
     .
     .
end EXAMPLE;
```

The architecture EXAMPLE contains two *alias declarations*. An alias declaration starts with the keyword **alias**, followed by the new name or alias, followed by the old, aliased name. The type of the new name or alias must be the same as the type of the old, aliased name. Following the alias declarations, we can simply use the shorter aliases to reference the longer input/output signals.

Aliases can also be used to define names for parts of a bit vector. For example, consider the 9-bit signal PARITY_SIG shown below. The most significant bit, PARITY_SIG(8), is the parity bit; the lower-order 8 bits carry the data. As explained in the

signal PARITY_SIG: BIT_VECTOR(8 **downto** 0);

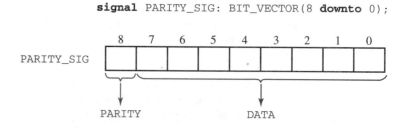

previous section, we can use slices to refer to parts of the bit vector PARITY_SIG. Alternatively, we can declare aliases to define names to refer to parts of PARITY_SIG.

```
-- Alias declarations
alias PARITY : BIT is PARITY_SIG(8);
alias DATA : BIT_VECTOR(7 downto 0) is PARITY_SIG(7 downto 0);
```

Sometimes, aliases are a better approach to referencing parts of a bit vector than slices because aliases can provide descriptive labels.

13.7.4 Operations

VHDL provides several operators for arrays, in general, and bit vectors, in particular. The logic and relational operators discussed in Section 13.3 can be applied to bit vectors as well as to scalar objects such as individual signals.

For example, logic operators can accept bit vectors; the logic operations are applied on an element-by-element basis. For binary logic operators, the two bit vector operands must be of the same length. The following example illustrates the binary **nor** operator accepting two 8-bit vectors.

```
signal SIG_A, SIG_B, SIG_C : BIT_VECTOR(7 downto 0);

    SIG_C <= SIG_A nor SIG_B;
```

If SIG_A equals B"1010_0110" and SIG_B equals B"1101_0011", then SIG_C equals B"0000_1000". That is, SIG_C(7) equals SIG_A(7) **nor** SIG_B(7), SIG_C(6) equals SIG_A(6) **nor** SIG_B(6), and so on.

Relational operators can also accept bit vectors and are also applied on an element-by-element basis. Unlike the logic operators, the bit vector operands need not be of the same length and the result is a *scalar* value of type BOOLEAN, in other words, either TRUE or FALSE.

For the equality relational operator, two bit vectors are equal if the vectors are of the same length and all corresponding element values are equal. For example, if SIG_A equals B"0000_1111" and SIG_B equals B"0000_1111", then the relational expression "SIG_A = SIG_B" yields the value TRUE because SIG_A and SIG_B are both of length 8 and SIG_A(7) = SIG_B(7), SIG_A(6) = SIG_B(6), and so on.

For the ordering relational operators, two cases must be considered: equal length operands and unequal length operands. If the left-hand and right-hand operands are of equal length, then corresponding elements of the two bit vectors are compared until a difference is found or all elements are checked. Corresponding elements are compared, starting with the leftmost elements and ending with the rightmost elements. For instance, assume SIG_A equals B"0100_1111" and SIG_B equals B"0110_0000". To evaluate the relational expression "SIG_A < SIG_B," corresponding array elements are compared, starting with the leftmost elements SIG_A(7) and SIG_B(7). Since SIG_A(7) equals SIG_B(7), SIG_A(6) and SIG_B(6) are then compared. Again, since SIG_A(6) equals SIG_B(6), SIG_A(5) and SIG_B(5) are then compared. Since SIG_A(5) is less than SIG_B(5), SIG_A is less than SIG_B and the relational expression "SIG_A < SIG_B" evaluates to TRUE. (Recall that for the predefined type BIT, '0' is less than '1'.)

If the left-hand and right-hand operands are of unequal length, then corresponding elements of the two bit vectors are compared left to right until a difference is found or all elements of the shorter array are checked. In the latter case, the shorter bit vector is *less than* the longer bit vector. For instance, consider the two bit vectors SIG_4 and SIG_8.

```
signal SIG_4 : BIT_VECTOR(3 downto 0);
signal SIG_8 : BIT_VECTOR(7 downto 0);
```

First, let SIG_4 equal B"1000" and SIG_8 equal B"0110_0000". To evaluate the relational expression "SIG_4 < SIG_8," corresponding array elements are compared, starting with the leftmost elements SIG_4(3) and SIG_8(7). Since SIG_4(3) is greater than SIG_8(7), SIG_4 is greater than SIG_8 and the relational expression "SIG_4 < SIG_8" evaluates to FALSE. Second, let SIG_4 equal B"0110" and SIG_8 equal B"0110_0000". To evaluate the relational expression "SIG_4 < SIG_8," corresponding array elements are again compared, starting with the leftmost elements SIG_4(3) and SIG_8(7). Since SIG_4(3 **downto** 0) equals SIG_8(7 **downto** 4), all elements of SIG_4 are checked and the relational expression "SIG_4 < SIG_8" evaluates to TRUE because SIG_4 is shorter than SIG_8.

Finally, the *concatenation* operator, "&," also accepts bit vectors. The concatenation operation catenates or joins "pieces" of arrays together to form a single larger array. The "pieces" can be individual elements, arrays, or combinations of elements and arrays, as shown below.

```
signal BVEC_8 : BIT_VECTOR(7 downto 0);

BVEC_8 <= B"0000" & B"1111";    -- B"0000_1111"

BVEC_8 <= '0' & B"0001111";    -- B"0000_1111"

BVEC_8 <= '0' & '0' & '0' & '0' & '1' & '1' & '1' & '1';
```

The first example joins two 4-bit vectors to form an 8-bit vector. The elements of BVEC_8 are formed by taking the elements of the left operand, B"0000", followed by the elements of the right operand, B"1111". The second example joins an individual element '0' with a 7-bit vector B"0001111" to form an 8-bit vector. The last example joins eight bit literals to form an 8-bit vector. The last example also shows that three or more operands are catenated together using multiple binary concatenation operators.

13.7.5 Predefined Attributes

The final aspect of our discussion of bit vectors concerns *predefined attributes*. Attributes specify "extra" information about some aspect of a VHDL model. VHDL provides several predefined attributes for querying information about arrays, in general, and bit vectors, in particular. For example, for the 8-bit string BVEC_8 declared in the previous section, the attribute BVEC_8'LEFT yields the left bound of the indexing scheme, namely, 7. The general form of a bit vector attribute is the name of the bit vector, followed by an apostrophe, "'" (called a "tick"), followed by the name of the attribute.

Predefined attributes for bit vectors are listed on the next page.

Attribute	Definition
name'LEFT	Left bound of indexing scheme
name'RIGHT	Right bound of indexing scheme
name'HIGH	Upper bound of indexing scheme
name'LOW	Lower bound of indexing scheme
name'RANGE	"name'LEFT **to** name'RIGHT" if the indexing scheme is ascending. "name'LEFT **downto** name'RIGHT" if the indexing scheme is descending.
name'REVERSE_RANGE	"name'RIGHT **downto** name'LEFT" if the indexing scheme is ascending. "name'RIGHT **to** name'LEFT" if the indexing scheme is descending.
name'LENGTH	Number of elements in the array

Most of the predefined attributes are self-explanatory. The values of the predefined attributes for the example bit vector BVEC_8 are listed below:

```
signal BVEC_8 : BIT_VECTOR(7 downto 0);
```

- BVEC_8'LEFT = 7
- BVEC_8'RIGHT = 0
- BVEC_8'HIGH = 7
- BVEC_8'LOW = 0
- BVEC_8'RANGE = 7 **downto** 0
- BVEC_8'REVERSE_RANGE = 0 **to** 7
- BVEC_8'LENGTH = 8

The designer does not have to be concerned about setting or maintaining predefined attributes. Once a bit vector is declared, its predefined attributes are automatically provided to the designer.

Attributes are often used to eliminate "hardcoding" specific numbers into a VHDL model. Instead of using "7" to refer to the leftmost index of BVEC_8 throughout an entire VHDL model, it is a better modeling practice to use the predefined attribute BVEC_8'LEFT. If attributes are judiciously used and the indexing scheme of BVEC_8 ever has to be modified, then only the signal declaration has to be changed. This is much simpler than the tedious process of scouring a large VHDL model looking for references to the indexing scheme of BVEC_8. We will see specific examples of the usefulness of attributes as we build our knowledge of VHDL; attributes will be covered in more detail in Chapter 16.

13.8 VHDL-93: SHIFT OPERATORS

VHDL-87 does not provide predefined shift and rotate operations on vectors. However, shift and rotate operations on vectors are so frequently used in VHDL modeling that VHDL-93 provides built-in or predefined shift and rotate operators to eliminate the tedious task of writing custom routines. Shift and rotate operators are collectively called *shift operators* (see Table 13.4); admittedly, the terminology could be better. Shift and rotate operators move around elements of an array and have the meanings listed on the next page.

Table 13.4

VHDL shift operators	Class	Operators					
	Shift	**sll**	**srl**	**sla**	**sra**	**rol**	**ror**

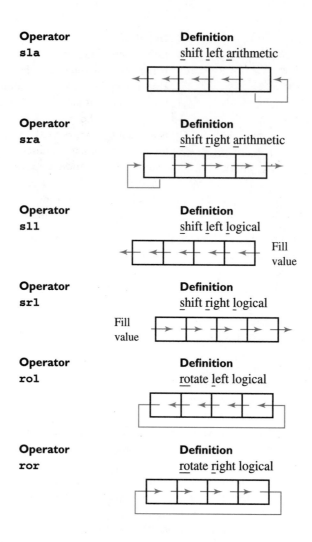

Operator	Definition
sla	shift left arithmetic

Operator	Definition
sra	shift right arithmetic

Operator	Definition
sll	shift left logical

Fill value

Operator	Definition
srl	shift right logical

Fill value

Operator	Definition
rol	rotate left logical

Operator	Definition
ror	rotate right logical

Examples of shift and rotate operators are given next.

```
-- Shift left arithmetic
B"1100" sla 1 = B"1000"    B"0011" sla 1 = B"0111"
```
(a)

```
-- Shift right arithmetic
B"1100" sra 1 = B"1110"    B"1100' sra -1 = B"1000"
```
(b)

```
-- Shift left logical
B"1111" sll 1 = B"1110"    B"1100" sll 4 = B"0000"
```
(c)

```
-- Shift right logical
B"1111" srl 1 = B"0111"    B"1100" srl 5 = B"0000"
```
(d)

```
-- Rotate left logical
B"1100" rol 1 = B"1001"    B"1100" rol -1 = B"0110"
```

(e)

```
-- Rotate right logical
B"1100" ror 1 = B"0110"    B"1100" ror -1 = B"1001"
```

(f)

The shift and rotate operators accept two operands. The left operand is a one-dimensional array having elements of type BIT or BOOLEAN. The right operand is an integer specifying the number of shifts or rotation steps. The number of shifts may exceed the length of the array, as shown in example (d). A negative number of shifts reverses the shift/rotate direction, as shown in examples (b), (e), and (f). In other words, B"1100" ror -1 is the same as B"1100" rol 1. The fill value is '0' for arrays having elements of type BIT and FALSE for arrays having elements of type BOOLEAN. Any values shifted out of an array are discarded.

Note that arithmetic shifts do not always yield expected numerical results. For instance, example (a) above shows that the shift left arithmetic **sla** operator does not always realize a multiply-by-two operation: The shift left arithmetic of B"0011" (3_{10}) by one position yields B"0111" (7_{10}). Also, example (b) shows that the shift right arithmetic **sra** operator does not always realize a divide-by-two operation: The shift right arithmetic of B"1100" (12_{10}) by one position yields B"1110" (14_{10}). Table 13.5 updates the operator precedence information given in Table 13.3 to include shift operators.

Table 13.5

Logic, relational, and shift operator precedence

Precedence	Operators					
Highest				not		
↓	sla	sra	sll	srl	rol	ror
↓	=	/=	<	<=	>	>=
Lowest	and	or	nand	nor	xor	xnor*

Note: (*) The logic operator **xnor** is provided only in VHDL-93.

13.9 SUMMARY

Hardware modeling can be broadly divided into two styles: *structural* and *behavioral*. Structural modeling *implicitly* defines the input/output function by describing components and their interconnections; the input/output function can be derived from the structural implementation. Conversely, behavioral modeling *explicitly* defines the input/output function by describing signal transformations. The concurrency of the behavioral specification may suggest an implementation.

In VHDL, behavioral models containing concurrent signal assignment statements are further classified as data flow models because the concurrent signal assignment statements describe the movement or flow of data. There are two types of concurrent signal assignment statements: conditional and selected. Both types of statements execute based on *stimulus-response* semantics, meaning that a concurrent signal assignment statement executes whenever any signal on the right-hand side changes value.

It is often a matter of individual preference whether to use a conditional or selected signal assignment statement. The conditional concurrent signal assignment statement can enumerate actions for only a select set of conditions, which can be given an ordered priority. The selected concurrent signal assignment statement must enumerate actions for all possible values of the discriminant, and all possible values have equal priority.

In examining the construction of concurrent signal assignment statements, we discussed logic, relational, concatenation, and shift (VHDL-93) operators in VHDL and how these operators can be used to compute values assigned to target signals. Table 13.6 summarizes the VHDL operators studied thus far; the table will be updated as we build our understanding of VHDL.

We also learned how the concurrency of data flow descriptions could be exploited to indicate the parallelism of hardware. Such information can be used to guide the design process toward a particular implementation. Multiple implementations of a design can be described in VHDL by *multiple alternative architectures* for a single design entity.

Finally, we examined how to declare and use bit vectors in VHDL using the predefined type BIT_VECTOR. BIT_VECTOR is a one-dimensional array type; each element of the array is of type BIT. VHDL provides an extensive set of capabilities for arrays in general, and bit vectors, in particular, including array values called bit vector literals, operations, and attributes. VHDL also provides convenient array accessing mechanisms, including indexed names, slices, and aliases.

Table 13.6

Logic, relational, and shift VHDL operators	Class	Operators						
	Logic[1]	and	or	nand	nor	xor	not	xnor
	Relational		=	/=	<	<=	>	>=
	Shift[2]		sll	srl	sla	sra	rol	ror
	Arithmetic				&			

Note:
- Note 1: The logic operator **xnor** is provided only in VHDL-93.
- Note 2: The shift operators are provided only in VHDL-93.

13.10 PROBLEMS

1. For the following logic expressions,

$$F_1(A, B, C, D) = \sum m(4, 5, 6, 7, 11, 12, 13)$$
$$F_2(A, B, C) = \sum m(1, 2, 4, 7)$$

   a. Minimize the logic expressions using any of the minimization techniques discussed in Chapter 4.
   b. Write VHDL structural models of the minimized logic. Declare the required components and write the associated design entities.

    c. Write VHDL behavioral models of the minimized logic.

    d. Verify that the structural and behavioral models describe the minimized logic by simulating the models for all possible inputs.

    e. Explain the main difference between structural and behavioral VHDL models.

2. Consider the following conditional *concurrent* signal assignment statements.

```
C <= A or B;   -- Signal assignment "C"
D <= C or B;   -- Signal assignment "D"
E <= D or A;   -- Signal assignment "E"
```

Assume that executing a concurrent signal assignment statement always changes the value of the associated target signal.

    a. Describe the sequence in which the concurrent signal assignment statements will execute in response to a change on signal A.

    b. Describe the sequence in which the concurrent signal assignment statements will execute in response to a change on signals A and B.

    c. Do the execution sequences given in parts (a) and (b) change if the textual listing of the above concurrent signal assignment statements changes by swapping the order of the first two statements?

3. Consider the following digital system.

```
entity COMB_LOGIC is
  port
    (A, B, C, D : in BIT;
     Z : out BIT);
end COMB_LOGIC;


architecture DATA_FLOW of COMB_LOGIC is
begin
  Z <= (not A and not B and not C and not D) or
       (A and not B and not C) or
       (A and not B and C) or
       (not A and not C and (B xor D));

end DATA_FLOW;
```

    a. Generate the associated truth table by mapping the logic expression onto a Karnaugh map.

    b. Verify the results of part (a) by simulation.

4. Consider the following digital system.

```
entity SWAP_BITS is
  port
    (OLD_SIG : in BIT_VECTOR(1 downto 0);
     SWAP    : in BIT;
     NEW_SIG : out BIT_VECTOR(1 downto 0));
end SWAP_BITS;
```

```
architecture DATA_FLOW of SWAP_BITS is
begin
  NEW_SIG <= OLD_SIG when (SWAP = '0') else
             OLD_SIG(0) & OLD_SIG(1);

end DATA_FLOW;
```

a. Generate the associated truth table.
b. Verify the results of part (a) by simulation.

5. Consider the following combinational digital system, called a light-emitting diode (LED) driver.

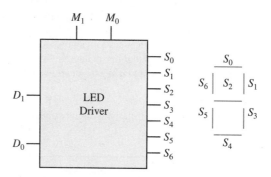

The LED driver converts a 2-bit binary number (D_1D_0) into an LED-displayed numeral. For example, $D_1D_0 = 10_2$ is displayed by asserting S_0, S_1, S_2, S_4, and S_5. The input signals M_1 and M_0 are control signals that test the LED display and determine the active output state according to the following truth table.

M_1M_0	Action
00	Asserted all segments, active-0
01	Asserted all segments, active-1
10	Display numeral, active-0
11	Display numeral, active-1

a. Generate the truth table for the LED driver. Note that $D_1D_0 = 01_2$ is displayed by asserting S_1 and S_3.
b. Generate a VHDL model of the LED driver using seven conditional concurrent signal assignment statements.
c. Generate a VHDL model of the LED driver using only one conditional concurrent signal assignment statement.

6. Evaluate each of the following expressions with A = '0', B = '1', C = '0', and D = '1'. List the order of operator evaluation.
   a. **not** A **and** B **or** **not** C **xor** D
   b. (**not** A **and** B) **or** (**not** C **xor** D)

7. Evaluate each of the following expressions with A = '0', B = '1', C = '0', and D = '1'. List the order of operator evaluation.
   a. **not** A < B **and** C = D
   b. **not** (A < B **and** C = D)

8. Consider the following digital system.

```
entity GEN_LOGIC is
  port
    (A, B : in BIT;
     OP_SEL : in BIT_VECTOR(1 downto 0);
     Z : out BIT);
end GEN_LOGIC;

architecture DATA_FLOW of GEN_LOGIC is
begin
 with OP_SEL select
 Z <= A and B  when B"00",
      A nand B when B"01",
      A or  B   when B"10",
      A nor B   when B"11";

end DATA_FLOW;
```

  a. Generate the associated truth table.
  b. Verify the results of part (a) by simulation.

9. Consider a logic operator that can be either a buffer or an inverter. If a mode select line called MODE_CNTRL is set to '0', then the output SIG_OUT equals the inverse of the input SIG_IN. If MODE_CNTRL is set to '1', then the output SIG_OUT equals the input SIG_IN.

  a. Generate a VHDL model of the digital system using the simplified form of the conditional concurrent signal assignment statement.

```
SIG_OUT <= logical expression;
```

  b. Generate a VHDL model of the digital system using the more general form of the conditional signal assignment statement.

```
target signal <= value1 when condition1 else
                 value2 when condition2 else
                            .
                 valueN-1 when conditionN-1 else
                 valueN;
```

  c. Generate a VHDL model of the digital system using a selected concurrent signal assignment statement.
  d. Compare and contrast the three data flow descriptions.

10. Consider the following signals.

```
signal VECA : BIT_VECTOR(7 downto 0) := X"E5";

signal VECB : BIT_VECTOR(10 to 15) := O"36";
```

a. Draw a diagram showing the elements and their values for VECA and VECB. List left to right the elements VECA'LEFT/VECB'LEFT to VECA'RIGHT/VECB'RIGHT.

b. Draw a diagram showing the elements and their values for each of the following expressions.

```
VECA(6 downto 2)

VECB(10 to 13)

alias MSB : BIT is VECA(7);

alias OCTET : BIT_VECTOR(1 to 3) is VECB(10 to 12);
```

---

c. Give the values of the predefined array attributes listed in Section 13.7.5 for VECA and VECB.

11.

```
signal VECA : BIT_VECTOR(7 downto 0) := X"E5";
signal VECB : BIT_VECTOR(10 to 15) := O"36";

VECA and B"1111_0000"

VECA xor X"FF"

VECA >= VECB
```

---

a. Evaluate each of the above expressions.

b. Explain the effects of the first and second logic operations and how the operations might be used in a digital design.

REFERENCES AND READINGS

[1] Armstrong, J. *Chip Level Modeling with VHDL.* Englewood Cliffs, NJ: Prentice-Hall, 1989.

[2] Bhasker, J. "Process-Graph Analyser: A Front-End Tool for VHDL Behavioural Synthesis," *Software—Practice and Experience,* vol. 18, no. 5, pp. 469–483, May 1988.

[3] Carlson, S. *Introduction to HDL-Based Design Using VHDL.* Mountain View, CA: Synopsys Inc., 1990.

[4] Dewey, Allen and A. de Geus. "VHDL: Toward a Unified View of Design," *IEEE Design & Test of Computers,* vol. 9, no. 2, pp. 8–17, June 1992.

[5] Ecker, W. and M. Hofmeister. "The Design Cube—A Model for VHDL Designflow Representation," *Proceedings of the European Design Automation Conference,* pp. 752–757, September 1992.

[6] Gajski, D. and R. Khun. "Guest Editors' Introduction: New VLSI Tools," *IEEE Computer,* vol. 16, no. 12, pp. 11–14, December 1983.

[7] Gilman, A. "VHDL—The Designer Environment," *IEEE Design & Test of Computers,* vol. 3, no. 2, pp. 42–47, April 1986.

[8] Mazor, S. *Guide To VHDL.* Boston, MA: Kluwer Academic Publishers, 1992.

[9] Reintjes, P. "A Set of Tools for VHDL Design," *Proceedings of the Eighth International Conference on Logic Programming,* pp. 549–562, June 1991.

[10] Ries, W. and K. Just. "VHDL in Logic Synthesis: An Applications Perspective," *EURO ASIC '91,* pp. 78–82, May 1991.

[11] van Eijndhoven, J. and L. Stok. "A Data Flow Graph Exchange Standard," *Proceedings of the European Conference on Design Automation,* pp. 193–199, March 1992.

14

STRUCTURAL MODELING IN VHDL: PART 2

In Chapter 12, we introduced structural modeling in VHDL by describing the interconnection of logic operators comprising a schematic. Recall that a structural description is generally composed of three VHDL statements: component declarations, signal declarations, and component instantiations.

```vhdl
entity DSGN_ENTITY is
  port
    (P1 : in BIT; ...);
end DSGN_ENTITY;

architecture STRUCTURE of DSGN_ENTITY is
  -- Component declarations
  component DSGN_SUBENTITY
    port (C1 : in BIT; C2 : out BIT; ...);
  end component;

  -- Signal declarations
  signal S1, ...., : BIT;

begin
  -- Component instantiations using named association
  X1: DSGN_SUBENTITY port map (C1 => P1, C2 => S1, ...);

end STRUCTURE;
```

Component instantiations create instances or copies of the declared components; the component instances are wired together using declared signals and/or ports of the enclosing design entity. The *association list* of the **port map** defines which signals are connected to which component ports using either *named* or *positional* association. Design entities providing the behavior of the subelements are associated with component instances via default configuration rules that pair entities and components having matching interface declarations.

In this chapter, we will investigate additional aspects of structural modeling in VHDL. We will discuss *port modes* and their use in more detail. We will also look at how to model *constant-valued* and *unconnected* ports. Then, we will explain how to use *generate statements* and *generics* to easily and succinctly describe structures that exhibit some degree of a regular geometric pattern. Generate statements and generics will introduce *arithmetic* operators and *integers* and *floating point* numbers. We will conclude with a discussion of new structural modeling capabilities provided in VHDL-93: *instance names* and *foreign architectures*.

14.1 PORT MODES AND THEIR PROPER USE

Recall that a design entity port defines an input/output signal and has the following three properties:

1. name,
2. port mode, and
3. type.

The name identifies the port, the mode defines the direction of information flow across the port, and the type defines the set of values that may flow across the port. VHDL provides five port modes.

Port Mode	Direction of Information Flow
in	Information flows only *into* the design entity.
out	Information flows only *out of* design entity.
inout	Information flows *into* and *out of* design entity.
buffer	Information flows *into* and *out of* design entity.
linkage	Information flows *into* and *out of* design entity.

For the present, we will examine only the **in**, **out**, and **inout** ports because they are the most commonly used port modes in typical VHDL models. The **buffer** and **linkage** ports tend to be less frequently used because they have additional properties that address special hardware connectivity situations.

To understand the rules regarding **in**, **out**, and **inout** port modes, we need to define the terms *reading* and *writing* with respect to signals. Reading a signal means accessing the value of a signal. Writing a signal means assigning a value to a signal. Writing a signal is also called *driving* or *updating* a signal.

14.1.1 The **in** Port

As stated above, information flows only *into* a design entity through an **in** port. An **in** port cannot be used in any manner that violates its defined direction of information flow. This rule, for example, implies that an input/output signal of mode **in** may not be the target (in other words, appear on left-hand side) of a concurrent signal assignment statement. Writing to a port from within a design entity implies that information flows out of the design entity, which contradicts the definition of an **in** port.

The properties of an **in** port also imply that an **in** port of a design entity may not be connected to a port of mode **out** or **inout** of a component of the design entity, as shown in Figure 14.1. Such port-to-port connections are illegal because the component is writing to the **in** port, again incorrectly implying that information flows out of an **in** port.

14.1.2 The **out** Port

Information flows only *out of* a design entity through an **out** port. Like the **in** port, an **out** port cannot be used in any manner that violates its defined direction of information flow. For example, an input/output signal of mode **out** may not appear on the right-hand side of a concurrent signal assignment statement. Reading from a port from within a design entity implies that information flows into the design entity, which contradicts the definition of an **out** port.

Figure 14.1

Illegal port connections—in port

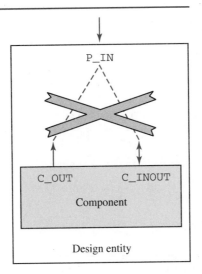

The VHDL model described in Figure 14.2(b) of the simple combinational logic shown in Figure 14.2(a) is incorrect because it attempts to read from the **out** port, X. A correct VHDL model of the combinational logic shown in Figure 14.2(a) is given in Figure 14.3 (p. 420).

The properties of an **out** port also imply that an **out** port of a design entity may not be connected to a port of mode **in** or **inout** of a component of the design entity, as shown in

Figure 14.2

Illegal use of an out port

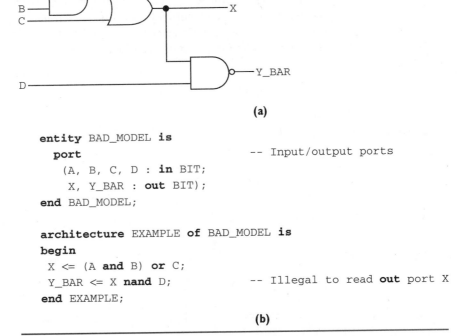

(a)

```
entity BAD_MODEL is
  port                              -- Input/output ports
    (A, B, C, D : in BIT;
     X, Y_BAR : out BIT);
end BAD_MODEL;


architecture EXAMPLE of BAD_MODEL is
begin
  X <= (A and B) or C;
  Y_BAR <= X nand D;                -- Illegal to read out port X
end EXAMPLE;
```

(b)

Figure (a) gives the schematic diagram of a combinational digital system. Figure (b) presents an incorrect VHDL model of the combinational system that attempts to read from an **out** port.

Figure 14.3

A correct VHDL model for Figure 14.2(a)

```
entity GOOD_MODEL is
  port                              -- Input/output ports
    (A, B, C, D : in BIT;
     X, Y_BAR : out BIT);
  end GOOD_MODEL;

architecture EXAMPLE of GOOD_MODEL is
  signal INT : BIT;
begin
  -- Use an internal signal as an intermediary to avoid reading
  -- from an out port.
  INT <= (A and B) or C;
  X <= INT;
  Y_BAR <= INT nand D;
  end EXAMPLE;
```

An internal signal is introduced to avoid reading from the **out** port.

Figure 14.4. Such port-to-port connections are illegal because the component is reading from the **out** port, incorrectly implying that information flows into an **out** port.

14.1.3 The **inout** Port

For an **inout** port, information can flow either *into* and/or *out of* a design entity. We are free to read from and/or write to an input/output signal of mode **inout**. Consequently, an input/output signal of mode **inout** may appear on the left-hand and/or the right-hand side of a concurrent signal assignment statement. Also, an **inout** port of a design entity may be connected to an **in**, **out**, or **inout** port of an enclosed component.

Figure 14.4

Illegal port connections— out port

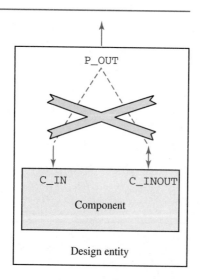

14.2 CONSTANT-VALUED AND UNCONNECTED PORTS

A *constant-valued* port is a port that is tied to a single value that does not change over time. As an example, consider the combinational system shown in Figure 14.5(a). The combinational system uses a 4-input multiplexer to realize the logic expression $F_1(A, B, C, D) = \sum m(1, 2, 10, 11)$. The truth table on the next page shows how the four multiplexer inputs are generated. The I_1 and I_3 multiplexer inputs are constant-valued ports, tied to the logic value 0.

Figure 14.5

VHDL example involving constant-valued ports

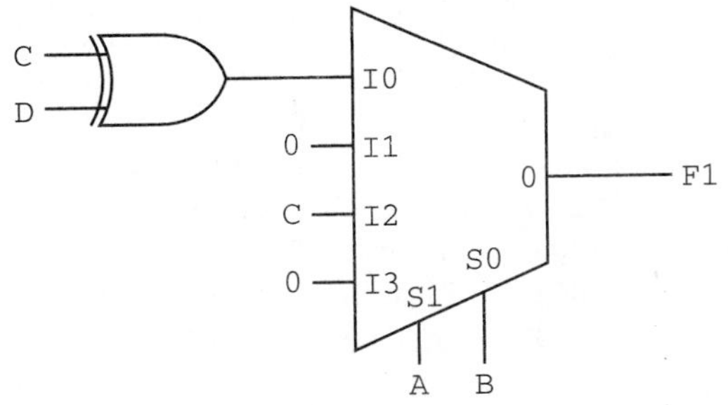

(a)

```
entity MUX_EX is
  port                                    -- Input/output ports
    (A, B, C, D : in BIT;
     F1 : out BIT);
end MUX_EX;

architecture XOR2_MUX4 of MUX_EX is
 component XOR2_GATE
   port (A, B : in BIT; Z : out BIT);
 end component;
 component MUX
   port (I0, I1, I2, I3, S1, S0 : in BIT;
         01 : out BIT);
 end component;

 signal INT1 : BIT;
 signal TIE_ZERO : BIT := '0';

 begin
   -- Connect components to describe schematic.
   X1: XOR2_GATE port map (C, D, INT1);
   M1: MUX        port map (INT1, TIE_ZERO, C, TIE_ZERO, A, B, F1);
 end XOR2_MUX4;
```

(b)

Figure (a) illustrates a schematic diagram of a combinational digital system using an **xor** gate and a multiplexer. Figure (b) presents a structural VHDL model.

ABCD	F_1	Multiplexer input
0000	0	
0001	1	$I_0 = F_1 = C \oplus D$
0010	1	
0011	0	
0100	0	
0101	0	$I_1 = F_1 = 0$
0110	0	
0111	0	
1000	0	
1001	0	$I_2 = F_1 = C$
1010	1	
1011	1	
1100	0	
1101	0	$I_3 = F_1 = 0$
1110	0	
1111	0	

Figure 14.5(b) shows a structural VHDL model of the F_1 combinational system. Ports I1 and I3 are tied to the constant binary value '0' by connecting them to the signal TIE_ZERO that is initialized to the value '0'. Since TIE_ZERO is not ever subsequently updated, the signal retains its initial value of '0' over time, thus holding the ports I1 and I3 at a constant value. The initialization expression ":= '0';" is an optional part of a signal declaration.

Another application of constant-valued ports is to tie off unused ports, such as connecting a '0' to unused **or** gate inputs or a '1' to unused **and** gate inputs. Instead of using constant-valued signals, Figure 14.6 shows an alternative approach of tying off unused ports to a constant value by using the keyword **open** *and default port values.*

Figure 14.6 is a variation of Figure 14.5, describing the logic expression $F_1(A, B, C, D) = \sum m(1, 2, 10, 11)$ using a 4-input multiplexer and a 3-input **xor** gate

Figure 14.6

Using the keyword open to model unconnected ports

```
architecture XOR3_MUX4 of MUX_EX is
component XOR3_GATE
  port(A, B, C : in BIT; Z : out BIT);
end component;
component MUX
  port (I0, I1, I2, I3, S1, S0 : in BIT;
        01 : out BIT);
end component;

signal INT1 : BIT;
signal TIE_ZERO : BIT := '0';

begin
 -- Indicate unconnected port by open
 X1: XOR3_GATE port map (open, C, D, INT1);
 M1: MUX        port map (INT1, TIE_ZERO, C, TIE_ZERO, A, B, F1);

end XOR3_MUX4;
```

rather than a 2-input **xor** gate. One of the inputs of the 3-input **xor** gate is not used. The A input port of the X1 instance of the XOR3_GATE component is not used and is left unconnected by associating the port with the keyword **open**.

However, we are not yet finished writing our VHDL model because the design entity representing the **xor** gate must still be given values for *all* inputs to compute an output response. Figure 14.7 shows the design entity representing the **xor** gate. A value for the **open** input port A is supplied via a *default* port value.

The design entity XOR3_GATE in Figure 14.7 is associated with or represents the component XOR3_GATE in Figure 14.6 via default configuration rules discussed in Chapter 12 and illustrated in Figure 14.8 (p. 424). The design entity XOR3_GATE shown on the far left supplies the needed port value for the unconnected port shown on the far right by specifying a default port value. The initialization expression ":= '0';" in the port declaration states that the input signals A, B, and C will take on the default value '0' if they are left unconnected by a component instantiation. Since the instantiation of component XOR3_GATE in Figure 14.6 leaves port A unconnected, this port takes on its default value '0' supplied by the associated design entity XOR3_GATE in Figure 14.7. An initialization expression applies to all ports listed in the associated port declaration and must specify a legal value of the port type.

Similar to unconnected *input* ports, unconnected *output* ports are also designated by using the keyword **open**. However, the associated design entity does not have to supply a default port value. Examples of input and output unconnected ports are given in Figure 14.9 (p. 425), which presents a VHDL model of the binary decoder implementation of the temperature controller discussed in Chapter 5. The O3 output port of the DECODE3_8 binary decoder is left unconnected by associating the port with the keyword **open**.

Figure 14.9 also shows that indexed names are allowed in **port map** association lists to connect signals to individual elements of ports. The three elements of port I of component DECODE3_8 are connected by the three associations shown here.

```
port map (I(2)=>TEMP2, I(1)=>TEMP1, I(0)=>TEMP0, ...);
```

When splitting up a port connection across several associations, *all* elements of the port must be specified in a *contiguous* sequence of *named* associations in the **port map** association list. Thus,

```
port map (I(2)=>TEMP2, I(1)=>TEMP1, TEMP0, ...);
```

Figure 14.7

VHDL example involving default input port values

```
entity XOR3_GATE is
  port
    (A, B, C : in BIT: '0'; -- Default value for ports A, B, and C
     Z : out BIT);
end XOR3_GATE

architecture BEHAVIOR of XOR3_GATE is
begin
  Z <= A xor B xor C;
end BEHAVIOR;
```

Figure 14.8

Default port values

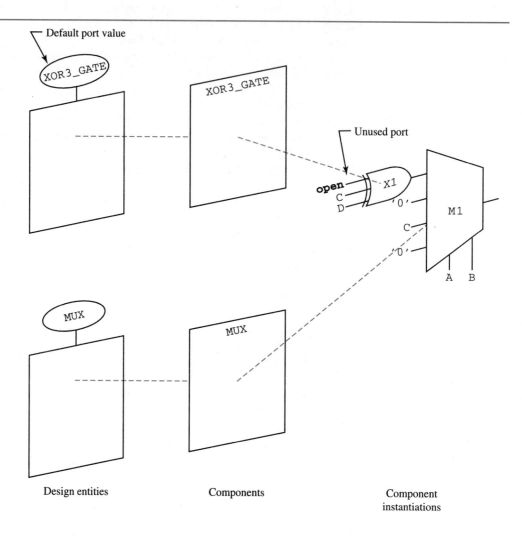

Design entities   Components   Component
                                                         instantiations

is illegal because all associations for the port I must be named associations, and

```
port map (I(2)=>TEMP2, 00=>INT0, I(1)=>TEMP1, I(0)=>TEMP0, ...);
```

is also illegal because the three associations for port I are not contiguous; in other words, they are not textually adjacent to one another.[1] Slice names are also allowed in **port map** association lists to connect signals to groups of contiguous elements of ports.

To summarize, constant-valued ports are modeled in VHDL by hooking up the ports to signals having constant values. A constant-valued signal is easily generated by adding an initialization expression to a signal declaration and, thereafter, not updating the signal. Unused ports are modeled in VHDL by hooking up the ports to the keyword **open**. If an *input* port

---

[1] VHDL-93: In VHDL-93, individual elements of a port may not be selectively associated with the keyword **open**.

Figure 14.9

VHDL example involving unconnected input and output ports

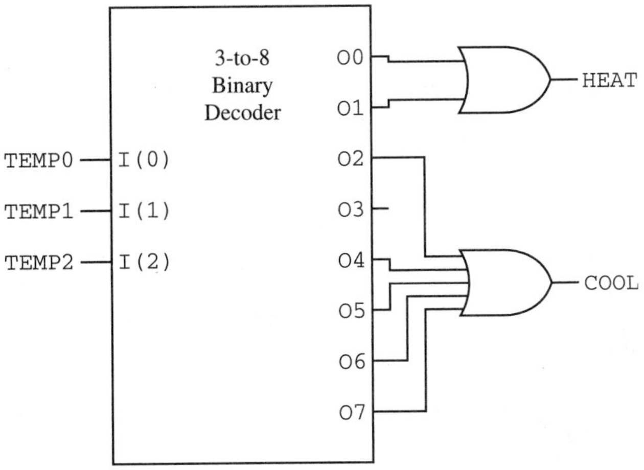

(a)

```
entity DECODER_EX is
 port
   (TEMP2, TEMP0, TEMP1 : in BIT;
    HEAT, COOL : out BIT);
end DECODER_EX;


architecture STRUCTURE of DECODER_EX is
 component OR5_GATE
   port (A, B, C, D, E : in BIT; Z : out BIT);
 end component;
 component DECODE3_8
   port (I : in BIT_VECTOR(2 downto 0);
        O0, O1, O2, O3, O4, O5, O6, O7 : out BIT);
 end component;

 signal INT0, INT1, INT2, INT4, INT5, INT6, INT7 : BIT;
begin
 -- Connect components to describe schematic.
 D1: DECODE3_8
   port map (I(2)=>TEMP2, I(1)=>TEMP1, I(0)=>TEMP0, O0=>INT0,
             O1=>INT1, O2=>INT2, O3=>open, O4=>INT4, O5=>INT5,
             O6=>INT6, O7=>INT7);
 O1: OR5_GATE port map (INT0, INT1, open, open, open, HEAT);
 O2: OR5_GATE port map (INT2, INT4, INT5, INT6, INT7, COOL);
 end STRUCTURE;
```

(b)

Figure (a) illustrates a schematic diagram of a combinational digital system using a 3-to-8 binary decoder. Figure (b) presents a structural VHDL model. The design entity representing the 5-input **or** gate is assumed to provide input port default values.

of a component is associated with the keyword **open**, then the design entity representing the component must define a default value for the unused port.

14.2.1 VHDL-93: Constant-Valued Input Ports

Although it might seem reasonable to tie off an input port to a constant value by associating the port with a bit literal, for example, "0" or "1", instead of a constant-valued signal or the keyword **open**, such structural descriptions are illegal in VHDL-87. In VHDL-87, only *signals* can be associated with component ports in a component instantiation.

To facilitate structural modeling, VHDL-93 extends the capabilities of association lists and allows *expressions* to be associated with input ports. The value of the expression, computed just before starting simulation, becomes the constant value for the input port. Thus, in VHDL-93, the MUX_EX design entity in Figure 14.5(b) can be rewritten as follows.

```
entity MUX-EX is
 port
   (A, B, C, D : in BIT;
    F1 : out BIT);
end MUX_EX;

architecture XOR2_MUX4 of MUX_EX is
 component XOR2_GATE
  port (A, B : in BIT; Z : out BIT);
 end component;
 component MUX
  port (I0, I1, I2, I3, S1, S0 : in BIT;
        01 : out BIT);
 end component;

 signal INT1 : BIT;
 begin
 -- Connect components to describe schematic.
 X1: XOR_GATE port map (C, D, INT1);
 M1: MUX       port map (INT1, '0', C, '0', A, B, F1);
 end XOR2_MUX4;
```

The M1 component instantiation statement ties off the I1 and I3 input ports of the MUX component to the constant value "0" by associating the bit literal directly with the input ports. The structural description is more readable and uses one less signal.

Since expressions associated with input ports are evaluated just before starting simulation, they must not contain any runtime or dynamic objects, such as signals. The expressions must contain only constant values, such as literals and generics. (Generics are discussed in Section 14.6.)

14.3 REGULAR STRUCTURES

Many digital designs are composed of components connected in a regular geometric pattern. For example, iterative circuits (discussed in Chapter 5) are composed of a series of "bit slices" cascaded together to form a one-dimensional vector pattern.

Consider, for instance, the iterative implementation of the comparator discussed in Chapter 5 and repeated in Figure 14.10 that possesses a regular structure of the cascade of identical bit-wise comparators. The comparator accepts two N-bit binary numbers, A $(a_{N-1}a_{N-2} \ldots a_1a_0)$ and B $(b_{N-1}b_{N-2} \ldots b_1b_0)$, as operands and yields a result that shows their relative magnitudes according to the encodings in Table 14.1.

Each bit-wise comparator accepts the results of the previous, higher-order bit-wise comparator, adds in its results, and passes the resulting information along to the next, lower-order bit-wise comparator. If the previous, higher-order bits of A and B are equal $(x_iy_i = 00)$, then the outputs x_{i+1} and y_{i+1} are set according to the relative values of a_i and b_i. If, on the other hand, the previous, higher-order comparisons have determined that $A < B$ or $A > B$, $(x_iy_i = 01$ or $x_iy_i = 10)$, then the result is simply passed on irrespective of the values of a_i and b_i.

Figure 14.11 (p. 428) shows a structural VHDL model of an 8-bit comparator. Notice that the component instantiation statements tend to be very similar, exhibiting the same degree of regularity found in the actual hardware. For instance, the only differences between the C6 and C5 instantiations of the BIT_COMPARE component are the indices of the connecting signals. The C7 and C0 instantiations are, however, slightly different from the other component instantiation statements, as C7 and C0 account for connecting to the input/output signals of

Table 14.1

Cascade signals encoding for a bit-wise comparator

x_iy_i	Encoded result
00	$A = B$
01	$A < B$
10	$A > B$
11	Don't care

Figure 14.10

Iterative implementation of a comparator

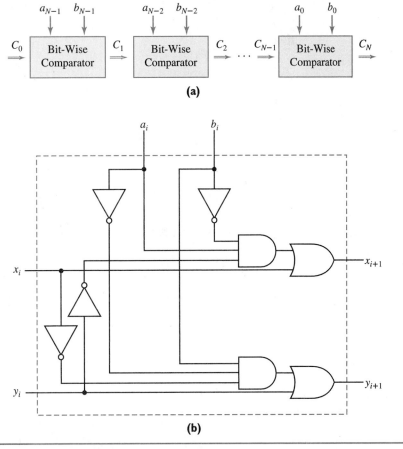

(a)

(b)

Figure (a) shows an N-bit comparator implemented using a series of 1-bit comparators. Figure (b) shows the 1-bit comparator.

Figure 14.11

**A VHDL model of an
8-bit comparator**

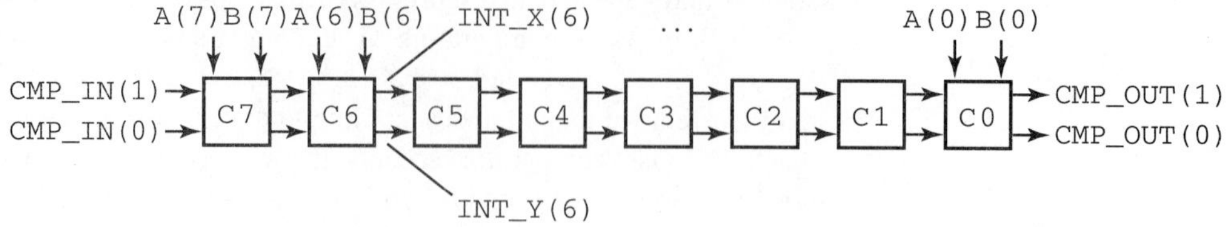

(a)

```
entity COMPARE8 is
 port
   (A, B     : in BIT_VECTOR(7 downto 0);
    CMP_IN  : in BIT_VECTOR(1 downto 0);
    CMP_OUT : in BIT_VECTOR(1 downto 0));
 end COMPARE8;


architecture MAN_STRUC of COMPARE8 is
 component BIT_COMPARE
   port (A, B, X_IN, Y_IN : in BIT;
         X_OUT, Y_OUT : out BIT);
 end component;
 signal INT_X, INT_Y : BIT_VECTOR(7 downto 1);


begin
 -- Connect components to describe iterative structure.
 C7: BIT_COMPARE port map (A(7), B(7), CMP_IN(1), CMP_IN(0),
                           INT_X(7), INT_Y(7));
 C6: BIT_COMPARE port map (A(6), B(6), INT_X(7), INT_Y(7),
                           INT_X(6), INT_Y(6));
 C5: BIT_COMPARE port map (A(5), B(5), INT_X(6), INT_Y(6),
                           INT_X(5), INT_Y(5));

                         .
                         .

 C1: BIT_COMPARE port map (A(1), B(1), INT_X(2), INT_Y(2),
                           INT_X(1), INT_Y(1));
 C0: BIT_COMPARE port map (A(0), B(0), INT_X(1), INT_Y(1),
                           CMP_OUT(1), CMP_OUT(0));
 end MAN_STRUC;
```

(b)

Figure (a) gives a schematic diagram of an iterative implementation of an 8-bit comparator. Figure (b) presents a structural VHDL model; only five of the eight component instantiation statements are shown.

the enclosing design entity COMPARE8. The C7 and C0 instantiations of the BIT_COMPARE component represent slight irregularities of the general regular iterative vector structure. Such irregularities often occur at the extremities of a regular structure and are called *boundary conditions*.

Clearly, describing large, regular structures of components by writing a component instantiation statement for each component can be a laborious and tedious task. In the following section, we will learn how to exploit the regularity in digital system schematics and the associated commonality in component instantiation statements via the generate statement to rewrite the structural model of the 8-bit comparator in a more compact and succinct manner.

14.4 GENERATE STATEMENTS

As the name implies, *generate* statements in VHDL provide the powerful ability to succinctly describe regular and/or slightly irregular structures by automatically generating component instantiations instead of manually writing each instantiation. As an example, Figure 14.12 (p. 430) shows an alternative structural description of the 8-bit comparator, COMPARE8, using generate statements. The architecture GEN_STRUC contains two kinds of generate statements: *iteration* and *conditional*. An iteration generate statement is also called a **for generate** statement; a conditional generate statement is also called an **if generate** statement. The net effect of "executing" the generate statements in COMPARE8 is to generate the required eight component instantiation statements that define the structure of the 8-bit comparator.

The **for generate** statement, repeated below, is similar to a **for** loop found in most common software programming languages.[2]

```
CASCADE:   -- Iteration generate
for I in 7 downto 0 generate

   -- Enclosed statements

end generate CASCADE;
```

[2] VHDL-93: In VHDL-93, a **for generate** statement may include an enclosing declarative region and the keyword **begin**.

```
CASCADE: -- Iteration generate
for I in 7 downto 0 generate
   -- Declarations
begin
   -- Statements
end generate CASCADE;
```

Figure 14.12

A VHDL model of an 8-bit comparator using generate statements

```
architecture GEN_STRUC of COMPARE8 is
 component BIT_COMPARE
  port
   (A, B, X_IN, Y_IN : in BIT;
    X_OUT, Y_OUT : out BIT);
 end component;
 signal INT_X, INT_Y : BIT_VECTOR(7 downto 1);

begin
 -- Connect components to describe iterative structure.
 CASCADE: -- Iteration generate
 for I in 7 downto 0 generate
  INPUT_CASE: -- Conditional generate
  if (I=7) generate
   C7: BIT_COMPARE port map (A(I), B(I), CMP_IN(1), CMP_IN(0),
                                 INT_X(I), INT_Y(I));
  end generate INPUT_CASE;

  NORM_CASE: -- Conditional generate
  if (I<=6 and I>=1) generate
   CX: BIT_COMPARE port map (A(I), B(I), INT_X(I+1), INT_Y(I+1),
                                 INT_X(I), INT_Y(I));
  end generate NORM_CASE;

  OUTPUT_CASE: -- Conditional generate
  if (I=0) generate
   C0: BIT_COMPARE port map (A(I), B(I), INT_X(I+1), INT_Y(I+1),
                                 CMP_OUT(1), CMP_OUT(0));
  end generate OUTPUT_CASE;
 end generate CASCADE;
end GEN_STRUC;
```

The interface for COMPARE8 is given in Figure 14.11(b).

The **for generate** statement starts with a label, CASCADE, that is repeated at the close of the statement. The label can be any legal identifier and is required. "I" is the *loop index*, having the *range* 7 **downto** 0. The **for generate** statement "executes" the enclosed statements for every possible value of the loop index. On the first pass through the enclosed statements, $I = 7$; on the second pass, $I = 6$; and so on. The loop index I is implicitly declared; one of the few instances in VHDL of an implicit declaration. The scope of the loop index is only the body of the **for generate** statement. That is, the loop index does not exist before or after the **for generate** statement. Moreover, the loop index is a *constant* inside the **for generate** statement. We can only access the value of the loop index; we are not allowed to change the value of the loop index.

Embedded inside the **for generate** statement are three **if generate** statements, identified by the three labels INPUT_CASE, NORM_CASE, and OUTPUT_CASE. The **if generate** statement, repeated below, is similar to an **if** statement found in most common software programming languages.[3]

```
NORM_CASE:  -- Conditional generate
if (I<=6 and I>=1) generate
                         .
    -- Enclosed statements
                         .
end generate NORM_CASE;
```

Again, the label can be any legal identifier and is required. The **if generate** statement "executes" the enclosed statement(s) if the Boolean expression given in parentheses, (I<=6 **and** I>=1), evaluates to TRUE. The Boolean expression (I<=6 **and** I>=1) evaluates correctly without the need for any additional parentheses because relational operators have higher precedence than the logic **and** operator (see Chapter 13).

The term "execute" is written in quotes because generate statements do not "execute" in the same fashion as signal assignment statements. The net effect of "executing" a generate statement is *code generation*, also sometimes called *macro expansion*. Generate statements do not directly participate in the definition of the input/output transform of a design entity. Rather, the code generated by "executing" generate statements defines the input/output transformation. Thus, generate statements serve as a shorthand notation for VHDL code. For the COMPARE8 design entity, the generate statements serve as a shorthand notation for the eight component instantiation statements.

For example, the first time through the **for generate** statement with I = 7, the INPUT_CASE **if generate** statement "executes" and generates the following component instantiation statement.

```
C7: BIT_COMPARE port map (A(7), B(7), CMP_IN(1), CMP_IN(0),
                          INT_X(7), INT_Y(7));
```

The second time through the **for generate** statement with I = 6, the NORM_CASE **if generate** statement "executes" and generates the following component instantiation statement.

```
CX: BIT_COMPARE port map (A(6), B(6), INT_X(7), INT_Y(7),
                          INT_X(6), INT_Y(6));
```

---

[3] VHDL-93: In VHDL-93, an **if generate** statement may include a declarative region and the keyword **begin**.

```
NORM_CASE: -- Conditional generate
if (I<=6 and I>=1) generate
    -- Declarations
begin
    -- Statements
end generate NORM_CASE;
```

Hence, each pass through the **for generate** loop generates one of the eight required component instantiation statements.

Admittedly, for the simple 8-bit comparator, the generate statements do not yield appreciable code savings, compared to manually coding each component instantiation statement. However, for more complex examples, substantial reductions in model code size can be obtained by exploiting generate statements.

14.4.1 General Use

Besides modeling regular *structural* descriptions, the basic macro expansion capability of a generate statement can also be used to model regular *behavioral* descriptions. Instead of generating a collection of similar component instantiation statements, we can generate a collection of similar concurrent signal assignment statements.

Let us reconsider the VHDL model of the cascade or serial implementation of the even parity generator discussed in Figure 13.10(b) in Chapter 13. Figure 14.13 shows how to use generate statements to generate the required seven concurrent signal assignment statements. Again, a **for generate** statement is used to set up a loop and **if generate** statements are embedded to generate concurrent signal assignment statements for particular loop iterations. Seven passes through the **for generate** statement generate the seven concurrent signal assignment statements shown on the next page that were manually written in Figure 13.10(b) in Chapter 13.

Figure 14.13

Using generate statements for regular behavioral models

```
entity EVEN_PARITY is
 port
   (BVEC : in BIT_VECTOR(7 downto 0);
    PARITY : out BIT);
end EVEN_PARITY;

architecture GEN_CASCADE of EVEN_PARITY is
 signal INT : BIT_VECTOR(1 to 6);
begin
 CASCADE: -- Iteration generate
 for I in 0 to 6 generate
  INPUT_CASE
  if (I=0) generate
   INT(I+1) <= BVEC(I) xor BVEC(I+1);
  end generate INPUT_CASE;
  NORM_CASE:
  if (I>=1 and I<=5) generate
   INT(I+1) <= INT(I) xor BVEC(I+1);
  end generate NORM_CASE;
  OUTPUT_CASE:
  if (I=6) generate
   PARITY <= INT(I) xor BVEC(I+1);
  end generate OUTPUT_CASE;
  end generate CASCADE;
end GEN_CASCADE;
```

```
INT(1) <= BVEC(0) xor BVEC(1);  INT(2) <= INT(1) xor BVEC(2);
INT(3) <= INT(2) xor BVEC(3);  INT(4) <= INT(3) xor BVEC(4);
INT(5) <= INT(4) xor BVEC(5);  INT(6) <= INT(5) xor BVEC(6);
PARITY <= INT(6) xor BVEC(7);
```

14.5 UNCONSTRAINED PORTS

Ports of a design entity can be unconstrained arrays, which can be exploited to describe regular behaviors and structures. For example, a port

```
entity UNIT is
 port
  (INPUT_PORT : in BIT_VECTOR; -- Unconstrained array
```

INPUT_PORT can be of type BIT_VECTOR, which is an unconstrained array of elements of type BIT. The range of the port INPUT_PORT is left unconstrained until the design entity UNIT is used in a component instantiation statement by a higher-level design entity. When the design entity UNIT is instantiated, INPUT_PORT takes on the range of the signal that is associated with or connected to the port.

Figure 14.14 (p. 434) shows how to use an unconstrained port to generate a structural description of an *N*-bit even parity generator, in other words, a design entity that generates even parity for an arbitrary number of bits.

BVEC is an unconstrained port; its range is set when the design entity EVEN_PARITY_N is used. For instance, the following sample component instantiation statement for EVEN_PARITY_N connects the lower-order eight bits of the signal TRANSMIT_DATA to the input port BVEC, setting BVEC to be an 8-bit vector having the range 7 **downto** 0. Note the use of the slice name TRANSMIT_DATA(7 **downto** 0) and the indexed name TRANSMIT_DATA(8) to connect parts of a signal to two ports. Design entity and component association is again accomplished by default configuration rules.

```
-- Component declaration
component EVEN_PARITY_N
 port
  (BVEC : in BIT_VECTOR; -- Unconstrained array
   PARITY : out  BIT);
end component;

-- 9-bit vector, MSB is even parity bit
signal TRANSMIT_DATA : BIT_VECTOR(8 downto 0);

-- Component instantiation
U1 : EVEN_PARITY_N port map (BVEC=>TRANSMIT_DATA(7 downto 0),
                             PARITY=>TRANSMIT_DATA(8));
```

Inside the architecture GEN_CASCADE, various hardcoded numbers are replaced by expressions involving the predefined array attribute LENGTH to define the internal signal

Figure 14.14

Using unconstrained ports to model *N*-bit even parity logic

```
entity EVEN_PARITY_N is
 port
   (BVEC : in BIT_VECTOR; -- Unconstrained array
    PARITY : out BIT);
end EVEN_PARITY_N;

architecture GEN_CASCADE of EVEN_PARITY_N is
 component XOR2_GATE
  port (A, B : in BIT; Z : out BIT);
 end component;

 signal INT : BIT_VECTOR(1 to BVEC'LENGTH-2);
begin
 CASCADE:
 for I in 0 to BVEC'LENGTH-2 generate
  INPUT_CASE:
  if (I=0) generate
   X1: XOR2_GATE port map (A=>BVEC(I), B=>BVEC(I+1),
                                       Z=>INT(I+1));
  end generate INPUT_CASE;

  NORM_CASE:
  if (I>=1 and I<=(BVEC'LENGTH-3)) generate
   XX: XOR2_GATE port map (A=>BVEC(I+1), B=>INT(I), Z=>INT(I+1));
  end generate NORM_CASE;

  OUTPUT_CASE:
  if (I=BVEC'LENGTH-2) generate
   XN: XOR2_GATE port map (A=>BVEC(I+1), B=>INT(I), Z=>PARITY);
  end generate OUTPUT_CASE;
 end generate CASCADE;
end GEN_CASCADE;
```

The **xor** gates are connected in a series configuration.

declaration and generate statements consistent with the range of BVEC. Recall that BVEC'LENGTH yields the number of elements in BVEC.

14.6 GENERICS AND PARAMETERIZED DESIGN ENTITIES

Let us extend the 8-bit comparator model given in Figure 14.12 to handle *N* bits. We could declare the input ports A and B to be unconstrained arrays of type BIT_VECTOR as discussed in the previous section, but we would have to impose somehow an additional restriction that the signals associated with the ports A and B have the *same* range. Inadvertently attempting to compare an 8-bit vector connected to port A with a 16-bit vector connected to port B could lead to a lot of problems.

A better approach to modeling an *N*-bit comparator is to make *N* a *parameter* that is set each time the design entity is instantiated. In other words, we seek a *parameterized* design entity. With a parameterized design entity, parameters are passed to a design entity each time

Problem-Solving Technique

Making a solution programmable greatly enhances the flexibility and general applicability of the solution.

it is used in a component instantiation, similar to the concept in software programming languages of passing arguments to a function each time it is invoked. The values of the parameters customize the design entity's input/output transformation, making the input/output transformation programmable.

Thus far in our study of VHDL, the only thing we can do with design entities in a structural description is use them, *as is*, by instantiating and interconnecting them. We have not been able to reach into a design entity and change things around a little to suit our particular needs, but design entities parameters essentially provide this capability. Through parameters, we can reach into a design entity and set various values that, in turn, can alter the behavior of the design entity. In other words, design entity parameters are like knobs on an oven; setting the knobs controls oven operation and setting the parameters controls design entity operation. However, there are typically limits to what operations can be controlled. Knobs on an oven can set temperature, but typically cannot turn an oven into a washing machine. Similarly, design entity parameters might set input bit string length or output active state, but typically would not turn a controller into an adder.

In VHDL, design entity parameters are called *generics* and are declared by adding a **generic** clause to the entity interface declaration. Figure 14.15 shows how to declare and use generics to model an *N*-bit comparator. The generic parameter, N, defines the length of the ports A and B, thereby constraining input ports to have the same length. The generic parameter also defines the length of the internal signals INT_X and INT_Y, the range of the **for generate** statement, and Boolean conditions of the **if generate** statements. If $N = 8$, COMPAREN describes an 8-bit comparator; if $N = 64$, COMPAREN describes a 64-bit comparator; and so on.

Figure 14.15

A VHDL model of an
N-bit comparator using
generics

```
entity COMPAREN is
  -- Introduce generic clause in interface for parameters.
  generic
    (N : POSITIVE);
  port
    (A, B : in BIT_VECTOR(N-1 downto 0);
     CMP_IN : in BIT_VECTOR(1 downto 0);
     CMP_OUT : out BIT_VECTOR(1 downto 0));
end COMPAREN;

architecture GEN_STRUC of COMPAREN is
  component BIT_COMPARE
    port (A, B, X_IN, Y_IN : in BIT;
          X_OUT, Y_OUT : out BIT);
  end component;
  signal INT_X, INT_Y : BIT_VECTOR(N-1 downto 1);

begin
  CASCADE: -- Iteration generate
  for I in N-1 downto 0 generate
    INPUT_CASE: -- Conditional generate
    if (I = N-1) generate
      CN: BIT_COMPARE port map (A(I), B(I), CMP_IN(1), CMP_IN(0),
                                INT_X(I), INT_Y(I));
    end generate INPUT_CASE;
```

Figure 14.15 continued

```
NORM_CASE: -- Conditional generate
if (I<=(N-2) and I>=1) generate
  CX: BIT_COMPARE port map (A(I), B(I), INT_X(I+1), INT_Y(I+1),
                            INT_X(I), INT_Y(I));
end generate NORM_CASE;

OUTPUT_CASE: -- Conditional generate
if (I = 0) generate
  C0: BIT_COMPARE port map (A(I), B(I), INT_X(I+1), INT_Y(I+1),
                            CMP_OUT(1), CMP_OUT(0));
end generate OUTPUT_CASE;
end generate CASCADE;
end GEN_STRUC;
```

The **generic** clause in the entity declaration of COMPAREN precedes the **port** clause and defines the parameter N. The parameter N is of type POSITIVE, which is a predefined *integer type* representing the numbers 1, 2, 3, ... The name of a generic can be any legal identifier. Any number of generics can be declared within the enclosing parentheses of the **generic** clause. The declarations are separated by semicolons and each declaration has the form "*parameter name : parameter type.*"

Within the design entity, a generic is considered a constant. We can only access the value of a generic; we cannot change the value of a generic.

How do we set the value of a generic parameter? As explained earlier, generics are given values each time the associated design entity is used in a structural description via a component instantiation statement. The following portion of a VHDL structural model shows how the parameterized design entity COMPAREN could be used in a higher-level design.

```
entity SYSTEM_DSGN is
  port (P1, P2 : in BIT_VECTOR(15 downto 0) ...);
end SYSTEM_DSGN;

architecture STRUCTURE of SYSTEM_DSGN is
  -- Component declaration for comparator
  component COMPAREN
   generic
     (N : POSITIVE);
   port
     (A, B : in BIT_VECTOR(N-1 downto 0);
      CMP_IN : in BIT_VECTOR(1 downto 0);
      CMP_OUT : out BIT_VECTOR(1 downto 0));
  end component;

  -- Signal declarations
  signal SIG_IN, SIG_OUT : BIT_VECTOR(1 downto 0);
begin
  -- Create a 16-bit comparator using named association.
  CMP1: COMPAREN
   generic map (N=>16)
   port map (A=>P1, B=>P2, CMP_IN=>SIG_IN, CMP_OUT=>SIG_OUT);

end STRUCTURE;
```

The component declaration for COMPAREN mirrors the associated entity declaration, declaring the generic parameter(s) and port(s) via **generic** and **port** clauses, respectively. The component instantiation of COMPAREN adds a **generic map** clause to set the value of the parameter N to 16. Just as the **port map** clause assigns values (signals) to ports, the **generic map** clause assigns values to generics. The example **generic map** clause uses named association; positional association is also allowed. The rules regarding the use of named and/or positional association for ports also apply to generics (see Chapter 12).

A generic parameter may be left unassociated, in other words, may not be given a value at component instantiation, by associating the generic with the keyword **open** or by omitting the generic from the **generic map** clause. If a generic is not given a value by a using design entity, the design entity providing the generic must supply a default value.

```vhdl
entity COMPAREN is
 generic
   (N : POSITIVE := 16); -- Default value
 port
   ( ..... );
end COMPAREN;

entity SYSTEM_DSGN is
 port
   ( ..... );
end SYSTEM_DSGN;

architecture STRUCTURE of SYSTEM_DSGN is
 component COMPAREN
  generic
    (N : POSITIVE);
  port
    .....
 end component;

begin
 -- Create a 16-bit comparator using named association.
 CMP1: COMPAREN
   generic map (N=>open) -- Use default value of 16
   port map ( ... );

                       .
                       .
end STRUCTURE;
```

Providing default values for generics is generally a good modeling practice because it suggests reasonable nominal values and saves users from having to write **generic map** clauses.

Like ports, generics provide channels of communication. With generics, a design entity no longer represents one hardware unit. Rather, a parameterized design entity defines a template, representing a family of hardware units. A particular hardware unit model is selected by setting the generics. We will study more about generics in future chapters, when we look at parameterizing technology-dependent aspects of VHDL models.

14.7 ARITHMETIC OPERATORS

The VHDL models demonstrating generate statements and generics in the previous sections introduced *arithmetic operators*. For instance, a portion of the COMPAREN design entity given in Figure 14.15 and repeated below uses the *subtraction*, "(N-2)," and *addition*, "(I+1)," operators.

```
        .
        .
   if (I<=(N-2) and I>=1) generate
      CX: BIT_COMPARE port map (A(I), B(I), INT_X(I+1), INT_Y(I+1),
                                INT_X(I), INT_Y(I));
        .
        .
```

Arithmetic operators form one of the major classes of operators in VHDL (recall that logic, relational, and shift operators were studied in Chapter 13) and are listed in Table 14.2. Arithmetic operators have the conventional algebraic definitions summarized below. To under-

Operator	Definition
+	positive sign or addition
–	negative sign or subtraction
*	multiplication
/	division
mod	modulus
rem	remainder
abs	absolute value
**	exponentiation
&	concatenation

stand VHDL arithmetic operators, we first need to discuss the associated operands because most of the rules governing the proper use of VHDL arithmetic operators concern an understanding of the differences between integer and real mathematics.

Table 14.2

VHDL arithmetic operators	Class	Operators
	Arithmetic	+ – * / ** & **mod** **rem** **abs**

Note: The "+" and "–" operators include the unary sign operators and binary addition and subtraction operators.

Integer mathematics is described by using VHDL arithmetic operators accepting operands of an *integer type*. An operand of an integer type is any number (positive or negative) not containing a radix point. VHDL provides three predefined integer types: INTEGER, NATURAL, and POSITIVE. The predefined type INTEGER represents a subset of the whole numbers, . . . −3, −2, −1, 0, 1, 2, 3, ... The range of the subset of whole numbers depends on the particular supporting software implementation. The predefined type NATURAL represents a subset of the natural numbers, 0, 1, 2, 3, ... The predefined type POSITIVE represents a subset of the positive numbers, 1, 2, 3, ... The upper bound of NATURAL and POSITIVE is the upper bound of INTEGER. The following section discusses in detail how to write integers, also called *integer literals*, in VHDL.

Real mathematics is described by using VHDL arithmetic operators accepting operands of a *floating point type*. An operand of a floating point type is any number (positive or negative) containing a radix point. VHDL provides one predefined floating point type: REAL. Like INTEGER, the range of REAL depends on the particular supporting software implementation. The following section also discusses in detail how to write floating point numbers, also called *floating point literals*, in VHDL.

Now, let us revisit the arithmetic operators listed above and explain associated restrictions on allowable operand types. First, the addition, subtraction, division, and multiplication operators require the operands to be of the *same* type. This requirement is consistent with the philosophy of a strongly typed language. Thus,

```
3 + 4
```

is correct, but

```
3 + 4.6
```

is incorrect because the first operand is an integer number and the second operand is a floating point number. Note that there is *no* implicit type conversion between integers and floating point numbers, again in the spirit of a strongly typed language — say what you mean and mean what you say.

Explicit type conversion between integer and floating point numbers is allowed, however, using *type conversion* notation. Type conversion notation uses the target type, followed by the expression to be converted enclosed within parentheses. For example,

```
REAL(3 + 4) = 7.0      INTEGER(4.6) = 5
```

Floating point to integer conversion rounds to the nearest integer; a floating point number halfway between two integers may be rounded up *or* down.

Next, the **rem** and **mod** operators are defined only for integer operands. These operators perform similar mathematical functions and are often confused. The **rem** operator yields the conventional integer remainder of dividing the left operand (dividend) by the right operand (divisor),

$$A \text{ rem } B = A - \left(\frac{A}{B}\right) \times B$$

where the absolute value of A **rem** B is less than the absolute value of B ($|A \text{ rem } B| < |B|$) and the sign of A **rem** B equals the sign of the dividend, A. Note that $\frac{A}{B}$ is an integer division. The **mod** operator, on the other hand, yields the value of A in the counting system having modulus B,

$$A \text{ mod } B = A - (B \times N)$$

where N is an integer such that the absolute value of A **mod** B is less than the absolute value of B ($|A \text{ mod } B| < |B|$) and the sign of A **mod** B equals the sign of the divisor, B.

Depending on the signs of the operands, the **rem** and **mod** operators can yield identical or different results, as shown in the following examples. The **rem** operations are shown on

(a)

(b)

the left-hand side and corresponding **mod** operations are shown on the right-hand side. The **rem** operations are performed using conventional division. The **mod** operations are performed using the cyclic counting sequences shown by traversing the directed arcs.

Finally, the exponentiation operator accepts both integer and floating point mantissas, but there are restrictions. A floating point mantissa may only be raised to an integer power

```
2.0 ** -2    -- Legal, yields 0.25
4.0 ** 0.5   -- Illegal
```

and an integer mantissa may only be raised to a positive integer power.

```
2 ** 10      -- Legal, yields 1024
2 ** -2      -- Illegal
```

An integer cannot be raised to a negative power in VHDL because the resulting type of the exponentiation operator must be the same as the type of the *left* operand, in other words, the mantissa. In the above example, $2^{-2} = 0.25$ yields a floating point number, but the left operand is an integer.

Table 14.3 updates operator precedence information, showing how the arithmetic operators relate to relational, logic, and shift operators. Table 14.3 contains all predefined VHDL operators.

14.8 INTEGER AND FLOATING POINT LITERALS

There are several ways to write integer and floating point numbers in VHDL using integer and floating point *literals*. Recall that values in VHDL are called *literals*; Chapter 13 explained how to specify literals of type BIT, such as '0' and '1', and type BIT_VECTOR, such as B"000" and B"010".

An integer is defined to be a number containing no decimal point. Several examples of specifying integer literals are shown below.

152000	152_000	1E3	1e+3
(a)	**(b)**	**(c)**	**(d)**

2#11111010#	8#372#	16#FA#	5#2#E3
(e)	**(f)**	**(g)**	**(h)**

Examples (a)–(d) implicitly assume a decimal (10) base. No spaces are allowed in an integer value. An underscore character may be used to improve readability, as shown in specifying 152,000 in examples (a) and (b). Examples (c) and (d) define the value 1000 using scientific notation, in other words, $1000 = 1 \times 10^3$. The exponent is denoted by "E" or "e," The "+" sign is optional; a negative exponent is not allowed for integer literals.

Table 14.3

Operator precedence	Precedence	Operators
	Highest	** abs not
	↓	* / mod rem
	↓	+ – (Note 1)
	↓	+ – & (Note 2)
	↓	sla sra sll srl rol ror (Note 3)
	↓	= /= < <= > >=
	Lowest	and or nand nor xor xnor (Note 4)

Note:
- Note 1: Unary sign operators.
- Note 2: Binary addition and subtraction operators.
- Note 3: Shift operators are provided only in VHDL-93.
- Note 4: The logic operator **xnor** is provided only in VHDL-93.

Examples (e)–(h) explicitly define a base and are called *based-integer literals*. The leading or leftmost integer defines the base; VHDL allows any base from 2 (binary) to 16 (hexadecimal). Example (e) is base 2 or binary; example (f) is base 8 or octal; example (g) is base 16 or hexadecimal; and example (h) is base 5. Following the base designator, the pound signs "#" enclose the based-integer value. The digits allowed within the pound signs depend on the base. For instance, only the numerals 0 and 1 can be used if the base is binary, only the numerals 0–7 can be used if the base is octal, and so on. If a based-integer literal is given using scientific notation, the exponent indicates the power of the *base* by which the mantissa is to be multiplied to obtain the integer value. In example (h), 5#2#E3 equals $2 \times 5^3 = 250_{10}$. Examples (e)–(h) denote the same integer value, 250_{10}.

A floating point value is defined to be a number containing a radix point. Several examples of specifying floating point literals are shown below. Many of the rules for constructing

$$
\begin{array}{cccc}
\texttt{0.0} & \texttt{0.000\_25} & \texttt{2.5E-4} & \texttt{5.6e+3} \\
\textbf{(a)} & \textbf{(b)} & \textbf{(c)} & \textbf{(d)}
\end{array}
$$

$$
\begin{array}{cccc}
\texttt{2\#110.01\#} & \texttt{8\#6.2\#} & \texttt{16\#6.4\#} & \texttt{5\#2.3\#E-1} \\
\textbf{(e)} & \textbf{(f)} & \textbf{(g)} & \textbf{(h)}
\end{array}
$$

integer literals also apply to constructing floating point literals. No spaces are allowed in a floating point value. The base may be either implicitly defined as decimal (10), shown in examples (a)–(d), or explicitly defined, shown in examples (e)–(h). Floating point literals may, however, have a negative exponent; examples (b) and (c) both denote 0.00025_{10}.

14.9 VHDL-93: FOREIGN ARCHITECTURES

We will conclude this chapter with a look at two new structural modeling features introduced in VHDL-93: *foreign architectures* and *instance names*. We will discuss foreign architectures in this section and instance names in the next section.

An architecture that is not written in VHDL is called a *foreign architecture*. A foreign architecture may be written in another computer language, such as C^{++}, or even realized by prototype hardware. It may be desirable to incorporate foreign architectures in a VHDL description to reduce modeling resources or accelerate design analysis. For instance, we might be working on a single-board computer design, and a model of the central microprocessor written in a computer language other than VHDL may already exist and be available from a vendor. Using such an "off-the-shelf" model could save considerable design time and costs compared to generating a custom VHDL model of the central microprocessor and verifying its accuracy.

VHDL-87 does not preclude using foreign architectures, but the coupling techniques are cumbersome and nonstandard. Thus, VHDL-93 provides a standard mechanism for integrating non-VHDL architectures into a VHDL model. In short, a foreign architecture is a "back door" through which we can bring in executable behaviors that are not written in VHDL.

As an example, the design entity implicitly configured with the BIT_COMPARE component in the VHDL comparator models given in Figure 14.11, Figure 14.12, and Figure 14.15 could be "stubbed out" and declared using a foreign architecture as follows.

```
entity BIT_COMPARE is
 port
   (A, B, X_IN, Y_IN : in BIT;
    X_OUT, Y_OUT     : out BIT);
end BIT_COMPARE;


architecture VENDOR_XYZ of BIT_COMPARE is
 -- Attach attribute to architecture to define foreign interface
  attribute FOREIGN of VENDOR_XYZ : architecture is
            "/lib/vendor_xyz/bcmp";


begin
end VENDOR_XYZ;
```

There is no behavior defined in the architecture VENDOR_XYZ for comparing bits. Instead, the behavior is provided by referencing a foreign model that, in this case, is a compiled C program located in the UNIX[4] directory "/lib/vendor_xyz/bcmp".

The architecture VENDOR_XYZ of BIT_COMPARE contains a single declarative statement, called an *attribute specification*, that tags or marks the architecture as a foreign architecture and gives interface or coupling information. This is the first time we have seen an attribute specification and the ability to define attribute values and associations. Thus far, we have examined only *predefined* attributes; information provided by predefined attributes is set and may only be queried. However, FOREIGN is a *user-defined* attribute, and its value and association with a part of a VHDL model may be specified. The attribute specification

```
attribute FOREIGN of VENDOR_XYZ : architecture is
            "/lib/vendor_xyz/bcmp";
```

specifies that an instance of the user-defined attribute FOREIGN is associated with the architecture VENDOR_XYZ and its value is the character string "/lib/vendor_xyz/bcmp".

The attribute specification marking a foreign architecture is always placed in the declarative section of the architecture. The attribute specification begins with the keyword **attribute,** followed by the attribute name, which must always be FOREIGN. Then, the name of the architecture is given after the keyword **of** and the value of the attribute FOREIGN is given after the keyword **is**. The value is a string of characters enclosed in double quotes that describes how to access or invoke the foreign architecture. A character string must fit on a single line; it cannot be split across multiple lines. (In Chapter 16 we discuss user-defined attributes in more detail.)

14.10 VHDL-93: NEW STRUCTURE ATTRIBUTE

In VHDL-87, design entities have very limited access to global information, which is information concerning an entire design encompassing many design entities. A design entity has access to only local information passed through ports and generics between its instantiating parents and instantiated children. This language feature enhances model modularity and

---

[4] UNIX® is a trademark of Bell Laboratories.

avoids nondeterministic behavior that can result from uncontrolled parallel accesses to global information. However, it would be helpful if a design entity had an easy way to access knowledge about the global hierarchy it is a part of, so individual instances can be identified within the context of the larger design. Design hierarchy information could be communicated via generics, but this method is cumbersome and nonstandard; every designer is free to devise a unique nomenclature.

VHDL-93 addresses this deficiency by providing the predefined attribute INSTANCE_NAME that records a string of characters defining the path from the root of a design hierarchy to a particular item within the hierarchy. Like the predefined attributes studied in Chapter 13 for bit vectors, the predefined attribute INSTANCE_NAME provides extra information about a VHDL model, namely, instance information. The value of the predefined attribute INSTANCE_NAME is automatically computed and made available for read-only use. To explain this new language feature, we will first study a few examples, and then we will discuss the construction and use of the new attribute in detail.

As an example, consider the two-level hierarchical structure of the MAJORITY_2X3 design entity presented in Chapter 12 and repeated in Figure 14.16. The particular **and** operator shaded in Figure 14.16(b) is identified by the following *instance name*.

Figure 14.16

A two-level hierarchical design entity structure

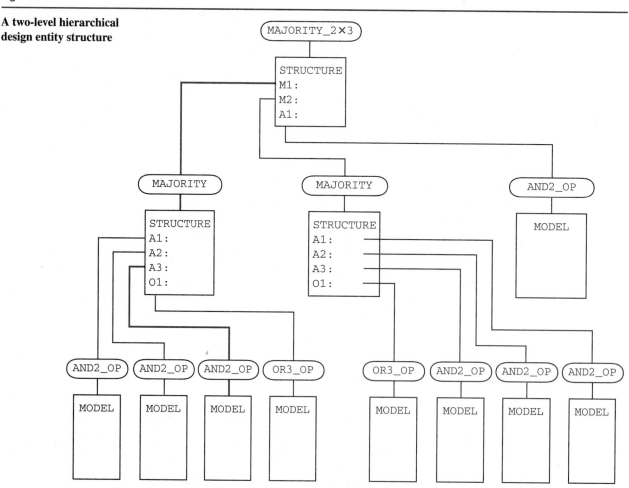

(a)

```
A3'INSTANCE_NAME =
  ":majority_2X3(structure):m1@majority(structure):a3"
```

The string of characters enclosed in double quotes identifies a particular **and** operator by defining the path between the operator and the top or root of the design hierarchy. Reading `A3'INSTANCE_NAME` from right to left, the instance of the entity/architecture `AND2_OP/MODEL` labeled `A3` is located in the instance of the entity/architecture `MAJORITY/STRUCTURE` labeled `M1`, which is, in turn, located in the root entity/architecture `MAJORITY_2X3/STRUCTURE`. This path is shown by the heavy line in Figure 14.16.

All basic identifiers are given in lowercase, regardless of how they were originally written. A colon ":" separates hierarchical boundaries within a design entity; an "at" symbol "@" separates hierarchical boundaries between design entities. For instance, the single @ sign in the instance name above denotes the hierarchical boundary between the `MAJORITY_2X3` and `MAJORITY` design entities.

As another example, consider the structural model of the 8-bit comparator described earlier in Figure 14.12. The two-level hierarchical structure of the `COMPARE8` design entity is shown in Figure 14.17 (p. 446), assuming a default configuration of the component `BIT_COMPARE` with a design entity having the same interface and an architecture named `MODEL`. The particular port, in other words, signal, highlighted in Figure 14.17(b) is identified by the following *instance name*.

Figure 14.16 Continued

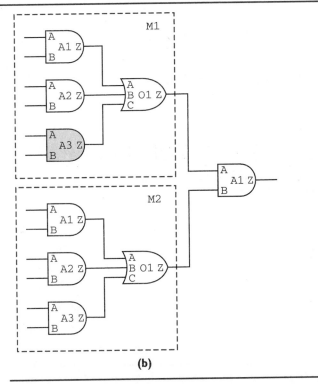

(b)

Figure (a) shows the VHDL design entity hierarchy of the logic schematic shown in Figure (b).

```
X_IN'INSTANCE_NAME =
":compare8(gen_struc):cascade(7):input_case:c7@bit_compare(model):x_in"
```

Again, the string of characters identifies a path from the particular X_IN port to the top of the hierarchy; the instance name is a little more complicated than the previous example because the path traverses a couple of hierarchical boundaries involving generate statements.

Figure 14.17

A two-level hierarchical design entity structure

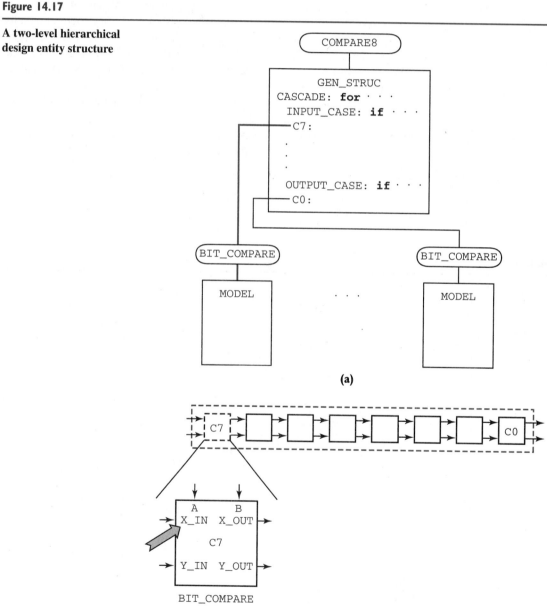

(a)

(b)

Figure (a) shows the VHDL design entity hierarchy of the logic schematic shown in Figure (b).

Reading A_IN'INSTANCE from right to left, the instance of the port X_IN is located in the instance of the entity/architecture BIT_COMPARE(MODEL) labeled C7, which is, in turn, instantiated within the **if generate** statement labeled INPUT_CASE of the seventh iteration of the **for generate** statement labeled CASCADE contained in the root entity/architecture COMPARE8(GEN_STRUC). Whew!

In summary, the value of an INSTANCE_NAME attribute is available only after a complete model has been assembled, in other words, elaborated, just before simulation and is accessible anywhere the named item is accessible. An instance name is a series of path segments or boundaries that collectively define a unique path from the top of the design hierarchy down to a particular named item. Path segments between design entities are separated by @ signs; these path segments involve VHDL component instantiation statements. Path segments contained within a design entity are separated by colons; these path segments involve VHDL statements that introduce nested scopes (Chapter 12), such as architectures, iteration generate statements, conditional generate statements, block statements, process statements, loop statements, and subprograms. The first four path segments are illustrated above; the last four path segments refer to VHDL statements that will be covered in future chapters.

14.11 SUMMARY

In this chapter, we have expanded our understanding of structural VHDL modeling, building on the basic concepts introduced in Chapter 12. We started by studying the definition and use of the commonly used ports **in**, **out**, and **inout**.

Port mode definitions imply certain restrictions on properly using ports in behavioral and structural descriptions. A port cannot be used in any way that violates the defined direction of the port's information flow. Writing to an **in** port or reading from an **out** port are not allowed. Thus, an **in** port may not appear on the left-hand side of a concurrent signal assignment statement and an **out** port may not appear on the right-hand side of a concurrent signal assignment statement. Also, when connecting an instantiated component port to a port of the enclosing design entity, the participating ports must have consistent modes. An **in** port may not be connected to an **out** or **inout** port and an **out** port may not be connected to an **in** or **inout** port.

We presented two approaches for modeling unconnected ports. An unconnected port can either be tied off to a constant value or associated with the keyword **open**. In VHDL-87, the constant value must be a signal; in VHDL-93, the constant value may be a signal or an expression. If the keyword **open** is associated with an unconnected input port, the associated design entity representing the particular component must have a default value defined for the unconnected port.

Following the discussion of ports, we then studied regular structures and how such structures can be easily described using generate statements and generics. Generate statements provide a text expansion capability, automatically generating VHDL statements that exhibit a degree of regularity or commonality. Generate statements can be used to create regular behavioral descriptions, as well as regular structural descriptions. Generics provide a powerful capability to parameterize a design entity, allowing the input/output transformation to be customized for each use or instantiation of the design entity.

VHDL provides a standard set of arithmetic operators accepting integer and floating point operands. We presented the ways of specifying integers and floating point numbers in VHDL.

Finally, we discussed that VHDL-93 provides foreign architectures and instance names to facilitate structural modeling. A foreign architecture is an architecture not written in VHDL and can be integrated into a VHDL model by using the FOREIGN attribute. Like all attributes, the attribute FOREIGN gives additional information about a VHDL construct, namely, architectures; the additional information defines how to access and exercise the foreign behavior. Instance names are provided by the predefined attribute INSTANCE_NAME and provide a means of identifying and reporting specific locations and items within a hierarchy of VHDL design entities. A colon ":" separates hierarchy boundaries within a design entity; an "@" symbol separates hierarchical boundaries between design entities.

14.12 PROBLEMS

1. Complete Table 14.4. Enter a "Y" if the component port to enclosing design entity port connection is allowed; otherwise, enter a "N."

Table 14.4

Port connectivity mode compatibility

Component port	Design entity port		
	in	out	inout
in			
out			
inout			

2. Consider the following structural VHDL model.

```
entity SMODEL is
 port
   (P1 : in BIT;
    P2 : out BIT;
    P3 : inout BIT);
 end SMODEL;


architecture STRUCTURE of SMODEL is
  component UNIT
     port (C1, C2 : in BIT; C3 : out BIT);
   end component;


begin
   U1: UNIT port map (C1=> ?, C2=> ?, C3=> ?);
 end STRUCTURE;
```

a. Complete the structural description by giving a legal set of port-to-port connections for entity ports P1, P2, and P2 and component ports C1, C2, and C3.

b. Is there more than one possible set of legal port-to-port connections?

3. Implement the following combinational system

$$F(A, B, C, D) = \sum m\,(0, 2, 8, 11, 12, 13, 14, 15)$$

using a 4-input multiplexer and 2-input **nand** gates.

a. Draw the logic diagram.
b. With the following package,

```
package LOGIC_COMPONENTS is
 component MUX
  port (I1, I2, I3, I4, S1, S0 : in BIT;
        O1 : out BIT);
 end component;


 component NAND2_GATE
  port (A, B : in BIT; Z : out BIT);
 end component;
end LOGIC_COMPONENTS;
```

describe the logic diagram in part (a) in VHDL as a structural model. Connect any unused input ports to constant-valued signals initialized to an appropriate value.

c. Generate the VHDL design entities for the MUX and NAND2_GATE components and verify by simulation that the model in part (b) describes the combinational system F.

d. Repeat part (b) designating any unused input ports with the keyword **open**.

e. Modify the design entities written in part (c) to accept unused ports and verify by simulation that the model in part (d) describes the combinational system F.

4. Implement a combinational system that detects Fibonacci numbers using a 4-to-16 decoder. The sequence of Fibonacci numbers ($F(n)$) is defined by Equation 14.1.

$$F(n) = \begin{cases} n, & n = 0, n = 1, n = 2 \\ F(n-1) + F(n-2), & n > 2 \end{cases}$$
$$F(n) = 0, 1, 2, 3, 5, 8, 13, 21, 34, \ldots \tag{14.1}$$

Each number in the Fibonacci sequence is the sum of the two preceding Fibonacci numbers. The input to the combinational system is a 4-bit binary number. The output is active-1.

a. Generate the truth table and draw the logic diagram.
b. With the following package,

```
package LOGIC_COMPONENTS is
 component DECODE4_16
  port (I : in BIT_VECTOR(3 downto 0);
        O0,O1,O2,O3,O4,O5,O6,O7,O8,O9,O10,O11,O12,O13,O14,O15
           : out BIT);
 end component;


 component OR3_GATE
  port (A, B, C : in BIT; Z : out BIT);
 end component;
end LOGIC_COMPONENTS;
```

describe the logic diagram in part (a) in VHDL as a structural model. Associate any unused ports with the keyword **open**.

c. Assuming default configuration rules, generate the design entities associated with the components DECODE4_16 and OR3_GATE.

d. Verify by simulation that the Fibonacci number detector generated in parts (b) and (c) describes the truth table generated in part (a).

5. Describe in VHDL an 8-bit ripple-carry adder using the following full adder component declaration; reference Chapter 5 for a discussion of ripple-carry adders.

```
package LOGIC_COMPONENTS is
 component FULL_ADDER
  port (A, B, CIN : in BIT; S, COUT : out BIT);
 end component;
end LOGIC_COMPONENTS;
```

   a. Use eight component instantiation statements.
   b. Use generate statements.
   c. Assuming implicit configuration, write the design entity for the FULL_ADDER component and use the design entity to simulate the VHDL model generated in part (b) adding 1. X"04" + X"04", 2. X"1D" + X"1D", and 3. X"FF" + X"FF".

6. The following VHDL model illustrates embedded **for generate** statements to describe a two-dimensional structure.

```
entity LOGIC is
 port
  (A, B : in BIT_VECTOR(3 downto 0);
   X    : out BIT);
end LOGIC

architecture STRUCTURE of LOGIC is
 component AND2_GATE
  port
   (A, B : in BIT; Z : out BIT);
 end component;
 component OR2_GATE
  port
   (A, B : in BIT; Z : out BIT);
 end component;
 signal ASIG, OSIG : BIT_VECTOR(3 downto 0) := X"0";
begin
 R: for COL in 1 to 2 generate
  C: for ROW in 3 downto 0 generate
   R1: if (COL=1) generate
     AX: AND2_GATE port map (A(ROW), B(ROW), ASIG(ROW));
    end generate R1;
   R1C: if (COL=2 and ROW/=0) generate
     OX: OR2_GATE port map (ASIG(ROW), OSIG(ROW), OSIG(ROW-1));
    end generate R1C;
   R1C0: if (COL=2 and ROW=0) generate
     O0: OR2_GATE port map (ASIG(ROW), OSIG(ROW), X);
    end generate R1C0;
  end generate C;
 end generate R;

end STRUCTURE;
```

    a. Draw the logic diagram described by the VHDL model.
    b. Generate the logic expression for $X = F(A, B)$.
    c. Under what set of input conditions is the output X asserted active-1?

7. To show that generate statements can describe regular behaviors as well as regular structures,
    a. Rewrite the entity LOGIC given in Problem 6 using concurrent signal assignment statements instead of component instantiation statements.
    b. List the concurrent signal assignment statements that are generated by elaborating the model in part (a).

8. VHDL allows recursive structures where a design entity instantiates itself. The series of recursive component instantiations can be stopped by a generate statement.
    a. Write a VHDL model using structural recursion of an **and** gate that accepts an arbitrary number of inputs.
    b. Write a VHDL model that uses the structural recursive model written in part (a) to create a 4-input **and** gate and simulate the 4-input **and** gate using the inputs 0101_2 and 1111_2.

9. Use unconstrained ports and a generate statement to write a design entity that accepts a bit vector of arbitrary length and reverses the bit ordering. The most significant bit of the output is the least significant bit of the input and the least significant bit of the output is the most significant bit of the input. For example, reversing the bit ordering of the 4-bit input and output signals

```
A : in BIT_VECTOR(3 downto 0);
B : out BIT_VECTOR(3 downto 0);
```

    yields

```
B(3) <= A(0); B(2) <= A(1); B(1) <= A(2); B(0) <= A(3);
```

    a. Write a VHDL design entity that uses the bit reversing design entity for 4-bit vectors.
    b. Simulate the design entity developed in part (b) using the inputs 0011_2, 1000_2, and 1100_2.

10. Rewrite the entity LOGIC given in Problem 6 to support input bit vectors of arbitrary length.
    a. Use unconstrained inputs.
    b. Use a generic.
    c. Discuss possible relative advantages and/or disadvantages of the models written in parts (a) and (b).

11. Give the values of the following type conversions.
    a. INTEGER(-10.7)   **b.** POSITIVE(-10.7)   **c.** INTEGER(-10.3)
    d. INTEGER(10.5)   **e.** INTEGER(4.8/REAL(2))

12. Using the number system conversions discussed in Chapter 2,
    a. Express 165 in VHDL in base 5 and base 6.
    b. Express 190 in VHDL in base 2, 8, and 16.
    c. Express 100,000 in VHDL with and without scientific notation.

13. Using the number system conversions discussed in Chapter 2,
    a. What is 5#44#E-2 in base 10?
    b. What is 8#34.6# in base 10?
    c. What is 2#1000_1000.1# in base 10?

14. For each of the following expressions, state if the expression is legal in VHDL and, if legal, give the resultant value.

    a. 3 ** -1   **b.** 30 ** 2   **c.** 16.0 ** 2   **d.** 18 **mod** 3
    e. 16.0 ** -2   **f.** 16.0 **mod** 4   **g.** 16.0 ** -0.5

15. <u>VHDL-93:</u> Consider the following VHDL structural model.

```
entity UNIT is
  port (A : in BIT;  B : out BIT);
end UNIT;


architecture UNIT of UNIT is
begin

                    . . . .

end UNIT;


entity ELEMENT is
  port (A : in BIT; B : out BIT);
end ELEMENT;
architecture STRUCTURE of ELEMENT is
  component UNIT
    port (A : in BIT; B : out BIT);
  end component;
  signal INT1, INT2 : BIT;
begin
  -- Connect components to describe schematic.
  CASCADE: -- Iteration generate
  for I in 0 to 2 generate
   INPUT_CASE: -- Conditional generate
   if (I=0) generate
    U1: UNIT port map (A=>A, B=>INT1)
    end generate INPUT_CASE;

    MIDDLE_CASE: -- Conditional generate
    if (I=1) generate
     U2: UNIT port map (A=>INT1, B=>INT2);
    end generate MIDDLE_CASE;

    OUTPUT_CASE: -- Conditional generate
    if (I=2) generate
     U3: UNIT port map (A=>INT2, B=>B);
    end generate OUTPUT_CASE;
   end generate CASCADE;
  end STRUCTURE;
```

    a. Draw the structural hierarchy.
    b. List the instance name for the port A, in other words, A' INSTANCE_NAME, for each design entity UNIT in the structural hierarchy.

REFERENCES AND READINGS

[1] Acosta, R., S. Smith, and J. Larson. "Mixed-Mode Simulation of Compiled VHDL Programs," *Proceedings of the IEEE International Conference on Computer Aided Design*, pp. 176–179, November 1989.

[2] Coelho, D. *The VHDL Handbook*. Boston, MA: Kluwer Academic Publishers, 1989.

[3] Hines, J. "Where VHDL Fits Within the CAD Environment," *Proceedings of the IEEE/ACM 24th Design Automation Conference,* pp. 491–494, June 1987.

[4] Leong, Y. and W. Birmingham. "Modeling and Simulation of a Partial Behavior of Intel 80286 in AHPL and VHDL," *Proceedings of the Summer Computer Simulation Conference,* pp. 212–216, July 1992.

[5] Leong, Y. and W. Birmingham. "The Automatic Generation of Bus Interface Models," *Proceedings of the 29th Design Automation Conference,* pp. 634–637, June 1992.

[6] Lipsett, R., C. Schaefer, and C. Ussery. *VHDL: Hardware Description and Design.* Boston, MA: Kluwer Academic Publishers, 1989.

[7] Saunders, L. "The IBM VHDL Design System," *Proceedings of the IEEE/ACM 24th Design Automation Conference,* June 1987.

[8] Spillane, J. and Z. Navabi. "Describing Controlling Hardware in VHDL," *Proceedings of the 4th IEEE International ASIC Conference and Exhibit,* September 1991.

Part 5

VHDL:

MANUFACTURING

TECHNOLOGIES

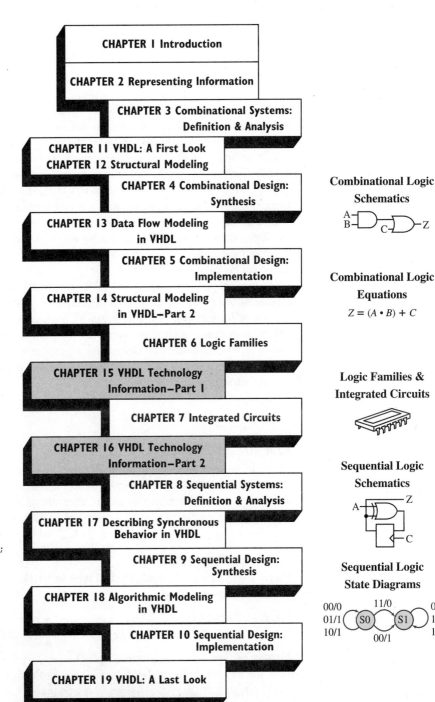

Structural Modeling

```
I1: NOT_OP
  port map (A, B);
```

Asynchronous Data Flow Modeling

```
B <= not A;
```

Technology Modeling

```
B <= not A after 5 ns;
```

Synchronous Data Flow Modeling

```
B1: block (C = '1')
begin
 B <= guarded not A after 5 ns;
end block B1;
```

Algorithmic Modeling

```
P1: process (PS)
begin
 NS <= STATE_TABLE(PS);
end process P1;
```

CHAPTER 1 Introduction

CHAPTER 2 Representing Information

CHAPTER 3 Combinational Systems: Definition & Analysis

CHAPTER 11 VHDL: A First Look
CHAPTER 12 Structural Modeling

CHAPTER 4 Combinational Design: Synthesis

CHAPTER 13 Data Flow Modeling in VHDL

CHAPTER 5 Combinational Design: Implementation

CHAPTER 14 Structural Modeling in VHDL–Part 2

CHAPTER 6 Logic Families

CHAPTER 15 VHDL Technology Information–Part 1

CHAPTER 7 Integrated Circuits

CHAPTER 16 VHDL Technology Information–Part 2

CHAPTER 8 Sequential Systems: Definition & Analysis

CHAPTER 17 Describing Synchronous Behavior in VHDL

CHAPTER 9 Sequential Design: Synthesis

CHAPTER 18 Algorithmic Modeling in VHDL

CHAPTER 10 Sequential Design: Implementation

CHAPTER 19 VHDL: A Last Look

Combinational Logic Schematics

Combinational Logic Equations

$$Z = (A \cdot B) + C$$

Logic Families & Integrated Circuits

Sequential Logic Schematics

Sequential Logic State Diagrams

15
VHDL TECHNOLOGY INFORMATION: PART 1

The VHDL descriptions that we have examined thus far can be classified as *technology-independent* models because they do not infer a particular implementation technology or logic family; the models describe only the logic function in a structural or behavioral fashion. Developing technology-independent VHDL descriptions improves the portability of models, just as developing operating-system-independent software improves the portability of programs. A technology-independent VHDL model can be "ported" to different implementations by different organizations without becoming outdated. There are, however, situations that warrant developing *technology-dependent* VHDL descriptions that imply a particular logic family. For example, technology-dependent information may be required to improve modeling accuracy so that a VHDL model will more closely reflect the behavior of the actual hardware.

In this chapter, we introduce VHDL constructs that model the electrical circuit-related aspects of logic families discussed in Chapter 6, in other words, digital design technology-dependent information. We will begin by discussing how to represent dimensional quantities, such as time, voltage, and capacitance. Then we will introduce the important aspect of *timing* in VHDL models by discussing *propagation delay*. Next, we will look at modeling the properties of wired logic. Finally, we conclude the chapter with a discussion of *objects*: *signals*, *variables*, and *constants*.

15.1 SPECIFYING PHYSICAL VALUES

To model technology-dependent information in VHDL, we need to be able to specify physical quantities, such as time, power, distance, and temperature. A physical quantity or number has a value *and* a dimension or unit of measurement. Physical quantities are represented in VHDL via *physical data types*. The value of a physical type is called a *physical literal*.

15.1.1 Specifying Time
There is only one predefined physical type in VHDL: TIME. Examples of specifying values of TIME are illustrated below.

7 ns	2#111# ns	1 min	min	10.65 us	10.65 fs
(a)	(b)	(c)	(d)	(e)	(f)

A value of type TIME is composed of a numeric value followed by a dimension or unit of time; the numeric value and dimension must be separated by a space. The units of time defined by type TIME are listed on the facing page.

Unit	Definition
fs	femtoseconds (10^{-15} seconds)
ps	picoseconds (10^{-12} seconds)
ns	nanoseconds (10^{-9} seconds)
us	microseconds (10^{-6} seconds)
ms	milliseconds (10^{-3} seconds)
sec	seconds
min	minutes (60 seconds)
hr	hours (3600 seconds)

A femtosecond is referred to as the *base unit*; all other units are referred to as *derived units*.

The numeric value can be an integer or a floating point literal (see Chapter 14). The first two examples both specify 7 nanoseconds: Example (a) uses an integer literal and example (b) uses a based-integer literal. Examples (c) and (d) both specify 1 minute, demonstrating that a numeric value is optional. If a numeric value is not given, a value of 1 is implied. Finally, the last two examples use floating point literals.

It is important to note that the value of a physical literal is defined in terms of *integral* multiples of the base unit. Physical literals of type TIME are represented in terms of integral multiples of the base unit (femtoseconds). For instance, the literal 10.65 us is actually represented as 1065×10^7 fs. Also, the literal 10.65 fs is actually represented as 10 femtoseconds; fractions of femtoseconds are truncated. Thus, be aware that the smallest available resolution of the predefined type TIME is 1 femtosecond to avoid inadvertently losing timing information. The range of type TIME, the low and high limits of the number of femtoseconds, is installation defined but guaranteed to be no smaller than $-2,147,483,647$ to $2,147,483,647$. (Yes, VHDL allows for negative values of time, which, as one might guess, are not commonly used.[1])

If a model does not require the precision of femtoseconds, then the resolution of type TIME can be set to a derived unit to allow for longer periods of runtime or simulation. The resolution of type TIME is set outside the language, typically as a simulator command or parameter.

15.1.2 User-Defined Physical Types

Thus far in our study of VHDL we have used *predefined* types, such as BIT, BIT_VECTOR, POSITIVE, and REAL. These predefined types provide ready-to-use sets of values convenient for modeling digital systems. In cases where the predefined data types do not adequately support modeling applications, VHDL provides the powerful capability of *user-defined* data types. As the name implies, user-defined data types allow a user to define a custom set of data abstractions. User-defined data types make VHDL more flexible and extensible, molding the hardware description language to fit an application rather than contorting an application to fit a rigid or confining language. For the present, we will look at only user-defined physical types; in future sections we will address other user-defined types.

Figure 15.1 (p. 458) shows an example of how to define a physical type to represent capacitance. The keyword **type** starts every user-defined type declaration. CAPACITANCE is the name of the user-defined physical type; the name can be any legal identifier (see Section 11.3). Following the type name, the clause

---

```
is range 0 to INTEGER'HIGH
```

---

[1] <u>VHDL-93</u>: VHDL-93 provides the predefined type DELAY_LENGTH that represents only positive values of type TIME. DELAY_LENGTH will be discussed in more detail in Chapter 18.

Figure 15.1

User-defined physical type
for capacitance

```
type CAPACITANCE is range 0 to INTEGER'HIGH
  units
    fF;                    -- Femtofarads
    pF = 1000 fF;          -- Picofarads
    nF = 1000 pF;          -- Nanofarads
  end units;
```

specifies the range of base units. The range may be ascending (**to**) or descending (**downto**). The limits can, in general, be arithmetic expressions, but must be computable at analysis time; no elaboration time information, such as generics, or runtime information, such as signals, are allowed. Moreover, the limits must be *integers*. INTEGER'HIGH is an example of a predefined attribute for the predefined type INTEGER. In Chapter 13, HIGH provided the upper bound of a bit vector indexing scheme; in this chapter HIGH provides the upper bound of the type INTEGER, in other words, the largest available integer. The largest available integer generally depends on the computing installation.

The first name following the keyword **units** is defined to be the base unit; fF (femtofarads) is the base unit for CAPACITANCE. Any number of additional derived or secondary units may be defined following the definition of the base unit. Derived units must be defined as integral multiples of the base unit. The identifiers pF (picofarads)[2] and nF (nanofarads) denote two derived units for CAPACITANCE. The keywords **end units** close the declaration of the secondary units and the physical type declaration.[3]

15.1.3 Arithmetic Operations

Physical numbers can participate in arithmetic expressions involving *scalar analysis*, but not *dimensional analysis*. To understand this distinction, consider the following sample arithmetic operations involving physical literals of type TIME.

$$7 \text{ ns} + 10 \text{ ns} = 17 \text{ ns}$$
$$1.2 \text{ ns} - 12.6 \text{ ps} = 1187400 \text{ fs}$$
$$5 \text{ ns} \times 4.3 = 21.5 \text{ ns}$$
$$\frac{20 \text{ ns}}{5 \text{ ns}} = 4$$

Two values of time can be added or subtracted to yield another value of time. A value of time can be multiplied (scaled) by an integer or floating point number to yield another value of time. Also, two values of time can be divided; the units cancel and the result is an integer. However, two values of time cannot be multiplied because VHDL does not possess a predefined dimensional analysis capability to understand that the resulting value should have units of time.

---

[2] A picofarad is commonly called a "puff."

[3] VHDL-93: In VHDL-93, a physical type declaration may close by repeating the name of the physical type.

```
type CAPACITANCE is range 0 to INTEGER'HIGH
  units

    .....

  end units CAPACITANCE;
```

The subtraction example emphasizes that arithmetic operations are executed in terms of the integral, base unit representation of the operands. Executing arithmetic operations on objects of type TIME in terms of the base unit (femtoseconds) provides an absolute accuracy in timing calculations. Picoseconds can be reliably subtracted from nanoseconds without numerical roundoff error because the arithmetic operation is conducted on operands that are each resolved into the common base units of femtoseconds. Thus, when a VHDL model is executed or simulated, we can be sure that the results reflect the actual behavior of the model and are not mistakenly corrupted by computational numerical errors.

15.1.4 Predefined Attributes

VHDL provides several predefined attributes for physical types.

- LOW
- HIGH
- LEFT
- RIGHT
- POS

- VAL
- PRED
- SUCC
- LEFTOF
- RIGHTOF

The range of base units may be queried by the attributes LOW/HIGH and LEFT/RIGHT. For example, the minimum and maximum number of microvolts (base unit) of the physical type VOLTAGE declaration

```
type VOLTAGE is range 0 to 2**32-1
  units
   uV;                    -- Microvolt
   mV = 1000 uV;          -- Millivolt
    V = 1000 mV;          -- Volt
  end units;
```

can be queried as

```
VOLTAGE'LOW  = VOLTAGE'LEFT  = 0
VOLTAGE'HIGH = VOLTAGE'RIGHT = 2**32-1
```

For an ascending range, LEFT equates to LOW, and RIGHT equates to HIGH. The reverse association holds for a descending range; LEFT equates to HIGH, and RIGHT equates to LOW.

The number of base units representing a physical number can be queried by the predefined attribute POS. For example, VOLTAGE'POS(10.6 mV) yields the integer 10,600, denoting 10,600 microvolts. The predefined attribute VAL performs the reverse mapping of base units to associated physical number; VOLTAGE'VAL(10_600) yields the voltage literal 10_600 uV or, equivalently, 10.6 mV.

Finally, VHDL also provides the predefined attributes PRED/SUCC and LEFTOF/RIGHTOF for physical types. Given a physical number, the predefined attribute PRED (predecessor) can be used to query the physical number represented by one less base unit; the predefined attribute SUCC (successor) can be used to query the physical number represented by one more base unit. For example, VOLTAGE'PRED(10.6 mV) yields the voltage 10_599 uV and VOLTAGE'SUCC(10.6 mV) yields the voltage 10_601 uV. The LEFTOF and RIGHTOF attributes provide a similar function, except that they are sensitive to the range direction. For an ascending range, LEFTOF equates to PRED, and RIGHTOF equates to SUCC.

The reverse association holds for a descending range; LEFTOF equates to SUCC, and RIGHTOF equates to PRED.

15.2 PROPAGATION DELAY

Having learned how to describe physical quantities in VHDL, we can now use this modeling skill to describe technology-dependent information, such as propagation delay. In Chapter 6, we learned that logic functions have a finite input/output delay due to the physical properties of electronic devices and circuits. In this section, we will discuss how to describe propagation delay in VHDL, building on behavioral statements introduced in previous chapters.

Figure 15.2 shows how the behavioral VHDL model of the combinational majority function shown in Figure 13.2 can be easily extended to account for an input/output delay of 20 nanoseconds. The clause **after** 20 ns is added to the conditional concurrent signal assignment statement driving Z_OUT to specify an input/output delay. From a hardware point of view, the concurrent signal assignment statement can be interpreted as follows.

Twenty nanoseconds after a change on any of the input signals A_IN, B_IN, or C_IN, Z_OUT will take on the value determined by the logic expression defining the majority function.

Equivalently, from a language point of view, the same concurrent signal assignment statement can be interpreted as follows.

In response to a change on any of the input signals A_IN, B_IN, or C_IN, the concurrent signal assignment statement will execute. The result of evaluating the logic expression is assigned as a *future* value of Z_OUT to occur after 20 nanoseconds.

From either perspective, the delay between input stimulus and output response is described by simply appending an **after** clause to a signal transformation. Figure 15.3 shows a *timing diagram* illustrating the behavior of the MAJORITY design entity. Signals are listed vertically; time progresses horizontally from left to right. Signal transitions appearing on the

Figure 15.2

Adding input/output delay to the majority function

```
entity MAJORITY is
   -- Input/output ports
  port
    (A_IN, B_IN, C_IN : in BIT;
     Z_OUT            : out BIT);
end MAJORITY;

architecture DATA_FLOW of MAJORITY is
begin
  -- Use concurrent signal assignment statement to describe
  -- input/output function.
  Z_OUT <= (not A_IN and B_IN and C_IN) or
           (A_IN and not B_IN and C_IN) or
           (A_IN and B_IN and not C_IN) or
           (A_IN and B_IN and C_IN) after 20 ns;
end DATA_FLOW;
```

Figure 15.3

The timing diagram for
MAJORITY

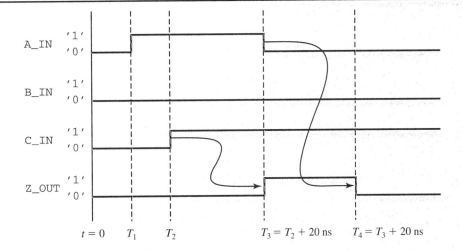

left precede, or occur before, signal transitions appearing on the right. Arrows indicate cause and effect, in other words, which signal transitions cause other signal transitions.

With $t = 0$ an initial condition, assume that all input signals, A_IN, B_IN, and C_IN are '0' and, consequently, Z_OUT is '0'. At $t = T_1$, the input A_IN changes value from '0' to '1' and triggers execution of the conditional concurrent signal assignment statement to compute a new value for Z_OUT. The new value '0' is the same as the old value, so nothing happens to Z_OUT. At $t = T_2$, the input C_IN changes value from '0' to '1' and also triggers execution of the conditional concurrent signal assignment statement to compute a new value for Z_OUT. Since there are now a majority of inputs asserted, the new value '1' differs from the old value, so Z_OUT is scheduled to transition to '1' after 20 nanoseconds have elapsed, in other words, after an input/output delay of 20 nanoseconds. This output rising transition occurs at time $t = T_3$. Also, at time $t = T_3$, the input A_IN returns to '0'. The falling transition on A_IN again triggers execution of the conditional concurrent signal assignment statement. Since there is no longer a majority of inputs asserted, the new value '0' again differs from the old value, so Z_OUT is scheduled to return to '0' after 20 nanoseconds have elapsed, which occurs at time $t = T_4$.

Both the conditional and selected concurrent signal assignment statements can accept **after** clauses. For instance, Figure 15.4 (p. 462) shows that an **after** clause can be similarly added to the concurrent selected signal assignment statement in the VHDL description of the even Gray code detector given in Figure 13.6. Figure 15.5 shows a timing diagram illustrating the behavior of the even Gray code detector. With $t = 0$ an initial condition, assume that CNTRL is asserted active-1, enabling the detector. Also assume that the data inputs GRAY(1) = '1' and GRAY(0) = '0' encode the number 3_{10}. Since 3_{10} is an odd number, the output signal EVEN = '0'. At $t = T_1$, the data input GRAY(0) changes value from '0' to '1' and triggers execution of the selected concurrent signal assignment statement to compute a new value for EVEN. Since GRAY(1) = '1' and GRAY(0) = '1' now encode the even number 2_{10}, the new value '1' differs from the old value, so EVEN is scheduled to transition to '1' after 7 nanoseconds have elapsed, in other words, after an input/output delay of 7 nanoseconds. This output rising transition occurs at time $t = T_2$. At $t = T_3$, the control signal CNTRL changes value from '1' to '0'. Again, the selected concurrent signal assignment executes and EVEN is scheduled to take on the value '0' after 5 nanoseconds have elapsed, in other words, after an input/output delay of 5 nanoseconds. This output falling transition occurs at time $t = T_4$.

Figure 15.4

Adding input/output delay to the even gray code detector

```
entity EVEN_GRAY is
  port
    (GRAY  : in BIT_VECTOR(1 downto 0);
     CNTRL : in BIT;
     EVEN  : out BIT);
end EVEN_GRAY;


architecture DATA_FLOW of EVEN_GRAY is
begin
  -- Use selected concurrent signal assignment statement
  -- to describe input/output function.
  with CNTRL select
  EVEN <= not (GRAY(1) xor GRAY(0)) after 7 ns when '1',
                '0'                  after 5 ns when '0';
  end DATA_FLOW;
```

Notice that, in general, the input/output delay may depend on the particular input stimulus.

15.2.1 Inertial Delay Model

The VHDL examples given in the previous section implicitly assume an *inertial delay model* that reflects the inertia of physical systems. Since electronic devices do not respond instantaneously to input stimuli, logic gates exhibit a degree of *lowpass frequency filtering*. That is, as the rate is increased at which input signals change value, there comes a point when a logic gate can no longer respond to changes to input stimuli. High frequency input signal changes that are not appropriately reflected in output changes are effectively "filtered out"; it is as if they never occurred. These high frequency input signal changes are often called *spikes*.

For the VHDL inertial delay model, any input signal change that does not persist for at least the propagation delay of the device is not reflected at the output. Figure 15.6 illustrates the behavior of the VHDL inertial delay model for an inverter. The **after** clause specifies the

Figure 15.5

Timing diagram for even gray code detector

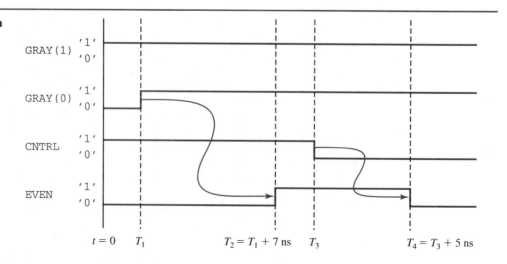

Figure 15.6

Behavior of the VHDL inertial delay model

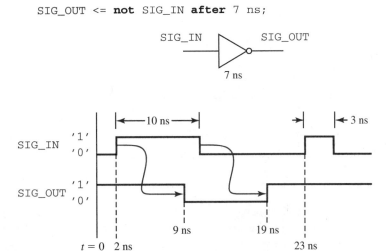

```
SIG_OUT <= not SIG_IN after 7 ns;
```

The inertial delay model filters short-duration input signal changes.

inverter propagation delay to be 7 nanoseconds. The 10-nanosecond input pulse is properly reflected at the inverter output because the pulse lasts more than the inverter propagation delay of 7 nanoseconds. However, the 3-nanosecond input spike is not reflected at the inverter output because it did not persist for at least 7 nanoseconds.

15.2.2 VHDL-93: Enhanced Inertial Delay Model

To improve language consistency and symmetry, VHDL-93 allows the inertial delay model to be *explicitly*, as well as *implicitly*, declared by allowing concurrent signal assignment statements to include the keyword **inertial** following the signal assignment operator. Thus, the conditional concurrent signal assignment statement in Figure 15.2

```
Z_OUT <= (not A_IN and B_IN and C_IN) or
         (A_IN and not B_IN and C_IN) or
         (A_IN and B_IN and not C_IN) or
         (A_IN and B_IN and C_IN) after 20 ns;
```

can be rewritten as follows

```
Z_OUT <= inertial (not A_IN and B_IN and C_IN) or
                  (A_IN and not B_IN and C_IN) or
                  (A_IN and B_IN and not C_IN) or
                  (A_IN and B_IN and C_IN) after 20 ns;
```

Likewise, the selected concurrent signal assignment statement given in Figure 15.4

```
with CNTRL select
EVEN <= not(GRAY(1) xor GRAY(0)) after 7 ns when '1',
        '0'                       after 5 ns when '0';
```

can be rewritten as

```
with CNTRL select
EVEN <= inertial(GRAY(1) xnor GRAY(0)) after 7 ns when '1',
             '0'                        after 5 ns when '0';
```

(Recall that VHDL-93 provides the **xnor** operator.)

VHDL-93 also allows inertial delay, sometimes called a *pulse rejection limit*, to be *different from* the propagation delay. For example, what if the inverter in Figure 15.6 has an inertial delay different from 7 nanoseconds, such as 5 nanoseconds? Such behavior would be difficult to model using concurrent signal assignment statements in VHDL-87. However, Figure 15.7 shows that VHDL-93 allows for the specification of separate inertial and propagation delays.

The clause **reject** 5 ns following the signal assignment operator declares the pulse rejection limit to be 5 nanoseconds. The pulse rejection limit must be less than the propagation delay. The keyword **inertial** is required when declaring an inertial delay different from the propagation delay.

15.2.3 Transport Delay Model

VHDL also provides a *transport delay model*. With a transport delay model, *all* input signal changes are reflected at the output, regardless of how long the signal changes persist. Figure 15.8 shows the behavior of an inverter with a transport delay model.

Transport delay is denoted by the keyword **transport** placed just after the signal assignment operator. The timing diagram shows that all input signal transitions, including the 3-nanosecond input signal spike, are reflected at the inverter output.

The inertial delay model is the default delay model because it is more commonly used than the transport delay model. However, the simpler transport delay model is used for high-level modeling applications, such as architecture or performance modeling, that are not concerned with the low-level inertial aspects of hardware.

15.2.4 Delta Delay

Thus far, we have artfully avoided discussing the timing behavior when a signal assignment statement contains a propagation delay of 0 time units, but this is legal and needs to be dis-

Figure 15.7

Enhanced inertial delay model

```
SIG_OUT <= reject 5 ns inertial not SIG_IN after 7 ns;
```

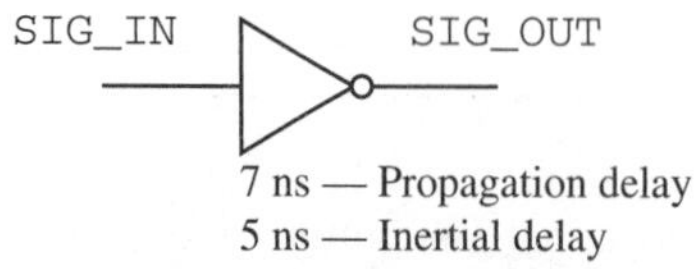

7 ns — Propagation delay
5 ns — Inertial delay

Figure 15.8

Transport delay responses to all input signal changes

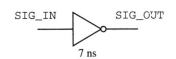

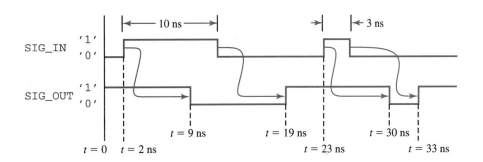

cussed. A propagation delay of 0 time units is equivalent to omitting the **after** clause and is called a *delta delay*.

To understand delta delay, it is helpful to define the VHDL *simulation cycle*. A simplified diagram of the VHDL simulation cycle is given in Figure 15.9 (p. 466). The simulation cycle defines how to mentally "play out" the execution of a VHDL model to understand its input/output behavior over time. In more formal terms, the simulation cycle defines VHDL dynamic semantics. We implicitly used the simulation cycle when explaining the timing diagrams of the MAJORITY design entity given in Figure 15.2 and the EVEN_GRAY design entity given in Figure 15.5.

The easiest way to explain the VHDL simulation cycle is via an example. Figure 15.10 illustrates the execution of the simulation cycle for a simple 4-input, cascade odd parity detector circuit using no delta delays. The odd parity detector combinational logic generates a '1' whenever an odd number of 1's are present at the inputs. (See Chapter 13 for a discussion of parity coding and note that an odd parity detector is equivalent to an even parity generator.)

At the start of simulation, $t = 0$, we assume that all signals are initialized to '0'. At $t = 5$ ns, we assume that the input signals A, C, and D change from '0' to '1'. Simulation time advances to $t = 5$ ns and the input signals A, C, and D are updated. These signal changes cause the three conditional concurrent signal assignment statements to be evaluated during the "execute model" step. As a result of evaluating the signal assignment statements, INT1, INT2, and ODD are all *projected* to take on the value '1' after 5 nanoseconds or, equivalently, 5 nanoseconds into the future relative to the present simulation time.

The next simulation cycle advances time to the next event(s) scheduled to occur. Time is advanced by 5 nanoseconds, $t = 5$ ns + 5 ns = 10 ns, and the projected values for INT1, INT2, and ODD become the present values. The changes on INT1 and INT2 cause the second and third concurrent signal assignment statements to again be evaluated, and INT2 and ODD are projected to change back to '0' after 5 nanoseconds.

The next simulation cycle again advances time by 5 nanoseconds, $t = 10$ ns + 5 ns = 15 ns, and the projected values for INT2 and ODD become present values. The change on INT2 causes the third concurrent signal assignment statement to be evaluated, and the output signal ODD is projected to take on the value '1' after 5 nanoseconds.

Problem-Solving Skill
Every solution has static and dynamic aspects.

Figure 15.9

**Simplified diagram of
VHDL simulation cycle**

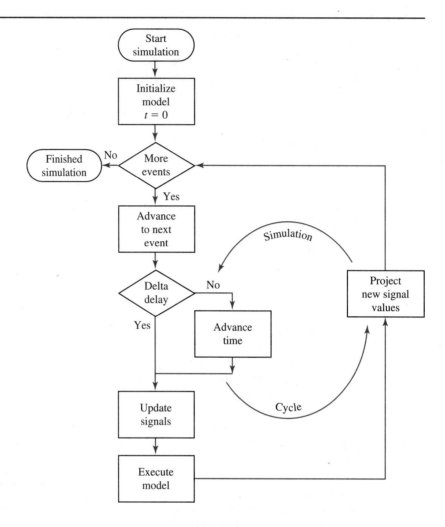

Finally, the last simulation cycle is exercised by advancing simulation time to 15 ns + 5 ns = 20 ns and updating the value of ODD to '1'. No signal assignment statements are evaluated and no new events are generated, so the simulation ends. The result is that the parity detector combinational logic correctly generates the value of '1', indicating an odd number of input 1's, with a total propagation delay of 15 nanoseconds. (Whew!)

Figure 15.11 shows the revised architecture body of the 4-input, cascade parity detector circuit using delta delays. How does the behavior of the new parity detector model with delta delays compare to the behavior of the parity detector model with finite delays? For the inputs given in Figure 15.10(c), both models execute the same number of simulation cycles, and the same concurrent signal assignment statements are evaluated during each simulation cycle. Thus, the same events, or signal transitions, occur in the same order or sequence. However, the parity detector model with finite delays describes an *absolute* ordering of events, whereas the parity detector model with delta delays describes only a *relative* ordering of events.

Figure 15.10

An example of a simulation cycle with no delta delay

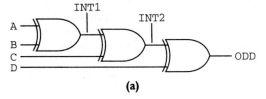

```
entity ODD_PARITY_DETECT is
  port
    (A, B, C, D : in BIT;
     ODD        : out BIT);
end ODD_PARITY_DETECT;

architecture DATA_FLOW of ODD_PARITY_DETECT is
  signal INT1, INT2 : BIT;
begin
  INT1 <= A xor B after 5 ns;
  INT2 <= INT1 xor C after 5 ns;
  ODD  <= INT2 xor D after 5 ns;
end DATA_FLOW;
```

(a)

(b)

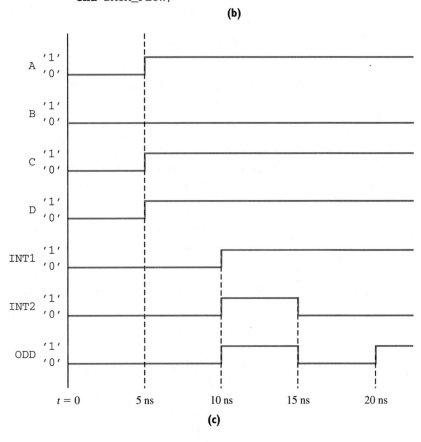

(c)

Figure 15.11

Parity model using delta delays

```
architecture DATA_FLOW of ODD_PARITY_DETECT is
  signal INT1, INT2: BIT;
begin
  INT1 <= A xor B;
  INT2 <= INT1 xor C after 0 ns; -- Same as no timing clause
  ODD  <= INT2 xor D;
end DATA_FLOW;
```

Notice that **after** 0 ns is equivalent to no timing clause.

467

The parity detector model with delta delays describes a relative ordering of events because executing a concurrent signal assignment statement with a delta propagation delay *does not advance simulation time* between simulation cycles. As an example, evaluating

---

```
INT1 <= A xor B;
```

---

means that the *next* value of INT1—in other words, the value of INT1 on the *next* simulation cycle—is projected to be A **xor** B. Simulation time is not advanced between the present and next simulation cycles. For the two parity detector models, both simulations initially advance time to $t = 5$ ns, which is the time the input signals change value. However, the time at the end of simulation for the parity detector model with finite delays is $t = 20$ ns, whereas the time at the end of simulation for the parity detector model with delta delays is still $t = 5$ ns.

Figure 15.12 shows a helpful way to relate finite and delta delays: Both can be thought of as forming a two-dimensional time space. Along the horizontal axis, time advances in finite steps. At any point on the horizontal axis, time may "advance" up the vertical axis in delta steps. A delta delay is denoted by "δ." Notice that time can advance in a relative sense any number of delta steps up the vertical axis and never advance in an absolute sense on the horizontal axis.

Delta delays are often used during the early stages of digital design when we are principally concerned with function, in other words, the right events occurring in the right sequence. Having obtained a design that yields the desired function, we can then refine the design to address performance by replacing delta delays with finite propagation delays.

15.2.5 Timing Expressions

The examples of propagation delay shown thus far have used literal physical values, such as 5 ns. Propagation delay can, in general, be an arbitrarily complex expression to model timing as a function of several factors, such as the environment or logic family.

Figure 15.12

Two-dimensional aspect of time

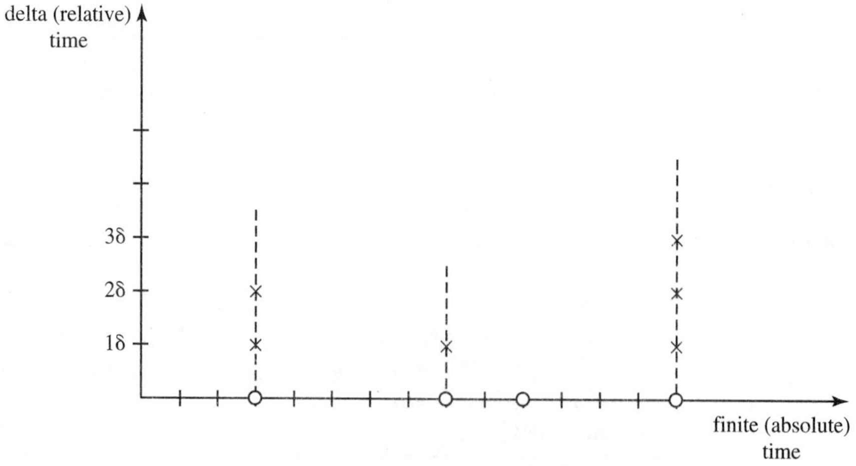

As a simple example, Figure 15.13(a) shows how to modify the exclusive-or gate described in Figure 14.7 to include a programmable propagation delay via a generic parameter. (See Chapter 14 for a discussion of generic parameters.)

By making the propagation delay a generic parameter, the XOR3_GATE design entity represents an **xor** gate having a variable input/output delay. Figure 15.13(b) shows how the component declaration and instantiation statements are appropriately modified to account for the generic parameter and to allow each use of the XOR3_GATE to potentially have a unique propagation delay.

15.3 FUNCTIONS

In Chapter 6, we introduced *wired logic*, which refers to the implicit logic that can sometimes be realized by simply connecting or wiring together several gate outputs. Wired logic is described in VHDL via a *resolution function* that resolves multiple sources driving a particular node into a single value. Since there are a lot of rules concerning the composition and use of functions in VHDL, we will depart from our standard "learn-by-example" format and first discuss functions and then discuss their application in describing wired logic.

Figure 15.13

Using generic parameters to specify programmable propagation delay

```
entity XOR3_GATE is
  generic
    (IO_DELAY : TIME); -- Generic parameter
  port
    (A, B, C : in BIT := '0'; -- Default value for A, B, and C
     Z       : out BIT);
end XOR3_GATE;

architecture BEHAVIOR of XOR3_GATE is
begin
  Z <= A xor B xor C after IO_DELAY; -- Generic is I/0 delay;
end BEHAVIOR;
```

(a)

```
component XOR3_GATE;
  generic (IO_DELAY : TIME);
  port (A, B, C : in BIT; Z : out BIT);
end component;

X1: XOR3_GATE
  generic map (IO_DELAY => 5 ns)
  port map (A=> .... );
```

(b)

Figure (a) shows how to use a generic parameter to make the **xor** gate delay programmable. Figure (b) shows how to set the value of the generic parameter when using an instance of the **xor** gate.

Functions in VHDL are analogous to functions in common software programming languages; they represent abstract operators. Just as the logic operator **and** accepts two operands and returns the conjunction, a function accepts operands as *parameters*, performs a computation, and returns a result. Functions support describing signal transformations by serving a secondary role of describing abstract computations or operations. To that end, VHDL functions execute in zero time, access only locally declared objects, and cannot directly update signals.

As an example, Figure 15.14(a) shows how to declare a function that performs the logic and-or-invert (A-O-I) operation introduced in Chapter 5 and shown in Figure 5.13. The and-or-invert switching algebra operation is not a VHDL predefined logic operation, so it is a good candidate for describing with a VHDL function if we intend to frequently use and-or-invert logic. The various parts involved in constructing functions, in general, and the A_O_I function, in particular, will be explained in detail in the following sections.

15.3.1 Specification

A function declaration starts with the keyword **function**, followed by an identifier giving the name of the function, followed by the *formal parameter list*. A formal parameter list of a function is similar to a port list of a design entity. The formal parameter list specifies the information communication channel(s) into a function, listing the function's operands. The type of the value returned by the function is declared following the keyword **return**. Repeating the function name at the closing of a function declaration is optional; however, this modeling practice is recommended to improve readability.

Formal parameters receive their values when the function is invoked. Figure 15.14(b) shows an example of a function call or invocation, assigning to the target signal X_SIG the logic and-or-invert function of signals A_SIG, B_SIG, C_SIG, and D_SIG after a delay of 20 nanoseconds. The signals A_SIG, B_SIG, C_SIG, and D_SIG are called *actual arguments* or *actuals*. Using positional association, the values of A_SIG, B_SIG, C_SIG, and D_SIG are

Figure 15.14

Function to perform the and-or-invert (A-O-I) operation

```
function A_O_I (constant A_ARG,B_ARG,C_ARG,D_ARG : in BIT)
                    return BIT is
 -- Declarations
begin
 -- Sequential statements
 return((A_ARG and B_ARG) nor (C_ARG and D_ARG));
end A_O_I;
```
                              (a)

```
signal A_SIG, B_SIG, C_SIG, D_SIG, X_SIG : BIT;
                              .
                              .
X_SIG <= A_O_I(A_SIG, B_SIG, C_SIG, D_SIG) after 20 ns;
```
                              (b)

Figure (a) presents a function declaration. Figure (b) shows how to use the function in a concurrent signal assignment statement.

assigned respectively to the formal parameters A_ARG, B_ARG, C_ARG, and D_ARG; named association (see Chapter 12) may also be used.

```
X_SIG <= A_O_I(A_ARG => A_SIG,
               B_ARG => B_SIG,
               C_ARG => C_SIG,
               D_ARG => D_SIG) after 20 ns;
```

Be careful not to confuse function calls and type conversions, introduced in Chapter 14. Although the syntax may be similar, the semantics are different. For example, "CONVERT(ARG)" could be an invocation of the function CONVERT with a single actual ARG or a conversion of an object ARG to a closely related type CONVERT.

A formal parameter has three properties: *class*, *mode*, and *type*. The properties of mode and type were introduced in Chapter 11 in the discussion of port declarations. The property of class is introduced for function formal parameter declarations because unlike ports, which can be only signals, function formal parameters can be either signals or *constants*. Hence, class specifies the kind of formal parameter and can be either **signal** for signals or **constant** for constants. Only signals may be associated with formal parameters that are signals, whereas general expressions may be associated with formal parameters that are constants. A class designator may be omitted, in which case the class **constant** is assumed.

Mode specifies the direction of information flow. Only mode **in** is allowed for function formal parameters. *All* function formal parameters are *always* of mode **in**. Values of formal parameters are set when a function is called and cannot be changed within the function. A mode designator may be omitted, in which case mode **in** is assumed.

Type specifies the set of values a formal parameter may take on. Formal parameters of the same class and type may be collectively defined in a single declaration (Figure 15.14(a)) or individually defined in separate declarations.

```
function A_O_I (constant A_ARG : in BIT;
               constant B_ARG : in BIT;
               constant C_ARG : in BIT;
               constant D_ARG : in BIT) return BIT is;
  -- Declarations
begin
  -- Sequential statements
  return((A_ARG and B_ARG) nor (C_ARG and D_ARG));
end A_O_I;
```

A formal parameter type can denote an unconstrained array. For example, Figure 15.15 (p. 472) shows formal parameters A_ARG and B_ARG of the predefined unconstrained array type BIT_VECTOR. The formal parameters A_ARG and B_ARG are constrained at the time VECTOR_OPR is called, taking on the range of the associated actual arguments. For instance, in Figure 15.15(b) A_ARG and B_ARG respectively take on the range of A_SIG and B_SIG, which is 7 **downto** 0.

A final point about function formal parameter lists concerns default values. Just as ports and generics can have default values, formal parameters can have default values by including an optional initial value expression.

```
(constant A_ARG, B_ARG, C_ARG, D_ARG : in BIT := '0')
```

Figure 15.15

Function with uncon-strained arrays as formal parameters

```
function VECTOR_OPR (signal A_ARG, B_ARG, : BIT_VECTOR)
                          return BIT is
  -- Declarations
begin
  -- Sequential statements
                                    .
                                    .
  end VECTOR_OPR;
```

(a)

```
  signal A_SIG, B_SIG : BIT_VECTOR(7 downto 0);
  signal C_SIG : BIT;
                                    .
                                    .
  C_SIG <= VECTOR_OPR(A_SIG, B_SIG) after 20 ns;
```

(b)

Figures (a) and (b) respectively show an example of the declaration and invocation of a function VECTOR_OPR having unconstrained arrays as formal parameters.

Formal parameters that are not set at function invocation are assigned associated default values. The default value applies to all associated formal parameters and must be a legal value of the formal parameter type.

15.3.2 Body

The function name, formal parameter list, and return type declaration form "header" information, more formally called the *function specification*. Following this information, the function "body" is described, which defines the computation to be performed on the input operands and the result to be returned to the calling VHDL statement. A function body is organized into two parts: *declarative part* and *statement part*. The keywords **is** and **begin** bracket the declarative part; the allowed declarations are listed below.

- function
- procedure
- type
- subtype
- constant

- variable
- file
- alias
- attribute
- use clause

Some of these declarations have been discussed, such as type, alias, and function declarations; the latter declaration allows for nesting of functions. Other declarations, such as file, subtype, and variable declarations, will be covered in later sections. Note that signal declarations are *not* allowed in functions. The scope of the declarations extends to the end of the function; in other words, the declared items can only be used inside the enclosing function.

The statement part of a function contains the executable statements that perform the desired user-defined operation. Functions support *only sequential statements*. Recall from Chapter 13 that there are two general classes of executable statements in VHDL: concurrent and sequential. The order in which concurrent statements are written has no bearing on the order in which they execute, whereas sequential statements are executed from top to bottom in the textual order they are written.

Thus far, we have used only concurrent statements. VHDL sequential statements will be briefly introduced in this chapter and covered in more detail in Chapter 18, which addresses algorithmic hardware modeling. In short, the example function A_O_I contains only one sequential statement: **return**. The **return** statement uses the logic operators **and** and **nor** to compute the and-or-invert operation and return the result to the calling parent. Every function must contain at least one **return** statement.

15.3.3 Overloading

Two or more functions may have the same name; such functions are called *overloaded*. It is permissible to declare overloaded functions, provided that a VHDL compiler can determine which function the user intends to use for a particular function call. In other words, overloaded functions must be able to be differentiated or *disambiguated*.

Overloaded functions are disambiguated based on the formal parameters. For example, consider declaring two functions having the same name MEAN. One version of MEAN computes the arithmetic mean of two integers, while the other version of MEAN computes the arithmetic mean of two floating point numbers.

```
function MEAN (A_ARG, B_ARG : INTEGER) return INTEGER is
                           .
                           .
                           .
function MEAN (A_ART, B_ARG : REAL) return REAL is
                           .
                           .
```

A function call MEAN(5, 11) refers to the first function because the operands are integers.

A common use of overloaded functions is to overload predefined operators. We will see examples of overloaded logic operators when we discuss modeling multi-valued logic in Chapter 16.

15.3.4 Functions and Packages

Now that we know how to write functions, how can we make them available for other designers to use? An easy way to make functions available for general use is to place the functions inside a package, as shown in Figure 15.16 for the A_O_I function.

Figure 15.16 (p. 474) illustrates two new VHDL constructs: *function specification* and *package body*. The function specification for A_O_I,

```
-- Function specification
function A_O_I (constant A_ARG,B_ARG,C_ARG,D_ARG : in BIT)
                return BIT;
```

given in the package declaration defines the header information; namely, function name, parameters, and return type. Placing only the function specification in the package declaration exports interface information, but not implementation information. Enough information is made public to enable function use, but, at the same time, enough information is kept private to enable flexible function implementation(s).

Figure 15.16

Placing functions inside packages

```
-- Package declaration
package SPECIAL_OPRS is
  -- Function specification
  function A_O_I (constant A_ARG,B_ARG,C_ARG,D_ARG : in BIT)
                 return BIT;
end SPECIAL_OPRS;

-- Package body
package body SPECIAL_OPRS is
  -- Function declaration
  function A_O_I (constant A_ARG,B_ARG,C_ARG,D_ARG : in BIT)
                 return BIT is
  begin

    return((A_ARG and B_ARG) nor (C_ARG and D_ARG));
  end A_O_I;

end SPECIAL_OPRS;
```

The function implementation or *body* associated with a function specification is placed in the package body. Any package may have a package body. The relationship between a package declaration and its body is similar to the relationship between an entity declaration and its architecture; both separate specification from implementation. Information contained in the package body is *not* made available via a **use** statement; in other words, the information is hidden from the user of the package. Thus, alternative implementations for the A_O_I function can be developed over time without impacting models using the package SPECIAL_OPRS, provided the function interface or specification remains unchanged.

A package body starts with the keywords **package body**, followed by the name of the associated package declaration, followed by the keyword **is**. In its role of completing the declarations defined in the associated package declaration, a package body may contain any of the following declarations.

- function
- procedure
- type
- alias
- subtype
- constant
- file
- use clause

Again, some of these declarations have been discussed; others will be introduced in future chapters. A package body definition closes with the keyword **end**, followed optionally by the name of the associated package declaration. We will adopt the practice of repeating the name of the associated package declaration to improve readability.[4]

---

[4] VHDL-93: In VHDL-93, a package body may include *group* and *group template* declarations and close with the keywords **end package body**.

```
package body SPECIAL_OPRS is
  .....
end package body SPECIAL_OPRS;
```

15.4 VHDL-93: PURE AND IMPURE FUNCTIONS

VHDL-93 provides several enhancements to functions. First, formal parameters have been extended to include files denoted by class **file**. Files store information on the host computing platform. Formal parameters of class **file** have no mode and may only be associated with files. (Files are discussed in detail in Chapter 18.) Second, functions allow *group template* and *group* declarations, which are also discussed in later chapters. Third, function declarations may end with the keywords **end function**.

```
function A_O_I ( ..... ) return BIT is
  -- Declarations
begin
  -- Sequential statements
end function A_O_I;
```

Finally, VHDL-93 refines and clarifies the use of functions by classifying functions as either *pure* or *impure*; this is a significant language modification. To understand the rationale for this classification, recall that VHDL functions serve the principal role of describing abstract operations. Consequently, VHDL-87 functions have several restrictions designed to ensure that they return the same value for a given set of actuals. In other words, functions exhibit *deterministic* behavior or, equivalently, have no *side-effects*. However, the restrictions are unfortunately not fool-proof because VHDL-87 functions can access locally declared files. Information stored in such files persists between invocations and can be used to generate different results for identical invocations.

To clarify the role of functions and the restrictions enforcing such roles, VHDL-93 classifies functions as either pure or impure. A pure function has no side-effects and is the default function classification. A function declaration can explicitly denote a pure classification by adding the keyword **pure**.

```
pure function A_O_I ( ..... ) return BIT is
  -- Declarations
begin
  -- Sequential statements
end function A_O_I;
```

A pure function is similar to VHDL-87 functions with the added restriction that local file declarations and accesses are disallowed, in other words, no more "hanky-panky."

An impure function can exhibit nondeterministic behavior; in other words, it can return different values when invoked several times with the same set of actuals. A function declaration must explicitly denote an impure classification by adding the keyword **impure**.

```
impure function A_O_I ( ..... ) return BIT is
  -- Declarations
begin
  -- Sequential statements
end function A_O_I;
```

An impure function can access objects declared outside the function as well as locally declared objects. Also, an impure function allows local file declarations and accesses. An impure function can even have a foreign implementation.

```
-- Package declaration
package SPECIAL_OPRS is
  -- Function specification
impure function A_O_I (constant A_ARG,B_ARG,C_ARG,D_ARG : in BIT)
                       return BIT;

  -- Attach attribute to function to define foreign interface
  attribute FOREIGN of A_O_I : function is
             "/lib/vendor_xyz/and_or_invert";
end SPECIAL_OPRS;
```

Foreign interfaces for architectures were introduced in Chapter 14. The attribute specification attaches an instance of the attribute FOREIGN to the function A_O_I and gives the attribute a character string value describing interface or coupling information.

15.5 MODELING WIRED LOGIC

We can now discuss how to use functions to describe wired logic. Chapter 6 introduced *wired logic*, where wiring together the outputs of several logic gates performs an implicit logic function. Figure 15.17 shows, for example, wired-**and** logic, supported by integrated injection logic (I^2L), and wired-**or** logic, supported by emitter-coupled logic (ECL). Wired logic is properly viewed as implementation technology information because its properties depend on a particular logic family.

15.5.1 Resolution Functions

The condition of multiple gates driving a single node is modeled as multiple sources, or signal transformations, driving a single signal. The resolution of the multiple gate outputs into a single node value is modeled in VHDL by a *resolution function* that resolves multiple sources into a single signal value. The behavior of the resolution function determines the

Figure 15.17

Wired logic

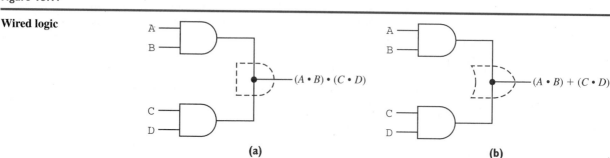

(a)   (b)

Figures (a) and (b) illustrate wired-**and** and wired-**or** logic, respectively.

wired logic properties. A user defines a resolution function and associates it with signals having multiple sources, but does not explicitly invoke a resolution function. Resolution functions are automatically invoked during execution whenever the value of a multiple-source signal is needed.

As an example of modeling wired-**and** logic, consider the logic schematic given in Figure 15.18, which takes advantage of the implicit **and** gate to implement the majority function in a product-of-sums form having an active-0 output. The output ZOUT_BAR is '0' if the majority (two or more) of the inputs are '1'.

Figure 15.19 (p. 478) presents structural and behavioral models of the majority function using wired logic. To understand wired logic modeling, we must first define the concept of "signal sources." Basically, any VHDL construct that can directly or indirectly affect the value of a signal, in other words, write to a signal, qualifies as a *source*, also called a *driver*, for that signal. Figure 15.19(a) shows how component instantiation statements can be sources; the outputs of the three OR2_GATEs all drive the common node (signal) INT_SIG. Alternatively, Figure 15.19(b) shows how concurrent signal assignment statements can be sources; the targets of the three **or** operators all drive the common node (signal) INT_SIG.

INT_SIG is called a *resolved signal* because it has multiple sources. All resolved signals must have an associated *resolution function* that defines how the multiple sources contribute or are resolved into a single value for the signal.

The resolution function for INT_SIG is WIRED_AND, which is specified in the declaration of INT_SIG

```
signal INT_SIG : WIRED_AND BIT;
```

by adding the name of the resolution function just before the type name BIT. If the resolved signal is a port, the resolution function is added in a similar fashion to the port declaration. The definition of the resolution function WIRED_AND and the component declarations for OR2_GATE and INVERTER are contained in the package LOGIC_PKG shown in Figure 15.20 (p. 479).

Any time the value of INT_SIG is needed, such as to drive ZOUT_BAR through the inverter, the resolution function WIRED_AND is automatically invoked to compute a single value based on the output values of the three sources, in other words, three **or** gate component instances or three **or** operator signal assignments. Figure 15.21 (p. 480) illustrates signal resolution. The values of the three sources are passed into the resolution function WIRED_AND as three elements of the formal parameter SOURCES, which is a bit vector. The body of the resolution function WIRED_AND then executes and returns a single value, which is the resolved value for INT_SIG.

Figure 15.18

The majority function with wired logic

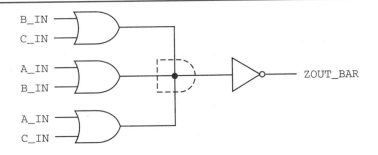

Figure 15.19

Modeling wired logic

```
entity MAJORITY is
  port
    (A_IN, B_IN, C_IN : in  BIT;
     ZOUT_BAR          : out BIT); -- Active-0 output
end MAJORITY;

use WORK.LOGIC_PKG.all;
architecture STRUCTURE of MAJORITY is
  -- Declare resolved signal to interconnect gates
  signal INT_SIG : WIRED_AND BIT;

begin
  -- Product-of-sums form—active-0 output
  O1: OR2_GATE port map (B_IN, C_IN, INT_SIG);   -- Source
  O2: OR2_GATE port map (A_IN, B_IN, INT_SIG);   -- Source
  O3: OR2_GATE port map (A_IN, C_IN, INT_SIG);   -- Source
  I1: INVERTER port map (INT_SIG, ZOUT_BAR);

end STRUCTURE;
```

(a)

```
use WORK.LOGIC_PKG.all;
architecture DATA_FLOW of MAJORITY is
  -- Declare resolved signal to interconnect gates
  signal INT_SIG : WIRED_AND BIT;

begin
  INT_SIG <= B_IN or C_IN;          -- Source
  INT_SIG <= A_IN or B_IN;          -- Source
  INT_SIG <= A_IN or C_IN;          -- Source
  ZOUT_BAR <= not INT_SIG;

end DATA_FLOW;
```

(b)

Models (a) and (b) present structural and behavioral descriptions of the majority logic given in Figure 15.18. The package LOGIC_PKG is given in Figure 15.20.

A resolution function accepts only one input parameter, which *must be* an unconstrained array so the resolution function has the flexibility to address any number of sources. The type of the sources determines the type of elements of the unconstrained array. WIRED_AND accepts any number of sources of type BIT and returns a single value of type BIT.

Like functions in general, the body of a resolution function contains only sequential statements. WIRED_AND contains two sequential statements: a **for-loop** statement and a **return** statement. The **for-loop** supports classical iteration-based control. The **for-loop** parameter I iterates through each element of SOURCES, accumulating the logic **and** of all sources in the *variable* RESULT. The **return** statement returns the final value of RESULT to the calling VHDL statement.

Figure 15.20

The supporting package
for wired logic

```
-- Package declaration
package LOGIC_PKG is
 -- Component declaration
 component OR2_GATE
  port (A, B : in BIT; Z : out BIT);
 end component;
 component INVERTER
  port (A : in BIT; A_BAR : out BIT);
 end component;

 -- Function specification
 function WIRED_AND (SOURCES : BIT_VECTOR) return BIT;
end LOGIC_PKG;

-- Package body
package body LOGIC_PKG is
 -- Function declaration
 function WIRED_AND (SOURCES : BIT_VECTOR) return BIT is
  -- Variable declaration
  variable RESULT : BIT := '1';
 begin
  -- Sequential statements
  for I in SOURCES'RANGE loop
   RESULT := RESULT and SOURCES(I);
  end loop;
  return( RESULT );
 end WIRED_AND;
end LOGIC_PKG;
```

Recall from Chapter 13 that SOURCES'RANGE is an example of a predefined attribute for bit vectors. The indexing range of the formal parameter SOURCES cannot be hardcoded in the **for-loop** because the length of SOURCES depends on the number of sources of the associated resolved signal, which can vary between invocations of the resolution function. To specify an indexing range that is not statically known before simulation or runtime, we simply use the predefined attribute RANGE.

The object RESULT is the first example we have seen of *variables* in VHDL. Variables represent another class of objects in VHDL, in addition to signals and constants. Variables hold values and have properties similar to variables traditionally used in most software programming languages. Within the declaration section of WIRED_AND, RESULT is declared

```
variable RESULT : BIT := '1';
```

to be of type BIT and initialized to the value '1'. RESULT is then assigned a value

```
RESULT := RESULT and SOURCES(I);
```

Figure 15.21

Signal resolution

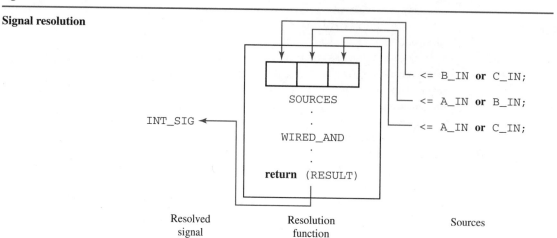

using the assignment operator := rather than the signal assignment operator <=. We will discuss in more detail the properties of signals, variables, and constants at the end of this chapter.

15.6 PROHIBITING WIRED LOGIC

We have just seen how to describe logic families that support wired logic. What about logic families that *do not* support wired logic? Recall from Chapter 6, for instance, that transistor-transistor logic (TTL) gates with totem pole configured outputs cannot be connected together because the active pull-up/pull-down structures can cause high currents and/or invalid voltage levels. For such cases of digital systems that do not support wired logic, we need to know how to use **buffer** ports to describe them.

15.6.1 Buffer Ports

Recall that ports have modes; the properties of **in**, **out**, and **inout** modes were discussed in Chapter 14. **Buffer** ports are similar to **inout** ports in that information can flow either into and/or out of **buffer** ports. However, unlike an **inout** port, a **buffer** port must be the *only* source for its associated port or signal. In other words, a port or signal associated with a **buffer** port must not be driven by any other source.

The term *buffer* is used because the value driven by a **buffer** port is guaranteed to be the value read by the **buffer** port. To understand this property, we will consider first the properties of multiple sources shown in Figure 15.22. Figure 15.22(a) shows a signal DATA associated with the two P1 **inout** ports of design entity A and design entity B. The signal DATA is a resolved signal because each design entity **inout** port serves as a source. Figure 15.22(b) shows how the two port values are resolved into a single value for DATA. The two port values, or sources, are routed into a resolution function, and the resolved result is routed back to the design entities. With the intervening resolution function, the source value driven by a design entity out through its **inout** port may *not* be the resolved value read by the design entity in through the same **inout** port.

This situation does not apply, however, to **buffer** ports because a **buffer** port must be the only source driving its associated signal. Hence, there is no intervening resolution function

Figure 15.22

Resolved signal

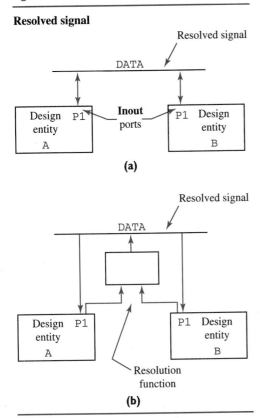

(a)

(b)

Figure 15.23

Illegal port connections—buffer port

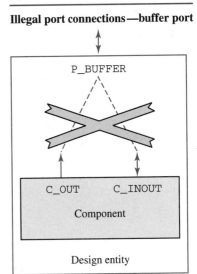

Figure (a) shows multiple signal sources. Figure (b) shows signal resolution.

and, consequently, the value driven by a **buffer** port is guaranteed to be same as the value read by the **buffer** port. The property of identical read (input) and write (output) values is characteristic of a buffer gate (Chapter 3), hence the term **buffer** port.

The properties of **buffer** ports imply the connectivity restrictions shown in Figure 15.23. A **buffer** port of a design entity can be associated with an **in** or **buffer** port of an instantiated component or a declared signal having no other source. A **buffer** port of a design entity cannot be connected to a port of mode **out** or **inout** of a component of the design entity because there is no way of guaranteeing that the component ports are driven by only one source. This restriction ensures that **buffer** ports are properly propagated or connected throughout a design hierarchy.

15.6.2 Application of **buffer** Ports

To illustrate the application of **buffer** ports in prohibiting wired logic, Figure 15.24(b) shows the proper use of TTL, totem pole logic gates to describe the simple logic schematic shown in Figure 15.24(a) (p. 482). **Buffer** ports are used to model the restriction that the outputs of these gates must not be connected. The restriction that the outputs of the TTL, totem pole **and** gate and **or** gate cannot be wired to the output of any other gate is described by declaring the respective output ports to be **buffer** ports in the component declarations for AND2_TTL_TP and OR2_TTL_TP. The signals INT1 and INT2 can be connected to these **buffer** ports because they have no other sources. Moreover, connecting the output of the

Figure 15.24

Using buffer ports to model TTL, totem pole logic

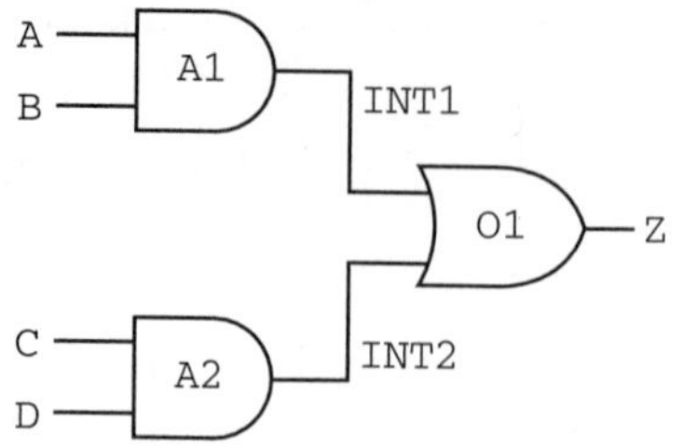

(a)

```
entity SIMPLE_LOGIC is
  port
    (A, B, C, D : in BIT;
     Z              : buffer BIT);   -- Buffer port
end SIMPLE_LOGIC;

architecture STRUCTURE of SIMPLE_LOGIC is
  -- TTL_TP denotes TTL, totem pole logic
  component AND2_TTL_TP
    port (A, B : in BIT; Z : buffer BIT);
  end component;
  component OR2_TTL_TP
    port (A, B : in BIT; Z : buffer BIT);
  end component;
  signal INT1, INT2 : BIT;

begin
  A1: AND2_TTL_TP port map (A, B, INT1);
  A2: AND2_TTL_TP port map (C, D, INT2);
  O1: OR2_TTL_TP  port map (INT1, INT2, Z);

end STRUCTURE;
```

(b)

component OR2_TTL_TP to the output of the enclosing design entity SIMPLE_LOGIC places a connectivity restriction on the proper use of SIMPLE_LOGIC by other design entities. That is, the output of the design entity SIMPLE_LOGIC may not be connected to the output of any other gate. Thus, the output port, Z, of the design entity SIMPLE_LOGIC is declared to be a **buffer** port, which propagates the connectivity restriction out of the enclosing design entity SIMPLE_LOGIC and "up the hierarchy."

As a counter example, Figure 15.25 shows an incorrect use of **buffer** ports. The VHDL model is incorrect because the rules for properly using **buffer** ports have been violated. Since information can flow into and out of a **buffer** port, the first component instantiation statement (A1) creates a source for the signal INT1. Even though INT1 is a resolved signal and can therefore have multiple sources, INT1 cannot be associated with the output **buffer** port of the A2 **and** gate in the second component instantiation because INT1 already has a source. The structural description is illegal, reflecting the constraint that the outputs of the two TTL, totem pole **and** gates cannot be tied together.

Figure 15.25

Incorrect use of buffer ports

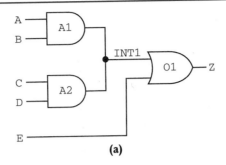

(a)

```
entity BAD_MODEL is
  port
    (A, B, C, D, E : in BIT;
     Z                : buffer BIT);   -- Buffer port
end BAD_MODEL;

use WORK.LOGIC_PKG.all;
architecture STRUCTURE of BAD_MODEL is
  -- TTL_TP denotes TTL, totem pole logic
  component AND2_TTL_TP
   port (A, B : in BIT; Z: buffer BIT);
  end component;
  component OR2_TTL_TP
   port (A, B : in BIT; Z: buffer BIT);
  end component;
  signal INT1 : WIRED_AND BIT;

begin
  A1: AND2_TTL_TP port map (A, B, INT1); -- Creates source for INT1
  A2: AND2_TTL_TP port map (C, D, INT1); -- Illegal!
  O1: OR2_TTL_TP  port map (INT1, E, Z);

end STRUCTURE;
```

(b)

The resolution function WIRED_AND is declared in the package LOGIC_PKG shown in Figure 15.20.

15.7 SIGNALS, VARIABLES, AND CONSTANTS

The above discussions on describing technology-dependent information in VHDL have introduced several important aspects of VHDL objects that bear reviewing in more detail before we move on to other topics. An object is something that can hold a value. Consistent with the strongly typed philosophy of VHDL, objects must be declared before they are used. An object declaration defines its name, class (kind), type (values it may hold), and possibly an initial value. VHDL supports the following three classes of objects.[5]

---

[5] VHDL-93: VHDL-93 provides a fourth class of objects: *files*. Files are discussed in detail in Chapter 18.

1. signals
2. variables
3. constants

Signals, variables, and constants have different properties, which will be described in the following sections.

15.7.1 Signals and Waveforms

Several examples of signal declarations having the following general form have already been presented.

```
signal name_list : resolution_function type := initial_value;
```

The `name_list` is composed of one or more signal names. A signal name is any legal identifier. If more than one signal name is given, the names are separated by commas. The `resolution_function` is optional and, if present, declares the name of the function to use to resolve the values of multiple sources into a single value. The `type` defines the set of values the declared object(s) may take on. The `initial_value` is optional and, if present, defines the *default* or *initial* value for the signal. Note that from a language perspective ports in the interface of a design entity are considered to be signals having the additional property of mode or direction.

A signal has several unique properties. Unlike variables and constants that represent a single present value, signals represent a time-ordered list of values denoting past, present, and future values. This time history of values is also called a *waveform*. A value/time pair is called a *waveform element* or a *transaction*. If a transaction changes the value of a signal, it is also called an *event*.

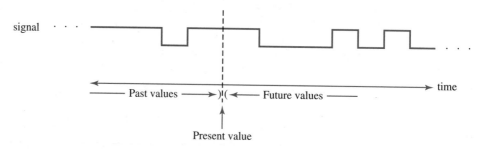

There are several rules concerning the proper use of signals. As with life, one cannot alter the past or see into the future. Past and present signal values can be accessed, but not changed. Future signal values can be changed, but cannot be accessed. When a value/time pair is assigned to a signal, we are projecting the signal's *future* values, in other words, values the signal will take on as time progresses. A signal assignment *never* affects the present value of a signal. There is always a propagation delay, even with delta delays.

The time component is always specified as a time relative to the present time, signified by the keyword **after**. VHDL provides the flexibility to specify more than one future value for a signal, defining several waveform elements comprising a waveform. For example, consider the selected concurrent signal assignment waveform shown in Figure 15.26.

The timing diagram illustrates the waveform, consisting of three waveform elements, assigned to X_SIG when the selected signal assignment statement executes with SYNC = '0'. Hence, in any of the forms of concurrent signal assignment statements studied thus far, the

Figure 15.26

Specifying a waveform

```
signal X_SIG, SYNC : BIT;
                          .
                          .
                          .
with SYNC select
X_SIG <= '1' after 1 ms, '0' after 5 ms, '1' after 10 ms when '0',
         '0' after 1 ms, '1' after 5 ms, '0' after 10 ms when '1';
```

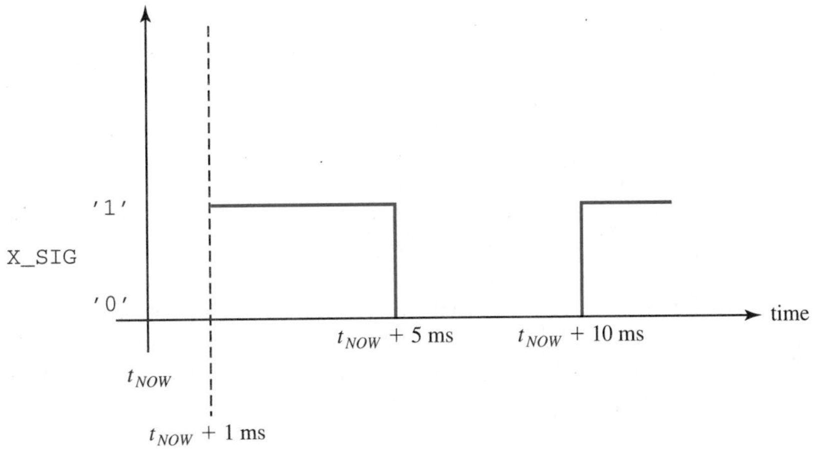

single value/time pair can be replaced by any number of value/time pairs comprising a waveform.

15.7.2 Variables

In the discussion of resolution functions and wired logic in Section 15.5, we introduced another class of objects: variables. The resolution function WIRED_AND in Figure 15.20 uses the variable RESULT to accumulate the logic **and** of multiple sources. The general format for a variable declaration is similar to a signal declaration.

```
variable name_list : type := initial_value;
```

Variables do not have resolution functions. Again, the initial_value is optional and, if present, defines the default or initial value for the variable.

Unlike signals, a variable does not represent a time history of values. Rather, a variable represents a single value, which is its present value. Also, the semantics of variable assignment differs from signal assignment, which is why two different assignment operator symbols are used: <= is used for signal assignment and := is used for variable assignment. A variable assignment always takes affect immediately, changing its present value to a new value.

It is important to understand where variables can and cannot be used in VHDL and the rationale for the restrictions. Generally, signals are viewed as being more representative of hardware than variables, evidenced by the extensive use of signals in design entity interfaces, concurrent statements, and the simulation cycle. Variables, on the other hand, are generally used in describing abstract computations, supporting the procedural or algorithmic aspects of high-level hardware modeling tasks. Thus, variable declarations are *not* allowed in design entities, architectural bodies, and packages. We will study the use of variables in more detail in future chapters on algorithmic modeling.

15.7.3 Constants

Constants are the simplest class of VHDL objects. The general format for a constant declaration is given below.

```
constant name_list : type := initial_value;
```

Constants do not have resolution functions and the `initial_value` is required. Once a constant is initialized, its value cannot be changed. We can only access the value of a constant; we cannot assign a new value to it.

The liberal use of constants has been traditionally advocated as a good software engineering practice to make code more readable; constants take the "funny numbers" out of programs. If a specific number, such as "5," is encountered in the middle of a complex program, it is difficult to understand what "5" represents. However, if the literal "5" is replaced by the constant `MAX_COUNT`,

```
constant MAX_COUNT : INTEGER := 5;
```

then the constant name is more descriptive of the literal's definition. Constants can also improve code maintainability. Changing the value of a literal that is used many times throughout a program can be tedious and error prone. If the literal is replaced by a constant, then only the constant declaration need be modified.

15.8 SUMMARY

In Chapter 6, we presented an overview of several popular ways of implementing combinational digital systems via logic families. We also examined several new physical properties of digital systems, derived from the limitations of electronic devices in serving as switches. In this chapter, we investigated how to describe these physical properties, also called *technology-dependent information*, in VHDL. It is important to be aware of technology-independent and -dependent information when developing VHDL models. Technology-dependent information should be made "programmable" and isolated from the technology-independent information. These practices will make VHDL models more portable across a variety of implementation technologies and will facilitate updating VHDL models in response to changes to implementation technology.

We introduced *physical types* to represent dimensional quantities or numbers having a value and a unit, such as time or voltage. We examined *predefined* and *user-defined* physical types. The predefined physical type TIME models time in VHDL. User-defined physical types can build additional data abstractions that uniquely support a particular modeling task or application.

Having learned how to describe values of time, we then examined how to describe *propagation delay*. The general subject of timing is important in VHDL and motivates many of the fundamental differences between hardware description languages and software programming languages.[6] VHDL provides two propagation delay models: *inertial* and *transport*. Inertial delay is the default propagation delay model, reflecting the lowpass frequency filtering property of most physical devices.

Section 15.5 and Section 15.6 discussed how to model logic families that do and do not support *wired logic*, respectively. Instead of predefining wired logic properties in the language or in a supporting simulator, wired logic properties in VHDL are *user-definable* via a resolution function because different logic families exhibit different wired logic properties. Buffer ports are used to describe logic families that do not support wired logic. A **buffer** port has the special connectivity/usage restriction that it must not participate in signal resolution.

Since the discussion of propagation delay introduced the new timing aspect of *signals* and the discussion of resolution functions introduced *variables*, the final section of this chapter presented a comparative overview of *signals*, *variables*, and *constants*. Signals, variables, and constants are all considered *objects* in VHDL that can hold values of a certain type. However, the three objects have different update/access properties and uses.

15.9 PROBLEMS

1. Convert the following physical literals of type TIME to seconds.
   a. 6#240# us _____ sec   **b.** 2#1010# min _____ sec
   c. 152_000 ms _____ sec   **d.** 5E+6 ns _____ sec
   e. 2.5E-3 hr _____ sec   **f.** 8#6.4# min _____ sec
   g. 8#6.4# fs _____ sec

2. Perform the following arithmetic operations involving physical literals of type TIME.

   a. $\left(\dfrac{30\ \text{ns}}{5\ \text{ns}}\right) \times (1.2\ \text{min} - \text{min})$

   b. $\left(\dfrac{5\ \text{ns}}{30\ \text{ns}}\right) \times (1.2\ \text{min} - \text{min})$

   c. 1 ns − 1 fs

3. Define the following physical types.
   a. Current
      • base unit - microamperes (uA)
      • derived units - milliamperes (mA) and amperes (A)
   b. Length
      • base unit - micron (u)
      • derived units - millimeter (mm), centimeter (cm), milli-inch (mil), and inch (in.)
   c. Temperature
      • base unit - Celsius (C)

4. Consider the following combinational logic.

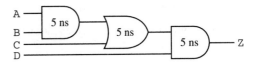

   a. Develop a VHDL model describing input/output function and timing. Assume inertial delays.
   b. Verify by simulation that the VHDL model describes the logic schematic for all possible inputs.

---

[6] Note that future research in hardware description languages and realtime programming languages may resolve these differences.

5. The following truth table defines the 9's complement operation for binary coded decimal (BCD) numbers.

Decimal value	N4 N3 N2 N1	C4 C3 C2 C1	Decimal value
0	0 0 0 0	1 0 0 1	9
1	0 0 0 1	1 0 0 0	8
2	0 0 1 0	0 1 1 1	7
3	0 0 1 1	0 1 1 0	6
4	0 1 0 0	0 1 0 1	5
5	0 1 0 1	0 1 0 0	4
6	0 1 1 0	0 0 1 1	3
7	0 1 1 1	0 0 1 0	2
8	1 0 0 0	0 0 0 1	1
9	1 0 0 1	0 0 0 0	0
	1 0 1 0	– – – –	
	1 0 1 1	– – – –	
	1 1 0 0	– – – –	
	1 1 0 1	– – – –	
	1 1 1 0	– – – –	
	1 1 1 1	– – – –	

a. Design a minimized combinational system. Show your work.
b. Describe input/output function and timing in VHDL. Assume inverters have a 5-nanosecond inertial delay and all other gates have a 7-nanosecond inertial delay.
c. Verify by simulation that the VHDL model describes the BCD 9's complement operation.

6. Consider the following combinational logic.

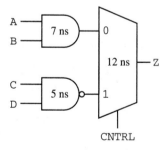

a. Develop a VHDL model describing input/output function and timing. Assume inertial delays.
b. Verify by simulation that the VHDL model describes the logic schematic for all possible inputs.

7. Consider the cascaded buffers shown below.

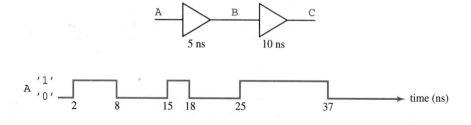

a. Generate a VHDL model describing input/output function and timing. Assume inertial delays.

b. Complete the timing diagrams for signals B and C.

8. Repeat Problem 7 using transport delays.

9. <u>VHDL-93:</u> Repeat Problem 7 using an inertial delay of 5 nanoseconds and propagation delay of 10 nanoseconds for the second buffer.

10. Describe the behavior of the cascaded buffers in Problem 7 on a simulation cycle basis. Identify each simulation cycle and the associated signal transactions.

11. Describe the behavior of the cascaded buffers in Problem 8 on a simulation cycle basis. Identify each simulation cycle and the associated signal transactions.

12. Describe in VHDL a **nor** gate having a programmable propagation delay sensitive to a rising (t_{PLH}) and falling (t_{PHL}) output transition.

13. Describe in VHDL an oscillator having a programmable period or clock cycle time.

14. Using the physical type for temperature declared in Problem 3(c),

a. Describe in VHDL a **nor** gate having a programmable propagation delay sensitive to the ambient temperature. Electronic circuits generally perform faster at lower ambient temperatures. Assume a baseline or nominal delay of 10 ns at a temperature of 20°C and a gradient of 5 ns per 10°C.

b. Simulate all four combinations of inputs of the **nor** gate at 10°C, 20°C, and 30°C.

15. An abstract operator is needed to detect illegal binary coded decimal (BCD) values, $10_{10}-15_{10}$.

a. Write a function that performs this abstraction operation. The input is a 4-bit vector and the output is a single bit, active-1. (*Hint*: In addition to the **for-loop** and **return** sequential statements, VHDL provides the conventional **if** statement.)

```
if (condition) then
  --Sequential statements
else
  -- Sequential statements
end if;
```

b. Place the function in a package and write a VHDL design entity that uses the function.

c. Simulate the VHDL design entity created in part (b) for input values 2_{10}, 9_{10}, 11_{10}, and 15_{10}.

16. Write a function that accepts a bit vector of arbitrary length and returns an integer that is the total number of 1's contained in the bit vector, and place the function in a package.

17. Like component instantiations, VHDL allows recursive function invocations, where a function can call itself.

a. Write a function that recursively computes the **and** of an arbitrary number of inputs, and place the function in a package.

b. Write a VHDL model that uses the recursive function, and create an instance of the model to realize a 4-input **and** gate.

c. Simulate the 4-input **and** gate using the inputs 0101_2 and 1111_2.

18. Use the following basic I^2L gate to do the following tasks.

```
entity IIL_GATE is
  port
    (A_IN : in BIT;
     Z1_OUT, Z2_OUT, Z3_OUT : out BIT);
end IIL_GATE;
```

```
architecture DATA_FLOW of IIL_GATE is
begin
  Z1_OUT <= not A_IN after 5 ns;
  Z2_OUT <= not A_IN after 5 ns;
  Z3_OUT <= not A_IN after 5 ns;
end DATA_FLOW;
```

a. Describe the logic given in Figure 6.16(a) in VHDL. Simulate the VHDL model using all possible combinations of inputs.

b. Describe the logic given in Figure 6.16(b). Simulate the VHDL model using all possible combinations of inputs.

19. List the properties of signals, variables, and constants and the rules concerning their proper use.

REFERENCES AND READINGS

[1] Harr, R., ed. *Applications of VHDL to Circuit Design.* Boston, MA: Kluwer Academic Publishers, 1991.

[2] Augustin, L. "Timing Models in VAL/VHDL," *Proceedings of the IEEE International Conference on Computer Aided Design*, pp. 122–125, November 1989.

[3] Hanson, D. "Structural Level Simulation of Analog/Microwave Circuits in VHDL," *Proceedings of the VHDL User's Group Conference*, pp. 135–141, Spring 1991.

[4] Leung, S. "Behavioral Modeling of Transmission Gates in VHDL," *Proceedings of the 26th ACM/IEEE Design Automation Conference*, pp. 746–749, June 1989.

[5] Luckham, D., A. Stanculescu, Y. Huh, and S. Ghosh. "The Semantics of Timing Constructs in Hardware Description Languages," *Proceedings of the IEEE International Conference on Computer Design*, pp. 10–14, October 1986.

[6] Sama, A. and J. Armstrong. "Behavioral Modeling of RF Systems with VHDL," *Proceedings of the VHDL User's Group Conference*, pp. 167–173, Spring 1991.

[7] Stanculescu, A. et al. "Switch-Level VHDL Descriptions," *Proceedings of the IEEE International Conference on Computer-Aided Design*, pp. 180–183, November 1989.

[8] Stanisic, B. and M. Brown. "VHDL Modeling for Analog-Digital Hardware Design," *Proceedings of the IEEE International Conference on Computer-Aided Design*, pp. 184–187, November 1989.

[9] Zhou, W. and H. Carter. "AnaVHDL: Mixed-Mode Simulation Using VHDL," *Proceedings of the VHDL User's Group Conference*, pp. 125–128, Spring, 1991.

16
VHDL TECHNOLOGY INFORMATION: PART 2

In this chapter, we will continue our discussion of describing technology-dependent information in VHDL by studying how to describe multi-valued logic and manufacturing-related information. Multi-valued logic will continue to build our understanding of *types* by introducing *enumeration* and *array* types. Manufacturing-related information will introduce *user-defined attributes* and *groups*.

16.1 MULTI-VALUED LOGIC

Multi-valued logic refers to a logic system that uses more than the conventional two states of "0" and "1." For example, tri-state logic (Chapter 6) can be viewed as supporting a *ternary* logic system composed of three states: "0," "1," and "Z" (high impedance). Compared to binary logic, multi-valued logic can describe a wider range of digital logic family behavior with greater accuracy.

There are many possible multi-valued logic systems, and each candidate logic system has its own advantages and disadvantages. Although different modeling requirements may warrant different multi-valued logic systems, multiple logic systems, unfortunately, tend to inhibit VHDL model interoperability, in other words, the ability to exchange and share VHDL models. VHDL models using different logic systems essentially speak different dialects of VHDL. To avoid the proliferation of incompatible multi-valued logic systems, a standard 9-valued logic system, given in Table 16.1, has been defined. This standard 9-valued logic system is referred to as *IEEE 1164-1993* [9].

To understand the standard 9-valued logic system listed in Table 16.1, it is convenient to view a signal value as having two possible properties: *state* and *strength*. A signal's state denotes its logic level, whereas a signal's strength denotes the quality of the source establishing the logic level, in other words, driving the signal. Stronger sources can override weaker sources. A signal connected directly to a power supply is an example of strong signal, whereas a signal connected via a resistance to a power supply is an example of a weak signal.

Table 16.1 lists the state and strength associated with each logic value. Some logic values have only an associated state, some logic values have only an associated strength, and some logic values have an associated state and strength. The 9-valued logic system is based on the following five states and three strengths.

Logic States
1. *0*
2. *1*
3. *Unknown*
4. *Uninitialized*
5. *Don't care*

Table 16.1

A 9-valued logic system

Value	State	Strength
U	Uninitialized	—
0	0	Forcing
1	1	Forcing
X	Unknown	Forcing
L	0	Resistive
H	1	Resistive
W	Unknown	Resistive
Z	—	High impedance
–	Don't care	—

Logic Strengths

1. *Forcing*—sometimes called *active, driving,* or *strong*
2. *Resistive*—sometimes called *passive* or *weak*
3. *High impedance*

Starting at the top of Table 16.1, the first logic value, U, represents an *uninitialized* value. The term *uninitialized* is somewhat of a misnomer because the U value is often used to initialize signal values at the beginning of simulation to model the "prepower-up" state of hardware. As simulation progresses and system inputs propagate through stages of logic to system outputs, uninitialized signal values are replaced by other "active" logic values. Signals that retain their initial U value at the end of simulation denote portions of logic that were not exercised. Without an uninitialized value, it would be difficult to tell whether a signal was driven to a certain value or just happened to retain its initial value. Thus, giving a signal a special "not-yet-active" value at the beginning of simulation serves as a helpful aide in understanding simulation results and tracking down possible design errors. An uninitialized signal value has no associated strength, or quality of source, because there is no source driving an uninitialized signal.

The values 0 and 1 respectively represent the logic states "0" and "1" with a forcing strength. These two values correspond to the conventional binary logic system. The values L and H also respectively represent the logic states "0" and "1," but with a resistive strength.

To understand forcing and resistive strengths, consider modeling two static NMOS inverters driving the same output node, as shown in Figure 16.1. If the top inverter (I1) is driven by a 0 and the bottom inverter (I2) is driven by a 1, what is the resolved value of the output node? Recall from Chapter 6 that the transistor resistances are set to enable the pull-down device (enhancement-mode NMOS transistor) to override the countering efforts of the pull-up device (depletion-mode NMOS transistor), thus forcing the output low when the input is high. In our 9-valued logic scheme, an NMOS inverter yields a weak 1 (H) output in response to a 0 or L input and a strong 0 (0) output in response to a 1 or H input. In Figure 16.1, the stronger output signal from inverter I2 (0) dominates the weaker output from inverter I1 (H); thus, the output node is correctly resolved to a strong 0 (0) value.

The values X and W represent *unknown* or *indeterminate* logic values with forcing and resistive strengths, respectively. In an unknown or indeterminate state, a signal voltage may be within a range denoting a logic 0, a range denoting a logic 1, or an intermediate range denoting neither a 0 or 1. The X value often denotes two gates driving the same output node with conflicting forcing values, creating the electrical equivalent of a short circuit between two power supplies. The W value often denotes two gates driving the same output node with conflicting resistive values, creating the electrical equivalent of a voltage divider between two power supplies. We generally do not like to see a lot of X and W values in model simula-

Figure 16.1

Modeling wired logic for static NMOS technology

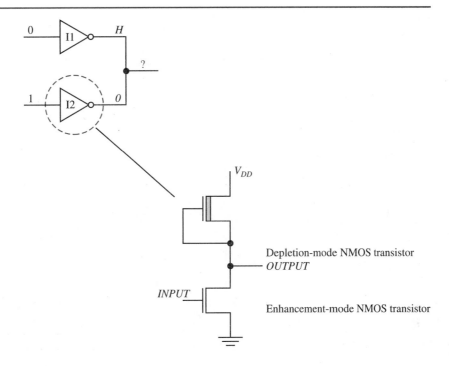

tions because they often denote connectivity "design bugs," in other words, cases where gates are connected in a manner that violates logic family electrical usage rules.

The Z value represents a high impedance or disconnected strength, denoting a source that is not driving its output. If a gate is not driving its output, the output state is not relevant; thus, a high impedance value does not have an associated logic state. To show the concept of a high impedance (Z) logic value, Figure 16.2 (p. 494) shows a tri-state CMOS inverter. When the control input signal ENABLE is deasserted, the two inside PMOS and NMOS transistors turn off and the inverter does not drive the output.

Finally, the value – represents a don't care condition, where we are not concerned about the input/output behavior. Maybe the output is not being monitored under certain input conditions or maybe certain input conditions are never expected to occur. (See Chapter 4 for detailed discussion of don't care values.) A – value is sometimes confused with X and W values because these values all denote some degree of uncertainty, but their roles are different. A don't care value ($-$) is a top-down specification concept, whereas an unknown value (X or W) is a bottom-up implementation concept.

For instance, consider the following truth table of a partial 2-to-4 decoder. Assume that the inputs ($I_1 I_0$) will never both be asserted active-1 at the same time, so the last row of the truth table yields don't care outputs.

I_1	I_0	O_3	O_2	O_1	O_0
0	0	0	0	0	1
0	1	0	0	1	0
1	0	0	1	0	0
1	1	–	–	–	–

Figure 16.2

A tri-state CMOS inverter

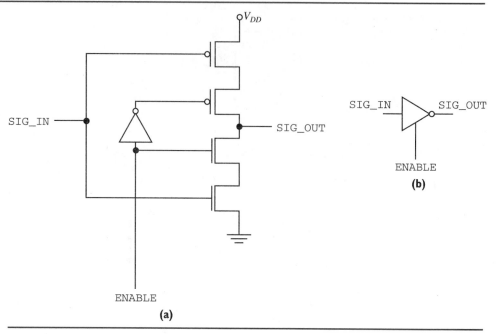

Figures (a) and (b) present the circuit schematic and logic symbol, respectively.

Recalling the earlier discussion of unknown values, it would not be appropriate to enter X's or W's for the last row of outputs. Unknown values are gate-related concepts and, at this point, we are not considering specific gates. Rather, we are considering specifying a logic behavior and thus the last row of outputs should be entered as don't care $(-)$ values.

The definition of the 9-valued logic system will become clearer as we present examples of VHDL models using the logic scheme. The following sections will explain the procedure for defining multi-valued logic, in general, and the standard 9-valued logic, in particular, in VHDL.

16.1.1 Definition

To understand how multi-valued logic can be modeled in VHDL, let us first examine how binary logic is modeled in VHDL. The declaration of the predefined type BIT provided in the package STANDARD is given below.

```
type BIT is (''0''0', '1''0');
```

Type BIT is called an *enumeration type* because the collection of values represented by BIT are simply enumerated or listed. The enumerated values for type BIT are the two character literals '0' and '1'. Enumeration types will be introduced in this section and discussed in more detail in Section 16.2.

Note that there is nothing magical or secret about the declaration of the predefined type BIT. Anyone could have written the above declaration of type BIT. However, the designers of VHDL judged the declaration of type BIT so fundamental to hardware modeling that they decided to include the enumeration type declaration in the package STANDARD and designate it as a predefined type. Recall from Chapter 12 that every design unit is assumed to have access to the information contained in STANDARD, which is, in turn, contained in the predefined library STD.

To define a logic system having more than two values, we start by declaring an appropriate *user-defined enumeration type*, delineating the symbols that will denote the individual logic values. For example, the following declares an enumeration type defining a tri-state or ternary logic system.

```
type TRI-STATE is ('0', '1', 'Z');
```

As another example, the enumeration type declaration for the standard 9-valued logic system defined in Table 16.1 is given below.

```
type STD_ULOGIC is ('U', 'X', '0', '1', 'Z', 'W', 'L', 'H', '-');
```

Character literals '0' and '1' are used to avoid confusion about whether 0 and 1 denote integers or logic values. Since character literals are used for '0' and '1', they tend to also be used for other logic values.

Instead of declaring signals to be of the predefined type BIT that can take on only the values '0' or '1', signals can be declared, for example, to be of the enumeration type STD_ULOGIC that can take on the nine values 'U', 'X', '0', '1', 'Z', 'W', 'L', 'H', and '-'. The "U" in STD_ULOGIC refers to "unresolved," meaning that a signal declaration using STD_ULOGIC needs to include a resolution function if the associated signals have multiple sources.

16.1.2 Logic Properties

Now that we have defined symbols to denote logic values, we next need to define the logic properties of these symbols. Looking, for example, at the standard 9-valued logic system, what does the conjunction of 'L' and 'Z' ($L \cdot Z$) mean? The predefined logic operators in VHDL accept operands of only the predefined types BIT and BOOLEAN. Thus, the standard logic operators in VHDL need to be redefined to accept operands of a user-defined, multi-valued logic type.

Figure 16.3 (p. 496) shows an example of how to redefine the logic **not** operator for the 9-valued logic system defined by STD_ULOGIC. Figure 16.3 contains several new VHDL constructs concerning arrays that will be introduced in this section and explained in detail in Section 16.3.

We will examine Figure 16.3 from the bottom to top, starting first with the function declaration. The double-quoted character string function name notation, "not", designates that the VHDL predefined logic operator **not** is being redefined. In other words, the predefined definition of **not** describing the complement operation for 2-valued logic is being overloaded

Figure 16.3

Defining complement operation for standard 9-valued logic

```
-- Constrained array type declaration
-- One-dimensional array - index and element values of type
-- STD_ULOGIC
   type STDLOGIC_1D is array (STD_ULOGIC) of STD_ULOGIC;

-- Constant declaration
-- Declare constant array and initialize with appropriate
-- complement values using an aggregate
   constant NOT_TABLE : STDLOGIC_1D :=
-- -------------------------------------------------
-- | 'U'   'X'   '0'   '1'   'Z'   'W'   'L'   'H'   '-'
-- -------------------------------------------------
   ('U', 'X', '1', '0', 'X', 'X', '1', '0', 'X');

-- Function declaration
-- Overload predefined logic not operator
 function "not" (ARG : STD_ULOGIC) return STD_ULOGIC is
 begin
    -- Compute complement of formal parameter ARG by table
    -- lookup
    return( NOT_TABLE(ARG) );
 end "not";
```

with an alternate definition describing the complement operation for 9-valued logic. The enclosing double quotes are used to differentiate the special case of *operator* overloading from the general case of *function* overloading. For example,

```
function "+" (L_ARG, R_ARG : STD_ULOGIC) . . . .
```

Problem-Solving Skill
Overloading is a convenient mechanism for defining different implementations of the same conceptual property or operation.

defines addition for 9-valued logic, creating an alternate or overloaded definition for the predefined addition for integer, floating point, and physical numbers.

In line with function overloading discussed in Chapter 15, operator overloading requires that a compiler can determine which operator definition applies to each operator reference. Moreover, only information about the number of operands and their type(s) may be used in selecting an operator definition. For example, the following code segment implies the use of the new, overloaded **not** definition given in Figure 16.3, because the operand is of the 9-valued logic type STD_ULOGIC.

```
signal MVL_SIG : STD_ULOGIC;
MVL_SIG <= not 'L';   -- Infix notation
```

Although overloaded operators are typically written like predefined operators using the *infix* notation shown above, they can also be written like conventional function calls using the following *prefix* notation.

```
signal MVL_SIG : STD_ULOGIC;
MVL_SIG <= "not" ('L');              -- Prefix notation
```

Within the body of the "not" function, the complement operation is implemented using a lookup table. The operand value, bound to the formal parameter ARG, serves as the index into the lookup table, which is the one-dimensional array NOT_TABLE. The content of the accessed array element NOT_TABLE(ARG) is returned as the complement of ARG, or $\overline{\text{ARG}}$.

NOT_TABLE is a constant of type STDLOGIC_1D, which is defined by the following *user-defined, constrained array type* declaration.

```
type STDLOGIC_1D is array (STD_ULOGIC) of STD_ULOGIC;
```

The notation (STD_ULOGIC) specifies that the array type is indexed by the nine values of the enumeration type STD_ULOGIC; hence, STDLOGIC_1D represents a one-dimensional array or vector of length nine. Notice that array indices do not have to be integers. The notation **of** STD_ULOGIC specifies that the nine individual elements of the array are of type STD_ULOGIC.

NOT_TABLE is declared as a constant because we only need to read from the lookup table; we never need to write to the lookup table. The contents of NOT_TABLE are initialized by an *array literal*, also called an *aggregate*. Aggregates will be explained in Section 16.3.3. For our present discussion, it is sufficient to know that the aggregate defines the simple data structure given in Table 16.2. The NOT_TABLE array encodes a truth table. The array indices (top row) denote the nine possible inputs; the array elements (bottom row) denote the associated nine outputs. The bottom row is the complement of the top row.

16.1.3 Tri-State Logic Example

Let us now examine a sample modeling application of a user-defined, multi-valued logic system in VHDL. The modeling application will use the 9-valued logic defined in the previous sections so we may gain experience using this important VHDL-related IEEE standard.

Figure 16.4 (p. 498) shows a VHDL model of the tri-state CMOS inverter shown in Figure 16.2. The package STD_LOGIC_1164 contains the complete definition of the standard 9-valued logic system and is part of the associated IEEE publication. The package STD_LOGIC_1164 defines the 9-valued logic symbols via enumeration type declarations, such as STD_ULOGIC. The package STD_LOGIC_1164 also defines the 9-valued logic properties via overloaded definitions of VHDL logic operators, such as **not**. IEEE Standard 1164 requires that STD_LOGIC_1164 be located in a library named IEEE. Thus, the INVERTER_TRI design entity accesses STD_LOGIC_1164 by declaring the IEEE library via

Table 16.2

	Array index	'U'	'X'	'0'	'1'	'Z'	'W'	'L'	'H'	'-'
Array NOT_TABLE for the 9-valued logic complement operation	Array element	'U'	'X'	'1'	'0'	'X'	'X'	'1'	'0'	'X'

Figure 16.4

VHDL model of a tri-state
CMOS inverter

```
library IEEE;
use IEEE.STD_LOGIC_1164.all;
entity INVERTER_TRI is
 port
   (SIG_IN, ENABLE : in  STD_ULOGIC;
    SIG_OUT        : out STD_ULOGIC);
end INVERTER_TRI;

architecture BEHAVIOR of INVERTER_TRI is
begin
  SIG_OUT <= 'Z' when (ENABLE = '0' or ENABLE = 'L') else
             not SIG_IN;
end BEHAVIOR;
```

the library clause and then using the selected name IEEE.STD_LOGIC_1164 in the use clause.

The architecture BEHAVIOR uses the 9-valued logic declarations provided by STD_LOGIC_1164 to describe the tri-state inverter. The single conditional concurrent signal assignment statement assigns a high impedance value, 'Z', to the output signal SIG_OUT if the control signal ENABLE is deasserted. Otherwise, the output signal SIG_OUT is assigned the complement of the input signal SIG_IN. The **not** operator refers to the overloaded 9-valued declaration shown in Figure 16.3 and given in the package STD_LOGIC_1164.

16.2 ENUMERATION TYPES

The previous discussion on defining multi-valued logic in VHDL introduces user-defined *enumeration types*, which we will now discuss in more detail. The general format for an enumeration type declaration is given below.

```
type name is ( list_of_values );
```

All type declarations start with the keyword **type**. The enumeration type name can be any legal identifier. The values of the enumeration type are listed within the parentheses, separated by commas. A value can be an *identifier* or a *character literal*. A character literal is any *graphic* character enclosed in single quotes.

For a particular enumeration type, the listed values, also called *enumeration literals*, must be unique, in other words, have no repeats. However, enumeration literals may be overloaded. That is, the same enumeration literal may be used in more than one enumeration type declaration, such as the following:

```
-- Character literals '0' and '1' are overloaded
type BIT is ('0', '1');
type TERNARY is ('0', '1', 'Z');
```

To comply with the strongly typed philosophy of VHDL, a compiler must be able to determine the type of an overloaded enumeration literal from the context in which the literal is used.

16.2.1 Predefined Enumeration Types

VHDL provides several predefined enumeration types. We have already seen the definition of type BIT.

```
type BIT is ('0', '1');
```

Type BOOLEAN is declared in a similar fashion.

```
type BOOLEAN is (FALSE, TRUE);
```

Type CHARACTER defines the 128-character set for VHDL taken from the International Standards Organization (ISO) 646 Standard.[1]

```
type CHARACTER is (NUL, . . .
                   '0', '1', . . .
                   'A', 'B', . . .
                   'a', 'b', . . .);
```

Finally, there is the predefined enumeration type SEVERITY_LEVEL, which will be used in Chapter 17 when assertions are discussed.[2]

```
type SEVERITY_LEVEL is (NOTE, WARNING, ERROR, FAILURE);
```

---

[1] VHDL-93: In VHDL-93, the predefined enumeration type CHARACTER defines a 256-character set for VHDL taken from the ISO 8859-1 Standard.

[2] VHDL-93: VHDL-93 defines the enumeration types

```
type FILE_OPEN_KIND is (READ_MODE, WRITE_MODE, APPEND_MODE);
type FILE_OPEN_STATUS is (OPEN_OK, STATUS_ERROR, NAME_ERROR,
     MODE_ERROR);
```

that will be used in Chapter 18 when files are discussed.

16.2.2 Ordering and Predefined Attributes

A *position number* is automatically assigned to each enumeration literal based on the *textual order* in which the literal is listed in the declaration. Moving from left to right, the first enumeration literal is assigned a position number of 0, the next literal is assigned a position number of 1, and so on.

The concept of a position number enables the predefined attributes discussed in Chapter 15 for physical types to apply to enumeration types. The predefined attributes POS and VAL convert between enumeration literals and the corresponding position numbers. For example, using the enumeration type STD_ULOGIC,

```
type STD_ULOGIC is ('U', 'X', '0', '1', 'Z', 'W', 'L', 'H', '-');

STD_ULOGIC'POS('U') = 0 -- Literal to position number
STD_ULOGIC'VAL(0) = 'U' -- Position number to literal
```

Position numbers are also used to define *predecessor* and *successor* values. Given an enumeration value and its associated position number P, its predecessor, given by the predefined attribute PRED, has position number $P - 1$ and its successor, given by the predefined attribute SUCC, has position number $P + 1$.

```
STD_ULOGIC'PRED('L') = 'W'
STD_ULOGIC'SUCC('L') = 'H'
```

The attributes LEFTOF and RIGHTOF also yield similar predecessor and successor values. STD_ULOGIC is considered to have an "ascending range," so LEFTOF equates to PRED and RIGHTOF equates to SUCC.

The attributes LEFT/RIGHT and LOW/HIGH may be used to query the range of enumeration values. Again, since STD_ULOGIC is considered to have an ascending range, STD_ULOGIC'LEFT = STD_ULOGIC'LOW = 'U' and STD_ULOGIC'RIGHT = STD_ULOGIC'HIGH = '-'.

Finally, position numbers form the basis for comparing objects of an enumeration type via relational operators (Section 13.3), in other words, "=," "/=," "<," and so on. For example, the textual ordering of the letters in the CHARACTER enumeration type declaration allows the following simple test for uppercase characters. Relations are determined using associated position numbers.

```
variable LETTER : CHARACTER;

if (LETTER >= 'A' and LETTER <= 'Z') then
  -- LETTER is uppercase
  . . . . .
```

16.3 ARRAYS

The previous discussion on modeling multi-valued logic also introduced user-defined *array types*. Arrays provide data structures for representing and naming a collection of data, where each element of the collection has the *same* properties. The following sections will discuss the declaration and use of *unconstrained* and *constrained* arrays. Also, we will examine *array literals* or *aggregates*.

16.3.1 Constrained Array Types

A *constrained array type* declaration defines the following information.

1. Array type name
2. Array type indexing scheme
   - Type of each array index
   - Range of each array index
3. Array element type

Defining the range of each index fixes or "constrains" the size of the array.

The general format for a constrained array type declaration is given below.

```
type name is array ( indexing_scheme ) of element_type;
```

For example, the following illustrates a constrained array type declaration representing a group of eight bits.

```
type BYTE_TYPE is array (7 downto 0) of BIT;
```

The constrained array type declaration begins with the keyword **type**, followed by the type name, BYTE_TYPE. An array type name can be any legal identifier.

Next, the array type *indexing scheme*, also called an *index constraint*, is given within parentheses. An indexing scheme defines a finite, ordered set of values, also called a *discrete range*, for each index. The index constraint (7 **downto** 0) defines a discrete range having the integer values 7, 6, . . . , 1, 0. Thus, BYTE_TYPE represents an array having eight elements, as shown in Table 16.3.

Following the indexing scheme, the definition of the element type completes the constrained array type declaration. The clause **of** BIT specifies that every array element is of the predefined enumeration type BIT.

The array type declaration given in Figure 16.3

```
type STDLOGIC_1D is array (STD_ULOGIC) of STD_ULOGIC;
```

shows another example of a constrained array type, illustrating the flexibility in specifying indexing schemes. Index constraints do not always have to be integer ranges.

Table 16.3

The form of the constrained array type BYTE_TYPE

Array index	7	6	5	4	3	2	1	0
Array element								

STDLOGIC_1D uses the enumeration type STD_ULOGIC to define an index constraint. The discrete range is the ordered set of the nine STD_ULOGIC enumeration literals. Thus, STDLOGIC_1D represents an array having nine elements, as shown in Table 16.4.

Multidimensional array types are declared by listing multiple discrete ranges; individual ranges are separated by commas. The following STDLOGIC_TABLE array type declaration represents a 9×9 array having elements of type STD_ULOGIC, as shown in Table 16.5.

```
-- 9x9 array or table - indices and element values of type -- -
-- STD_ULOGIC
  type STDLOGIC_TABLE is array (STD_ULOGIC, STD_ULOGIC) of
          STD_ULOGIC;
```

The STD_LOGIC_1164 package uses STDLOGIC_TABLE to declare two-dimensional lookup tables for defining binary logic operations, such as **and** and **nor**.

16.3.2 Unconstrained Array Types

An unconstrained array type declaration defines the array element type and the type of each index, but *not* the range of each index. In other words, legal values for the array elements and indices are specified, but the array size is not specified; in other words, it is left *unconstrained*. Array size definition is postponed until a later time, such as when an object is declared or when a function is invoked. Hence, an unconstrained array type represents a "family" of arrays having the same element type and number of dimensions, but different sizes.

The format for an unconstrained array type declaration is similar to the format of a constrained array type declaration, except an unconstrained array type declaration does not include discrete ranges. As an example, the declaration for the unconstrained array type STD_ULOGIC_VECTOR is given on the next page.

Table 16.4

The form of the constrained array type STDLOGIC_1D	**Array index**	'U'	'X'	'0'	'1'	'Z'	'W'	'L'	'H'	'-'
	Array element									

Table 16.5

The form of the constrained array type STDLOGIC_TABLE	Left index	Right index								
		'U'	'X'	'0'	'1'	'Z'	'W'	'L'	'H'	'-'
	'U'									
	'X'									
	'0'									
	'1'									
	'Z'									
	'W'									
	'L'									
	'H'									
	'-'									

```
type STD_ULOGIC_VECTOR is array (NATURAL range <>) of
    STD_ULOGIC;
```

STD_ULOGIC_VECTOR is the 9-valued logic equivalent of BIT_VECTOR, providing an easy way to declare a signal or port to be a group of 9-valued "bits." The indexing scheme given within the parentheses, (NATURAL range <>), specifies that the array type is indexed by natural numbers and that the length of the array or, equivalently, the allowed index range, is not defined. The type NATURAL (Chapter 14) is a predefined integer type representing the numbers—0, 1, 2, 3, ... The special symbol "<>" is called a *box*.

STD_ULOGIC_VECTOR can be easily generalized to two dimensions, as shown below.

```
type STD_ULOGIC_ARRAY is array (NATURAL range <>,
    of STD_ULOGIC;              NATURAL range <>)
```

Each index must be a natural number, but the specific range (bounds *and* direction) of each index is not defined.

For convenience, VHDL provides two predefined unconstrained array types: BIT_VECTOR and STRING.

```
type BIT_VECTOR is array (NATURAL range <>) of BIT;
type STRING is array (POSITIVE range <>) of CHARACTER;
```

BIT_VECTOR is a one-dimensional array of elements of type BIT, introduced in Chapter 13. STRING is a one-dimensional array of elements of type CHARACTER. Type POSITIVE is a predefined integer type, representing the numbers 1, 2, 3, ...

Eventually, unconstrained array types need to be constrained so objects of finite size can be declared. The size of an unconstrained array type is set by simply supplying the missing index constraint. The following object declarations for STORE_REGISTER and INTERCONNECT show examples of index constraints for the unconstrained array types STD_ULOGIC_VECTOR and STD_ULOGIC_ARRAY, respectively.

```
-- Constrain STD_ULOGIC_VECTOR to be 8 elements
variable STORE_REGISTER : STD_ULOGIC_VECTOR(0 to 7);

-- Constrain STD_ULOGIC_ARRAY to be 8 x 8
signal INTERCONNECT : STD_ULOGIC_ARRAY(7 downto 0, 7 downto 0);
```

The constraint limits must be of the designated index type, and the direction can be ascending *or* descending.

For the special case of a *constant* array object, an index constraint need not be explicitly defined; the index constraint is inferred from the initial value. As an example, the following ALARM_STATE constant declaration constrains the unconstrained array type BIT_VECTOR to be of length 4, based on the initial value specified by the *4-bit string literal* B"1001". Bit string literals (Chapter 13) are considered to have an ascending range, so the implied range constraint for ALARM_STATE is 0 to 3.

```
constant ALARM_STATE : BIT_VECTOR := B"1001";
```

As another example, the DSGNR_NAME constant declaration constrains the unconstrained array type STRING to be of length 10, based on the initial value specified by the *10-character string literal* "Mr. Wizard".

```
constant DSGNR_NAME : STRING := "Mr. Wizard";
```

Character string literals are formed by enclosing a list of characters within double quotes. The characters may be any graphic character. A double quote within a character string is denoted by two adjacent double quotes. A character string must fit on a single line; multiple character strings can be joined using the concatenation operator.

```
constant DSGNR_NAME : STRING := "The Cat" & " " & "In The Hat";
```

16.3.3 Specifying Values with Aggregates

There are three ways to refer to the contents of an array: an individual element, a set of contiguous elements, or all elements. An individual array element is referred to by an *indexed name*. For instance,

```
type STD_ULOGIC is ('U', 'X', '0', '1', 'Z', 'W', 'L', 'H', '-');
type STD_ULOGIC_VECTOR is array (NATURAL range <>) of STD_ULOGIC;

variable STORE_REGISTER : STD_ULOGIC_VECTOR(0 to 7);
```

STORE_REGISTER(4) is an indexed name referring to an individual element of the array STORE_REGISTER. Alternatively, a set of contiguous array elements is referred to by a *slice name*; STORE_REGISTER(4 to 7) is a slice name referring to the four elements having the contiguous indices 4, 5, 6, and 7. Finally, the set of all array elements is referred to by just the array name, such as STORE_REGISTER. A set of all element values of an array is called an *array value*.

For the two special cases of objects of the predefined array types BIT_VECTOR and STRING, VHDL respectively provides compact array value notations via bit string literals, such as B"1110_0011", and character string literals, such as "Hello World!".[3] For the general case of objects of any array type, array values are specified via *aggregates*. The following variable assignment statement uses an aggregate to provide a value for every element of the array STORE_REGISTER.

---

[3] <u>VHDL-93</u>: In VHDL-93, bit string literals (discussed in Section 16.4) are not restricted to just arrays of type BIT_VECTOR.

```
STORE_REGISTER :=
    ('0',                      -- STORE_REGISTER(0)
     '1',                      -- STORE_REGISTER(1)
     2 to 5 | 7 => '1',        -- STORE_REGISTER(2) to STORE_REGISTER(5)
                               -- and STORE_REGISTER(7)
     others => '0'             -- STORE_REGISTER(6)
    );
```

The contents of STORE_REGISTER are shown in Table 16.6.

An aggregate must define once and only once a value for every array element. Also, the type of the values must agree with the type of the array elements. Values may be associated with array elements using either positional or named association. However, positional associations must precede all named associations if the two association styles are mixed in a single aggregate. For instance, the first two elements of STORE_REGISTER are assigned values using positional association; the remaining array elements are assigned values using named association.

With positional association, values are associated with array elements by aligning left to right the textual ordering of values listed in the aggregate with the ordering of array elements. For instance, the first two values of the example aggregate ('0' and '1') are associated with the first two elements of the target array (STORE_REGISTER(0) and STORE_REGISTER(1)), respectively. With named association, values are associated with array elements by explicitly naming or identifying the element indices.

There are several ways to name element indices. They can be individually specified, a contiguous set of indices can be specified by an ascending or descending range, or some combination of these two forms can be used. A vertical bar, |, combines forms. For instance, the aggregate clause

```
2 to 5 | 7 => '1',
```

assigns the value '1' to the five array elements STORE_REGISTER(2) through STORE_REGISTER(5) and STORE_REGISTER(7). Indices can also be named using the keyword **others**, which denotes all remaining indices not specified by preceding aggregate clauses. The final clause for the STORE_REGISTER array aggregate

```
others => '0'
```

assigns the value '0' to all remaining array elements that have not been assigned a value, namely, STORE_REGISTER(6). If the keyword **others** is used, it must be the last aggregate clause.

Table 16.6

Contents of STORE_REGISTER	Array index	0	1	2	3	4	5	6	7
	Array element	'0'	'1'	'1'	'1'	'1'	'1'	'0'	'1'

Aggregates can be embedded or nested to construct values for multidimensional arrays. The following constant declaration uses embedded aggregates to initialize the contents of the 4×4 DELAY_TABLE array.

```
type TIME_TABLE is array (1 to 4, 4 downto 1) of TIME;

variable DELAY_TABLE : TIME_TABLE := ( (others => 1 ms),
                                       (others => 2 ms),
                                       (others => 3 ms),
                                       (others => 4 ms)
                                     );
```

The aggregate for the two-dimensional array DELAY_TABLE can be viewed as being composed of four one-dimensional array aggregates. Each one-dimensional array aggregate, or list, corresponds to a row in the array. The contents of DELAY_TABLE are shown in Table 16.7.

16.3.4 Predefined Attributes

The following predefined attributes introduced in Chapter 13 for bit vectors also apply to arrays, in general.[4]

- LEFT
- RIGHT
- HIGH
- LOW
- RANGE
- REVERSE_RANGE
- LENGTH

The predefined attributes provide information about array indexing. To query information about multidimensional arrays, an argument is added to the predefined attributes to indicate the particular dimension.

Table 16.7

Contents of DELAY_TABLE		Right index			
	Left index	4	3	2	1
	1	1 ms	1 ms	1 ms	1 ms
	2	2 ms	2 ms	2 ms	2 ms
	3	3 ms	3 ms	3 ms	3 ms
	4	4 ms	4 ms	4 ms	4 ms

[4] VHDL-93: VHDL-93 provides the predefined attribute ASCENDING for constrained arrays.

Attribute	Definition
name'ASCENDING	TRUE if one-dimensional array has an ascending index scheme, otherwise FALSE.
name'ASCENDING(N)	TRUE if the Nth dimension of array has an ascending index scheme, otherwise FALSE.

As an example, the values of the predefined array attributes for the 4×4 array variable DELAY_TABLE declared in the previous section are given below.

1. First Dimension
   - DELAY_TABLE'LEFT(1) = DELAY_TABLE'LEFT = 1
   - DELAY_TABLE'RIGHT(1) = 4
   - DELAY_TABLE'HIGH(1) = 4
   - DELAY_TABLE'LOW(1) = 1
   - DELAY_TABLE'RANGE(1) = 1 **to** 4
   - DELAY_TABLE'REVERSE_RANGE(1) = 4 **downto** 1
   - DELAY_TABLE'LENGTH(1) = 4
2. Second Dimension
   - DELAY_TABLE'LEFT(2) = 4
   - DELAY_TABLE'RIGHT(2) = 1
   - DELAY_TABLE'HIGH(2) = 4
   - DELAY_TABLE'LOW(2) = 1
   - DELAY_TABLE'RANGE(2) = 4 **downto** 1
   - DELAY_TABLE'REVERSE_RANGE(2) = 1 **to** 4
   - DELAY_TABLE'LENGTH(2) = 4

DELAY_TABLE'LEFT shows that if the argument is omitted, a default of 1, or the first dimension, is assumed.

16.3.5 Array Operations

Several operations can be performed on entire arrays. Two arrays can be compared using the relational equality (=) and inequality (/=) operators. For two arrays to be equal, all corresponding element values must be equal.

Relational ordering operations ($<, <=, >, >=$) can also be used to compare arrays, but the operands must be one-dimensional arrays. Starting with the leftmost elements, corresponding elements are compared until a difference is found or all corresponding elements have been checked. In the former case, the element difference determines the array relational result. In the latter case, the arrays are equal if they are of equal length, otherwise the shorter array is less than the longer array.

Logic operators are also defined for one-dimensional array operands. The array operands must be of equal length and have elements of either type BIT or BOOLEAN. The result is a one-dimensional array of the same type as the operands. The value of each element of the result is obtained by performing the logic operation on corresponding elements of the array operands, starting with the leftmost elements. Finally, the concatenation operator can be used to construct a one-dimensional array from smaller one-dimensional arrays and/or aggregates.

16.4 VHDL-93: BIT STRING LITERALS

In VHDL-87, bit string literals, such as B"1010" or X"AE", specify values of only type BIT_VECTOR. Thus, user-defined logic systems, such as IEEE 1164, cannot exploit the shorthand notational convenience of bit string literals to specify binary, octal, and hexadecimal values and must instead use character string literals.

```
signal TEST : STD_ULOGIC_VECTOR(7 downto 0);


TEST <= X"F4";        -- Illegal use of bit string literal
TEST <= "11110100";   -- Legal use of character string literal
```

VHDL-93 enhances the utility of bit string literals by allowing them to be used for *any* array having the character literals `'0'` and `'1'` as legal element values. Thus, bit string literals can be used for the 9-valued logic system IEEE 1164 because the array element type, `STD_ULOGIC`, includes the character literals `'0'` and `'1'`.

VHDL-93 bit string literals still allow only bit values of `'0'` and `'1'`; other bit values, such as `'W'` or `'Z'` cannot be used. To specify bit values other than `'0'` and `'1'`, we must use a character string literal or some concatenation of bit string literals and character string literals.

16.5 USER-DEFINED ATTRIBUTES

We have been careful to use the term *predefined* attributes because VHDL also provides *user-defined* attributes. The motivation for predefined and user-defined attributes mirrors the motivation for predefined and user-defined data types. User-defined data types augment predefined data types, providing users a structured mechanism for defining custom data abstractions. Similarly, user-defined attributes augment predefined attributes, providing users a structured mechanism for defining custom model information. Custom model information typically involves design information not directly associated with input/output behavior/structure, such as administrative data, design automation tool data, or manufacturing data.

User-defined attributes are defined by the following pair of statements:
- an *attribute declaration* and
- an *attribute specification*.

These two statements collectively create an attribute, associate the attribute with some part of a VHDL model, and give the attribute a value. Perhaps the easiest way to introduce user-defined attributes is by a simple example. Figure 16.5 shows how a user-defined attribute can document that Mr. Wizard designed the `INVERTER_TRI` VHDL model given in Figure 16.4.

> **Problem-Solving Technique**
>
> *User-defined attributes illustrate another application of the general design principle of providing a core set of capabilities that can be used to develop personalized solutions.*

Figure 16.5

Using a user-defined attribute to document the designer's name

```vhdl
library IEEE;
use IEEE.STD_LOGIC_1164.all;
entity INVERTER_TRI is
 port
   (SIG_IN, ENABLE : in  STD_ULOGIC;
    SIG_OUT        : out STD_ULOGIC);
end INVERTER_TRI;

architecture BEHAVIOR of INVERTER_TRI is
  -- Attribute declaration
  attribute DSGNR_NAME : STRING;

  -- Attribute specification
  attribute DSGNR_NAME of BEHAVIOR : architecture is
          "Mr. Wizard";

begin
  SIG_OUT <= 'Z' when (ENABLE = '0' or ENABLE = 'L') else
          not SIG_IN;
end BEHAVIOR;
```

The attribute declaration defines attribute DSGNR_NAME to be of type STRING, in other words, a character string. The attribute specification associates the attribute DSGNR_NAME with the architectural body BEHAVIOR of the entity INVERTER_TRI and gives the attribute the value "Mr. Wizard". The following sections will discuss user-defined attributes in more detail.

16.5.1 Attribute Declaration

An *attribute declaration* declares the data type of a user-defined attribute, in other words, the legal values the attribute may hold. An attribute declaration begins with the keyword **attribute**, followed by the attribute name. An attribute name can be any legal identifier. Attributes of the same type may be individually defined in separate declarations,

```
attribute DSGNR_NAME : STRING;
attribute DSGNR_DEPT : STRING;
attribute DSGNR_MNGR : STRING;
```

or collectively defined in a single declaration.

```
attribute DSGNR_NAME, DSNGR_DEPT, DSGNR_MNGR : STRING;
```

16.5.2 Attribute Specification

Once an attribute is declared, an attribute specification creates an instance of the attribute, associates the instance with some portion of a VHDL model, and gives the instance a value. The general format of an attribute specification is given below.

```
attribute name of item_specification is expression;
```

Like an attribute declaration, an attribute specification begins with the keyword **attribute**, followed by the attribute name. Then, the item_specification identifies the VHDL declared item(s) that will be associated with or "inherit" the attribute. Finally, the expression defines the attribute value. The expression is evaluated once before simulation during model elaboration and must yield a value of the attribute type.

There are many places to "hook" a user-defined attribute onto a VHDL model. A place is any implicitly or explicitly declared item, identified by its name and kind. The kind of a declared item is more formally called an *entity class*. Here, the term *entity* refers to a general concept of a declared item, not the specific concept of an interface and associated architecture. Admittedly, using the same term to denote two different meanings is somewhat confusing, but this is how VHDL is defined in the language reference manual. A listing of entity classes is given below.[5]

- An entity (**entity**)
- An architecture (**architecture**)
- A configuration (**configuration**)
- A package (**package**)

---

[5] Additional entity classes provided in VHDL-93 are discussed in Section 16.5.4.

- A subprogram (**function** or **procedure**)
- A type (**type** or **subtype**)
- An object (**signal**, **variable**, or **constant**)
- A component (**component**)
- A label (**label**)

We have already introduced most entity classes; the entity classes **configuration**, **procedure**, and **subtype** will be discussed in later chapters.

The attribute specification

```
attribute DSGNR_NAME of BEHAVIOR : architecture is "Mr. Wizard";
```

given in Figure 16.5 uses the entity class **architecture** to associate an instance of the attribute DSGNR_NAME having the value "Mr. Wizard" with the architecture BEHAVIOR.

Multiple declared items of the same entity class can be identified by listing the names, separated by commas. The following attribute specification associates the attribute WIRE_CAP with the signals ALARM and TEST, which is a convenient way to describe wire capacitance.

```
-- User-defined physical type declaration
type CAPACITANCE is range 0 to INTEGER'HIGH
  units
    fF;                    -- Femtofarads
    pF = 1000 pF;          -- Picofarads
    nF = 1000 pF;          -- Nanofarads
  end units;


-- Signal declaration
signal ALARM, TEST : BIT;


-- Attribute declaration
attribute WIRE_CAP : CAPACITANCE;


-- Attribute specification
attribute WIRE_CAP of ALARM, TEST : signal is 8 pF;
```

The item specification, ALARM, TEST : **signal**, associates an instance of the attribute WIRE_CAP with each of the signals ALARM and TEST.

Multiple declared items of the same entity class can also be identified by using the keywords **others** or **all**. The keyword **all** is like a "wildcard," identifying all declared items of the specified entity class declared within the *immediately* enclosing declarative part or scope. The keyword **others** identifies declared items of the specified entity class declared within the immediately enclosing declarative part that have not been identified by preceding attribute specifications. Attribute specifications using the keywords **others** and **all** serve to "pick up" all remaining instances of an entity class within a declarative part. Thus, an attribute specification using the keywords **all** and **others** for a particular entity class can be followed by neither (1) additional attribute specifications for the entity class nor (2) additional declarations of instances of the entity class.

The following example attempts to attach to signals an attribute that describes the preferred fabrication material and wiring plane or layer: polysilicon layer 1 (POLY1), polysilicon layer 2 (POLY2), metal layer 1 (METAL1), or metal layer 2 (METAL2). It may be necessary to define preferred wire fabrication options for performance critical signals because different materials and wiring layers have different physical and geometric properties. Such preference information could be used to guide physical design automation tools. (See Chapter 7 for an overview of integrated circuit fabrication.) The second attribute specification given below is illegal because it attempts to redefine the attribute WIRE_KIND for the signal ALARM, which is already defined by the first attribute specification using the keyword **all**. The last signal declaration for SENSOR is also illegal because it attempts to declare a signal after the first attribute specification has already accounted for all signals.

```
-- User-defined enumeration type declaration
type MATERIAL is (POLY1, POLY2, METAL1, METAL2);

-- Signal declaration
signal ALARM, TEST : BIT;

-- Attribute declaration
attribute WIRE_KIND : MATERIAL;

-- Attribute specifications
attribute WIRE_KIND of all : signal is METAL1;

attribute WIRE_KIND of ALARM : signal is POLY2; -- Illegal

-- Signal declaration
signal SENSOR : BIT;  -- Illegal
```

The following section presents a user-defined attribute example using the **others** notation.

Attribute specifications for subprograms, objects, components, and types must follow the declaration of these entity classes and appear within their immediately enclosing declarative region. Attribute specifications for entities, architectures, configurations, and packages must appear immediately within the design units' associated declarative parts. For example, the attribute specification for DSGNR_NAME for the architecture BEHAVIOR must appear immediately within the architecture's associated declarative part.

An attribute is considered a constant. Thus, an instance of an attribute can be assigned an initial value only once, and the initial value may not be changed thereafter. Notice that the attribute specification for DSGNR_NAME takes advantage of the fact that attributes are constants to infer from the initial character string literal "Mr. Wizard" the constraint on the unconstrained array STRING. Similar to predefined attributes, the value of a user-defined attribute may be queried using the "tick" notation. For instance, the value of the attribute WIRE_CAP associated with the signal ALARM may be queried by ALARM'WIRE_CAP.

16.5.3 CAD Tool Information Example

To illustrate the utility of user-defined attributes, let us reconsider the iterative adder discussed in Chapter 5. Recall that an iterative adder generates the sum of two multi-bit binary numbers column by column by connecting in a cascade or series structure several full adders. Each full adder accepts the corresponding column bit of each operand and a carry bit from the previous, lower-order full adder and generates a column sum bit and a carry bit for the next, higher-order full adder. The carry path from the least significant stage to the most significant stage is called the *critical path* because the time required for the carries to propagate through all full adder stages, or columns, determines the overall speed of the iterative adder. If speed is an important design objective, we may want to include critical path information as part of the VHDL model documentation. Such information can then be used to guide and/or adjust the manufacturing process.

Figure 16.6 shows an example of how user-defined attributes can describe the carry propagation critical path of an individual stage of an iterative adder, a full adder. The package LOGIC_PKG, containing the required component declarations for AND2_GATE, XOR2_GATE, and OR2_GATE and the attribute declaration for PD_PRIORITY, is given in Figure 16.7. Moving these declarations into a package to minimize the lines of code in the design entity FULL_ADDER model and to easily reuse the declarations for other design entities again illustrates the utility of packages.

The package LOGIC_PKG also contains the enumeration type declaration of CRITICAL_LEVELS used in the subsequent declaration of the attribute PD_PRIORITY. The component declarations given in the package LOGIC_PKG and the component instantiations given in the architecture STRUCTURE collectively define the logic schematic of the full adder shown in Figure 16.6(a).

Figure 16.6

User-defined attributes

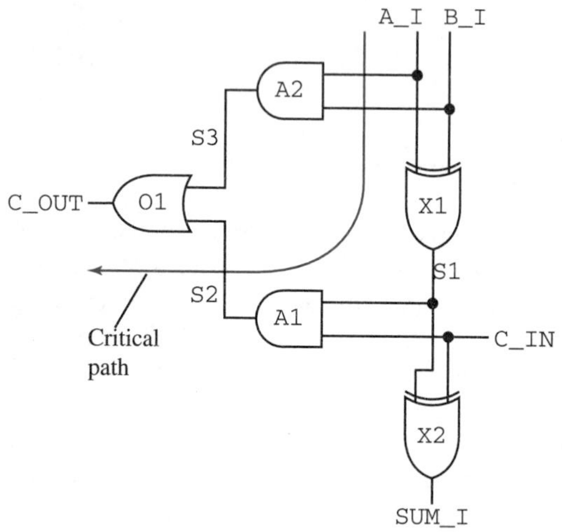

(a)

```
entity FULL_ADDER is
 port
   (A_I, B_I, C_IN : in    BIT;
    SUM_I, C_OUT   : out   BIT);
end FULL_ADDER;

use WORK.LOGIC_PKG.all;
architecture STRUCTURE of FULL_ADDER is
  signal S1, S2, S3 : BIT;

  -- Attribute specifications
  attribute PD_PRIORITY of X1, A1, 01 : label is HIGH;
  attribute PD_PRIORITY of others : label is LOW;

begin
  -- Structural description
  X1: XOR2_GATE port map (A_I, B_I, S1);
  X2: XOR2_GATE port map (S1, C_IN, SUM_I);
  A1: AND2_GATE port map (C_IN, S1, S2);
  A2: AND2_GATE port map (A_I, B_I, S3);
  01: OR2_GATE  port map (S2, S3, C_OUT);

end STRUCTURE;
```

(b)

Figure (a) gives a schematic diagram of a full adder. Figure (b) presents a VHDL structural model, illustrating user-defined attributes. The package LOGIC_PKG is given in Figure 16.7.

Figure 16.7

Enclosing component and attribute declarations in a package

```
package LOGIC_PKG is
 -- Component declarations
 component AND2_GATE
  port (A, B : in BIT; Z : out BIT);
 end component;
 component XOR2_GATE
  port (A, B : in BIT; Z : out BIT);
 end component;
 component OR2_GATE
  port (A, B : in BIT; Z : out BIT);
 end component;

 -- Enumeration type declaration
 type CRITICAL_LEVELS is (NONE, LOW, NORM, HIGH);

 -- Attribute declaration
 attribute PD_PRIORITY : CRITICAL_LEVELS;
end LOGIC_PKG;
```

The attribute PD_PRIORITY conveys physical design placement and routing priority information for each instantiated component, in other words, gate. Instances of the attribute PD_PRIORITY are associated with component instances and given appropriate values via the attribute specifications in the architecture STRUCTURE. The first attribute specification

```
attribute PD_PRIORITY of X1, A1, O1 : label is HIGH;
```

identifies the full adder carry propagation critical path by associating an instance of the attribute PD_PRIORITY with each gate in the critical path (X1, A1, and O1) and giving these instances the value HIGH. Physical design CAD tools can use this information to give these gates special attention and priority. The second attribute specification

```
attribute PD_PRIORITY of others : label is LOW;
```

associates instances of the attribute PD_PRIORITY with all component instantiations not already explicitly named (X2 and A2) and gives these instances the value LOW.

16.5.4 VHDL-93: Enhanced User-Defined Attributes

VHDL-93 provides two enhancements to user-defined attributes. First, VHDL-93 provides four additional entity classes: **literal**, **units**, **group**, and **file**. The **literal** entity class allows attributes to be associated with enumeration literals.

```
type STD_ULOGIC is ('U', 'X', '0', '1', 'Z', 'W', 'L', 'H', '-');

attribute FORCING : BOOLEAN;
attribute FORCING of '0', '1', 'X : literal is TRUE;
```

The **units** entity class allows attributes to be associated with base and secondary units of physical types.

```
type CAPACITANCE is range 0 to INTEGER'HIGH
  units
    fF;                          -- Femtofarads
    pF = 1000 fF;                -- Picofarads
    nF = 1000 pF;                -- Nanofarads
  end units;

attribute DEF : STRING;
attribute DEF of fF : units is "femtofarads";
```

The **group** and **file** entity classes allow attributes to be respectively associated with *groups* and *files*. Groups will be discussed in Section 16.6; files will be discussed in Chapter 18.

Another enhancement VHDL-93 provides for user-defined attributes is the ability to specify *signatures* to disambiguate overloaded subprograms and enumeration literals. In VHDL-87, if an attribute specification associates an attribute with a function that has overloaded, or multiple, declarations, then *all* overloaded function declarations receive the attribute. There is no way to associate an attribute with only one member of a family of overloaded functions. VHDL-93 addresses the problem of disambiguating overloaded names by providing signatures.

A signature identifies a particular member of a family of overloaded subprograms or literals by specifying the member's *profile*. For subprograms, the profile lists the formal parameter types and return type. The following example associates an instance of the attribute DSGNR_NAME with *only* the second MEAN function declaration applying to real numbers.

```
function MEAN (A_ARG, B_ARG : INTEGER) return INTEGER;
function MEAN (A_ARG, B_ARG : REAL) return REAL;

attribute DSGNR_NAME of MEAN[REAL, REAL return REAL] : function
          is "Tigger";
```

The signature

```
[REAL, REAL return REAL]
```

identifies the second declaration of the overloaded function MEAN by listing in square brackets the type of each formal parameter, followed by the keyword **return**, followed by the type of the returned value.

For literals, the signature profile lists the keyword **return**, followed by the appropriate enumeration type. For instance, the following example associates an instance of the attribute FORCING with *only* the enumeration literal 'Z' declared in the first TERNARY enumeration type declaration.

```
type TERNARY is ('0', '1', 'Z');
type STD_ULOGIC is ('U', 'X', '0', '1', 'Z', 'W', 'L', 'H', '-');

attribute FORCING : BOOLEAN;
attribute FORCING of 'Z'[return TERNARY] : literal is FALSE;
```

The signature

```
[return TERNARY]
```

identifies a particular enumeration literal by viewing the overloaded enumeration literal 'Z' declarations as the following overloaded function declarations.[6]

```
function 'Z' () return TERNARY;
function 'Z' () return STD_ULOGIC;
```

16.6 VHDL-93: GROUPS

VHDL-87 does not provide explicit mechanisms to rigorously define relationships between declared items. For instance, the connectivity relationship of two signals forming segments of a net can be inferred by analyzing a structural description, but cannot be explicitly declared. Other kinds of relationships among declared items may also be useful, such as variables that share the same physical register or entities that share the same physical integrated circuit. To address the limitation of expressing declared item relationships, VHDL-93 provides *groups*.

A group is a collection or set of declared items. A group is defined by a two-step process. First, the allowed membership of a group is defined via a *group template declaration*. Second, a specific group that complies with the allowed membership is defined via a *group declaration*. As an example, Figure 16.8 (p. 516) shows how the critical path of a full adder can be specified using groups, presenting an alternative to the user-defined attribute example shown in Figure 16.6.

The group template declaration

```
group PATH3_GRP is (label, label, label);
```

declares a particular group structure that defines group membership. The group template PATH3_GRP represents groups containing three labels. The group declaration

```
group ADDER_CRITICAL_PATH : PATH3_GRP(X1, A1, O1);
```

declares a group having the structure or template PATH3_GRP. The group ADDER_CRITICAL_PATH contains the three labels X1, A1, and O1.

It is instructive to emphasize the difference between Figure 16.6 and Figure 16.8. Relationships between declared items can be defined via user-defined attributes, but such relationships are often implicit. Associating instances of the same attribute with several declared items implies a relationship. Alternatively, relationships between declared items can be defined via groups, in which case the relationships are explicit. Defining qualifications for group membership and associated members clearly identifies related items.

---

[6] A little hoky, but it works.

Figure 16.8

Groups

```
entity FULL_ADDER is
 port
   (A_I, B_I, C_IN : in   BIT;
    SUM_I, C_OUT   : out  BIT);
end FULL_ADDER;

use WORK.LOGIC_PKG.all;
architecture STRUCTURE of FULL_ADDER is
 signal S1, S2, S3 : BIT;

   -- Group template declaration
   group PATH3_GRP is (label, label, label);

   -- Group declaration
   group ADDER_CRITICAL_PATH : PATH3_GRP(X1, A1, O1);

begin
 -- Structural description
 X1: XOR2_GATE port map (A_I, B_I, S1);
 X2: XOR2_GATE port map (S1, C_IN, SUM_I);
 A1: AND2_GATE port map (C_IN, S1, S2);
 A2: AND2_GATE port map (A_I, B_I, S3);
 O1: OR2_GATE  port map (S2, S3, C_OUT);

 end STRUCTURE;
```

Component declarations are provided by the package LOGIC_PKG given in Figure 16.7.

Any entity class, discussed in the previous section, can be a member of a group. Note that **group** is an entity class in VHDL-93, so groups can be members of groups, allowing group hierarchies. A group may contain any number of entity classes, defined in any order. A group may even leave the number of entity classes undefined by using the box (<>) notation. An entity class using the box notation must be the last member listed in the group template declaration. Figure 16.9 generalizes the group example shown Figure 16.8 by using the box notation in the group template declaration and placing the group template declaration in the package LOGIC_PKG, which is given in Figure 16.10.

16.7 SUMMARY

In this chapter, we concluded the discussion we started in Chapter 15 of describing technology-dependent data in VHDL. We first described modeling multi-valued logic. To accurately describe certain aspects of the behavior of logic families, such as ratioed logic or tri-state logic, it is convenient to develop a multi-valued logic system having more than just the traditional binary values of 0 and 1. Most multi-valued logic systems are based on a *state/strength* scheme; state denotes logic level and strength denotes the quality or source of the logic level. A multi-valued logic system is modeled in VHDL by first defining the logic value symbols via an enumeration type and then defining the associated logic operations via overloaded functions that redefine VHDL's predefined logic operators.

Figure 16.9

Generalized groups

```
entity FULL_ADDER is
 port
   (A_I, B_I, C_IN : in   BIT;
    SUM_I, C_OUT   : out  BIT);
end FULL_ADDER;

use WORK.LOGIC_PKG.all;
architecture STRUCTURE of FULL_ADDER is
 signal S1, S2, S3 : BIT;

 -- Group declaration
 group ADDER_CRITICAL_PATH : PATH_GRP(X1, A1, O1);

begin
 -- Structural description
 X1: XOR2_GATE port map (A_I, B_I, S1);
 X2: XOR2_GATE port map (S1, C_IN, SUM_I);
 A1: AND2_GATE port map (C_IN, S1, S2);
 A2: AND2_GATE port map (A_I, B_I, S3);
 O1: OR2_GATE  port map (S2, S3, C_OUT);

end STRUCTURE;
```

The component and group template declarations are provided by the package LOGIC_PKG given in Figure 16.10.

Figure 16.10

Enclosing attribute declarations in a package

```
package LOGIC_PKG is
 -- Component declarations
 component AND2_GATE
  port (A, B : in BIT; Z : out BIT);
 end component;
 component XOR2_GATE
  port (A, B : in BIT; Z : out BIT);
 end component;
 component OR2_GATE
  port (A, B : in BIT; Z : out BIT);
 end component;

 -- Group template declaration
 group PATH_GRP is (label <>);

end LOGIC_PKG;
```

In modeling multi-valued logic we introduced two new user-defined types: *enumeration* and *array* types. An enumeration type is classified as a *scalar* type because an object of an enumeration type holds only a single value. An array type is classified as a *composite* type because an object of an array type holds more than one value. Since we are gradually introducing VHDL typing as we learn digital system modeling concepts, it is useful to maintain a running log of the VHDL types we have discussed to date. The following list of VHDL types will be expanded and updated as we learn more about VHDL.

- *Scalar*
  - Enumeration
    - *Predefined:* BIT, BOOLEAN, CHARACTER, SEVERITY_LEVEL, FILE_OPEN_KIND*, *and* FILE_OPEN_STATUS*
    - *User-Defined: Section 16.2*
  - *Numeric*
    - Integer
      - *Predefined:* INTEGER, NATURAL, *and* POSITIVE
    - Floating Point
      - *Predefined:* REAL
    - Physical
      - *Predefined:* TIME *and* DELAY_LENGTH*
      - *User-Defined: Section 15.1.2*
- *Composite*
  - Array
    - *Constrained*
      - *Predefined: none*
      - *User-Defined: Section 16.3*
    - *Unconstrained*
      - *Predefined:* BIT_VECTOR *and* STRING
      - *User-Defined: Section 16.3*

* VHDL-93

In Section 16.5 we discussed *user-defined attributes*. It is important to realize the general utility of user-defined attributes. Although user-defined attributes were introduced to describe technology information, just about *anything* can be described via user-defined attributes. A user-defined attribute is a valuable mechanism for describing design-related information that is not conveniently modeled via the "core" set of VHDL constructs oriented toward describing basic signal transformations. To that end, user-defined attributes are often referred to as "*describing ancillary design information*" or "*decorating the design tree.*"

Finally, in Section 16.6 we introduced a new language feature provided in VHDL-93: *groups*. Groups provide a more rigorous definition of related declared items and make the definition explicit, thus giving the set of related declared items a name or handle that can be referenced elsewhere in a VHDL model.

16.8 PROBLEMS

1. For the IEEE Standard 1164,
   a. List and describe the possible logic values.
   b. Which logic values are required to model TTL?
   c. Which logic values are required to model TTL with open collector outputs?
   d. Which logic values are required to model static CMOS?

2. For the **not** operation declared in Figure 16.3,
   a. Does the **not** operator describe the complement behavior of a static NMOS inverter? Explain your answer.

b. Write a VHDL model of a static NMOS inverter. Declare the input and output signals to be of type STD_ULOGIC given in Section 16.1.1.

3. Consider the conjunction definition given in Table 16.8.

Table 16.8

The 9-valued logic conjunction operation

Left operand	Right operand								
	'=U'	'X'	'0'	'1'	'Z'	'W'	'L'	'H'	'-'
'U'	'U'	'U'	'0'	'U'	'U'	'U'	'0'	'U'	'U'
'X'	'U'	'X'	'0'	'X'	'X'	'X'	'0'	'X'	'X'
'0'	'0'	'0'	'0'	'0'	'0'	'0'	'0'	'0'	'0'
'1'	'U'	'X'	'0'	'1'	'X'	'X'	'0'	'1'	'X'
'Z'	'U'	'X'	'0'	'X'	'X'	'X'	'0'	'X'	'X'
'W'	'U'	'X'	'0'	'X'	'X'	'X'	'0'	'X'	'X'
'L'	'0'	'0'	'0'	'0'	'0'	'0'	'0'	'0'	'0'
'H'	'U'	'X'	'0'	'1'	'X'	'X'	'0'	'1'	'X'
'-'	'U'	'X'	'0'	'X'	'X'	'X'	'0'	'X'	'X'

a. Declare the **and** VHDL operator for the IEEE Standard 1164 multi-valued logic that does *not* support the short-circuit property. Show all type and object declarations.

b. Use the **and** operator written in part (a) to describe an IEEE 1164 **and** gate. Verify the IEEE 1164 **and** gate by simulating the design entity for all possible input combinations of 'U', '0', and '1'.

c. Declare the **and** VHDL operator for the IEEE Standard 1164 multi-valued logic that *does* support the short-circuit property. Show all type and object declarations. Note that VHDL provides the conventional sequential **if** statement having the form

```
if (boolean_condition) then

    . . . .

else

    . . . .

end if;
```

d. Use the **and** operator written in part (c) to describe an IEEE 1164 **and** gate. Verify the IEEE 1164 **and** gate by simulating the design entity for all possible input combinations of 'U', '0', and '1'.

4. Consider the disjunction definition given in Table 16.9.

Table 16.9

The 9-valued logic disjunction operation

Left operand	Right operand								
	'U'	'X'	'0'	'1'	'Z'	'W'	'L'	'H'	'-'
'U'	'U'	'U'	'U'	'1'	'U'	'U'	'U'	'1'	'U'
'X'	'U'	'X'	'X'	'1'	'X'	'X'	'X'	'1'	'X'
'0'	'U'	'X'	'0'	'1'	'X'	'X'	'0'	'1'	'X'
'1'	'1'	'1'	'1'	'1'	'1'	'1'	'1'	'1'	'1'
'Z'	'U'	'X'	'X'	'1'	'X'	'X'	'X'	'1'	'X'
'W'	'U'	'X'	'X'	'1'	'X'	'X'	'X'	'1'	'X'
'L'	'U'	'X'	'0'	'1'	'X'	'X'	'0'	'1'	'X'
'H'	'1'	'1'	'1'	'1'	'1'	'1'	'1'	'1'	'1'
'-'	'U'	'X'	'X'	'1'	'X'	'X'	'X'	'1'	'X'

a. Declare the **or** VHDL operator for the IEEE Standard 1164 multi-valued logic that does *not* support the short-circuit property. Show all type and object declarations.

b. Use the **or** operator written in part (a) to describe an IEEE 1164 **or** gate. Verify the IEEE 1164 **or** gate by simulating the design entity for all possible input combinations of `'U'`, `'0'`, and `'1'`.

c. Declare the **or** VHDL operator for the IEEE Standard 1164 multi-valued logic that *does* support the short-circuit property. Show all type and object declarations. Note that VHDL provides the conventional sequential **if** statement having the form

```
if (boolean_condition) then
    . . . .
else
    . . . .
end if;
```

d. Use the **or** operator written in part (c) to describe an IEEE 1164 **or** gate. Verify the IEEE 1164 **or** gate by simulating the design entity for all possible input combinations of `'U'`, `'0'`, and `'1'`.

5. In VHDL-87, a disadvantage of not using the predefined type BINARY is that we lose the general use of bit vector literals as convenient notations for bit strings.

   a. Explain why the general use of bit vector literals is lost when overloading logic operators to accept operands of a user-defined multi-valued logic system.

   b. Use the **and** operator defined in Problem 3 to write a VHDL model that accepts a 64-bit vector A_IN of type STD_ULOGIC_VECTOR(63 **downto** 0) and that generates $A\_IN \cdot 0123456789ABCDEF_{16}$.

6. Modeling input/output logic behavior based on truth tables is generally not a good idea when using IEEE 1164 because of the large number of possible inputs. Consider, for example, the following VHDL model based on the VHDL model given in Figure 13.2; the type information has been changed from the predefined type BIT to the user-defined type STD_ULOGIC.

```
library IEEE;
use IEEE.STD_LOGIC_1164.all;
entity MAJORITY is
  port
    (A_IN, B_IN, C_IN : in  STD_ULOGIC;
     Z_OUT            : out STD_ULOGIC);
end MAJORITY;


architecture DATA_FLOW of MAJORITY is
begin
  Z_OUT <= (not A_IN and B_IN and C_IN) or
           (A_IN and not B_IN and C_IN) or
           (A_IN and B_IN and not C_IN) or
           (A_IN and B_IN and C_IN);

end DATA_FLOW;
```

   a. How many entries are contained in the truth table describing the input/output behavior of the MAJORITY design entity?

    b. Generate a truth table describing the output response to all possible combinations of $'U'$, $'0'$, and $'1'$ values and verify by simulation.

    c. What is the value of Z_OUT at the beginning of simulation?

7. The following precedence graph can be used to describe the resolution property of the IEEE Standard 1164. Two different logic values of different precedence resolve to the logic value having the higher precedence. Two different logic values of the same precedence resolve to the logic value having the next higher precedence. Additionally, a don't care value $'-'$ resolved with an uninitialized value $'U'$ yields an uninitialized value $'U'$; a don't care value $'-'$ resolved with any other value yields a forcing unknown value $'X'$.

    a. Declare a resolution function called RESOLVED for the IEEE Standard 1164. Show all type and object declarations. (*Hint*: Note the special case where the resolution function is called with only one driver.)

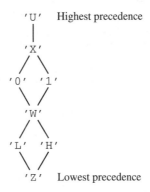

8. The following VHDL model is based on the VHDL model given in Figure 15.19; the type and resolution function information have been changed from the predefined type BIT to the user-defined type STD_ULOGIC.

```
library IEEE;
use IEEE.STD_LOGIC_1164.all;
entity HMMMMM is
  port
    (AIN, BIN, CIN : in STD_ULOGIC;
     ZOUT_BAR      : out STD_ULOGIC);    -- Active-0 output
end HMMMMM;


architecture DATA_FLOW of HMMMMM is
  -- Declare resolved signal to interconnect logic operators
  signal INT_SIG : RESOLVED STD_ULOGIC;

begin
  INT_SIG  <= BIN or CIN;        -- Source
  INT_SIG  <= AIN or BIN;        -- Source
  INT_SIG  <= AIN or CIN;        -- Source
  ZOUT_BAR <= not INT_SIG;
end DATA_FLOW;
```

    a. Generate a truth table describing the output response to all possible combinations of $'0'$ and $'1'$ values.

    b. Verify the truth table via simulation.

c. Does the VHDL model HMMMMM still describe the active-0 majority logic function? Explain your answer.

9. Consider a tri-state 3-to-8 decoder. When the enable signal is asserted active-0, the outputs yield a 3-to-8 decoder function. When the enable signal is deasserted, the outputs are disconnected from the inputs. (See Chapter 5 for a discussion of decoders.)
   a. Write a VHDL model using the IEEE 1164 9-valued logic system.
   b. Verify by simulation that the VHDL model written in part (a) performs the tri-state 3-to-8 decoder function for all possible inputs having values '0' and '1'.

10. For convenience, STD_LOGIC_1164 provides several commonly used subsets of the standard nine logic values. One subset is the 3-valued set 'X', '0', and '1' provided by the type X01. STD_LOGIC_1164 also provides edge detection functions.

```
function RISING_EDGE  (signal S : STD_ULOGIC) return BOOLEAN;
function FALLING_EDGE (signal S : STD_ULOGIC) return BOOLEAN;
```

---

   a. Use the type X01 and functions RISING_EDGE/FALLING_EDGE to write a VHDL model of a static CMOS inverter. Model propagation delay using t_{PLH} and t_{PHL} as generic parameters.
   b. Verify by simulation that the VHDL model written in part (a) performs the complement function for all possible inputs.

11. Improve the timing accuracy of the static CMOS inverter VHDL model developed in Problem 10 by adding a description of output rise and fall times using t_r and t_f.
   a. Generate a VHDL model having output rise and fall times using t_r and t_f as generic parameters.
   b. Verify by simulation that the VHDL model written in part (a) performs the complement function for all possible inputs.
   c. Draw the input/output waveforms for A transitioning from '0' to '1' and from '1' to '0'.

12. Consider the following enumeration type declaration.

```
type CRITICAL_LEVELS is (NONE, LOW, NORM, HIGH);
```

---

   a. What is CRITICAL_LEVELS'POS(CRITICAL_LEVELS'LOW)?
   b. What is CRITICAL_LEVELS'VAL(CRITICAL_LEVELS'POS(LOW))?
   c. What is CRITICAL_LEVELS'POS(CRITICAL_LEVELS'PRED(HIGH))?
   d. What is CRITICAL_LEVELS'SUC(CRITICAL_LEVELS'PRED(LOW))?

13. Would the following alternative enumeration type declarations for CRITICAL_LEVELS (Problem 12) be acceptable modeling practices? Explain your answers.
   a. type CRITICAL_LEVELS **is** (HIGH, NORM, LOW, NONE);
   b. type CRITICAL_LEVELS **is** (NONE, LOW, MOD, HIGH);

14. For each of the following array types, declare an object, draw a diagram showing the object array structure, and give values of the predefined object array attributes LEFT, RIGHT, HIGH, LOW, RANGE, REVERSE_RANGE, LENGTH, and ASCENDING (VHDL-93) for each dimension.
   a. type WORD_TYPE **is array** (10 **to** 17) **of** BIT:
   b. type MVL4 **is** ('X', '0', '1', 'Z');
      type ELEMENT_TYPE **is array** (MVL4) **of** MVL4;
   c. type TABLE_TYPE **is array** (3 **downto** 0, 0 **to** 3) **of** TIME;

15. Discrete ranges and element values can also be defined by specifying a subset of values of an enumeration or integer type using a *range constraint*, as shown in the following declaration.

```
type TERNARY_1D is array (STD_ULOGIC range 'X' to '1')
              of STD_ULOGIC range 'X' to '1';
```

Following the enumeration type STD_ULOGIC with the range constraint **range** 'X' **to** '1' defines the three-element subset or discrete range 'X', '0', '1'. Thus, TERNARY_1D represents an array having three elements; index and element values include 'X', '0', and '1'.

a. For the SUBLIST_TYPE and LIST_TYPE declarations, declare an object, draw a diagram showing the object array structure, and give values of the predefined object array attributes LEFT, RIGHT, HIGH, LOW, RANGE, REVERSE_RANGE, LENGTH, and ASCENDING (VHDL-93).

```
type MVL4 is ('X', '0', '1', 'Z');
type ELEMENT_TYPE is array (MVL4) of MVL4;
type SUBLIST_TYPE is array (MVL4 range '0' to '1') of
     ELEMENT_TYPE;
type LIST_TYPE is array (MVL4) of ELEMENT_TYPE;
```

16. Based on the types given in Problem 15, draw diagrams showing the structure and content of the following arrays.

a. constant ELEMENTA : ELEMENT_TYPE := ('X', 'Z' => '0'
                                               '0' **to** '1' => '1');

b. constant SUBLISTA : SUBLIST_TYPE := ((**others** => 'X'),
                                               (**others** => 'X'));

c. constant LISTA : LIST_TYPE := (MVL4'RANGE =>
                                        (MVL4'RANGE => '1'));

17. Describe the following combinational system in VHDL.

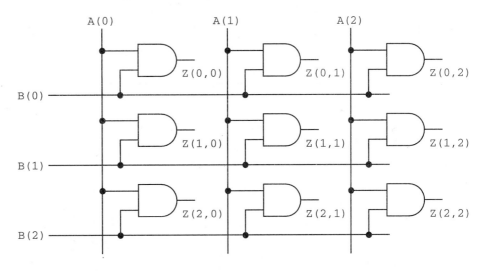

18. a. Use a lookup table to write a model of an MOS **and** gate using the 9-valued IEEE Standard logic having the propagation delay given in Table 16.10.

Table 16.10

Propagation delay (ns)	Gate Width (μm)	Gate Length (μm)		
		1	2	3
	1	10	12	15
	2	7	10	11
	3	5	8	10

b. Show simulations of the gate output transitioning from '0' to '1' for gate width of 2 μm and gate lengths of 1 μm, 2 μm, and 3 μm.

19. For the VHDL model written in Problem 18, add the following user-defined attributes.
   a. Describe the package type as dual-in-line, pin-grid array, or leadless chip carrier.
   b. Describe package pin names associated with port names.
   c. Describe input/output test vectors for verification.

20. Are the following code segments legal or illegal? Explain your answer.
   a. **attribute** METAL_LEVEL : POSITIVE;
   signal SIG_A, SIG_B, SIG_C : BIT;

   attribute METAL_LEVEL **of** SIG_A : **signal is** 2;
   attribute METAL_LEVEL **of others** : **signal is** 1;
   attribute METAL_LEVEL **of** SIG_C : **signal is** 2;
   b. **attribute** METAL_LEVEL : POSITIVE;
   signal SIG_A, SIG_B, SIG_C : BIT;

   attribute METAL_LEVEL **of** SIG_A : **signal is** 2;
   attribute METAL_LEVEL **of others** : **signal is** 1;
   signal SIG_D : BIT;
   c. **attribute** METAL_LEVEL : POSITIVE;
   signal SIG_A, SIG_B, SIG_C : BIT;

   attribute METAL_LEVEL **of all** : **signal is** 0;

21. Only the *value* of an object is passed in an association list; user-defined attributes of an object are *not*. In other words, user-defined attributes are not passed between design entities or between a subprogram and its parent. As an example, examine the following segments of a VHDL model.

```
package PD_PKG is
  -- Attribute declaration
  attribute WIRE_CAP : CAPACITANCE;
end PD_PKG;

-- Design entity
use WORK.PD_PKG.all;
entity SMALL_DESIGN is
  port  (X, Y, ... );
  attribute WIRE_CAP of X : signal is 10 pF;
end SMALL_DESIGN;

-- Design entity
use WORK.PD_PKG.all;
entity BIG_DESIGN is
  port  (A, B, ... );
  attribute WIRE_CAP of A : signal is 8 pF;
end BIG_DESIGN;

architecture STRUCTURE of BIG_DESIGN is
  component SMALL_DESIGN
    port (X, Y, ... );
  end component;
```

```
begin
  -- Structural description
  C1: SMALL_DESIGN port map (X => A, ... );
end STRUCTURE;
```

a. Assuming the *absence* of the attribute specification for X in SMALL_DESIGN, what is A'WIRE_CAP and X'WIRE_CAP in the structural description BIG_DESIGN? (Also assume default configuration bindings.)

b. Assuming the *presence* of the attribute specification for X in SMALL_DESIGN, what is A'WIRE_CAP and X'WIRE_CAP in the structural description BIG_DESIGN? (Also assume default configuration bindings.)

22. VHDL-93: Use groups to describe the relationship that input ports of an **and** gate are logically permutable.

REFERENCES AND READINGS

[1] Armstrong, J. "Tuning VHDL for Multivalue Logic Modeling," *IEEE Design & Test of Computers,* vol. 7, no. 3, pp. 8–10, June 1990.

[2] Carroll, M. "VHDL—Panacea or Hype?" *IEEE Spectrum,* vol. 30, no. 6, pp. 34–38, June 1993.

[3] Coelho, D. *The VHDL Handbook.* Boston, MA: Kluwer Academic Publishers, 1989.

[4] Coelho, D. "A VHDL Standard Package for Logic Modeling," *IEEE Design & Test of Computers,* vol. 7, no. 3, pp. 25–32, June 1990.

[5] d'Abreu, M. "Gate-Level Simulation," *IEEE Design & Test of Computers,* vol. 2, no. 6, pp. 63–71, December 1985.

[6] DeGroat, J. "Transparent Logic Modeling in VHDL," *IEEE Design & Test of Computers,* vol. 7, no. 3, pp. 42–48, June 1990.

[7] Gilman, A. "Logic Modeling in WAVES," *IEEE Design & Test of Computers,* vol. 7, no. 3, pp. 49–55, June 1990.

[8] Hanna, W., A. Kaskowitz, and L. Wallace. "Evaluating VHDL-Based ASIC Synthesis Tools," *Proceedings of the 5th IEEE International ASIC Conference and Exhibit,* pp. 253–256, September 1992.

[9] IEEE. IEEE Standard 1164 Multivalue Logic System for VHDL Model Interoperability. 1993.

[10] Menchini, P. "A Minimalist Approach to VHDL Logic Modeling," *IEEE Design & Test of Computers,* vol. 7, no. 3, pp. 12–23, June 1990.

[11] Ryan, R. "X-State Handling in VHDL," *Design Automation,* vol. 2, no. 4, pp. 32–35, May 1991.

[12] Wu, J., T. Young, E. Kawamoto, and W. Keutgens. "Accurate VHDL Libraries for ASIC Design," *Proceedings of the 5th IEEE International ASIC Conference and Exhibit,* pp. 327–330, September 1992.

Part 6

VHDL:

SEQUENTIAL

SYSTEMS

CHAPTER 1 Introduction

CHAPTER 2 Representing Information

CHAPTER 3 Combinational Systems: Definition & Analysis

CHAPTER 11 VHDL: A First Look
CHAPTER 12 Structural Modeling

CHAPTER 4 Combinational Design: Synthesis

CHAPTER 13 Data Flow Modeling in VHDL

CHAPTER 5 Combinational Design: Implementation

CHAPTER 14 Structural Modeling in VHDL–Part 2

CHAPTER 6 Logic Families

CHAPTER 15 VHDL Technology Information–Part I

CHAPTER 7 Integrated Circuits

CHAPTER 16 VHDL Technology Information–Part 2

CHAPTER 8 Sequential Systems: Definition & Analysis

CHAPTER 17 Describing Synchronous Behavior in VHDL

CHAPTER 9 Sequential Design: Synthesis

CHAPTER 18 Algorithmic Modeling in VHDL

CHAPTER 10 Sequential Design: Implementation

CHAPTER 19 VHDL: A Last Look

Structural Modeling

```
I1: NOT_OP
  port map (A, B);
```

Asynchronous Data Flow Modeling

```
B <= not A;
```

Technology Modeling

```
B <= not A after 5 ns;
```

Synchronous Data Flow Modeling

```
B1: block (C = '1')
begin
 B <= guarded not A after 5 ns;
end block B1;
```

Algorithmic Modeling

```
P1: process (PS)
begin
 NS <= STATE_TABLE(PS);
end process P1;
```

Combinational Logic Schematics

A
B ⟩—C ⟩—Z

Combinational Logic Equations

$Z = (A \cdot B) + C$

Logic Families & Integrated Circuits

Sequential Logic Schematics

A ⟩— Z
   C

Sequential Logic State Diagrams

00/0   11/0   01/0
01/1 (S0)   (S1)   10/0
10/1   00/1   11/1

17

DESCRIBING SYNCHRONOUS BEHAVIOR IN VHDL

In this chapter, we will investigate how to model in VHDL the sequential systems introduced in Chapter 8. We will develop VHDL models of latches, flip-flops, and general state machines. We will learn how to use *blocks* and *guards* to change the execution semantics of concurrent signal assignment statements from asynchronous to synchronous. *Predefined signal attributes* that provide information about the present and past values of signals will be introduced to model edge-sensitive behavior. *Assertions* will be introduced as a general mechanism for conveying design constraints, such as setup and hold times.

We will conclude this chapter with a discussion of *processes* and their role as the basic behavioral statement in VHDL. We will explain how concurrent signal assignment statements are actually shorthand notations for certain commonly used forms of the *process statement*. We will also take an expanded look at the simulation cycle and *postponed assertions* provided in VHDL-93.

17.1 LATCHES

Latches are easily described in VHDL using the concurrent signal assignment statements discussed in Chapter 13. For example, Figure 17.1 shows how conditional concurrent signal assignment statements can be used to model the behavior of a JK latch given by the following truth table. The signals Q and Q_BAR are read and updated because they appear on the

Row	JK	Q^+	Action
0	00	Q	No change
1	01	0	Reset
2	10	1	Set
3	11	$\overline{Q}$	Toggle

right-hand and left-hand sides of the concurrent signal assignment statements, respectively. Thus, Q and Q_BAR are declared as ports of mode **inout**, reflecting the output-to-input feedback characteristic of cross-coupled logic. The default values in the port declarations for Q and Q_BAR ensure that the ports correctly assume initial complementary values; model initialization will be discussed in more detail in Section 17.13.1. Note that when J='1' and K='1', the two concurrent signal assignment statements alternately wake each other up and the JK latch oscillates.

Figure 17.1

Data flow description of a
JK latch

```
entity JK_LATCH is
 port
    (J, K  : in BIT;
     Q     : inout BIT := '0';
     Q_BAR : inout BIT := '1');
 end JK_LATCH;


architecture TRUTH_TABLE of JK_LATCH is
begin
  -- Map truth table into conditional concurrent signal
  -- assignment statement
  Q <= Q     when (J='0' and K='0') else
       '0'   when (J='0' and K='1') else
       '1'   when (J='1' and K='0') else
       Q_BAR;

  Q_BAR <= not Q;
end TRUTH_TABLE;
```

Figure 17.2 shows an alternative description of a JK latch using its characteristic equation. Again, the two concurrent signal assignment statements are written in a cross-coupled fashion.

$$Q^+ = (J \cdot \overline{Q}) + (\overline{K} \cdot Q)$$

Figure 17.2

Alternative description of
a JK latch using its char-
acteristic equation

```
architecture CHAR_EQ of JK_LATCH is
begin
  -- Map characteristic equation into conditional concurrent
  -- signal assignment statement
  Q <= (J and Q_BAR) or (not K and not Q_BAR);

  Q_BAR <= not Q;
end CHAR_EQ;
```

17.2 VHDL-93: SIGNAL VALUE—UNAFFECTED

In VHDL-87, reassigning a signal to itself poses the potentially serious problem of subtly introducing design errors because only the *present value* is reassigned; any pending future values are discarded. In other words, the signal assignment SIG1 <= SIG1; does *not* reassign SIG1's entire set of time-ordered values to SIG1, leaving SIG1 unchanged. Rather, the signal assignment reassigns only SIG1's present value to SIG1 and discards any time-ordered values scheduled to occur in the future. Thus far, we have carefully avoided this potentially troublesome aspect of reassigning a signal to itself by ensuring that such signals did not accumulate pending future values.

VHDL-93 addresses the problem of assigning a signal to itself by providing a new signal "value": **unaffected**. Assigning the value denoted by the keyword **unaffected** to a target signal means the signal remains unchanged; in other words, its entire time-ordered set of values remains unaffected. Hence, the concurrent signal assignment statement driving the target signal Q in the JK latch model given in the previous section as

```
Q <= Q     when (J='0' and K='0') else
     '0'   when (J='0' and K='1') else
     '1'   when (J='1' and K='0') else
     Q_BAR;
```

can be more appropriately written as

```
Q <= unaffected when (J='0' and K='0') else
     '0'        when (J='0' and K='1') else
     '1'        when (J='1' and K='0') else
     Q_BAR;
```

Also, recall from Chapter 13 that the option of omitting the final **else** clause of a conditional concurrent signal assignment statement is VHDL-93 provides an alternative means of leaving a target signal unchanged.

```
Q <= '0'    when (J='0' and K='1') else
     '1'    when (J='1' and K='0') else
     Q_BAR  when (J='1' and K='1');
```

17.3 LEVEL-SENSITIVE SYNCHRONOUS BEHAVIOR

When we discussed gated latches in Section 8.2.2 we introduced the concept of a *control* signal. In general, a control signal determines whether a hardware entity will respond to or ignore transitions on certain inputs. For example, the clock signal for a gated JK latch controls whether the gated latch responds to or ignores transitions on its J and K inputs; this behavior is called *level-sensitive synchronous behavior.*

Level-sensitive synchronous behavior is easily described in VHDL via *blocks* and *guards.* As an example, the ungated or basic JK latch VHDL model given in Figure 17.1 is modified in Figure 17.3 to describe a gated JK latch.

The architecture contains a single **block** statement that contains the expression (CLK = '1'); this expression is called a *guard expression.* The guard expression controls the execution of the concurrent signal assignment statement containing the reserved keyword **guarded** enclosed within the block labeled CLKED. The conditional concurrent signal assignment statement driving Q,

```
Q <= guarded Q when (J='0' and K='0') else
             '0' when (J='0' and K='1') else
             '1' when (J='1' and K='0') else
             Q_BAR;
```

Figure 17.3

Data flow description of a gated JK latch

```
entity JK_GLATCH is
 port
    (J, K, CLK : in BIT;
     Q          : inout BIT := '0';
     Q_BAR      : inout BIT := '1');
 end JK_GLATCH;

architecture TRUTH_TABLE of JK_GLATCH is
begin
 CLKED: block (CLK = '1') -- Guard expression
 begin
  -- Guarded concurrent signal assignment
  Q <= guarded Q when (J='0' and K='0') else
                 '0' when (J='0' and K='1') else
                 '1' when (J='1' and K='0') else
                 Q_BAR;

   Q_BAR <= not Q;
  end block CLKED;
 end TRUTH_TABLE;
```

responds to changes on the J and K input signals and Q_BAR *only when the guard expression is true;* in other words, only when the clock signal CLK equals the logic BIT value '1'. When the guard expression is false, in other words, CLK = '0', the conditional concurrent signal assignment statement driving Q is temporarily disabled or "shut down."

Concurrent signal assignment statements not containing the keyword **guarded** ignore the controlling condition of the guard expression and retain asynchronous stimulus/response execution semantics. We will discuss blocks, guards, and synchronous behavior in more detail in the following section.

17.4 GUARD EXPRESSIONS AND GUARDED ASSIGNMENTS

A block statement provides the "scaffolding" to group-related concurrent statements within a name space. Moreover, the group of concurrent statements may also be placed under the control of a guard expression.

A block statement has the general form shown on the next page.[1]

---

[1] VHDL-93: In VHDL-93, the header of a block statement may close with the keyword **is**.

```
label: block (guard expression) is
 -- Declarative part
begin
 -- Statement part
end block label;
```

```
label: block (guard_expression)
 -- Declarative part
begin
 -- Statement part
end block label;
```

A block statement starts with a label identifying the block. A label is required and can be any legal identifier. Repeating the block label at the closing of a block statement is optional, however, this modeling practice is recommended to improve readability. The label is followed by the keyword **block**, which is in turn followed optionally by a guard expression. The declarative part contains declarations to be shared by the set of concurrent statements contained in the statement part. The scope of any declaration contained within a block statement extends to only the end of the block.

A guard expression, also more succinctly called a guard, is an optional part of a block statement. A guard must be a BOOLEAN expression that evaluates to either TRUE or FALSE. For instance, if we want a concurrent signal assignment statement to execute only when a control signal CONTROL_SIGNAL of type BIT is asserted active-1, in other words, '1', then the guard expression should be formed as

```
(CONTROL_SIGNAL = '1')
```

We will examine more complex guard expressions when edge-sensitive control behavior is modeled in the next section.

A guard expression is used to modify the execution semantics of concurrent signal assignment statements, in other words, to control when concurrent signal assignment statements may execute. Arbitrarily complex guard expressions can model control conditions ranging from simple interrelationships of a few signals to complicated interrelationships of many signals. A guard controls the concurrent signal assignment statements (enclosed within the associated block) that contain the keyword **guarded**.

A concurrent signal assignment statement that contains the keyword **guarded** is appropriately called a *guarded assignment*. A guarded assignment executes under either of the following two conditions.

1. The guard expression changes from FALSE to TRUE.
2. The guard expression is TRUE and one of the signals appearing on the right-hand side of the signal assignment statement changes value.

Conversely, a concurrent signal assignment statement that does not contain the keyword **guarded** is called an *unguarded assignment*. An unguarded assignment executes whenever any one of the signals appearing on the right-hand side of the signal assignment statement changes value. Thus, the keyword **guarded** and a guard expression collectively modify the asynchronous behavior of a concurrent signal assignment statement by defining a control condition that disables stimuli sensitivity.

Figure 17.3 shows that a block having a guard expression may contain a mix of guarded and unguarded assignments; the guarded assignments are controlled by the guard and the unguarded assignments ignore the guard. Typically, an unguarded assignment is placed within a particular block to improve model readability, grouping related signal transforms. However, unguarded assignments may be placed anywhere within the statement part of an architecture because they have no affinity to any block/guard. The architecture for the JK_GLATCH in Figure 17.3, for example, could equivalently be written as follows.

```
architecture TRUTH_TABLE of JK_GLATCH is
begin
 CLKED: block (CLK = '1') -- Guard expression
 begin
  -- Guarded concurrent signal assignment
  Q <= guarded Q when (J='0' and K='0') else
               '0' when (J='0' and K='1') else
               '1' when (J='1' and K='0') else
               Q_BAR;
 end block CLKED;

 Q_BAR <= not Q;

end TRUTH_TABLE;
```

To understand guarded and unguarded assignments, consider the following block statement. Assume all signals are of type BIT and have an initial value of '0'.

```
B1: block (CONTROL_SIGNAL = '1')
begin
 X <= guarded A or B after 5 min;
 Y <= A or B after 5 min;
end block B1;
```

Figure 17.4 presents the resulting timing diagram for the signals X and Y given the signals CONTROL_SIGNAL, A, and B.

Figure 17.4

Guarded and unguarded assignments

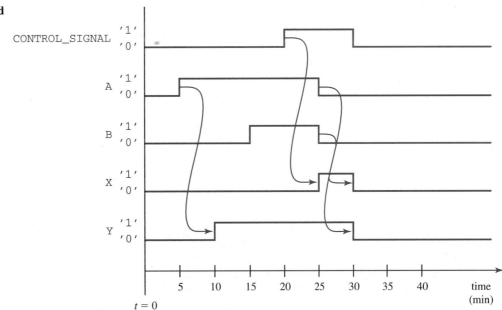

The unguarded concurrent signal assignment statement

```
Y <= A or B after 5 min;
```

---

executes in response to signal A changing value from '0' to '1' at 5 minutes, setting the signal Y to take on the value $A + B = 1 + 0 = 1$ at time 5 minutes + 5 minutes = 10 minutes. This unguarded assignment executes again in response to signal B changing value from '0' to '1' at 15 minutes; however, the value of Y does not change because $A + B = 1 + 1 = 1$. Finally, Y is reset to '0' at 30 minutes due to the changes on A and B at 25 minutes.

The guarded concurrent signal assignment statement

```
X <= guarded A or B after 5 min;
```

---

yields a different waveform for the target signal X because it is controlled by the guard expression (CONTROL_SIGNAL = '1'). Since the guard expression is FALSE during the change on signal A at 5 minutes and signal B at 15 minutes, the guarded assignment does not execute and the signal X remains at '0'. At 20 minutes, the control signal CONTROL_SIGNAL changes from '0' to '1', which causes the guard to change from FALSE to TRUE. The guarded assignment executes, setting X to take on the value $A + B = 1 + 1 = 1$ at time 20 minutes + 5 minutes = 25 minutes. Since the guard is TRUE at time 25 minutes, the guarded signal assignment responds to the changes on A and B at 25 minutes and resets X to '0' at 30 minutes.

The scope of a guard is the associated block. In other words, a guard expression applies only to the guarded assignments contained within the associated block. For nested block statements, the guard expression of an outer block may be accessed by the immediately enclosed block via the predefined signal GUARD, as illustrated below.

```
BLK_A: block (CNTRL_A = '1')
begin
 X <= guarded A or B after 5 min;

 BLK_B: block (CNTRL_B = '1' and GUARD) -- Nested block
 begin
  Y <= guarded A or B after 5 min;
 end block BLK_B;

end block BLK_A;
```

The guarded assignment driving the signal X is controlled by the guard expression associated with the outer block BLK_A, (CNTRL_A = '1'). Thus, this guarded assignment statement is enabled only when CNTRL_A = '1'. The guarded assignment driving the signal Y is controlled by the guard expression associated with the inner block BLK_B, (CNTRL_B = '1' and GUARD). GUARD is a signal of type BOOLEAN; the value of GUARD is the value of the *immediately* enclosing guard expression (CNTRL_A = '1'). Thus, the guarded assignment driving Y is enabled only when CNTRL_A = '1' *and* CNTRL_B = '1'.

A GUARD signal is implicitly declared for each block. The signal GUARD can be used as any other signal, except we cannot attempt to change the value of GUARD by using GUARD as the target of a signal assignment statement or associating GUARD with an **out**, **inout**, or **buffer** port.

17.5 EDGE-SENSITIVE BEHAVIOR

In Chapter 8, we learned about memory devices called edge-triggered flip-flops. Edge-triggered flip-flops are controlled by signal *transitions*, whereas the latches modeled in the previous sections are controlled by signal *values*. Edge-sensitive behavior can be easily described in VHDL by appropriately modifying the guard expression of a block statement.

Figure 17.5 shows a VHDL model of a positive edge-triggered JK flip-flop. The guard expression (CLK = '1' **and not** CLK'STABLE) identifies the rising edge of the signal CLK; in other words, the expression is TRUE when CLK transitions from '0' to '1'. STABLE is a *predefined attribute* of the signal CLK. The value of this attribute, accessed by the notation CLK'STABLE, is TRUE when CLK has *not* changed value. The expression **not** CLK'STABLE is TRUE when CLK *has* changed value. Thus, if CLK has just changed value (**not** CLK'STABLE) and its present value is '1' (CLK = '1'), then we have identified a rising edge. We will discuss predefined signal attributes in more detail in the following section.

As another example of edge-triggered behavior, Figure 17.6 (p. 536) shows a description of a 4-bit circular *shift register* or ring counter with synchronous reset. A *register* is a collection of memory devices used to store several bits of information. (See Chapter 10.)

Figure 17.5

Data flow description of a positive edge-triggered JK flip-flop

```
entity JK_FF is
 port
    (J, K, CLK : in BIT;
     Q          : inout BIT := '0';
     Q_BAR      : inout BIT := '1');
end JK_FF;


architecture DATA_FLOW of JK_FF is
begin
 CLKED: block (CLK = '1' and not CLK'STABLE)
 begin
   -- Guarded concurrent signal assignment
   Q <= guarded Q when (J='0' and K='0') else
               '0'    when (J='0' and K='1') else
               '1'    when (J='1' and K='0') else
               Q_BAR;

    Q_BAR <= not Q;
 end block CLKED;
end DATA_FLOW;
```

Figure 17.6

VHDL model of edge-triggered circular shift register

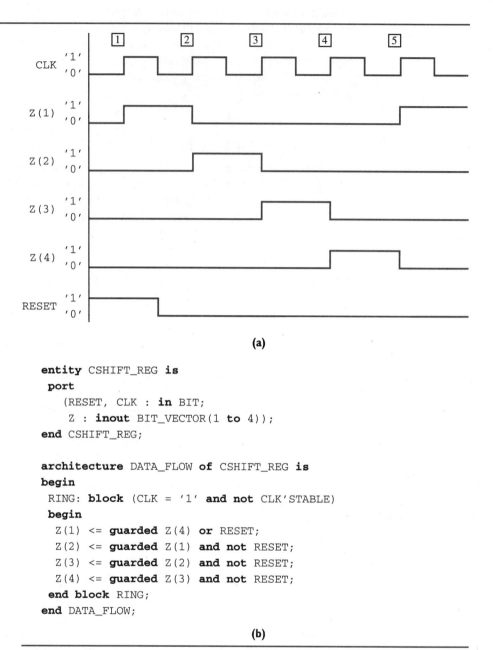

(a)

```
entity CSHIFT_REG is
 port
   (RESET, CLK : in BIT;
    Z : inout BIT_VECTOR(1 to 4));
end CSHIFT_REG;


architecture DATA_FLOW of CSHIFT_REG is
begin
 RING: block (CLK = '1' and not CLK'STABLE)
 begin
  Z(1) <= guarded Z(4) or RESET;
  Z(2) <= guarded Z(1) and not RESET;
  Z(3) <= guarded Z(2) and not RESET;
  Z(4) <= guarded Z(3) and not RESET;
 end block RING;
end DATA_FLOW;
```

(b)

Figure (a) shows a timing diagram of a single bit circulating in a shift register. Figure (b) shows a VHDL model of a circular shift register.

On the first rising clock edge labeled ①, the guard expression changes from FALSE to TRUE and all guarded concurrent signal assignments are executed. The circular shift register is initialized with a single bit at Z(1) by asserting the RESET signal.

On the next rising clock edge labeled ②, the guard expression again changes from FALSE to TRUE and all guarded concurrent signal assignments are again executed. The value of Z(1) is shifted one position and assigned to Z(2), Z(2) is shifted one position and assigned to Z(3), Z(3) is shifted one position and assigned to Z(4), and Z(4) wraps around and is assigned to Z(1). Thus, the single bit shifts one position from Z(1) to Z(2).

The single bit again shifts from Z(2) to Z(3) on the rising clock edge labeled $\boxed{3}$, shifts from Z(3) to Z(4) on the rising clock edge labeled $\boxed{4}$, and finally wraps around and shifts from Z(4) back to Z(1) on the rising clock edge labeled $\boxed{5}$.

17.6 PREDEFINED SIGNAL ATTRIBUTES

Recall from previous chapters that predefined attributes provide information about a VHDL model that a designer may use in describing behavior and/or structure. We have seen several examples of predefined attributes, such as RANGE for arrays and POS for discrete types. VHDL also provides a set of predefined attributes for signals.

Predefined signal attributes are defined below. In the following definitions, signal_name represents an identifier denoting a signal.

- **Transaction-Related Predefined Signal Attributes**

Attribute	Definition
signal_name'ACTIVE	A *value* of type BOOLEAN. If a transaction has occurred on signal_name, signal_name'ACTIVE is TRUE; otherwise it is FALSE.
signal_name'QUIET	A *signal* of type BOOLEAN. If a transaction has not occurred on signal_name, signal_name'QUIET is TRUE; otherwise it is FALSE.
signal_name'QUIET(T)	A *signal* of type BOOLEAN. If a transaction has not occurred on signal_name for the past T units of type TIME, signal_name'QUIET(T) is TRUE; otherwise it is FALSE.
signal_name'TRANSACTION	A *signal* of type BIT. If a transaction has occurred on signal_name, signal_name'TRANSACTION changes value; the new value is the complement of the old value.
signal_name'LAST_ACTIVE	A *value* of type TIME denoting the time elapsed since the last transaction occurred on signal_name. If a transaction has never occurred on signal_name, signal_name'LAST_ACTIVE is 0 seconds.

- **Event-Related Predefined Signal Attributes**

Attribute	Definition
signal_name'EVENT	A *value* of type BOOLEAN. If an event has occurred on signal_name, signal_name'EVENT is TRUE; otherwise it is FALSE.
signal_name'STABLE	A *signal* of type BOOLEAN. If an event has not occurred on signal_name, signal_name'STABLE is TRUE; otherwise it is FALSE.
signal_name'STABLE(T)	A *signal* of type BOOLEAN. If an event has not occurred on signal_name for the past T units of type TIME, signal_name'STABLE(T) is TRUE; otherwise it is FALSE.

signal_name'LAST_EVENT	A *value* of type TIME denoting the time elapsed since the last event occurred on signal_name. If an event has never occurred on signal_name, signal_name'LAST_EVENT is 0 seconds.
signal_name'LAST_VALUE	The *value* of signal_name before its present value. If an event has never occurred on signal_name, signal_name'LAST_VALUE is the present value of signal_name.

• **General Predefined Signal Attributes**

Attribute	Definition
signal_name'DELAYED	A signal identical to signal_name, but delayed one delta delay. (See Chapter 15.)
signal_name'DELAYED(T)	A signal identical to signal_name, but delayed T units of type TIME.

The predefined signal attributes are grouped into three categories: transaction-related, event-related, and general. Before discussing the individual attributes, let us first discuss the definitions of the terms *transaction* and *event*. When a signal is assigned a new value, in other words, is updated, a transaction occurs on the signal even if the new value is the same as the old value. However, if the new value is different from the old value, a special kind of transaction, called an event, occurs on the signal.

The predefined signal attributes ACTIVE, QUIET, TRANSACTION, and LAST_ACTIVE can be used to query information about *transaction* activity on a signal. The attributes ACTIVE, QUIET, and TRANSACTION provide essentially the same information in different forms. Whenever a signal, SIGNAL_X, is updated, SIGNAL_X'ACTIVE changes from FALSE to TRUE, SIGNAL_X'QUIET changes from TRUE to FALSE, and an event occurs on SIGNAL_X'TRANSACTION.

The attribute QUIET can optionally accept a parameter, QUIET(T), to query information about whether any transactions have occurred over the past T units of type TIME. The attribute LAST_ACTIVE provides the amount of time since the last transaction, if a transaction has occurred.

The predefined signal attributes EVENT, STABLE, LAST_EVENT, and LAST_VALUE can be used to query information about *event* activity on a signal. The attributes EVENT and STABLE provide essentially the same information in different forms. Whenever a signal, SIGNAL_X, is updated and changes value, SIGNAL_X'EVENT changes from FALSE to TRUE and SIGNAL_X'STABLE changes from TRUE to FALSE.

The attribute STABLE can optionally accept a parameter, STABLE(T), to query information about whether any events have occurred over the past T units of type TIME. The attribute LAST_EVENT provides the amount of time since the last event, if an event has occurred. The attribute LAST_VALUE provides the signal value just before its last event, again if an event has occurred.

It is important to note that the predefined signal attributes QUIET, TRANSACTION, and STABLE are *themselves* signals. For instance, SIGNAL_X'STABLE is a signal having values dependent on SIGNAL_X; during the simulation cycle in which SIGNAL_X is updated and changes value, SIGNAL_X'STABLE will be FALSE.

Finally, the predefined attribute DELAYED provides a means of "shifting" a signal in time. Figure 17.7 shows the relationship between the two signals, SIGNAL_X and SIGNAL_X'DELAYED(3 sec). The signal SIGNAL_X'DELAYED(3 sec) is SIGNAL_X delayed 3 seconds in time. In other words, all the events occurring on SIGNAL_X are delayed or shifted later in time by 3 seconds.

Figure 17.7

DELAYED predefined signal attribute

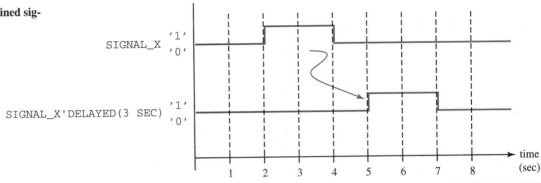

Assume SIGNAL_X is of type BIT.

17.7 SETUP AND HOLD TIMES

As we explained in Chapter 8, setup and hold times are timing restrictions placed on synchronous sequential systems to ensure proper operation of memory devices. Recall that signal values must remain stable for at least a time period denoted by the setup time before memory activation and a time period denoted by the hold time after memory activation to ensure correct memory device behavior and, consequently, sequential system behavior. Figure 17.8 illustrates positive edge-triggered setup and hold times.

Figure 17.9 (p. 540) shows an example of how predefined signal attributes and a new VHDL construct called an *assertion* describe setup and hold timing requirements for a D flip-flop. An assertion statement is used to assert a condition that must always exist for *correct* design entity behavior. If the asserted condition is ever violated during a given simulation, the assertion statement executes and issues the message given in the **report** clause, thereby notifying the designer that the model is not performing as desired. The designer does not have to be concerned with checking assertions during simulation. The proper handling of assertion statements is automatically taken care of by the VHDL simulation process.

Figure 17.10 shows how the condition of the assertion statement

---

not (CLK'DELAYED(HOLD)='1'   **and**
    not CLK'DELAYED(HOLD)'STABLE   **and**
    not D'STABLE(SETUP + HOLD))

Figure 17.8

Positive edge-triggered setup and hold times

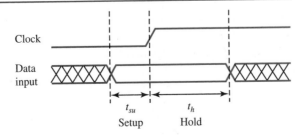

Figure 17.9

Describing setup and hold timing requirements in VHDL

```
entity D_FF is
  generic
    (SETUP, HOLD : TIME);
  port
    (D, CLK   : in BIT;
     Q, Q_BAR : inout BIT);
end D_FF;

architecture DATA_FLOW of D_FF is
begin
  -- Assertion statement defining setup and hold timing
  -- requirements
  assert not
      (CLK'DELAYED(HOLD)='1'               and
        not CLK'DELAYED(HOLD)'STABLE     and
        not D'STABLE(SETUP + HOLD))
  report "Setup/Hold Timing Violation";

  CLKED: block (CLK = '1' and not CLK'STABLE)
  begin
    -- Guarded concurrent signal assignment
    Q <= guarded D;
    Q_BAR <= not Q;
  end block CLKED;
end DATA_FLOW;
```

describes a positive edge-triggered setup and hold timing requirement. The top signal is the clock signal, CLK. Below the CLK signal is another signal derived from CLK, CLK'DELAYED(HOLD). The signal CLK'DELAYED(HOLD) is a copy of CLK shifted (delayed) in time by the period HOLD. The first two terms of the assertion expression

CLK'DELAYED(HOLD)='1' **and not** CLK'DELAYED(HOLD)'STABLE

detect the rising edge of the delayed clock signal.

Figure 17.10

Timing diagram illustrating setup and hold timing conditions

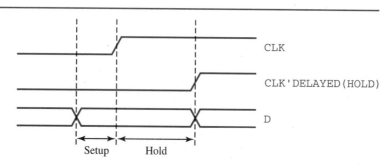

Having detected the rising edge of the delayed clock signal, we can then examine the history of the data input signal D with the last term of the assertion expression

```
not D'STABLE(SETUP + HOLD)
```

to check whether D has been stable for the period SETUP + HOLD. This term is TRUE if D has violated the setup and hold times.

If each of the three terms in the assertion expression is TRUE, then we have identified an error condition, in other words, a setup and hold timing failure. In this case, we need the assertion to fail, or to evaluate to FALSE, to trigger an assertion violation, so we negate the conjunction of the three terms to form the complete assertion expression. In other words, the complement of the asserted error condition is the desired asserted inviolate condition that must always exist for correct operation.

After a BOOLEAN expression that captures the desired conditions of setup and hold times is constructed, the **assertion** statement is completed by defining the **report** clause. The **report** clause

```
report "Setup/Hold Timing Violation";
```

contains a text string that will be issued to the designer if the setup and hold times are ever violated.

This example shows that the STABLE and DELAYED parameters need not be literal values of type TIME; the parameters can, in general, be expressions yielding values of type TIME. This example also shows that we can use a series of concatenated signal attributes. The signal CLK is used to generate a second signal CLK'DELAYED(HOLD), which is, in turn, used to generate a third signal CLK'DELAYED(HOLD)'STABLE.

17.8 ASSERTIONS

Now that we have introduced assertions to describe setup and hold timing requirements, we can examine the use and construction of **assertion** statements in more detail. There are several popular applications of assertions in VHDL models.

Assertions are often used to capture design constraints required for correct operation, also called *design intent*. Sometimes, specifying just the input/output transformation does not completely describe the proper behavior and operation of a component. For example, simply describing the synchronous behavior of a D flip-flop via the guarded assignments

```
CLKED: block (CLK = '1' and not CLK'STABLE)
begin
  -- Guarded concurrent signal assignment
  Q <= guarded D;
  Q_BAR <= not Q;
end block CLKED;
```

does not include the setup and hold timing requirements on D. Alternatively, describing an SR **nor** latch by its characteristic equation

```
Q <= S or (not R and Q);
Q_BAR <= not Q;
```

does not convey the important restriction that S and R must never be asserted at the same time. Thus, assertions provide a convenient mechanism to define caveats or special restrictions on particular signal values or timing conditions.

Assertions can also be used during model construction as a debugging aid. Assertions can be liberally inserted into the code to check for various conditions and report errors.[2] Finally, assertions can be used as high-level specifications or directives that must be satisfied as the digital system is designed.

The general form of a *concurrent assertion statement* is given below.

```
label : assert
          ( BOOLEAN expression )
          report STRING expression
severity SEVERITY_LEVEL expression;
```

The label is optional and may be omitted. The BOOLEAN expression defines the asserted condition that must always exist during normal operation; the expression should always evaluate to TRUE. Asserting *correct* behavior, in other words, behavior that must *always happen*, can sometimes be awkward because designers often focus on *incorrect* behavior, or behavior that must *never happen*. When focusing on correct behavior, we form the assertion expression as

```
(desired condition)
```

Alternatively, when focusing on incorrect behavior, we form the assertion expression as

```
not (undesired condition)
```

In other words, behavior that must always happen is simply the complement of the behavior that must never happen. The double negatives of the latter situation can admittedly be confusing at times.

A truth table often helps in formulating assertion expressions. For example, consider a truth table for the setup and hold timing assertion from the previous section. The first row identifies a falling transition or a '0' value for CLK, and the second row identifies a '1' value for CLK; these cases are not pertinent to our setup and hold timing check because the check is only concerned with the rising transition of CLK, so the assertion is TRUE. The third row identifies a rising clock transition and satisfied setup and hold times, so, again, the assertion is TRUE. The last row, however, identifies a setup and hold timing violation, so the

---

[2] Note that model debugging can also be accomplished by file input/output, which will be studied in Chapter 18.

CLK'DELAYED(HOLD)='1'	CLK'DELAYED(HOLD)'STABLE	D'STABLE(SETUP + HOLD)	Assertion
FALSE	–	–	TRUE
TRUE	TRUE	–	TRUE
TRUE	FALSE	TRUE	TRUE
TRUE	FALSE	FALSE	FALSE

assertion is FALSE. The assertion expression is formed as the complement of the must-never-occur condition, which is the conjunction of the first three column entries in the last row. If the asserted condition ever fails during simulation, in other words, evaluates to FALSE, an *assertion violation* occurs and the action taken depends on the **report** and **severity** clauses.

The **report** clause is optional and may be omitted. If the **report** clause is omitted, then the following default action occurs in response to an assertion violation:

- A default error message "Assertion violation" is issued.

If a **report** clause is supplied in an assertion statement, then the default error message is replaced by the user-defined text string, which must be of type STRING. (See Chapter 16 for a discussion of the predefined array type STRING and character string literals.)

The **severity** clause is also optional and may be omitted. As an example, the assertion statement describing the setup and hold timing requirements for the D flip-flop given in Figure 17.9 does not contain a **severity** clause. If the **severity** clause is omitted, then the following default action occurs in response to an assertion violation:

- A default severity of "ERROR" is assigned to the assertion violation.

If a **severity** clause is supplied in the assertion statement, then the default violation severity is replaced by the user-defined severity, which must be of type SEVERITY_LEVEL. The predefined type SEVERITY_LEVEL is an enumeration type composed of the four values: NOTE, WARNING, ERROR, and FAILURE. (See Chapter 16 for a discussion of the predefined enumeration type SEVERITY_LEVEL.) VHDL only defines the possible severity levels. VHDL *does not* define what actions are to be taken in response to a certain severity level; such actions are dependent on the user and/or computing environment.

The arguments for the **report** and **severity** clauses need not be static, literal values; they can in general be expressions that are dynamically computed during simulation to generate the appropriate report text string and severity level. For example, the following assertion checks that active-0 asynchronous set and clear signals (SET_BAR and CLR_BAR) should never be asserted at the same time. If an assertion violation occurs, a report text string is issued containing the particular values of SET_BAR and CLR_BAR that caused the problem.

```
assert not (SET_BAR='0' and CLR_BAR='0')
   report "Input Violation: SET_BAR = " & VALUE(SET_BAR) &
          ", CLR_BAR = "                 & VALUE(CLR_BAR);
```

The function VALUE yields a text string denoting the value of the signal given as its argument.

As a final topic on assertions, we will discuss *where* a concurrent assertion statement may be used in a VHDL model. There are three places a concurrent assertion statement may appear. As with all concurrent statements, a concurrent assertion statement may appear within the statement part of an architecture, as shown in Figure 17.9. A concurrent assertion statement may also appear within a block statement; however, the associated guard has no effect on the execution of the assertion unless the signal GUARD is explicitly referenced. Finally, a concurrent assertion statement may appear in the statement part of a design entity declaration, as shown in Figure 17.11.

Thus far in our study of VHDL, we have examined design entity declarations or interfaces that contain only declarations of generics and ports. In addition to *declarations*, Figure 17.11 shows that an interface may also contain *statements*. Thus, an entity declaration has the following general form.

```
entity EXAMPLE is
  -- Declarative part
    ....
begin
  -- Statement part
    ....
end EXAMPLE;
```

If there are no statements, the keyword **begin** is omitted.

Recall from the discussion on *scope* in Chapter 12 that anything declared within an entity declaration may be used within the entity declaration *and* all associated architec-

Figure 17.11

Assertions and design entities

```
entity D_FF is
  -- Declarative part
  generic
    (SETUP, HOLD : TIME);
  port
    (D, CLK    : in BIT;
     Q, Q_BAR : inout BIT);

begin
  -- Statement part
  assert not
    (CLK'DELAYED(HOLD)='1'            and
     not CLK'DELAYED(HOLD)'STABLE    and
     not D'STABLE(SETUP + HOLD))
  report "Setup/Hold Timing Violation";

end D_FF;

architecture DATA_FLOW of D_FF is
begin
  CLKED: block (CLK = '1' and not CLK'STABLE)
  begin
    -- Guarded concurrent signal assignment
    Q <= guarded D;
    Q_BAR <= not Q;
  end block CLKED;
end DATA_FLOW;
```

tures. Similarly, statements placed within an entity declaration apply to *all* associated architectures. Thus, placing the concurrent assertion statement for data setup and hold times in the entity declaration of D_FF is equivalent to writing the concurrent assertion statement in every associated architecture. Since the assertion on setup and hold timing requirements for the data input of the D flip-flop pertains to a property of a design entity interface, it is appropriately placed in the entity declaration rather than in a specific architecture. That is, one would expect the setup and hold timing requirements for the input signal D to apply to all alternative implementations.

17.9 VHDL-93: POSTPONED ASSERTIONS

Sometimes, assertions that check for steady-state conditions are mistakenly triggered by transient conditions. A transient logic value is called a *glitch*. A transient logic value lasting only a delta cycle is called a *delta glitch*. A false assertion triggering due to a glitch is called a *simulation-cycle sensitivity*. A false assertion triggering due to a delta glitch is called a *delta-cycle sensitivity*.

As a simple example of assertion delta-cycle sensitivities, consider the model shown in Figure 17.12 of parity detector logic. The parity detector combinational logic generates a '1' whenever an odd number of 1's are present at the inputs. The assertion statement is intended to check that the steady-state value of the output Z is always '1', indicating that the input bit vector A always has odd parity.

The timing diagram in Figure 17.13 (p. 546) shows that when the input signal A changes from B"10000" to B"11001" at $t = T_1$ and new logic values ripple through the **xor** stages, a glitch appears on the output Z as it transiently goes to '0' before reaching its steady-state value of '1'. This output glitch misfires the assertion statement at time $t = T_1 + \delta$. The change (falling transition) on the signal Z causes the assertion to execute to recheck the assertion condition. The BOOLEAN equality Z='1' fails, and the report clause is mistakenly issued.

Figure 17.12

Assertion delta-cycle sensitivity

```
entity ODD_PARITY_DETECT is
 port
    (A : in BIT_VECTOR(4 downto 0); -- MSB is parity bit
     Z : out BIT);

 begin
  assert (Z='1')
  report "Assertion Violation: Odd parity check failed.";
 end ODD_PARITY_DETECT;

architecture DATA_FLOW of ODD_PARITY_DETECT is
   signal INT1, INT2, INT3 : BIT;
 begin
  INT1 <= A(4) xor A(3);
  INT2 <= INT1 xor A(2);
  INT3 <= INT2 xor A(1);
    Z <= INT3 xor A(0);
  end DATA_FLOW;
```

Figure 17.13

Output delta glitch and assertion delta-cycle sensitivity

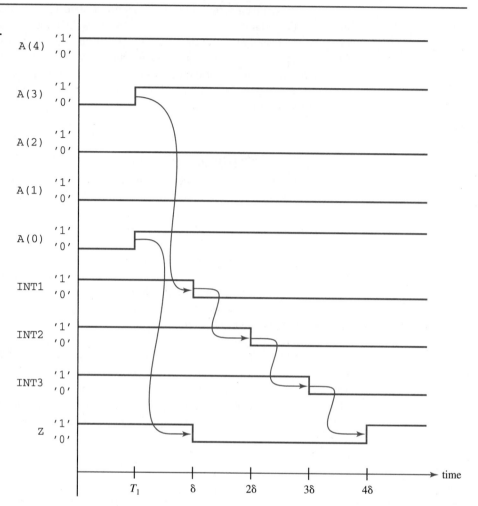

VHDL-93 addresses assertion delta-cycle sensitivity by introducing *postponed assertions*. Postponed assertions do not execute until the *last* delta cycle just prior to advancing real or absolute time. Waiting until the last delta cycle just prior to advancing absolute time ensures that all delta stages of logic have executed and that the combinational logic output values have reached a steady state.

Figure 17.14 shows how the delta-cycle sensitivity of the assertion in Figure 17.12 can be eliminated using a postponed assertion. The assertion in Figure 17.12 becomes a *postponed* assertion in Figure 17.14 by adding the leading keyword **postponed;** nothing else changes. Even though the postponed assertion is activated at time $t = T_1 + \delta$ when z transitions from '1' to '0', its execution is postponed until time $t = T_1 + 5\delta$; at this point, z has reached its steady-state value of '1' and the assertion does not misfire. We are assuming that after time $t = T_1 + 5\delta$, simulation time either ends or is advanced to a new absolute time $t = T_2$ for the next set of input values.

Figure 17.14

Postponed assertion, eliminating delta-cycle sensitivity	

```
entity ODD_PARITY_DETECT is
 port
    (A : in BIT_VECTOR(4 downto 0); -- MSB is parity bit
     Z : out BIT);

begin
 postponed assert (Z='1')
 report "Assertion Violation: Odd parity check failed.";
end ODD_PARITY_DETECT;

architecture DATA_FLOW of ODD_PARITY_DETECT is
   signal INT1, INT2, INT3 : BIT;
begin
 INT1 <= A(4) xor A(3);
 INT2 <= INT1 xor A(2);
 INT3 <= INT2 xor A(1);
    Z <= INT3 xor A(0);
end DATA_FLOW;
```

17.10 SYNCHRONOUS MACHINES

Thus far in this chapter we have mainly used simple memory devices to introduce VHDL constructs for modeling synchronous behavior. Let us now use these language constructs to model more complex sequential systems, namely, state machines.

As an example of a Mealy machine, Figure 17.15(b) (p. 548) shows a VHDL model of the sequential adder shown in Figure 17.15(a) and described earlier in Figure 8.21. The block DFF and associated guarded assignment describe the negative edge-triggered D flip-flop and next state combinational logic. The unguarded concurrent signal assignment describes the output combinational logic.

As an example of a Moore machine, Figure 17.16(b) (p. 549) shows a VHDL model of the binary counter shown in Figure 17.16(a) and described earlier in Figure 8.30. The DFF block and associated three guarded concurrent signal assignment statements model the three positive edge-triggered D flip-flops and associated next state logic. The last unguarded concurrent signal assignment statement models the output logic, which is simply the active-1, D flip-flop outputs. Note the declaration of the type of the signal Q holding the present state information, FF_TYPE. To align the signal names in the VHDL model with the signal names given on the logic schematic, we used the enumeration type FF_INDEX to define the indexing scheme for the array type FF_TYPE. Recall that an array may be indexed by any range constraint that defines a finite set of elements. Thus, the present state signal Q is an array of three elements of type BIT indexed by Q(A), Q(B), and Q(C).

17.11 RESOLVED AND GUARDED SIGNALS

What happens when multiple guarded assignments drive the same target signal? In this case, we need to resolve the contributions of the multiple guarded assignments into a single value for the common target signal, but the resolution is different from the resolution discussed in Chapter 15. We will review the properties of *resolved* signals discussed in Chapter 15 and then introduce *guarded* signals.

Figure 17.15

VHDL model of a sequential adder

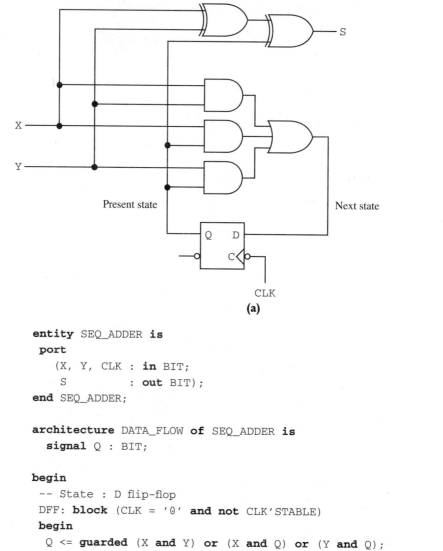

(a)

```
entity SEQ_ADDER is
 port
    (X, Y, CLK : in BIT;
     S          : out BIT);
 end SEQ_ADDER;


architecture DATA_FLOW of SEQ_ADDER is
   signal Q : BIT;

begin
 -- State : D flip-flop
 DFF: block (CLK = '0' and not CLK'STABLE)
 begin
  Q <= guarded (X and Y) or (X and Q) or (Y and Q);
 end block DFF;

 -- Output
 S <= X xor Y xor Q;
 end DATA_FLOW;
```

(b)

Figures (a) and (b) show the schematic and VHDL model of a sequential adder, respectively.

For the following discussion, it is useful to distinguish two conditions of a guarded concurrent signal assignment statement driving a resolved signal:

1. The condition when a guarded assignment is eligible/ineligible to execute and
2. The condition when a guarded assignment is serving/not serving as a source for its target resolved signal.

Figure 17.16

VHDL model of a binary counter

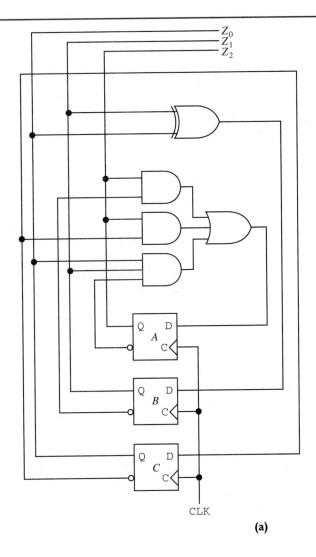

(a)

```
entity BIN_COUNTER is
 port
    (CLK : in BIT;
     Z    : out BIT_VECTOR(2 downto 0));
end BIN_COUNTER;

architecture DATA_FLOW of BIN_COUNTER is
  type FF_INDEX is (A, B, C);
  type FF_TYPE is array (FF_INDEX) of BIT;
  signal Q : FF_TYPE;
```

Figure 17.16 Continued

```
begin
  -- State : D flip-flops
  DFF: block (CLK = '1' and not CLK'STABLE)
  begin
    Q(A) <= guarded (Q(A) and not Q(B)) or (Q(A) and not Q(C)) or
                    (not Q(A) and Q(B) and Q(C));
    Q(B) <= guarded Q(B) xor Q(C);

    Q(C) <= guarded not Q(C);
  end block DFF;

  -- Output
  Z <= Q(A) & Q(B) & Q(C);
end DATA_FLOW;
```

(b)

Figures (a) and (b) show the schematic and VHDL model of a sequential binary counter, respectively.

The former condition is described by the terms *enabled/disabled*; the latter condition is described by the terms *connected/disconnected*.

To illustrate resolution of multiple guarded assignments, consider the following VHDL code segment.

```
signal DATA : WIRED_OR BIT;

B1: block ( MODE='1' ) -- Normal mode
begin
  S1 : DATA <= guarded NORMAL_DATA;
end block B1;

B2: block ( MODE='0' ) -- Test mode
begin
  S2 : DATA <= guarded TEST_DATA;
end block B2;
```

The signal declaration defines DATA as a resolved signal. Recall that a resolved signal may be driven by multiple sources, such as being the target of multiple concurrent signal assignment statements or being associated with multiple output ports. The values of the multiple sources are resolved into a single value via a resolution function. The signal DATA is of the predefined type BIT and has the resolution function WIRED_OR. Although not shown, we assume that the WIRED_OR function realizes the behavior of wired-**or** logic.

The VHDL code segment also contains two block statements. Block B1 controls the S1 concurrent signal assignment statement; block B2 controls the S2 concurrent signal assignment statement. S1 and S2 are user-defined labels (identifiers) that may be optionally added to concurrent signal assignment statements. We assume that NORMAL_DATA, TEST_DATA, and MODE are signals of type BIT.

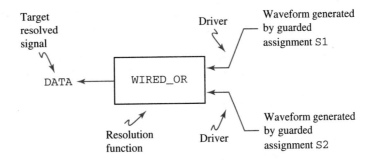

As discussed in Section 17.4, the guard expressions control when the guarded concurrent signal assignment statements driving the resolved signal DATA are eligible to execute. When MODE is '1', the guard expressions for blocks B1 and B2 are respectively TRUE and FALSE and the guarded assignments labeled S1 and S2 are respectively enabled and disabled. When MODE is '0', the guard expressions for blocks B1 and B2 are respectively FALSE and TRUE and the guarded assignments labeled S1 and S2 are respectively disabled and enabled.

It is important to note that regardless of whether the S1 and S2 guarded assignments are enabled or disabled, the guarded assignments are always *sources*, also called *drivers*, for DATA. In other words, the guarded assignments are always *connected* to the target signal DATA, as illustrated above. Any time the value of DATA is referenced, the resolution function WIRED_OR is automatically invoked to resolve the values contributed by the guarded assignments S1 and S2 into a single value.

17.11.1 Registers and Buses

There may be situations when we do not want a guarded assignment driving a particular resolved signal to always be connected as a source, in other words, to always contribute to the resolved signal value. In other words, we may want to selectively *disconnect* certain guarded assignments as sources to a resolved signal. For instance, if we require that the resolved signal DATA be the multiplexed value of *either* NORMAL_DATA when MODE='1' *or* TEST_DATA when MODE='0', then the above example is incorrect. Although the S2 guarded assignment is disabled when MODE='1', it is still serving as a source for DATA; hence, the value of DATA will be the logical **or** of the bit values generated by the S1 and S2 guarded assignments.

VHDL provides the ability to control whether a guarded assignment driving a resolved signal is connected or disconnected by declaring the target resolved signal to be a *guarded signal*. A guarded signal is a resolved signal with the keyword **register** or **bus** added to the signal declaration. Admittedly, the choice of the terminology "guarded signal" is a little confusing. Since any concurrent signal assignment statement having the keyword **guarded** is called a "guarded assignment," one might guess that any signal that is the target of a guarded assignment would be called a "guarded signal," but this is unfortunately not true. In VHDL, "guarded signals" refer to only those signals that are targets of guarded assignments *and* are declared as resolved signals having either the keyword **register** or **bus**. For a guarded signal, whenever the associated guarded assignment is not eligible to execute, the guarded assignment is also not serving as a source.

As an example, let us modify the VHDL code segment given earlier by changing DATA from a resolved signal to a guarded signal and reexamine the resulting behavior. Notice that the keyword **register** has been added to the declaration of DATA.

```
signal DATA : WIRED_OR BIT register;

B1: block ( MODE='1' ) -- Normal mode
begin
 S1 : DATA <= guarded NORMAL_DATA;
end block B1;

B2: block ( MODE='0' ) -- Test mode
begin
 S2 : DATA <= guarded TEST_DATA;
end block B2;
```

When MODE = '1', the guard expressions for blocks B1 and B2 are respectively TRUE and FALSE. In this case, the guarded assignment labeled S1 is *enabled* and *connected* and the guarded assignment labeled S2 is *disabled* and *disconnected*, as shown below. In other words, the S1 guarded assignment is eligible to execute in response to events occurring on

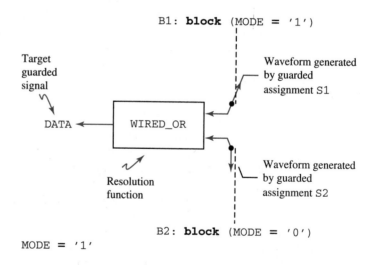

signal NORMAL_DATA and is serving as a source for DATA; the S2 guarded assignment is inel-igible to execute in response to events occurring on signal TEST_DATA and is not serving as a source for DATA. Since the guard expressions for blocks B1 and B2 are mutually exclusive, either S1 or S2 will be driving DATA at any given time, but never both S1 and S2. Thus, we have modeled the desired multiplexed behavior.

To understand the differences between resolved and guarded signals, study the timing di-agram given in Figure 17.17. The first three entries show example waveforms for the signals MODE, NORMAL_DATA, and TEST_DATA used in the VHDL code samples shown on page 550 and above.

The fourth waveform shows the driver for DATA generated by the S1 guarded assignment. Following the rules for guarded assignments, the rising transitions on MODE at times $t = T_1$ and $t = T_4$ change the B1 block guard from FALSE and TRUE, which, in turn, triggers the

Figure 17.17

Resolved and guarded signals

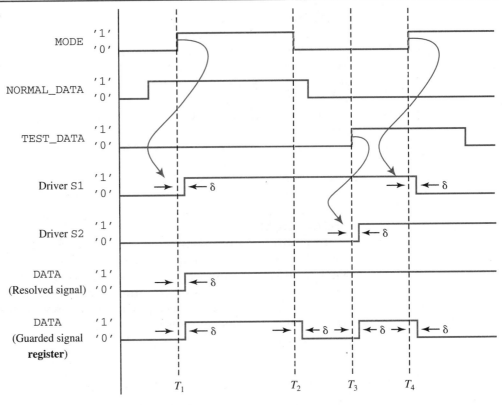

evaluation of the S1 guarded assignment. Hence, one delta cycle after $t = T_1$, the driver for S1 takes on the value of NORMAL_DATA, transitioning from '0' to '1'. Likewise, one delta cycle after $t = T_4$, the driver for S1 again takes on the value of NORMAL_DATA, transitioning from '1' to '0'. There are no other updates to the driver for the S1 guarded assignment because S1 is disabled while MODE is '0'.

The fifth waveform shows the driver for DATA generated by the S2 guarded assignment. Again, following the rules for guarded assignments, the falling transition on MODE at time $t = T_2$ changes the B2 block guard from FALSE and TRUE, which, in turn, triggers the evaluation of the S2 guarded assignment; the driver for S2 stays at '0'. However, since the S2 guarded assignment is enabled while MODE is '0', the rising transition on TEST_DATA at time $t = T_3$ triggers the reevaluation of S2. Consequently, one delta cycle later, the driver for S2 transitions from '·' to '1'.

The sixth waveform shows the target signal DATA, assuming DATA is a *resolved* signal as modeled on page 550. DATA is the resolved **or** of the drivers generated by the S1 and S2 guarded assignments. In particular, study the time period between $t = T_2$ and $t = T_3$. One might expect DATA to be '·' when TEST_DATA and NORMAL_DATA are both '0', but this is not the case. The S1 guarded assignment is disabled, so S1 does not respond to the falling transition on NORMAL_DATA and the associated driver retains its '1' value. Moreover, the S1 guarded assignment is always connected as a source for DATA and thus contributes to the resolution. The **or** of the drivers for S1 and S2 yield ′1′, so DATA remains at ′1′ between $t = T_2$ and $t = T_3$.

The last waveform shows the target signal DATA, assuming DATA is a *guarded* signal as modeled on page 552. Now, disabled guarded assignments are also disconnected and thus DATA is the multiplexed values of the drivers generated by the S1 and S2 guarded assignments. When MODE = ′0′, only driver S2 contributes to the resolution of DATA. When MODE = ′1′, only driver S1 contributes to the resolution of DATA. The resulting behavior of DATA as a guarded signal is notably different from the behavior of DATA as a resolved signal.

The only difference between a **register** signal and a **bus** signal is what happens when all guarded assignments associated with a guarded signal are disconnected. If we want a guarded signal to retain its previous value when all associated guarded assignments are disconnected, then the guarded signal is declared with the keyword **register**. When the inputs are "disconnected" from a register, the register continues to drive its output; likewise, when the sources are disconnected from a register guarded signal, the signal's value persists. Sometimes, a register guarded signal is simply called a *register*; registers can be viewed as a subclass of signals.

If we want a guarded signal to take on a default value when all associated guarded assignments are disconnected, then the guarded signal is declared with the keyword **bus**. When the inputs are disconnected from a bus, the bus often floats to a default value; likewise, when the sources are disconnected from a bus guarded signal, the signal takes on a default value supplied by the resolution function. Hence, resolution functions for bus guarded signals must account for the special case of supplying a resolved value even when called with no contributing sources.

17.12 DEFINITION OF CONCURRENT STATEMENTS

At this point, you should have a good understanding of concurrent statements. Over several chapters, we have explained how to use concurrent statements to describe a wide range of combinational and sequential digital system behavior. Although concurrent statements are clearly an important part of VHDL, you may be surprised to learn that from a *language point of view* concurrent statements are generally *not* considered to be basic behavioral constructs.

The *process* statement is considered to be the basic behavioral statement in VHDL. All time-dependent behavior in VHDL is defined in terms of the process statement.

Concurrent statements, such as concurrent signal assignment statements and concurrent assertion statements, are in fact shorthand notations for certain commonly used forms of the *process statement*.

For example, the conditional concurrent signal assignment statement used in Figure 17.1 to drive the Q output of a JK latch

```
Q <= Q     when (J='0' and K='0') else
     '0'    when (J='0' and K='1') else
     '1'    when (J='1' and K='0') else
     Q_BAR ;
```

is actually defined by the following process statement.

```
process
begin
 if (J='0' and K='0') then
    Q <= Q;
 elsif (J='0' and K='1') then
    Q <= '0';
 elsif (J='1' and K='0') then
    Q <= '1';
 else
    Q <= Q_BAR;
 end if;

 wait on J, K, Q, Q_BAR;
end process;
```

Process statements will be discussed in detail in Chapter 18. For the present, it is sufficient to know that a process statement supports *only sequential statements* that execute in a procedural fashion from top to bottom in the textual order they are written. When an event occurs on J, K, Q, or Q_BAR, the **if** statement executes, the appropriate sequential signal assignment statement updates the value of Q, and then process execution suspends until another event occurs on the J, K, Q, or Q_BAR by executing the **wait** statement. Such an event causes the process to exit the wait statement and resume execution; control implicitly returns to the beginning of the process statement, the **if** statement executes, and the sequence repeats. Thus, the combined behavior of the **if**, **wait**, and signal assignment sequential statements contained within the process statement is *equivalent* to the behavior of the conditional concurrent signal assignment statement. In other words, the process and conditional concurrent signal assignment statements will generate identical waveforms for the signal Q.

As another example of the equivalency between behavioral concurrent statements and process statements, the following concurrent assertion statement that checks that the S and R inputs of an SR **nor** latch must never be both asserted at the same time

```
assert not (S='1' and R='1')
   report "Input Violation: S and R simultaneously asserted.";
```

is equivalent to the following process statement.

```
process
begin
   assert not (S='1' and R='1')
      report "Input Violation: S and R simultaneously asserted.";

   wait on S, R;
end process;
```

The *sequential* assertion statement given within the body of the process statement is similar to the *concurrent* assertion statement, except that the sequential assertion statement is executed only when encountered by the procedural flow of control associated with executing the process statement. When the signals R and/or S change value, the body of the process statement is executed. The sequential assertion statement executes, and the report message is posted to the designer if both R and S are asserted active-1. Then, the wait statement executes, suspending process execution until another event occurs on R and/or S. When the process resumes execution, control implicitly returns to the beginning of the process statement, the sequential assertion statement executes, and the sequence repeats.

In summary, from a language perspective, behavioral concurrent statements can be viewed as "pseudo statements" because they are derived from the basic process statement. In principle, a concurrent signal assignment statement can always be replaced by its equivalent process statement without changing the modeled behavior. In practice, however, such a description might be overly verbose and difficult to understand—not good qualities of a hardware description language. Since data flow modeling is widely used in digital design, a set of behavioral concurrent statements are included in VHDL as a convenience for the designer. These behavioral concurrent statements serve as "macros" that expand into equivalent process statements. Thus, data flow modeling can be viewed as an abstraction built on top of the basic behavioral modeling provided via the process statement.

We will learn in future chapters that the process statement provides a general behavioral modeling capability that includes algorithmic, as well as data flow, abstractions. A designer need not understand the equivalency between behavioral concurrent statements and process statements to write data flow models, however, such knowledge leads to a better understanding of VHDL.

17.13 SIMULATION CYCLE

Since this chapter has been largely concerned with timing-related modeling issues, it is appropriate to return to a discussion of the *simulation cycle* that defines the dynamic semantics of VHDL. Recall that the simulation cycle (introduced in Chapter 15) defines the behavior of a VHDL model over time, in other words, how a VHDL model executes. Having introduced process statements as the basic behavioral statements of VHDL in the previous section, we can now reexamine the VHDL simulation cycle in more detail.

The simulation cycle is included as part of the language definition to prevent the situation where a VHDL model behaves differently on different simulators. If two different results are obtained by executing a VHDL model on two different simulators, the discrepancy should be resolvable based on the correct behavior defined by the simulation cycle.

The simulation cycle first illustrated in Figure 15.9 is repeated in Figure 17.18. The simulation cycle defines an *event-driven* simulation because actions are initiated based on events, or signals changing value. At any given time, a certain set of signals change value. The model response to these events is obtained by executing a set of processes, which compute new signal values, in other words, a set of events that will occur in the future. Simulation time is then advanced to the next event(s) and the model response is again obtained. The following sections discuss the parts of the simulation cycle, starting with how a model is *initialized*.

17.13.1 Initialization

To begin a simulation, a VHDL model must first be initialized, which means that all signals must be assigned initial waveforms. The steps involved in initializing a model are listed below.

1. The initial waveform of each declared signal is computed; this waveform is assumed to have persisted for an infinite length of time in the past.
2. Each guard expression is evaluated.
3. Each process is executed until it suspends.

Figure 17.18

The simulation cycle

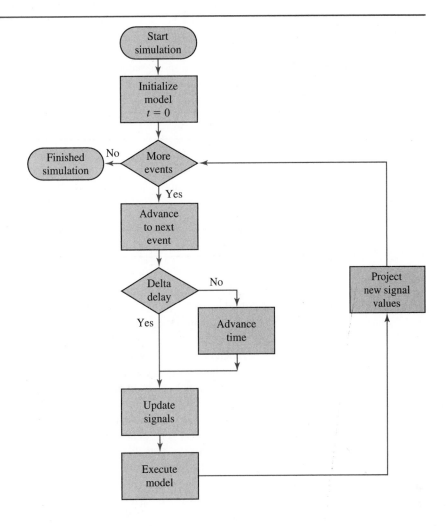

Initial signal values may be *explicitly* or *implicitly* defined. Initial signal values are explicitly defined by adding *default expressions* to signal declarations

```
-- Signal declarations with default expressions
signal TIE_ZERO : BIT := '0';
signal MASK : BIT_VECTOR(0 to 3) := B"1100";
```

or to port declarations.

```
-- Port declarations with default expressions
port
  (OPR_A : in   BIT := '0';
   DATA  : inout BIT := '1');
```

If an initial signal value is not explicitly defined, then an initial value is implicitly defined based on the signal or port type. If a signal or port is of a scalar type T, such as an enumeration type, integer type, floating point type, or physical type, then the implicit initial value is the value of the predefined attribute T'LEFT. For instance, signals or ports of the type BIT are implicitly initialized to the value '0' because BIT'LEFT, or the leftmost enumeration literal, for the enumeration type declaration for BIT

```
type BIT is ('0', '1');
```

---

is '0'. Similarly, signals and ports of the type STD_ULOGIC defining the standard IEEE 1164 9-valued logic system are implicitly initialized to the value 'U' because STD_ULOGIC'LEFT for the enumeration type declaration for STD_ULOGIC

```
type STD_ULOGIC is ('U', 'X', '0', '1', 'Z', 'W', 'L', 'H', '-');
```

---

is 'U', appropriately denoting the uninitialized value. (See Chapter 16 for a discussion of *IEEE 1164-1993* standard.)

If a signal or port is of a composite type, such as an array type, then the implicit initial value of each scalar element of type T is defined to be the value of the predefined attribute T'LEFT. For instance, each element of a signal or port of the array type BIT_VECTOR is implicitly initialized to the value '0' because the element type is BIT and BIT'LEFT = '0'.

When determining initial signal values using either the implicit or explicit definitions explained above, we must be careful with situations involving signal-to-signal associations in structural descriptions. As the name implies, a signal-to-signal association given in the port map of a component instantiation statement essentially links two signals as segments of a single network. The initial values of signals that are segments of a network cannot be computed independently; all signals associated with a particular network must have the *same* initial value, which is the initial value of the network.

For instance, consider the simple VHDL structural model outlined in Figure 17.19. The association A => HOOKUP in the I1 component instantiation statement in STRUCTURE architecture links the signals HOOKUP and A together as segments of a network. Notice that the default expression given in the port declaration of A in the DESIGN_SUBENTITY design entity explicitly defines A to have the initial value '0', assuming an implicit component configuration associating the DSGN_SUBENTITY component with the DSGN_SUBENTITY design entity. However, the default expression given in the signal declaration of HOOKUP in the DSGN_ENTITY design entity explicitly defines HOOKUP to have the initial value '1'. Since segments of a network cannot have different initial values, what is the initial value of HOOKUP and A? The answer is that A assumes the initial value of HOOKUP, which is '1'. An **in** port assumes an initial value according to the implicit and explicit rules discussed above only when the port is unconnected, in other words, is not participating in a larger network. When an **in** port is connected to a larger network, the **in** port takes on the initial value of its driving signal.

17.13.2 Event-Driven Simulation

Now that we know how a VHDL model is initialized, let us review the actions of the VHDL simulation cycle by examining the behavior of the simple clock generator given in Fig-

Figure 17.19

Initial value of networks

```
entity DSGN_SUBENTITY is
 port
    (A : in BIT := '0'; ... );
end DSGN_SUBENTITY;

architecture BEHAVIOR of DSGN_SUBENTITY is
begin
   ....
end BEHAVIOR;

entity DSGN_ENTITY is
 port
    (P1 : in BIT; ... );
end DSGN_ENTITY;

architecture STRUCTURE of DSGN_ENTITY is
 component DSGN_SUBENTITY
   port (A : in BIT; ...);
 end component;

 signal HOOKUP : BIT := '1';
begin
         ...
 I1 : DSGN_SUBENTITY port map (A => HOOKUP: ....);
         ...
end STRUCTURE;
```

ure 17.20 (p. 560). A timing diagram for the signal CLK will be developed by repeatedly executing the VHDL simulation cycle shown in Figure 17.18.

First, the VHDL model, CLOCK_GEN, is initialized. There is only one declared signal, CLK. The default value given in the port declaration explicitly initializes CLK to '1' at simulation time $t = 0$. Next, the guard expression (CLK = '1') is evaluated and, based on the initial value of CLK, the implicitly declared signal GUARD is set to TRUE. Then, the process statement

```
process
begin
 if GUARD then
  CLK <= not CLK after (CYCLE_TIME/2), CLK after CYCLE_TIME;
 end if;

 wait on GUARD, CLK;
end process;
```

Figure 17.20

**VHDL description of a
clock generator**

```
entity CLOCK_GEN is
  generic (CYCLE_TIME : TIME);
  port (CLK : inout BIT := '1');
end CLOCK_GEN;

architecture DATA_FLOW of CLOCK_GEN is
begin
  CYCLE: block (CLK = '1') -- Guard expression
  begin
    -- Guarded concurrent signal assignment
    CLK <= guarded not CLK after (CYCLE_TIME/2), CLK after CYCLE_TIME;
  end block CYCLE;
end DATA_FLOW;
```

representing the guarded concurrent assignment statement is executed, which generates the two events shown in Figure 17.21(a) at times $t = CYCLE\_TIME/2$ and $t = CYCLE\_TIME$. At this point, the clock generator design entity is initialized.

Now, the first simulation cycle begins. Time is advanced to the next event, $t_{NOW} = CYCLE\_TIME/2$. Signal values are updated; CLK changes from '1' to '0', and the guard expression changes from TRUE to FALSE. Next, the process representing the guarded assignment that is suspended on the wait statement "wakes up" in response to the events on CLK and GUARD, but no new values are assigned to CLK because the guard is FALSE. The first simulation cycle is now complete.

The second simulation cycle executes in a similar fashion. Time is advanced to the next event, $t_{NOW} = CYCLE\_TIME$. Signal values are again updated; CLK changes from '0' to '1' and the guard expression changes from FALSE to TRUE. In response to these two signal events, the process representing the guarded assignment again "wakes up" and executes. Since the guard is now TRUE, two new events shown in Figure 17.21(b) are posted to CLK at times $t = 1.5 \times CYCLE\_TIME$ and $t = 2 \times CYCLE\_TIME$. The second simulation cycle is now complete.

For this model, simulation never terminates because the clock generator always schedules new events. Under this situation, simulation is typically terminated after a user-specified period of time.

17.14 SUMMARY

Building on the basic propagation timing information introduced in Chapter 15, we discussed advanced timing issues in VHDL in this chapter. Level-sensitive and edge-sensitive synchronous behavior is described via *guard expressions* and *guarded assignments*. Guard expressions provide a general capability to describe an arbitrarily complex condition controlling whether certain outputs respond to certain inputs; in other words, whether certain signal assignments are enabled/disabled. If the keyword **guarded** is added to a concurrent signal assignment statement, then the resulting statement is called a *guarded assignment* and is enabled only when the associated guard expression is true.

Figure 17.21

Timing diagram for CLK

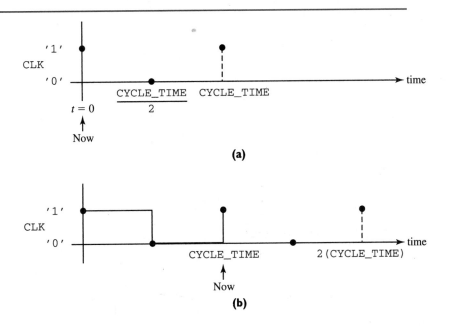

(a)

(b)

If a guarded assignment is driving a resolved signal, then it is important to distinguish when the guarded assignment is *connected/disconnected*, as well as *enabled/disabled*. If the resolved signal is declared as either a *register* or a *bus*, then associated guarded assignments are connected, or acting as sources, only when the guard expression is true. Thus, guarded assignments driving guarded signals provide the ability to describe selective synchronization.

To describe various timing relationships, we introduced predefined signal attributes and assertions. Predefined signal attributes provide an easy way to query information about the present and past values of signals. Assertions describe conditions that must always exist during the operation of a VHDL model and are used for conducting model debug, issuing synthesis directives, and describing design constraints. A design constraint typically supplements an input/output transformation description, addressing assumptions concerning illegal inputs, unused operating modes, and required timing conditions. Assertions also introduced the ability to include statements, as well as declarations, in an entity declaration. A concurrent assertion statement placed in an entity declaration applies to all associated architectural bodies. Postponed assertions provided in VHDL-93 eliminate transient delta-cycle misfirings of assertions checking for steady-state conditions.

A process is the basic behavioral statement in VHDL. It is often helpful to view VHDL as fundamentally a collection of processes communicating through signals. Behavioral concurrent statements, such as signal assignments and assertions, are merely shorthand notations for certain forms of process statements commonly used in data flow modeling. VHDL constructs like entities, architectures, packages, components, and blocks constitute a scaffolding provided to aid in conceptualizing and creating many intercommunicating processes. In line with this perspective, the VHDL simulation cycle is defined in terms of the repetitive execution of processes, posting new output signal events in response to input signal events.

Finally, we also discussed the VHDL-93 feature of the new signal value **unaffected**. The signal value **unaffected** effectively assigned nothing to the target signal, leaving the target signal unchanged.

17.15 PROBLEMS

1. Explain why the following model of a JK latch does *not* work.

```
entity JK_LATCH is
port
   (J, K  : in BIT;
    Q     : inout BIT := '0';
    Q_BAR : inout BIT := '1');
end JK_LATCH;


architecture TRUTH_TABLE of JK_LATCH is
begin
   -- Map truth table into conditional concurrent signal
   -- assignment statement
   Q <= Q   when (J='0' and K='0') else
        '0' when (J='0' and K='1') else
        '1' when (J='1' and K='0') else
        not Q;

   Q_BAR <= not Q;
end TRUTH_TABLE;
```

2. Describe a T latch in VHDL.
   a. Use the following truth table.

Row	T	Q^+	Action
0	0	Q	No change
1	1	$\overline{Q}$	Toggle state

   Verify the VHDL model by simulation for $T = 1$ and $T = 0$.
   b. Use the following characteristic equation.

$$Q^+ = T \oplus Q$$

   Verify the VHDL model by simulation for $T = 1$ and $T = 0$.

3. Describe a gated T latch in VHDL, using the following truth table.
   a. Use only concurrent signal assignment statements with no block statements.

Row	CT	Q^+	Action
0	0 –	Q	No change
1	1 0	Q	No change
2	1 1	$\overline{Q}$	Toggle state

b. Use concurrent signal assignment and block statements.

c. Compare and contrast the two models generated in parts (a) and (b).

4. We want to enable guarded signal assignments when a control signal named CONTROL_SIGNAL of type BIT equals '1'. Why is the following block statement incorrect?

```
B1: block (CONTROL_SIGNAL)
    begin
        . . . . . . . . .
    end block B1;
```

---

5. Consider the following concurrent signal assignment statements.

```
B1: block (SYNC_BAR = '0')
    begin
      W <= guarded A or B after 5 min;
      X <= A or B after 5 min;
      Y <= '1' after 10 min, '0' after 20 min;
      Z <= guarded '1' after 5 min;
    end block B1;
```

---

a. Complete the timing diagram below by generating the waveforms for signals W, X, Y, and Z. Label all signal updates.

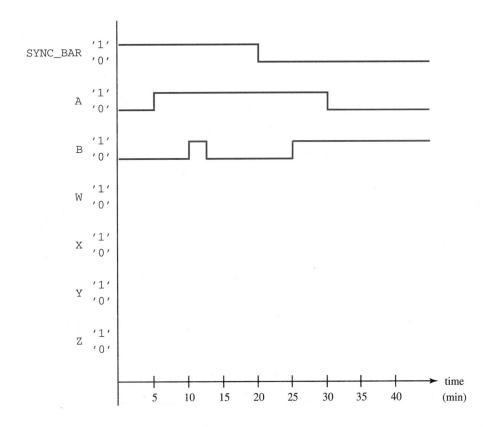

6. Consider the following concurrent signal assignment statements.

```
B1: block (SYNC_BAR = '0')
begin
  X <= A when (not GUARD) else
       B;
  Y <= guarded A when (not GUARD) else
       B;
end block B1;
```

a. Complete the timing diagram below by generating the waveforms for signals X and Y. Label all signal updates.

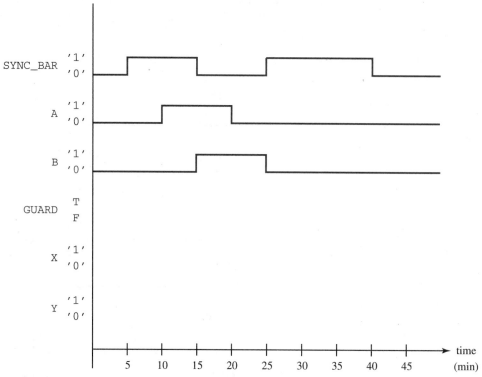

b. Identify which updates are transactions and events.

7. Consider the following concurrent signal assignment statements.

```
B1: block (CK_ONE = '1')
begin
  X <= guarded A or B after 5 min;
  Y <= A or B after 5 min;

  B2: block (GUARD and not CK_ONE'STABLE)
  begin
    Z <= guarded '1' after 5 min;
  end block B2;
end block B1;
```

a. Complete the timing diagram below by generating the waveforms for signals X, Y, and Z. Identify all transactions and events.

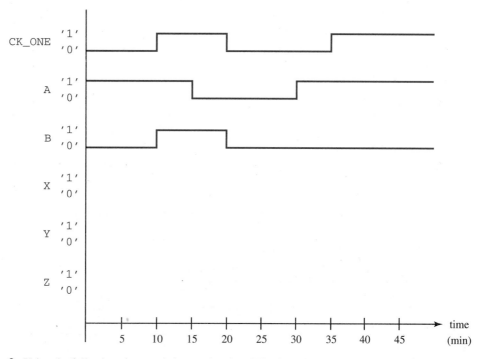

8. Using the following characteristic equation for a T latch, generate the following VHDL models.

$$Q^+ = T \oplus Q$$

a. A VHDL model for a positive edge-triggered T flip-flop
b. A VHDL model for a negative edge-triggered T flip-flop

9.

```
entity D_DEVICE is
 port
    (D, CLK    : in BIT;
     Q, Q_BAR : inout BIT);
end D_DEVICE;

architecture DATA_FLOW of D_DEVICE is
begin
 CLKED: block (CLK'EVENT and CLK='1')
 begin
  Q <= guarded D;
  Q_BAR <= not Q;
 end block CLKED;
end DATA_FLOW;
```

a. Why does the above VHDL model *not* describe a positive edge-triggered D flip-flop? (*Hint*: Think of the guard expression as a concurrent signal assignment statement driving the target signal GUARD.)
b. What kind of memory device does the above VHDL model describe?

10. For the signal waveform for Z generated in Problem 7, show the following predefined attributes.

  a. Z'ACTIVE   **b.** Z'QUIET
  c. Z'TRANSACTION **d.** Z'LAST_ACTIVE
  e. Z'EVENT   **f.** Z'STABLE(5 min)
  g. Z'LAST_VALUE   **h.** Z'DELAYED(10 min)

11. Consider the following timing diagram for signal SIG_X.

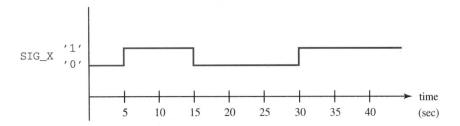

 a. Draw and explain the waveforms for the following signals X and Y.

```
X <= SIG_X'DELAYED(5 sec);
Y <= SIG_X after 5 sec;
```

 b. Draw and explain the waveforms for the following signals X and Y.

```
X <= SIG_X'LAST_VALUE;
Y <= SIG_X'LAST_VALUE after SIG_X'LAST_EVENT;
```

 c. Draw and explain the waveforms for the following signals X and Y.

```
X <= SIG_X and SIG_X'DELAYED(5 sec);
Y <= not (SIG_X xor SIG_X'DELAYED(5 sec));
```

12. Show an alternative way of writing the assertion for setup and hold times for the D flip-flop shown in Figure 17.9 without using the DELAYED predefined attribute.

13. Write the function VALUE shown in Section 17.8, place the function in a package, and use the function to write a model for an SR **nor** latch that includes an assertion statement that checks that S and R must not be asserted active-1 at the same time. Verify by simulation that the SR latch and assertion operate properly.

14. Write an assertion that checks for a programmable active-1 hold time and reports a severity based on how badly the active-1 hold timing requirement is violated.

Timing Violation	Severity
Percent Error <= 10%	NOTE
10% < Percent Error <= 30%	WARNING
30% < Percent Error <= 50%	ERROR
Percent Error > 50%	FAILURE

Verify by simulation that the assertion properly detects each level of severity.

15. Generate a VHDL model of the Moore machine analyzed in Problem 13(a) in Chapter 8 using concurrent signal assignment statements. The schematic is repeated on the following page.

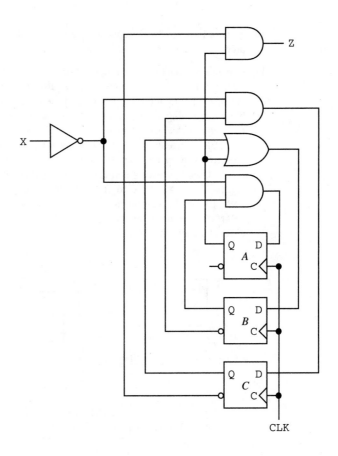

16. Generate a VHDL model of the Mealy machine analyzed in Problem 14 in Chapter 8 using concurrent signal assignment statements. The schematic is repeated below.

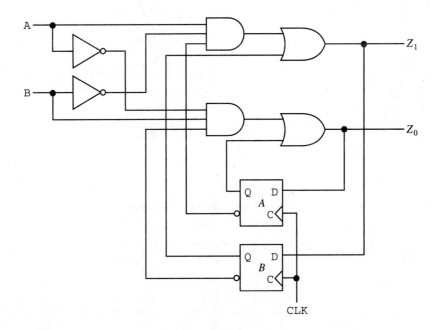

17. Rewrite Problem 16 by replacing the concurrent signal assignment statements by their equivalent process statement representations.

18. Generate a model of a communications arbiter. (Show all your work.) Based on a simple handshaking protocol, the communications arbiter connects one of two users to a communications channel. The input/output behavior of the communications arbiter is described by the following block and timing diagrams. When the communication channel STATUS is IDLE, user 1 or user 2 may request channel use by respectively changing the TR_1 or TR_2 signals from INACTIVE to ACTIVE. When TR_1 or TR_2 changes from INACTIVE to ACTIVE, the communications arbiter respectively connects DATA_1 or DATA_2 to CHANNEL and sets the channel STATUS to BUSY. User 1 or user 2 notifies the end of channel use by respectively changing TR_1 or TR_2 signals from ACTIVE to INACTIVE. When TR_1 or TR_2 changes from ACTIVE to INACTIVE, the communications arbiter respectively disconnects DATA_1 or DATA_2 from CHANNEL and resets the channel STATUS to IDLE. When neither user 1 nor user 2 is using the channel, the channel "floats" to X"00".

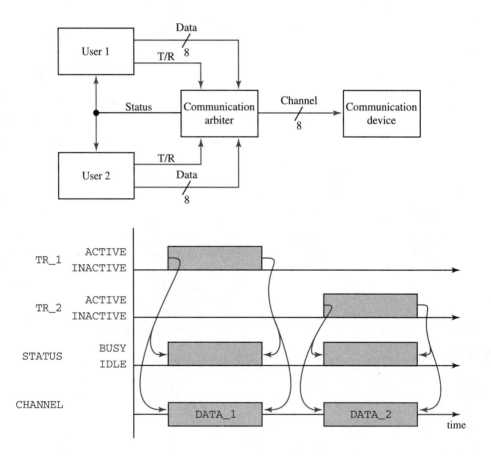

19. Describe the operation of the 4-bit circular shift register given in Figure 17.6 for five clock cycles, simulation cycle by simulation cycle. Assume the first clock cycle executes a reset.

20. VHDL-93: Add an assertion to the JK flip-flop model in Figure 17.5 that checks that the outputs Q and Q_BAR are always complementary. Eliminate any delta-cycle sensitivities, and verify by simulation.

21. VHDL-93: Design a combinational digital system that accepts 3-bit integers and generates the 3-bit two's complement. The input -4_{10} is considered an illegal input.

     a. Describe the design in VHDL using data flow modeling. Verify by simulation that the VHDL model describes the logic truth table.

     b. Add assertions that check for the illegal input -4_{10} and for correct two's complement output. Eliminate any delta-cycle sensitivities, and verify by simulation.

22. <u>VHDL-93:</u> Draw the waveforms for the following signals X and Y.

    a. `X <= '1' after 5 ns, '0' after 10 ns;`
       `Y <= '1' after 10 ns, '0' after 20 ns when (X='1') else`
          `Y;`

    b. `X <= '1' after 5 ns, '0' after 10·ns;`
       `Y <= '1' after 10 ns, '0' after 20 ns when (X='1') else`
          `unaffected;`

REFERENCES AND READINGS

[1] Belhadj, H., L. Gerbaux, M. Bertrand, and G. Saucier. "Specification and Synthesis of Communicating Finite State Machines," *IFIP Transactions A*, vol. A-22, pp. 91–102, 1993. Proceedings of WG10.2/WG10.5 Workshop.

[2] Debriel, A. and P. Oddo. "Synchronous Design in VHDL," *Proceedings of the European Design Automation Conference*, pp. 680–681, September 1992.

[3] Helbig, J., R. Schlor, W. Damm, G. Dohmen, and P. Kelp. "VHDL/S—Integrating Statecharts, Timing Diagrams, and VHDL," *19th Symposium on Microprocessing and Microprogramming, EUROMICRO*, pp. 571–580, September 1993.

[4] Huang, C., J. Lis, M. Quayle, and S. Shroff. "RTL Controller Synthesis," *IFIP Transactions A*, vol. A-22, pp. 3–17, 1993. Proceedings of WG10.2/WG10.5 Workshop.

[5] Markowitz, M. "VHDL Design Project, Part 1-6," *EDN*, vol. 38, no. 1-6, 1993.

[6] Navabi, Z. *VHDL: Analysis and Modeling of Digital Systems.* New York: McGraw-Hill, 1993.

[7] Spillane, J. and Z. Navabi. "Describing Controlling Hardware in VHDL," *Proceedings of the 4th IEEE International ASIC Conference and Exhibit*, September 1991.

[8] Tiedmann, W., S. Lenk, G. Grobe, and W. Grass. "Introducing Structure into Behavioral Descriptions Obtained from Timing Diagrams," *19th Symposium on Microprocessing and Microprogramming, EUROMICRO*, pp. 581–588, September 1993.

[9] Vahid, F. "Obtaining Functionally Equivalent Simulations Using VHDL and a Time-Shift Transformation," University of California, Irvine, Semiconductor Research Corporation (SRC) Publication C91830, November 1991.

[10] Wang, X., S. Grainger, and A. Cooper. "Behavioral VHDL Code Generation for Synchronous FSMs," *Proceedings of the 5th IEEE International ASIC Conference and Exhibit*, pp. 529–532, September 1992.

18
ALGORITHMIC MODELING IN VHDL

In Chapter 9, we discussed the use of state diagrams/tables as abstract representations of synchronous sequential systems. In this chapter, we will investigate how to describe these sequential machine abstractions in VHDL via *algorithmic* modeling techniques.

Recall from Chapter 13 that hardware modeling can be broadly divided into two styles: structural and behavioral. Behavioral modeling styles can be further divided into data flow and algorithmic. Data flow descriptions are *nonprocedural* and use *concurrent* statements, whereas algorithmic descriptions are *procedural* and use *sequential* statements. Algorithmic modeling in VHDL is similar to conventional software programming in Pascal, C, or Ada[1]; this similarity has advantages and disadvantages. An advantage is that you may already be familiar with many of the language constructs that we will discuss, such as **if** and **for-loop** statements. This familiarity can also be a disadvantage when one ignores the concurrency or parallel modeling aspects of VHDL and tries to turn all hardware models into software programs.

We have delayed introducing algorithmic modeling to emphasize the use of the more hardware-oriented descriptive styles of structural and data flow modeling techniques. We introduce algorithmic modeling at this point in the text to support high-level descriptions of digital systems aligned with sequential machine abstractions. Another reason for delaying algorithmic modeling is that the material can get a little dry, as we need to examine several new language constructs. Covering this material at the outset tends to be a rather unfriendly introduction to VHDL. However, we will try to introduce algorithmic modeling as gently as possible by interjecting illustrating examples.

As mentioned above, we will introduce several new VHDL statements. As a roadmap for this chapter, the VHDL statements and associated sections are summarized below.

VHDL Statement	Section
Process and Wait	18.2
Variable and Sequential Signal Assignment	18.3
If and Case	18.4
Sequential Assertion	18.5
Report	18.6
Loop, Next, and Exit	18.7
Procedure Call and Return	18.8
Null	18.9

In addition to sequential statements in VHDL, we will also introduce new data types to improve our information modeling skills. We will explain subtypes, user-defined integer types, and user-defined floating point types, as well as file types and file input/output.

---

[1] Ada® is a registered trademark of the U.S. Department of Defense.

Finally, we conclude the chapter with a few new VHDL-93 features. These features include sequential statement labels and new predefined attributes for processing character strings.

18.1 DESCRIBING STATE MACHINES: A FIRST LOOK

To introduce algorithmic modeling, we will revisit the binary counter discussed in Chapter 8; Figure 18.1 shows the state diagram. The counter is a Moore machine; state transitions are controlled by the clock. Figure 17.16 in Chapter 17 presented a data flow model of the binary counter. Figure 18.2 presents an alternative algorithmic model of the binary counter.

The binary counter is described by a single **process** statement contained within the architecture ALGORITHM. The **process** statement is the basic behavioral statement in VHDL. In Chapter 17, we studied certain forms of the **process** statement used to define behavioral concurrent statements. In this chapter, we will expand our discussion of the **process** statement and explain its general use in algorithmic modeling.

The **process** statement within the ALGORITHM architecture begins with the reserved keyword **process**, followed by the declaration of the variable PRESENT_STATE. PRESENT_STATE is a 3-bit vector and is initialized to the value 111_2 using the bit string literal B"111". After the variable declaration, three executable statements are contained in the body of the **process** statement following the keyword **begin**: a **case** statement, a **signal**

Figure 18.1

Binary counter state diagram

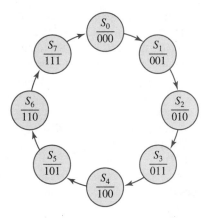

Figure 18.2

Algorithmic VHDL model of a binary counter

```
entity BIN_COUNTER is
 port
    (CLK : in  BIT;
     Z   : out BIT_VECTOR(2 downto 0));
 end BIN_COUNTER;


architecture ALGORITHM of BIN_COUNTER is
 begin
  process
   -- State information held in PRESENT_STATE
   variable PRESENT_STATE : BIT_VECTOR(2 downto 0) := B"111";
```
(Figure 18.2 continued on page 572)

Figure 18.2 continued

```
begin
   -- Update present state
   case PRESENT_STATE is
      when B"000" =>   PRESENT_STATE := B"001";
      when B"001" =>   PRESENT_STATE := B"010";
      when B"010" =>   PRESENT_STATE := B"011";
      when B"011" =>   PRESENT_STATE := B"100";
      when B"100" =>   PRESENT_STATE := B"101";
      when B"101" =>   PRESENT_STATE := B"110";
      when B"110" =>   PRESENT_STATE := B"111";
      when B"111" =>   PRESENT_STATE := B"000";
   end case;

   -- Update output
   Z <= PRESENT_STATE after 10 ns;

   -- Wait until next rising edge of clock
   wait until (CLK = '1');
 end process;
end ALGORITHM;
```

assignment statement, and a **wait** statement. These executable statements describe the input/output behavior of the counter. The **case**, **signal** assignment, and **wait** statements are called *sequential* statements because they execute in the textual sequence they appear, from top to bottom. All statements contained within a process are sequential statements.

To understand how the BIN_COUNTER design entity describes the state diagram in Figure 18.1, we will first look at the **wait** statement. The **wait** statement determines when the enclosing **process** statement executes. The **wait** statement suspends execution of the **process** statement until the clock signal, CLK, changes value and (CLK = '1'); in other words, until a rising clock edge occurs. When a rising edge occurs on CLK, the wait condition is satisfied and the process exits the **wait** statement, implicitly returns to the beginning of the process body, and executes the three sequential statements, starting with the **case** statement.

The **case** statement encodes the state table, defining the present-state-to-next-state transitions. The **when** clause matching the value of PRESENT_STATE determines which variable assignment statement executes to update PRESENT_STATE to the next state. For instance, if PRESENT_STATE equals 010_2, the third **when** clause

```
when B"010" =>   PRESENT_STATE := B"011";
```

matches and the next value of PRESENT_STATE is 011_2. This update describes the S_2 to S_3 transition in Figure 18.1. Recall from Chapter 15 that a variable assignment uses the := operator and that the assignment takes effect immediately.

After the **case** statement executes and updates PRESENT_STATE, the signal assignment statement executes. The output Z is updated based on the new state; the input/output delay is 10 nanoseconds. Be careful not to confuse the *sequential* signal assignment statement in Figure 18.2 with the *concurrent* signal assignment statements studied in previous chapters. A sequential signal assignment statement executes only when encountered by the procedural flow of control, regardless of whether any signals on the right-hand side change value. Sequential and concurrent signal assignment statements may look syntactically alike, but they appear in different contexts and have different execution semantics.

After the signal assignment statement executes and updates z, the **wait** statement executes, which suspends process execution until the next rising clock edge. When the next rising clock edge occurs, the process "wakes up" and again executes the sequential statements as described above.

18.2 PROCESS AND WAIT STATEMENTS

With this introduction to algorithmic modeling, we can now discuss associated VHDL statements in more detail, focusing first on the **process** statement. **Process** statements play a central role in VHDL because, as explained in Chapter 17, all time-based behavior in VHDL is defined in terms of the **process** statement. Concurrent statements, such as concurrent signal assignment statements and concurrent assertion statements, are actually shorthand notations for commonly used forms of the **process** statement. The **wait** statement will be discussed with the **process** statement because the **wait** statement defines when processes execute.

Like many other statements in VHDL, the **process** statement is composed of two parts: the declarative part and the statement part.[2]

```
process
  -- Declarative part

begin
  -- Statement part

end process;
```

A process may be given a label that can be used to identify the process; the label is any legal identifier.

```
PROC_LABEL: process
  -- Declarative part

begin
  -- Statement part

end process PROC_LABEL;
```

---

[2] <u>VHDL-93:</u> In VHDL-93, the header of a **process** statement may close with the keyword **is**.

```
process is
  -- Declarative part
begin
  -- Statement part
end process;
```

The declarative part defines names for objects that will be used in the statement part. The declarations allowed within a function (see Chapter 15) are also allowed within a process; these declarations are listed below.

- function
- procedure
- type
- subtype
- constant
- variable
- file
- alias
- attribute
- use clause
- group (VHDL-93)

Some of the listed declarations have already been discussed, such as type, constant, attribute, and group declarations. Other declarations, such as file and subtype, will be covered in later sections. Note that signal declarations are *not* allowed in processes. The scope of a declaration extends to the end of the process; in other words, declared objects can only be used inside the enclosing process.

The statement part contains *sequential* statements that describe time-based behavior via one or more signal transformations. The set of sequential statements in VHDL are listed below.

- wait
- variable assignment
- signal assignment
- if
- exit
- procedure call
- return
- case
- assertion
- loop
- next
- null
- report (VHDL-93)

We will discuss these sequential statements in detail in following sections.

18.2.1 Process Execution

Having explained the general construction of a **process** statement, let us now discuss when and how a process executes. The rules for process execution mimic hardware execution. A hardware device is turned on or "powered up," produces a series of outputs in response to a series of inputs, and finally is turned off or "powered down." In a similar fashion, a process "wakes up" or begins running at the start of simulation, executes a series of signal transforms in response to a series of input stimuli, and finally "dies" or stops running at the end of simulation. This general flow of control in executing a process is diagramed in Figure 18.3.

A process begins running by *elaborating* the declarations, in other words, by creating storage and setting up initial values. Then, the sequential statements are executed from top to bottom. When the last sequential statement executes, flow of control *implicitly* returns to the first sequential statement and the sequence of statements execute again. Thus, a process im-

Figure 18.3

Flow of control in executing a process

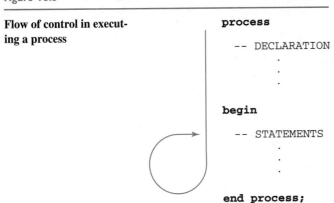

```
process

   -- DECLARATION
          .
          .
          .

begin

   -- STATEMENTS
          .
          .
          .

end process;
```

plicitly contains an infinite **do-loop** that repeatedly executes the sequential statements and terminates only at the end of simulation.

An important consequence of process execution is that objects are declared *once* at the beginning of simulation. Thus, all variables declared within a process are by definition *static variables*. Variables declared within a process are not reinitialized each time the process responds to input stimuli, rather the variables are initialized once at the beginning of simulation and retain values between process activations. For example, the variable declaration for PRESENT_STATE given in Figure 18.2

```
process
   -- State information held in PRESENT_STATE
   variable PRESENT_STATE : BIT_VECTOR(2 downto 0) := B"111";
   ....
```

is initialized to 111_2 only once, at the beginning of simulation.

18.2.2 Process Activation and Suspension

Between input stimuli, a hardware device does not turn off, rather it simply waits for the next input. Similarly, a process does not stop running between input stimuli, rather it temporarily suspends execution until the next input. Thus, a process activates in response to an input, produces an output, and then suspends and waits for the next input. Process activation and suspension are controlled by **wait** statements.

We will first discuss how to write a **wait** statement, then we will look at the rules for using **wait** statements. The **wait** statement has three basic forms:

1. *signal wait,*
2. *condition wait,* and
3. *time wait.*

A signal wait "waits on" an event to occur on one or more signals and has the following form:

```
wait on signal_a, signal_b, ...;
```

A signal wait suspends the enclosing process until one or more of the listed signals change value. As an example, the following signal wait

```
PROC_MONITOR: process
  -- Declarative part

begin
  . . .
  wait on ALARM_A, ALARM_B;
  . . .
end process PROC_MONITOR;
```

suspends the enclosing PROC_MONITOR process until an event occurs on signals ALARM_A and/or ALARM_B.

The second form of the **wait** statement, the condition wait, "waits until" a condition is satisfied and has the following form:

```
wait until condition;
```

The condition must be an expression yielding a value of type BOOLEAN, for example, FALSE or TRUE. When any signal appearing in the condition expression changes value, the condition is reevaluated. The process resumes execution if the condition reevaluates to TRUE, otherwise the process stays suspended. As an example, the condition wait in the binary counter model in Figure 18.2

```
process
  -- Declarative part

begin
  . . .
  wait until (CLK = '1');
end process;
```

suspends the enclosing process until an event occurs on the signal CLK and CLK equals 1, in other words, until a rising clock edge occurs.

The third form of the **wait** statement, the time wait, "waits for" a period of time and then resumes process execution and has the following form:

```
wait for period_of_time;
```

The period of time must be an expression yielding a value of type TIME. As an example, the following time wait

```
PROC_TIMER: process
  -- Declarative part

begin
  ...
  wait for 1 ms;
  ...
end process PROC_TIMER;
```

suspends the enclosing PROC_TIMER process for the time interval 1 millisecond.

The signal, condition, and time waits can be combined in a single **wait** statement by listing the signal wait, followed by the condition wait, followed by the time wait. Any of the eight possible combinations of the signal, condition, and time waits are allowed, but we must be careful to understand the interactions.

For example, the **wait** statement

```
wait on ALARM_A, ALARM_B until (CLK = '1');
```

combines signal and condition waits to suspend a process until an event occurs on the signals ALARM_A and/or ALARM_B and CLK is 1. In this case, an event on the signal CLK does *not* trigger reevaluation of the **wait** statement; only events on signals ALARM_A and/or ALARM_B trigger reevaluation. Notice that combining signal and condition waits slightly changes the semantics of the condition wait described earlier. When a **wait** statement contains only a condition wait, the condition is reevaluated when any signal appearing in the *condition expression* changes value. However, when a **wait** statement combines a condition wait with a signal wait, the condition is reevaluated only when any signal appearing in the *signal wait list* changes value; signals appearing in the condition expression do not trigger reevaluation. In this case, the signal wait defines an explicit sensitivity that dominates or overrides the implicit sensitivity defined by the condition wait.

As another example of combining **wait** statement forms, the **wait** statement

```
wait on ALARM_A, ALARM_B until (CLK = '1') for 1 ms;
```

combines signal, condition, and time waits to suspend a process until an event occurs on ALARM_A and/or ALARM_B and CLK equals 1 *or* the time interval of 1 millisecond has elapsed. When a time wait is combined with a signal and/or condition wait, the time wait specifies the *maximum* period of time a process may be suspended. A process may resume execution before the maximum time out interval expires if the signal and condition wait specifications become satisfied.

Now that we know how to construct **wait** statements, how do we use them? Any number of **wait** statements may be placed at any location within the statement part of a process to suspend/activate the enclosing process any number of times.

```
process
begin
  sequential statement
  sequential statement
  wait ...;
  sequential statement
  sequential statement
  wait ...;
  sequential statement
  sequential statement
  wait ...;
end process;
```

There is, however, one exception to this rule: When a process includes a *sensitivity list,* **wait** statements are not permitted within the process. A sensitivity list is an optional part of a **process** statement and is shown in Figure 18.4(a).

A sensitivity list is a convenient notation for defining process activation/suspension based on events occurring on one or more signals. Figure 18.4(a) shows that a sensitivity list follows the keyword **process** and lists within a set of parentheses one or more signals; multiple signals are separated by commas.

A sensitivity list implicitly defines a signal wait as the *last* statement within the process. For instance, the **wait** statement implied by the sensitivity list in Figure 18.4(a) is shown in Figure 18.4(b). The implied **wait** statement is the last statement to allow the enclosing process to participate in simulation initialization. Recall from Chapter 17 that during initial-

Figure 18.4

Process sensitivity list

```
PROC_LABEL:
process (ALARM_A, ALARM_B)   -- Sensitivity list
  -- Declarative part

begin
  -- Statement part

end process PROC_LABEL;
```
                                                        (a)
```
PROC_LABEL:
process
  -- Declarative part

begin
  -- Statement part

  wait on ALARM_A, ALARM_B; -- Signal wait
end process PROC_LABEL;
```
                                                        (b)

Processes in Figures (a) and (b) are equivalent. A sensitivity list is a shorthand notation for a signal wait.

ization, each process executes until it suspends. With the implied **wait** statement at the end of the process, the process executes one pass through the sequential statements during initialization to setup target signals.

Using a sensitivity list or explicitly coding the equivalent **wait** statement is a matter of personal preference. However, once a sensitivity list is used, it *alone* defines the activation/suspension behavior of the process and no other **wait** statement may be included in the process body.

Sensitivity lists are included in VHDL mainly for historical reasons. Early versions of VHDL defined process activation/suspension only by sensitivity lists. This limited process activation/suspension capability was later extended by adding a **wait** statement. Even though the **wait** statement subsumes the capabilities of a sensitivity list, sensitivity lists were left in the language.

Time for an example. Sensitivity lists and a condition wait are illustrated in the VHDL model given in Figure 18.5(b) that describes the sequential adder state diagram in Figure 18.5(a). (An alternative description of the sequential adder using data flow modeling techniques is given in Chapter 17.)

The ALGORITHM architecture contains three processes: STATE, NS, and OUTPUT. The STATE process models memory. The NS process models the next state combinational logic. The OUTPUT process models the output combinational logic.

A condition wait controls execution of the STATE process. When CLK changes from '0' to '1', the STATE process activates, executes the single sequential signal assignment statement to update PRESENT_STATE, loops back to the **wait** statement, and suspends execution until the next rising edge on the signal CLK.

Sensitivity lists control execution of the NS and OUTPUT processes. A new input (X or Y) or state value (PRESENT_STATE) activates the NS and OUTPUT processes. The next state (NEXT_STATE) and output (S) are respectively computed using the three-dimensional arrays NS_PS_TABLE and OUTPUT_TABLE as truth tables. The PRESENT_STATE denoting the carry-in and the inputs X and Y denoting the augend and addend serve as indices into the truth tables to yield the output S denoting the sum and the next state NEXT_STATE denoting the carry-out. Truth tables are initialized using aggregates (see Chapter 16).

Figure 18.5

VHDL model of a sequential adder

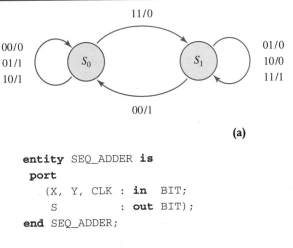

(a)

```
entity SEQ_ADDER is
 port
   (X, Y, CLK : in  BIT;
    S         : out BIT);
 end SEQ_ADDER;


architecture ALGORITHM of SEQ_ADDER is
   type STATE_TYPE is (S0, S1);
   signal PRESENT_STATE, NEXT_STATE : STATE_TYPE;
```

(Figure 18.5 continued on page 580)

Figure 18.5 continued

```
begin
 STATE: process
 begin
   wait until (CLK = '1');
   PRESENT_STATE <= NEXT_STATE;
 end process STATE;

 NS: process (PRESENT_STATE, X, Y)
  type STABLE_TYPE is array (STATE_TYPE, BIT, BIT) of
      STATE_TYPE;
  constant NS_PS_TABLE : STABLE_TYPE :=
   (S0 => ('0'=>(S0, S0), '1'=>(S0, S1)),
    S1 => ('0'=>(S0, S1), '1'=>(S1, S1)));
 begin
    NEXT_STATE <= NS_PS_TABLE(PRESENT_STATE, X, Y);
 end process NS;

 OUTPUT: process (PRESENT_STATE, X, Y)
  type BTABLE_TYPE is array (STATE_TYPE, BIT, BIT) of BIT;
  constant OUTPUT_TABLE : BTABLE_TYPE :=
   (S0 => ('0'=>('0', '1'), '1'=>('1', '0')),
    S1 => ('0'=>('1', '0'), '1'=>('0', '1')));
 begin
    S <= OUTPUT_TABLE(PRESENT_STATE, X, Y);
 end process OUTPUT;
end ALGORITHM;
```

(b)

Figures (a) and (b) show the state diagram and VHDL model of a sequential adder, respectively.

The SEQ_ADDER design entity introduces several new aspects of algorithmic modeling. First, algorithmic models are not limited to a single process. Algorithmic models of state machines often use three processes to describe the high-level concurrency of the memory devices, next state combinational logic, and output combinational logic. Second, algorithmic models are not limited to bit-level abstractions. In Figure 18.5(b), states have symbolic names, S0 and S1, instead of binary encodings. Third, next state and output values are determined by lookup or truth tables, NS_PS_TABLE and OUTPUT_TABLE, instead of switching algebra expressions.

18.3 VARIABLE AND SEQUENTIAL SIGNAL ASSIGNMENT STATEMENTS

Let us look next at the sequential statements that may be used within **process** statements to describe input/output signal transforms. Since the BIN_COUNTER and SEQ_ADDER design entities contain examples of variable and sequential signal assignment statements, we will discuss these statements first. Recall that variables and signals are objects that hold values of some type. Assignment statements provide the means to change variable and signal values.

A variable assignment uses the := operator. A variable assignment takes effect immediately. The value of the expression on the right-hand side of the := assignment operator must be of the type of the target variable on the left-hand side of the := operator. Examples of variable assignment statements are shown below.

```
variable LOGIC_A, LOGIC_B : BIT := '0';
        .
        .


LOGIC_A := '1';        -- Variable assignment
LOGIC_B := LOGIC_A;    -- Variable assignment
```

The first variable assignment statement assigns the value '1' to variable LOGIC_A. The value of LOGIC_A is immediately updated, so the second variable assignment statement assigns the new value of LOGIC_A to LOGIC_B, giving LOGIC_B the value '1'.

A sequential signal assignment uses the <= operator. As mentioned earlier, we must not confuse sequential and concurrent signal assignment statements. The two kinds of signal assignment statements have similarities, but also important differences.

Concerning syntax, both concurrent and sequential signal assignment statements can use transport or inertial propagation delay. Also, both concurrent and sequential assignment statements can assign a single waveform element or a collection of waveform elements comprising a waveform to a target signal.[3] However, a sequential signal assignment statement does not support the general conditional and selected forms of the concurrent signal assignment statement. Examples of sequential signal assignment statements are shown below.

```
signal LOGIC_A : BIT;
       .
       .


-- Inertial delay with waveform element and delta delay
LOGIC_A <= '0';

-- Inertial delay with waveform element
LOGIC_A <= '0' after 1 sec;

-- Inertial delay with waveform
LOGIC_A <= '0' after 1 sec, '1' after 3.5 sec;

-- Transport delay with waveform
LOGIC_A <= transport '0' after 1 sec, '1' after 3.5 sec;
```

Concerning semantics, concurrent and sequential signal assignment statements share the property that assignments *never* take effect immediately. Even if an **after** clause is not present, a delta propagation delay of one simulation cycle is assumed. However, unlike concur-

---

[3] VHDL-93: In VHDL-93, both concurrent and sequential signal assignment statements can specify an inertial delay that is different from the propagation delay; see Chapter 15 for discussion of pulse rejection.

rent statements that execute in a stimulus/response fashion, sequential statements execute only when encountered by the flow of control.

The following example illustrates the execution semantics of sequential signal assignment statements.

```
signal LOGIC_A, LOGIC_B : BIT := '0';
                         .
                         .

  LOGIC_A <= '1';        -- Sequential signal assignment
  LOGIC_B <= LOGIC_A;    -- Sequential signal assignment
```

The first sequential signal assignment statement assigns the value '1' to the target signal LOGIC_A with a delta delay; thus, the value of LOGIC_A is not updated until the next simulation cycle. Consequently, the second sequential signal assignment statement assigns the value '0' to the signal LOGIC_B, again with a delta delay.

A final difference between sequential and concurrent signal assignment statements concerns the equivalence between assignment statements and sources. Consider the following signal assignments, which assume that COMM_BUS is a signal of type BIT.

```
CONCURRENT:
block
begin
  COMM_BUS <= '1';    -- Concurrent signal assignment
  COMM_BUS <= '0';    -- Concurrent signal assignment
end block CONCURRENT;

SEQUENTIAL:
process
begin
  COMM_BUS <= '1';    -- Sequential signal assignment
  COMM_BUS <= '0';    -- Sequential signal assignment
end process SEQUENTIAL;
```

The first two signal assignment statements are concurrent statements because they appear within the context of a **block** statement. The two concurrent signal assignment statements contained within the block labeled CONCURRENT form *two* sources for COMM_BUS. In this case, COMM_BUS must be declared as a resolved signal.

In contrast, the last two signal assignment statements are sequential statements because they appear within the context of a **process** statement. The two sequential signal assignment statements contained within the process labeled SEQUENTIAL form only *one* source for COMM_BUS. The second sequential signal assignment statement overwrites the effects of the first sequential signal assignment statement. In general, each process that assigns to a given signal creates only *one* driver or source for that signal, regardless of the number of times the signal is assigned within the process.

18.4 CONDITIONAL STATEMENTS

Looking at other sequential statements, conditional statements include the **if** and **case** statements. The **if** statement allows various sets of sequential statements to be executed based on various conditions. The general form of the **if** statement is shown below.

```
if condition1 then
   sequential statements
elsif condition2 then
   sequential statements
elsif condition3 then
   sequential statements
         .
         .
         .

else
   sequential statements
end if;
```

The conditions must be expressions of type BOOLEAN. Starting with condition1, the conditions are evaluated in order and the statements associated with the first TRUE condition are executed. If there are no TRUE conditions, the statements associated with the final **else** clause are executed.

The **elsif** (notice it is not **elseif**) and **else** clauses are optional, so an **if** statement can also have the form

```
if condition1 then
   sequential statements
else
   sequential statements
end if;
```

or simply

```
if condition1 then
   sequential statements
end if;
```

Figure 18.6 (p. 584) shows how to use an **if** statement to describe the positive edge-triggered D flip-flop with asynchronous reset and clear inputs discussed in Chapter 8. The process uses a sensitivity list to define process activation and suspension; the process executes in response to an event on SET_BAR, CLR_BAR, or C. Process execution can be read directly from the **if** statement. The flip-flop output Q is assigned the value '1' if SET_BAR is asserted, or else Q is assigned the value '0' if CLR_BAR is asserted, or else Q is assigned the value of the input D if a rising edge has occurred on C, or else no changes are made to the flip-flop. After executing the **if** statement, the process suspends execution until the next event

Figure 18.6

Positive edge-triggered D flip-flop with reset and clear

```
entity D_FF is
 port
   (D, C              : in BIT;
    SET_BAR, CLR_BAR  : in BIT;
    Q, Q_BAR          : out BIT);
end D_FF;

architecture ALGORITHM of D_FF is
begin
 process (SET_BAR, CLR_BAR, C)
 begin
   if (SET_BAR='0') then          -- Set action
      Q <='1';   Q_BAR <= '0';

   elsif (CLR_BAR='0') then       -- Clear action
      Q <='0';   Q_BAR <= '1';

   elsif (C'EVENT and C='1') then  -- Flip-flop action
      Q <= D;   Q_BAR <= not D;
   end if;
 end process;
end ALGORITHM;
```

occurs on control signals SET_BAR, CLR_BAR, or C. The asynchronous set and clear inputs are given precedence over the synchronous clock input by checking the SET_BAR and CLR_BAR assert conditions before checking the rising edge of C condition.

A **case** statement may be a more straightforward way to model decision-based behavior when the alternatives all have *equal* priority and can be discriminated as different values of an expression. A **case** statement has the following general form.

```
case discriminant is
 when choice1 => sequential statements
 when choice2 => sequential statements
                   .
                   .
 when choiceN => sequential statements
end case;
```

The rules for writing a **case** statement are similar to the rules discussed in Chapter 13 for writing a selected concurrent signal assignment statement. The discriminant must be an expression that is of a type having a finite set of values that can be enumerated, or listed. The set of **when** clauses or choices must account for all possible discriminant values and each discriminant value must be accounted for only *once*; in other words, a discriminant value cannot be listed in more than one choice.

A choice can be a single value

```
when B"000" => ...
when 0 => ...
```

or an ascending range of contiguous values

```
when B"010" to B"100" => ...
when 2 to 4 => ...
```

or a descending range of contiguous values.

```
when B"100" downto B"010" => ...
when 4 downto 2 => ...
```

A choice can also be a combination of choices by using a vertical bar | .

```
when B"001" | B"101" => ...
when 1 | 5 => ...
```

Finally, a choice can use the keyword **others** to account for all remaining discriminant values not listed in previous **when** clauses or choices.

```
when others;
```

An **others** choice may not be combined with other choices and must be the final **when** clause within a **case** statement.

18.5 SEQUENTIAL ASSERTION STATEMENT

Recall from Chapter 17 that assertions supplement the modeling of input/output signal transformations by describing design intent or constraints. Constraints include illegal inputs, unsupported operating modes, and required synchronized signal transitions.

The syntax of the sequential assertion statement is basically the same as the concurrent assertion statement.

```
assert ( BOOLEAN expression )
    report STRING expression
    severity SEVERITY_LEVEL expression ;
```

The BOOLEAN expression defines the asserted condition that must always exist during normal operation; in other words, the expression should always evaluate to TRUE. If the asserted condition fails during simulation, in other words, evaluates to FALSE, an *assertion violation* occurs. The violation is recorded by reporting (1) the line of text given by evaluating the STRING expression in the **report** clause and (2) the severity given by evaluating the SEVERITY_LEVEL expression in the **severity** clause. The line of text provided by the **report** clause explains the reason for the violation and the severity level provided by the **severity** clause indicates the seriousness of the violation. Possible severity levels include NOTE, WARNING, ERROR, and FAILURE. Either the **report** or **severity** clauses may be omitted; in which case, the default error message "Assertion violation" is reported with the severity level ERROR.

Unlike syntax, the semantics of the sequential assertion statement is different from the concurrent assertion statement. A *concurrent* assertion statement executes whenever any signal in the asserted condition changes value, whereas the *sequential* assertion statement executes only when encountered by the procedural flow of control.

As an example, Figure 18.7 shows how a sequential assertion statement can check that the set and clear inputs of the D flip-flop described in Figure 18.6 are not simultaneously asserted. When the process executes in response to an event on SET_BAR, CLR_BAR, or C, the sequential assertion statement first checks the constraint on the asynchronous set and clear inputs. If SET_BAR and CLR_BAR are ever simultaneously asserted active-0, the BOOLEAN expression **not** (SET_BAR='0' **and** CLR_BAR='0') evaluates to FALSE, the assertion fails, and a violation is reported.

Figure 18.7

Sequential assertion statement to check set and clear

```
entity D_FF is
 port
    (D, C                : in BIT;
     SET_BAR, CLR_BAR : in BIT;
     Q, Q_BAR           : out BIT);
 end D_FF;

architecture ALGORITHM of D_FF is
begin
 process (SET_BAR, CLR_BAR, C)
 begin
    -- Sequential assertion statement
    assert not (SET_BAR='0' and CLR_BAR='0')
      report "Illegal Use: Set and Clear Asserted At Same Time."
      severity ERROR;

    if (SET_BAR='0') then              -- Set action
       Q <='1';    Q_BAR <= '0';

    elsif (CLR_BAR='0') then           -- Clear action
       Q <='0';    Q_BAR <= '1';

    elsif (C'EVENT and C='1') then     -- Flip-flop action
       Q <= D;   Q_BAR <= not D;
    end if;
 end process;
end ALGORITHM;
```

18.6 VHDL-93: REPORT STATEMENT

Although sequential assertion statements have a nice hardware-related eloquence, as a practical matter this eloquence is often overkill when we only wish to trace simulation flow of control. In this case, we require a simple print statement that always issues a notice when executed. Such a print statement can be provided by the following assertion statement that disables the assertion by setting the BOOLEAN condition to FALSE to always force an assertion violation and issue a report.

```
assert FALSE
    report "I'm now executing!"
    severity NOTE;
```

VHDL-87 models are sometimes laced with sequential assertion statements of the above form that essentially serve as debugging print statements.

Since using a sequential assertion statement can be cumbersome when a designer wants to report a message based solely on statement execution and not on an asserted condition, VHDL-93 provides a *report statement*. A **report** statement contains only **report** and **severity** clauses and always issues a report whenever executed.

A **report** statement has the following general form.

```
label: report   STRING expression
        severity SEVERITY_LEVEL expression ;
```

The **report** statement issues the text obtained by evaluating the STRING expression with the severity obtained by evaluating the SEVERITY_LEVEL expression. The label and **severity** clause are optional. If the **severity** clause is omitted, a severity of NOTE is assumed.

18.7 ITERATION STATEMENTS

The **loop** statement allows a set of sequential statements to be repeatedly executed. Some software programming languages have different statements for different iteration schemes, such as a **for** loop statement, a **while** loop statement, and an **until** loop statement. VHDL provides a *single* **loop** statement that can have one of the following three forms, depending on the desired iteration scheme:

1. infinite-loop,
2. while-loop, and
3. for-loop.

The infinite-loop is the simplest form of the **loop** statement; a set of sequential statements are repeatedly executed forever.

```
INF_LOOP:
loop
   sequential statement
   sequential statement

      .
      .

end loop INF_LOOP;
```

The loop label, `INF_LOOP`, is optional and may be omitted. The infinite-loop often uses **next**, **exit**, and/or **return** sequential statements to describe finite iteration schemes not easily described with the more structured while-loop and for-loop. The **next** and **exit** statements are discussed in Section 18.7.1; the **return** statement is discussed in Section 18.8.

The while-loop repeatedly executes a set of sequential statements while a condition is satisfied.

```
WHILE_LOOP:
while condition
loop
  sequential statement
  sequential statement
    .
    .
    .
end loop WHILE_LOOP;
```

The loop label, `WHILE_LOOP`, is optional and may be omitted. The condition must be an expression of type `BOOLEAN`. The `BOOLEAN` expression is checked before each loop iteration. The next loop iteration executes if the `BOOLEAN` expression evaluates to `TRUE`, otherwise the while-loop terminates.

Finally, a for-loop repeatedly executes a set of sequential statements a certain number of times.

```
FOR_LOOP:
for parameter in discrete_range
loop
  sequential statement
  sequential statement
    .
    .
    .
end loop FOR_LOOP;
```

The loop label, `FOR_LOOP`, is optional and may be omitted. The for-loop executes a loop iteration for each value of the discrete range. As the for-loop executes, the loop parameter successively takes on each value of the discrete range. The loop parameter can be any legal identifier, is implicitly declared to be of the type of the discrete range, and its scope is restricted to the for-loop statement. That is, the loop parameter can be used only within the body of the for-loop. Moreover, the loop parameter is *read only* within the loop body. The value of the loop parameter can be accessed, but cannot be updated by using the loop parameter as the target of a variable assignment statement.

A discrete range defines a finite set of values and can be specified using one of the following forms.

Discrete Range	Definition
Ascending range	expression **to** expression; for example, 3 **to** 10
Descending range	expression **downto** expression; for example, 10 **downto** 3
Discrete type	Enumeration and integer types; for example, `NATURAL`
Discrete type with constraint	Constrained enumeration and integer types; for example, `NATURAL` **range** 1 **to** 100
Attribute name	`array_name'RANGE`

If the discrete range evaluates to a null set having no values, no loop iterations are executed. For example, a for-loop containing the discrete range 2 **to** 1 never executes a loop iteration because the discrete range is a null set.

18.7.1 Next and Exit Statements

The **next** and **exit** statements can be used to modify or terminate the execution of a **loop** statement. The **next** statement terminates execution of a particular loop *iteration*, thereby providing a way to skip iterations or portions of iterations. Various forms of the next statement are given below.

1. **next**;
2. **next when** condition;
3. **next** label;
4. **next** label **when** condition;

The first form of the **next** statement unconditionally skips to the next iteration of the immediately enclosing loop, assuming there is another iteration. The second form of the **next** statement skips to the next iteration of the immediately enclosing loop if the condition is satisfied. The condition must be a BOOLEAN expression. As an example, the following for-loop executes only the first two sequential statements when the loop parameter I is 3.

```
for I in (1 to 10) loop
   sequential statement
   sequential statement
   next when (I=3);

      .
      .
end loop;
```

Instead of always terminating the iteration of the *immediately* enclosing loop, the third and fourth forms of the **next** statement contain labels that specify which enclosing loop iteration to terminate. This feature is useful for nested loops, as shown below.

```
LOOP_A:
for I in (1 to 10) loop
   sequential statement
   sequential statement
   LOOP_B:                      -- Nested loop
   for J in (1 to 10) loop
      sequential statement
      sequential statement
      next LOOP_A when (J=I);

         .
         .
   end loop LOOP_B;

      .
      .
end loop LOOP_A;
```

When J equals I, the **next** statement terminates the inner loop, LOOP_B, and skips to the next iteration of the outer loop, LOOP_A.

The **exit** statement terminates a loop entirely, not just a particular loop iteration. The syntax of the **exit** statement is similar to the **next** statement.

Problem-Solving Skill

Similar concepts should be given similar representations. Consistency facilitates simplicity.

1. **exit**;
2. **exit when** condition;
3. **exit** label;
4. **exit** label **when** condition;

The first and second forms of the **exit** statement respectively unconditionally and conditionally terminate the immediately enclosing loop. The third and fourth forms of the **exit** statement respectively unconditionally and conditionally terminate the loop identified by the given label.

18.7.2 Example

To illustrate the use of the iteration statements in an algorithmic modeling application, consider the Moore state machine shown in Figure 18.8 that describes a *polling* operation. The polling state machine is like a sequential priority encoder (see Chapter 5). The three inputs represent three devices, such as coprocessors or sensors, requesting some type of service, such as data transfers or interrupt checks. A device requests service by asserting its input. On every clock cycle, the polling machine checks the status of the three input devices and generates an output code that identifies the asserted input (device) to be serviced. The output encoding is given in Figure 18.8(a). If more than one device is requesting service, the device with the highest priority is selected. Each input is assigned a fixed priority denoted by its subscript; 3 is the highest priority and 1 is the lowest priority. To prevent "starving" the lower priority devices, the same asserted input is never selected on two successive pollings unless there are no other asserted inputs.

An algorithmic VHDL description of the 3-input sequential poller is given in Figure 18.9. The architecture ALGORITHM contains one process. Within the process, three variables are declared: LAST_CHOICE, PRESENT_CHOICE, and POSSIBLE_CHOICE. The statement part of the process first sets the variables POSSIBLE_CHOICE and PRESENT_CHOICE to 0. Then, a for-loop polls the inputs from highest to lowest priority. If there is a new asserted input, the **exit** statement immediately terminates for-loop execution because no additional inputs need be checked. If there are no asserted inputs or the input selected on the previous polling is again the only asserted input, the for-loop executes all iterations because all inputs need to be checked.

When the input polling is finished, the result is recorded in the variable LAST_CHOICE

```
LAST_CHOICE := PRESENT_CHOICE;
```

and assigned to the output signal ACK.

```
ACK <= ENCODE(PRESENT_CHOICE);
```

ENCODE is a function that generates a unique output code for each request input. The function can be defined in a package or in the declaration part of the enclosing process, architecture, or entity interface.

Figure 18.8

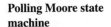

Polling Moore state machine

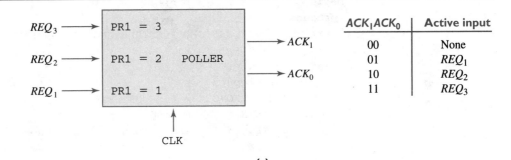

(a)

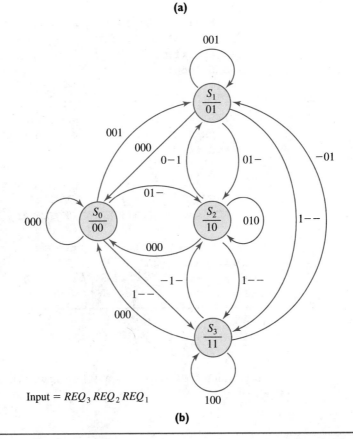

Input = $REQ_3\, REQ_2\, REQ_1$

(b)

Figures (a) and (b) show the input/output and state diagrams for a 3-input sequential poller, respectively.

Figure 18.9

Algorithmic description of a sequential polling system

```
entity POLLER is
 port
    (REQ : in BIT_VECTOR(3 downto 1);
     CLK : in BIT;
     ACK : out BIT_VECTOR(1 downto 0));
end POLLER;
```

(Figure 18.9 continued on page 592)

Figure 18.9 continued

```vhdl
architecture ALGORITHM of POLLER is
begin
 process
  variable LAST_CHOICE, POSSIBLE_CHOICE, PRESENT_CHOICE :
          NATURAL := 0;
 begin
     -- Poll inputs
     POSSIBLE_CHOICE := 0;   PRESENT_CHOICE := 0;
     for I in REQ'RANGE loop
       if (REQ(I)='1' and I/=LAST_CHOICE) then
         PRESENT_CHOICE := I;    -- New active input
         exit;                   -- Terminate for-loop
       elsif (REQ(I)='1' and LAST_CHOICE=I) then
         POSSIBLE_CHOICE := I;  -- Old choice still active
       end if;                  -- keep checking
     end loop;

     -- If last choice is only active input, choose it.
     if (PRESENT_CHOICE=0 and POSSIBLE_CHOICE/=0) then
       PRESENT_CHOICE := POSSIBLE_CHOICE;
     end if;

     -- Save the new choice and update output
     LAST_CHOICE := PRESENT_CHOICE;
     ACK <= ENCODE(PRESENT_CHOICE);

    -- Wait until next rising edge of clock
    wait until (CLK = '1');
 end process;
end ALGORITHM;
```

The declaration of the function ENCODE is shown below.

```vhdl
function ENCODE (CHOICE : NATURAL) return BIT_VECTOR is
begin
  case CHOICE is
    when 0       =>  return (B"00");
    when 1       =>  return (B"01");
    when 2       =>  return (B"10");
    when 3       =>  return (B"11");
    when others =>  return (B"00");
  end case;
end ENCODE;
```

The ENCODE function is only concerned with encoding the CHOICE values 0, 1, 2, and 3 because the sequential priority encoder POLLER has only three inputs. However, a final **when** clause is required to address CHOICE values 4, 5, 6, . . . because CHOICE is declared to be of type NATURAL (0, 1, 2, . . .) and a **case** statement must account for all values of the discriminant. Section 18.10 shows how to improve this somewhat awkward modeling situation.

It is instructive to compare the POLLER design entity in Figure 18.9 with the BIN_COUNTER design entity in Figure 18.2. The BIN_COUNTER describes input/output behavior from a state transition perspective, whereas the POLLER design entity describes input/output behavior from a more abstract functional perspective. Although the BIN_COUNTER and POLLER design entities employ slightly different levels of abstraction, both entities are considered examples of algorithmic modeling because they employ sequential statements. We will revisit the POLLER design entity in future sections.

18.8 PROCEDURES

A procedure is called a *subprogram* and is used to decompose behavioral descriptions. Just as a VHDL model can be hierarchically constructed using design entities, a behavioral description within a design entity can be hierarchically constructed using subprograms. There are two types of subprograms: *functions* and *procedures*. A function is an *expression* that returns a value, whereas a procedure is a *statement* that can return any number of values. The following sections will discuss how to declare and invoke procedures, paralleling the discussion given in Chapter 15 on how to declare and invoke functions.

18.8.1 Declaration

Procedures may be declared in procedures, functions, processes, blocks, architectures, entities, and packages. Declaring procedures within procedures allows procedures to be defined hierarchically. Declaring procedures within packages allows commonly used procedures to be encoded once and made available to many different applications.

A procedure declaration starts with the keyword **procedure**, followed by an identifier giving the name of the procedure, followed by a list of *formal parameters*.

```
procedure  name  (formal_parameters) is
   -- Declarative part

begin
   -- Statement part

end name;
```

Formal parameters define the information communication channel(s) between the calling or parent program and the procedure. The calling program passes information into a procedure via the formal parameters. This information is processed by the sequential statements contained in the *statement part* of the procedure and the results are returned to the calling program, again via the formal parameters.

The declarations and sequential statements allowed within a process, listed earlier in Section 18.2, also apply to procedures, except that a **wait** statement may not appear in a procedure that is called from a function or a process containing a sensitivity list (because neither a function nor a process containing a sensitivity list may contain **wait** statements).

Returning to the formal parameter list, a formal parameter declaration contains the following four items.

1. object class (**constant**, **variable**, or **signal**)
2. mode (**in**, **out**, or **inout**)
3. type
4. default expression

The object class specifies the kind of formal parameter and can be **constant** for constants, **variable** for variables, or **signal** for signals. Only signals may be associated with formal parameters that are signals. Likewise, only variables may be associated with formal parameters that are variables. General expressions may be associated with formal parameters that are constants. An object class designator may be omitted, in which case, the class **constant** is assumed if the parameter mode is **in** or the class **variable** is assumed if the parameter mode is **out** or **inout**.

Similar to port modes discussed in Chapter 14, formal parameter modes define the direction of information flow with respect to the procedure. Modes also define rules for accessing and updating formal parameters within the procedure. Formal parameters of mode **in** may be accessed, but not updated. Formal parameters of mode **out** may be updated, but not accessed. Formal parameters of mode **inout** may be accessed and updated. Consequently, a **constant** formal parameter cannot be declared with modes **out** or **inout** because a constant cannot be updated.

Type specifies the set of values a formal parameter may take on. Formal parameters of the same class and type may be collectively defined in a single declaration or individually defined in separate declarations.

The declaration of a **constant** or **variable** formal parameter of mode **in** may optionally include a *default expression*. The formal parameter takes on the value of the default expression if the formal parameter is not given a value by the calling program. The default value must be a legal value of the formal parameter type.

We will illustrate procedure declarations by modifying the POLLER design entity given in Figure 18.9. A procedure POLL_INPUTS will be used to encapsulate the polling action; the procedure declaration is given in Figure 18.10. Information is passed into the procedure POLL_INPUTS via the formal parameters INPUTS and LAST_CHOICE. The polling result is passed back to the POLLER design entity via the formal parameter PRESENT_CHOICE.

18.8.2 Invocation

A new version of the POLLER design entity using the procedure POLL_INPUTS declared in Figure 18.10 is given in Figure 18.11. POLL_INPUTS is invoked via a *procedure call statement*.

A procedure call statement associates objects/values in the calling program, called *actuals*, with the formal parameters of a procedure and then executes the procedure. Giving associated actuals and formals the same name is a good modeling practice. Actuals can be "matched up" with formals using either positional or named association. The procedure call in Figure 18.11 uses named association to associate the actuals REQ, LAST_CHOICE, and PRESENT_CHOICE in the calling process with the formal procedure parameters INPUTS,

Figure 18.10

Procedure declaration for a polling sequential system

```
procedure POLL_INPUTS
  (constant INPUTS        : in  BIT_VECTOR;
   constant LAST_CHOICE   : in  NATURAL := 0;
   variable PRESENT_CHOICE : out NATURAL) is

  variable POSSIBLE_CHOICE, CHOICE : NATURAL;
begin
    POSSIBLE_CHOICE := 0;   CHOICE := 0;
    for I in INPUTS'RANGE loop
      if (INPUTS(I)='1' and I/=LAST_CHOICE) then
       CHOICE := I;            -- New active input
       exit;                   -- Terminate for-loop
       elsif (INPUTS(I)='1' and LAST_CHOICE=I) then
         POSSIBLE_CHOICE := I;  -- Old choice still active
       end if;                 -- keep checking
    end loop;

    -- If last choice is only active input, choose it.
    if (CHOICE=0 and POSSIBLE_CHOICE/=0) then
      CHOICE := POSSIBLE_CHOICE;
    end if;

    PRESENT_CHOICE := CHOICE;
  end POLL_INPUTS;
```

LAST_CHOICE, and PRESENT_CHOICE, respectively. Alternatively, the positional association

```
POLL_INPUTS(REQ, LAST_CHOICE, PRESENT_CHOICE);
```

matches actuals with formals based on the textual order (left-to-right) in which the actuals are listed in the procedure call and the formals are listed in the header of the procedure declaration.

Figure 18.11

Algorithmic description with procedure of a sequential polling system

```
entity POLLER is
 port
    (REQ : in  BIT_VECTOR(3 downto 1);
     CLK : in  BIT;
     ACK : out BIT_VECTOR(1 downto 0));
 end POLLER;

architecture ALGORITHM of POLLER is
```

(Figure 18.11 continued on page 596)

Figure 18.11 continued

```
architecture ALGORITHM of POLLER is
begin
 process
  variable LAST_CHOICE, PRESENT_CHOICE : NATURAL := 0;
 begin
   -- Poll inputs
   POLL_INPUTS(INPUTS=>REQ,
              LAST_CHOICE=>LAST_CHOICE,
              PRESENT_CHOICE=>PRESENT_CHOICE);

   -- Save the new choice and update output
   LAST_CHOICE := PRESENT_CHOICE;
   ACK <= ENCODE(PRESENT_CHOICE);

   -- Wait until next rising edge of clock
   wait until (CLK = '1');
 end process;
end ALGORITHM;
```

The declarations of function ENCODE and procedure POLL_INPUTS are not shown.

18.8.3 Return Statement

A **return** statement causes control to return to the calling program and can only be used within a subprogram, in other words, a function or procedure. A **return** statement has the following two forms.

- **return** expression;
- **return**;

The first form is used within functions; the second form is used within procedures.

A **return** statement is required for a function to define the value returned to the calling program. A **return** statement is optional for a procedure. If a **return** statement is not included in a procedure, control implicitly returns to the calling program after the last sequential statement executes. If a **return** statement is included in a procedure, control explicitly returns to the calling program when the **return** statement is encountered in executing the sequential statements within the body of the procedure.

18.8.4 VHDL-93: Procedures

Many of the VHDL-93 enhancements discussed in Chapter 15 for functions also apply to procedures. For instance, formal parameters have been extended to include files denoted by class **file**. Formal parameters of class **file** have no mode and may only be associated with files. Files are discussed in Section 18.11.

Procedures allow *group template* and *group* declarations, which are discussed in Chapter 16. Also, procedure declarations may end with the keyword **procedure**.

```
procedure POLL_INPUTS ( ... ) is
  -- Declarations
begin
  -- Sequential statements
end procedure POLL_INPUTS;
```

Finally, procedures not written in VHDL may be used in a VHDL model by declaring the procedures as *foreign procedures*. A foreign procedure may be written in another computer language, such as C++ or Fortran. As an example, the procedure POLL_INPUTS written in VHDL in Figure 18.10 could be "stubbed out" and implemented as a foreign procedure by including the following in the declaration section of the process, architecture, or entity in Figure 18.11.

```
-- Procedure specification
procedure POLL_INPUTS
   (constant INPUTS          : in  BIT_VECTOR;
    constant LAST_CHOICE     : in  NATURAL := 0;
    variable PRESENT_CHOICE  : out NATURAL);

-- Attach attribute to procedure to define foreign interface
attribute FOREIGN of POLL_INPUTS : procedure is
            "/lib/assembly/polling";
```

The *attribute specification* following the procedure specification tags or marks the procedure POLL_INPUTS as a foreign procedure and gives interface or coupling information.

18.9 NULL STATEMENT

Last but not least in our tour of sequential statements, the **null** statement does nothing. A **null** statement is often used when at least one sequential statement is required, but no action is desired. For example, the ENCODE function discussed in Section 18.7 could have been written using a **null** statement.

```
function ENCODE (CHOICE : NATURAL) return BIT_VECTOR is
   variable BINARY_CODE : BIT_VECTOR(1 downto 0) := B"00";
begin
   case CHOICE is
     when 0      =>   BINARY_CODE := B"00";
     when 1      =>   BINARY_CODE := B"01";
     when 2      =>   BINARY_CODE := B"10";
     when 3      =>   BINARY_CODE := B"11";
     when others =>   null;                    -- No action
   end case;

   return (BINARY_CODE);
end ENCODE;
```

Even though the encoding function is only concerned with encoding the CHOICE values 0, 1, 2, and 3, a final **when** clause is required to address CHOICE values 4, 5, 6, . . . because CHOICE is declared to be of type NATURAL (0, 1, 2, . . .) and a **case** statement must account for all values of the discriminant. Since at least one sequential statement is required for each **when** clause, a **null** statement is given as the action of the final **when** clause to denote that no encoding action is to be taken when CHOICE is greater than 3.

18.10 MORE ON INFORMATION MODELING— TYPES AND SUBTYPES

To keep this discussion in perspective, let us review what we have learned about algorithmic modeling. Thus far, we have examined example models of finite state machines, including a Moore machine realizing a binary counter and a Mealy machine realizing a sequential adder. These examples introduced processes, which in turn introduced sequential statements. Thirteen sequential statements were presented. Next, we need to introduce a few new types because the sequential polling VHDL model presented in Figure 18.8 and Figure 18.9 illustrates a deficiency in the modeling skills we have developed to date.

The predefined type NATURAL, which includes integers in the range (0, 1, 2, . . . , INTEGER'HIGH), is used to define variables that take on only integers in the range 0, 1, 2, and 3. This use of data types is not a good modeling practice. The original designer may be able to mentally record the true integer range of the variables, but what happens if the original designer moves to other project or company? It may be very difficult for another designer to use the POLLER design entity or make modifications. A better modeling practice uses data types that more accurately reflect variable value ranges.

Problem-Solving Skill
Designing data structures to closely reflect the properties of objects being modeled improves model accuracy, understandability, and maintainability.

18.10.1 Integer Types

To improve the POLLER design entity, we need a data type that has not yet been discussed. VHDL provides the *predefined integer types* INTEGER, NATURAL, and POSITIVE. If these predefined integer types do not suit a particular modeling situation, then one can declare a custom integer type, called a *user-defined integer type*.

Examples of user-defined integer type declarations are given below.[4]

```
-- Ascending discrete range
type PRI_ENCODE_TYPE is range 0 to 3;


-- Descending discrete range
type PRI_ENCODE_TYPE is range 3 downto 0;


-- Attribute discrete range
type REQUEST_TYPE is range REQ'RANGE;
```

A user-defined integer type starts with the keyword **type**, followed by an identifier giving the type name. Then, the set of contiguous integers belonging to the type is specified by defining a discrete range.

The lower and upper bounds of the discrete range must not exceed the numeric limits supported by the local host. That is, the lower and upper bounds of the integer range cannot exceed INTEGER'LOW and INTEGER'HIGH, respectively. The first two integer type declarations for PRI_ENCODE_TYPE show that a discrete range can be ascending or descending. PRI_ENCODE_TYPE represents the integers 0, 1, 2, and 3. The third integer type declaration for REQUEST_TYPE shows that a discrete range can also be specified by an attribute. Recall that REQ is declared in the POLLER design entity in Figure 18.9 and Figure 18.11 as

---

[4] Integer literals were explained in Chapter 14.

```
REQ : in BIT_VECTOR(3 downto 1)
```

Thus, REQUEST_TYPE represents the integers 1, 2, and 3.

The predefined attributes discussed for physical types (Chapter 15) and enumeration types (Chapter 16) also apply to integer types. The attributes LEFT/RIGHT and LOW/HIGH query the bounds of an integer type. For instance,

```
REQUEST_TYPE'RIGHT = REQUEST_TYPE'LOW = 1
```

and

```
REQUEST_TYPE'LEFT = REQUEST_TYPE'HIGH = 3
```

The attributes PRED/SUCC and LEFTOF/RIGHTOF obtain integers having the value one less than and one greater than a given integer. For instance,

```
REQUEST_TYPE'LEFTOF(2) = REQUEST_TYPE'SUCC(2) = 3
```

and

```
REQUEST_TYPE'RIGHTOF(2) = REQUEST_TYPE'PRED(2) = 1
```

The attributes PRED and SUCC respectively give numeric predecessor and successor values. The attributes LEFTOF and RIGHTOF respectively give ordinal predecessor and successor values.

Returning to the POLLER design entity, the algorithmic description can be improved by replacing all occurrences of the predefined integer type NATURAL with the more descriptive user-defined integer type PRI_ENCODE_TYPE. A new version of the ENCODE function is shown below.

```
function ENCODE (CHOICE : PRI_ENCODE_TYPE) return BIT_VECTOR is
begin
  case CHOICE is
    when 0  =>   return (B"0");
    when 1  =>   return (B"01");
    when 2  =>   return (B"10");
    when 3  =>   return (B"10");
  end case;
end ENCODE;
```

Notice that the awkward final **when** clause of the **case** statement containing the **others** choice is removed because CHOICE's data type PRI_ENCODE_TYPE includes only the integers 0 to 3.

18.10.2 Floating Point Types

Floating point types are similar to integer types, so now is a good time to talk about them. VHDL provides the *predefined* floating point type REAL. If this predefined floating point type does not suit a particular modeling situation, then one can declare a custom floating point type, called a *user-defined floating point type*.

Examples of user-defined floating point type declarations are given below.[5]

```
--Ascending range
type RELIABILITY_TYPE is range 0.0 to 100.0;

-- Descending range
type PIN_CNT_TYPE is range 1.0E2 downto 0.0;
```

A floating point type declaration is similar to an integer type declaration, except the range limits are floating point numbers instead of integers. A user-defined floating point type starts with the keyword **type**, followed by an identifier giving the type name. Then, the set of floating point numbers belonging to the type is specified by defining a range.

The lower and upper bounds of the floating point range must not exceed the numeric limits supported by the local host. That is, the lower and upper bounds of the floating point range cannot exceed REAL'LOW and REAL'HIGH, respectively. The example floating point type declarations show that the range can be ascending or descending.

Floating point types support the predefined attributes LEFT/RIGHT and LOW/HIGH, but do *not* support the attributes POS/VAL, PRED/SUCC, and LEFTOF/RIGHTOF because floating point types are not *discrete* types. That is, the values of a floating point type cannot be exhaustively enumerated.

18.10.3 Subtypes

To finish this discussion on information modeling, we will examine *subtypes*. A type defines a set of values; a subtype defines a subset of values. For example, consider the following subtype declaration.

```
subtype NEGATIVE is INTEGER range INTEGER'LOW to -1;
```

INTEGER is the "parent" type, more formally called the *base* type, and represents the set of integers INTEGER'LOW to INTEGER'HIGH. NEGATIVE is itself a type and is also a subtype of INTEGER. NEGATIVE represents a subset of integers containing only the negative numbers. A subtype declaration starts with the keyword **subtype**, followed by an identifier giving the subtype name. Then, after the keyword **is**, the name of the parent (base) type is specified; the parent type can be a subtype. Finally, following the parent type, a constraint is given that describes the subset of values of the parent type that define the subtype. The constraint can be a range constraint, as shown above for NEGATIVE, or an index constraint, as shown on the following page for BIT_V3. The constraint is optional; if no constraint is given, the subtype represents all the values of the parent type.

---

[5] Floating point literals were explained in Chapter 14.

Common applications of subtypes in VHDL are listed below.
- Derive a type from another type.
- Provide a shorthand notation for constraining types.
- "Relax" type checking rules.
- Associate a resolution function with a type instead of a signal.

To understand the first subtype application of deriving a type from another type, consider the 9-value logic system discussed in Chapter 16 and defined by the enumeration type STD_ULOGIC.

```
type STD_ULOGIC is ('U', 'X', '0', '1', 'Z', 'W', 'L', 'H',
                    '-');
```

A binary (2-valued) logic system based on the 9-valued logic system can be defined by deriving the type STD_UBINARY from the type STD_ULOGIC via the following subtype declaration.

```
subtype STD_UBINARY is STD_ULOGIC range '0' to '1';
```

To understand the second subtype application, note that a constrained use of an unconstrained array type *implicitly* declares a subtype. For instance, consider the predefined unconstrained array type BIT_VECTOR in the BIN_COUNTER design entity in Figure 18.1. The type BIT_VECTOR(2 downto 0), used to declare the output port and a local process variable, implicitly declares a subtype. The subtype is called *anonymous* because it has no name. The base type is BIT_VECTOR and the constraint is the index constraint (2 downto 0). If a particular bit vector is used several times throughout a model, it may be better to explicitly declare a subtype

```
subtype BIT_V3 is BIT_VECTOR(2 downto 0);
```

and use the subtype name instead of repeatedly using the complete base type/index constraint notation. In short, we save some typing.

Relaxing VHDL's type checking rules is perhaps the most important use of subtypes. To understand this application, consider the difference between the following two type declarations.

```
--Integer type declaration
type PRI_ENCODE_TYPE is range 0 to 3;

-- Integer subtype declaration
subtype PRI_ENCODE_SUBTYPE is INTEGER range 0 to 3;
```

PRI_ENCODE_TYPE and PRI_ENCODE_SUBTYPE are both types representing the set of integers 0 to 3. However, PRI_ENCODE_TYPE is completely unrelated to INTEGER, whereas PRI_ENCODE_SUBTYPE is a subtype of INTEGER. The difference is important. Objects of types PRI_ENCODE_TYPE and INTEGER *cannot* be freely mixed in expressions and statements because PRI_ENCODE_TYPE and INTEGER are different types, as shown in the following example.

```
variable ENCODE : PRI_ENCODE_TYPE;
variable DATA   : INTEGER;

DATA := ENCODE;        -- Illegal!
```

Even though the variables ENCODE and DATA both hold integers, an integer held by ENCODE is not considered to be a legal value for DATA because PRI_ENCODE_TYPE and INTEGER are different types. Alternatively, objects of type PRI_ENCODE_SUBTYPE and INTEGER can be mixed because PRI_ENCODE_SUBTYPE is a subtype of INTEGER.

```
variable ENCODE : PRI_ENCODE_SUBTYPE;
variable DATA   : INTEGER;

DATA := ENCODE;        -- Legal!
```

More generally, objects of subtypes of the same base type can be mixed in VHDL expressions and statements. This rule allows a convenient degree of interoperability between bit vectors that are declared using index constraints on the same base type, enabling portions of bit vectors to be assigned to other bit vectors.

The final use of subtypes concerns resolution functions. Recall from Chapter 15 that a resolution function resolves multiple sources into a single value and can be associated with a signal by including the name of the resolution function in the signal's declaration. For example,

```
signal INT_SIG : WIRED_AND BIT;
```

declares INT_SIG to be a resolved signal of type BIT; the associated resolution function is WIRED_AND. For a large model involving many signals, repeatedly including the resolution function in each signal declaration can be tedious. As an alternative, a resolution function can be associated with a type via a subtype declaration and then the subtype can be used in signal declarations. For example,

```
subtype WIRED_AND_BIT is WIRED_AND BIT;

signal INT_SIG : WIRED_AND_BIT;
```

also declares INT_SIG to be a resolved signal of type BIT having the WIRED_AND resolution function. The subtype declaration "constrains" the base type BIT by associating the resolution function WIRED_AND.

To conclude our discussion of subtypes, VHDL provides the following three predefined subtypes.

```
subtype NATURAL is INTEGER range 0 to INTEGER'HIGH;

subtype POSITIVE is INTEGER range 1 to INTEGER'HIGH;

subtype DELAY_LENGTH is TIME range 0 fs to TIME'HIGH;
```

The subtype DELAY_LENGTH is provided only in VHDL-93.

18.11 FILE INPUT/OUTPUT

Another common aspect of algorithmic modeling involves using the storage facilities of the host operating system, in other words, *file input/output*. As a historical note, the original design of VHDL did not include file input/output. The language designers argued that file input/output is used primarily for model debug and proposed the assertion statement as a more "hardware-related" mechanism for addressing this need. Later language design activities argued that assertion statements do not fully support file input/output applications, such as processing test vectors or modeling a large memory. Consequently, file types and operations were added to VHDL. The following sections explain how to declare and use files.

18.11.1 File Types, Objects, and Operations

Files are objects in VHDL that hold values, lots of values. Thus, like all VHDL objects, we need to declare a *file type* and then declare objects (files) of this type.

VHDL provides predefined and user-defined file types. We will discuss the predefined file type called TEXT for files containing "text" or ASCII character strings in the next section. We discuss user-defined file types in this section. User-defined file type declarations may appear in entities, architectures, subprograms, packages, blocks, and processes. Example user-defined file type declarations are listed below.

```
-- General form
type name is file of type_designation;

-- Examples
type DATA_FILE_TYPE is file of BIT_VECTOR(7 downto 0);

type COUNT_FILE_TYPE is file of NATURAL;
```

A file type declaration includes the type name (any legal identifier) and the type of the values contained in the file. In our examples, files of type DATA_FILE_TYPE can hold only bytes and files of type COUNT_FILE_TYPE can hold only natural numbers. A file may not contain multidimensional arrays, files, or pointers. Pointers, or objects of an access type, are discussed in Chapter 19.

Having declared a file type, we can then declare objects of that type. Objects of file types, called *files*, may be declared in entities, architectures, subprograms, packages, blocks, and processes. Example file declarations are listed below.

```
-- General form
file name  : file_type is mode logical_name;

-- Examples
file TEST_FILE  : DATA_FILE_TYPE  is "c:\verify\test.dat";

file DATA_FILE  : DATA_FILE_TYPE  is in "/u/microcode/rom.dat";

file COUNT_FILE : COUNT_FILE_TYPE is out "COUNT FILE A";
```

The name of a file object may be any legal identifier. The mode may be either **in** or **out**. The contents of a file of mode **in** can be read, but cannot be updated. The contents of a file of mode **out** can be updated, but cannot be read. Mode keywords are optional. If a mode keyword is omitted, as shown in the TEST_FILE declaration, mode **in** is assumed. The last part of a file declaration specifies the logical name of the file, which is the host operating system file name. The logical name is character string. The first, second, and third file declaration examples give string literals illustrating DOS, Unix,[6] and VM[7] file system naming conventions, respectively.

Three predefined subprograms are provided for file input/output operations.

1. READ
2. WRITE
3. ENDFILE

READ and WRITE are procedures and ENDFILE is a function.

The procedure READ inputs data from a file. The procedure READ requires two arguments; the first argument is a file and the second argument is a variable that holds the data read from the file. File reads are sequential. The first READ operation retrieves the first data value in the file, the second READ operation retrieves the next (second) data value, and so on.

The procedure WRITE outputs data to a file. The procedure WRITE also requires two arguments; the first argument is a file and the second argument is an expression yielding the data to be added to the file. WRITE always appends to the end of a file.

Finally, the function ENDFILE checks for "end-of-file." This function accepts a file and returns the BOOLEAN value TRUE if the next READ operation will not retrieve any more data, otherwise the returned value is FALSE.

Figure 18.12 illustrates an application of file input/output: recording in a file the frequency of detecting a sequence of bits. The design entity SEQ_DETECTOR describes the Mealy machine state diagram given in Figure 9.7. The output, PATTERN, is asserted active-1 whenever the input sequence 1100_2 or 1000_2 is detected. The sequence detector is a nonresetting state machine, meaning that the pattern 1000_2 may be partially contained within the pattern 1100_2.

On each clock cycle, the **if** statement checks whether the present input combined with the last three inputs stored in the variable LAST_3 equals the pattern 1100_2 or 1000_2. If there is a

---

[6] Unix® is a registered trademark of AT&T/Bell Laboratories.

[7] VM® is a registered trademark of International Business Machines.

Figure 18.12

Algorithmic description of
a nonresetting sequence
detector

```
entity SEQ_DETECTOR is
 port
    (DATA    : in  BIT;
     CLK     : in  BIT;
     PATTERN : out BIT);
 end SEQ_DETECTOR;

architecture ALGORITHM of SEQ_DETECTOR is
begin
 process
   -- Stores last three inputs, least recent ... most recent
   variable LAST_3 : BIT_VECTOR(1 to 3) := B"000";

   -- Stores times of pattern matches
   type RECORD_FILE_TYPE is file of TIME;
   file RECORD_FILE    : RECORD_FILE_TYPE is out
                              "S_DETECT TIMES A";
  begin
   if ((LAST_3 & DATA = B"1100") or (LAST_3 & DATA = B"1000"))
 then
       PATTERN <= '1';              -- Assert output
       WRITE(RECORD_FILE, NOW);   -- Record time
    else
       PATTERN <= '0';
    end if;

    LAST_3 := LAST_3(2 to 3) & DATA;

    wait until (CLK = '1');
  end process;
 end ALGORITHM;
```

pattern match, the output signal PATTERN is set to '1' and the simulation time is recorded in the file RECORD_FILE. NOW is a predefined function in VHDL that yields the present simulation time. After a simulation with a particular input data set, the RECORD_FILE file could be examined to study how often the two bit sequences were detected.

Notice that VHDL does not provide explicit file open and close operations. Files are implicitly opened when they are elaborated at the invocation of subprograms or at the start of simulation of entities, architectures, packages, blocks, and processes. Files are implicitly closed when they are destroyed at the completion for subprograms or at the end of simulation of entities, architectures, packages, blocks, and processes.

18.11.2 Text File Input/Output

File input/output described in the previous section stores on the local host data in *binary* format. For instance, an integer is typically stored as 2 or 4 bytes, depending on the host computer. Binary format is memory space-efficient, but is not portable across different hosts and cannot be easily examined, for example, by scrolling the file on a terminal or viewing the file with a text editor. To address these disadvantages, data is sometimes stored in *text* format instead of binary format. VHDL supports *text* file input/output by providing a predefined file type called TEXT and a set of associated subprograms.

Files of type TEXT are called *text files*. A text file contains lines of text. Each line of text is of the predefined type LINE and is, in turn, composed of data values or *fields*. Text files are declared using the general format described in the previous section and the predefined type TEXT. For example, the declaration

---

```
file ROM_FILE : TEXT is in "/u/firmware/rom";
```

---

defines a text file named ROM_FILE having the local host name /u/firmware/rom. VHDL provides two predefined text files INPUT and OUTPUT respectively having the local host file names STD_INPUT and STD_OUTPUT. These local host file names can be associated with a terminal or display device to enable user interaction during simulation.

Six predefined subprograms are provided for text file input/output operations.

1. READLINE
2. READ
3. WRITELINE
4. WRITE
5. ENDLINE
6. ENDFILE

READ, READLINE, WRITE, and WRITELINE are procedures and ENDLINE and ENDFILE are functions.

Text file input/output is performed line by line. For input, a line of text is read from a file with the READLINE procedure and then individual data values are extracted from the line of text with the READ procedure. For output, individual data values are appended to a line of text with the WRITE procedure and then the entire line of text is appended to a file with the WRITELINE procedure.

The READLINE procedure requires two arguments. The first argument is a text file and the second argument is a variable of type LINE to hold the line of text read from the file. For example, assume the file ROM_FILE declared above contains the following line of text containing three fields.

```
Memory_Address    X"FB"    16#07#
```

This line of text can be input by reading the line into the variable INSTRUCTION using the READLINE procedure.

```
variable INSTRUCTION : LINE;

READLINE(ROM_FILE, INSTRUCTION);
```

The READ procedure requires at least two arguments and can accept three arguments. The first argument is the line of text. The second argument is a variable of type BIT,

BIT_VECTOR, BOOLEAN, CHARACTER, INTEGER, REAL, STRING, or TIME to hold data read from a field of the line. The optional third argument is a variable of type BOOLEAN that is set to TRUE if the READ procedure is successful, otherwise the variable is set to FALSE. For example, the data fields in the line of text held in the INSTRUCTION variable can be input by reading the data fields into the variables LABEL, ADDRESS, and DATA using the READ procedure.

```
variable LABEL    : STRING(1 to 14);
variable ADDRESS  : BIT_VECTOR(7 downto 0);
variable DATA     : INTEGER;
variable VALID    : BOOLEAN;

READ(INSTRUCTION, LABEL);          -- LABEL = Memory_Address
READ(INSTRUCTION, ADDRESS);        -- ADDRESS = X"FB"
READ(INSTRUCTION, DATA, VALID);    -- DATA = 16#07#
                                   -- VALID = TRUE
```

The WRITE procedure constructs a line of text. WRITE requires at least two arguments and can accept up to six arguments. The first argument is the line of text. The second argument is an expression of type BIT, BIT_VECTOR, BOOLEAN, CHARACTER, INTEGER, REAL, STRING, or TIME yielding the data to be added as the next field in the line. The optional four remaining arguments control the format of the data field and are listed below.

Formal Parameter	Description
FIELD	Specifies field width in terms of characters. Value can be any integer greater than or equal to 0. Default field width is the minimum width required to contain data.
JUSTIFIED	Specifies whether the data value is left or right justified, or aligned, within the field. Value can be LEFT or RIGHT. Default is right-justified.
DIGITS	Specifies the number of digits to the right of the decimal point for floating point numbers. Value can be any integer greater than or equal to 0. Default format is normalized scientific notation.
UNIT	Specifies units for values of type TIME. Value can be any unit of type TIME. Default is ns (nanoseconds).

The WRITELINE procedure requires two arguments. The first argument is a text file and the second argument is an expression yielding the line of text to add to the file. WRITELINE always appends lines to the end of a file.

Finally, the ENDLINE and ENDFILE functions check for "end-of-line" and "end-of-file," respectively. ENDLINE accepts a line and returns the BOOLEAN value TRUE if the next READ operation will not retrieve any more data, otherwise the returned value is FALSE. ENDFILE accepts a file and returns the BOOLEAN value TRUE if the next READLINE operation will not retrieve any more data, otherwise the returned value is FALSE.

Figure 18.13 illustrates an application of text file input/output by modifying the SEQ_DETECTOR in Figure 18.12 to use text files instead of binary files. The text file type and input/output subprograms are accessed via the package TEXTIO contained in the library STD. The WRITE procedure constructs the line of text containing the time value. The

Figure 18.13

Nonresetting sequence detector with text input/output

```vhdl
use STD.TEXTIO.all;
architecture ALGORITHM of SEQ_DETECTOR is
begin
 process
  variable LAST_3 : BIT_VECTOR(1 to 3) := B"000";

    -- Text file to store times of pattern matches
  file RECORD_FILE : TEXT is out "S_DETECT TIMES A";
  variable TIME_STAMP : LINE;
 begin
  if ((LAST_3 & DATA = B"1100") or (LAST_3 & DATA = B"1000"))
  then
     PATTERN <= '1';                              -- Assert output
     WRITE(L=>TIME_STAMP, VALUE=>NOW, UNIT=>ms); -- Record time
     WRITELINE(RECORD_FILE, TIME_STAMP);
  else
     PATTERN <= '0';
  end if;

  LAST_3 := LAST_3(2 to 3) & DATA;

    -- Wait until next rising edge of clock
  wait until (CLK = '1');
 end process;
end ALGORITHM;
```

WRITELINE procedure appends the line of text to the file RECORD_FILE. A sample contents of RECORD_FILE is given below.

```
3.4 ms
7.8 ms
12.2 ms
```

18.12 VHDL-93: FILE INPUT/OUTPUT

VHDL-93 involves substantial changes to file input/output that are *not* upward compatible. That is, VHDL-87 models using file input/output will probably not recompile and simulate successfully using VHDL-93 compliant design tools. Hence, it is important to understand differences between VHDL-87 and VHDL-93 file input/output and the supporting rationale.

File type declarations have not changed; file types are declared in VHDL-93 the same way file types are declared in VHDL-87.

```vhdl
type DATA_FILE_TYPE is file of BIT_VECTOR(7 downto 0);
```

File declarations, however, have changed. VHDL-93 provides explicit as well as implicit file open and close operations. Example VHDL-93 file declarations are listed below.

```
-- No file open
file TEST_FILE  : DATA_FILE_TYPE;

-- Implicit file open in READ_MODE
file TEST_FILE  : DATA_FILE_TYPE is "c:\verify\test.dat";

-- Explicit file open in READ_MODE
file TEST_FILE  : DATA_FILE_TYPE open READ_MODE is
                  "c:\verify\test.dat";
```

The first file declaration defines a file TEST_FILE, but leaves the binding of TEST_FILE with a host or external file for a later, separate file open operation. The second file declaration defines a file TEST_FILE, binds TEST_FILE with the host file c:\verify\test.dat, and implicitly opens the file for access or read only. Note that only the form of the second VHDL-93 file declaration is upward compatible with VHDL-87 file declarations. The third file declaration defines a file TEST_FILE, binds TEST_FILE with the host file c:\verify\test.dat, and explicitly opens the file for access or read only.

The VHDL-87 file mode designations of **in** and **out** are no longer used in VHDL-93 and have been replaced by the file mode designations READ_MODE, WRITE_MODE, and APPEND_MODE.

File Mode	Definition
READ_MODE	Access or read only. Reads are sequential, starting with first value in the host or external file.
WRITE_MODE	Update or write only. Writes are sequential, starting at the beginning of the host or external file. Previous file contents are destroyed.
APPEND_MODE	Update or write only. Writes are sequential, starting at the end of the host or external file. Previous file contents are retained.

In addition to READ, WRITE, and ENDFILE subprograms. VHDL-93 provides FILE_OPEN and FILE_CLOSE procedures. The FILE_OPEN procedure binds a file object with a host file and opens the file. FILE_OPEN has two forms. The first form requires at least two arguments and can accept three arguments. The first argument is a file and the second argument is an expression yielding a character string specifying the host file. The optional third argument specifies the file mode; if the third argument is omitted, READ_MODE is assumed.

```
-- Default READ_MODE
FILE_OPEN(TEST_FILE, "c:\verify\test.dat");

FILE_OPEN(TEST_FILE, "c:\verify\test.dat", APPEND_MODE);
```

The second form of FILE_OPEN provides an additional argument supplying status information about the file open operation.

```
variable STATUS : FILE_OPEN_STATUS;

-- Default READ_MODE
FILE_OPEN(STATUS, TEST_FILE, "c:\verify\test.dat");

FILE_OPEN(STATUS, TEST_FILE, "c:\verify\test.dat",
          APPEND_MODE);
```

The additional argument must be of type FILE_OPEN_STATUS, defined in Chapter 16. The status of the file open operation is returned as the value of the additional argument; possible values are listed below.

File Open Status	Definition
OPEN_OK	File open operation successful.
STATUS_ERROR	File object is already open.
NAME_ERROR	The host file cannot be associated with the file object.
MODEL_ERROR	The host file cannot be associated with the file object with requested mode.

The procedure FILE_CLOSE closes a file. The procedure FILE_CLOSE accepts a single argument that is a file.

```
FILE_CLOSE(TEST_FILE);
```

The final change in VHDL-93 files concerns object classes. In VHDL-87, files are special kinds of variables, just as ports are special kinds of signals. However, since files cannot be assigned to or compared, files actually have little affinity to variables. In VHDL-93, files are defined as their own object class, **file**. Thus, formal parameter lists for subprograms that include files must declare the files using the new object class **file** instead of the old object class **variable**. Moreover, such formal parameter declarations have *no* mode. Concepts and rules of VHDL modes **in**, **out**, and **inout** do not apply to files because updating a file implies reading (file index) and reading a file implies updating (file index). Thus, for consistency, mode properties are removed from file parameters in VHDL-93. The creation of a **file** object class also affects VHDL constructs using entity classes, such as user-defined attributes and groups.

18.13 VHDL-93: SIMPLE_NAME, IMAGE, AND VALUE ATTRIBUTES

The **report** statement and TEXT files discussed in previous sections provide character string input/output capabilities, so now is a good time to talk about new character string features in VHDL-93. The predefined attribute SIMPLE_NAME provides the character string representation of a named item, such as an entity, architecture, variable, function, type, or group. For instance, referring to Figure 18.5, SEQ_ADDER'SIMPLE_NAME yields "seq_adder" and STATE_TYPE'SIMPLE_NAME yields "state_type". Text representations of named items using VHDL-87 identifiers are given in lowercase; text representations of named items using VHDL-93 extended identifiers preserve case.

The predefined attribute IMAGE provides the character string representation of the value of a scalar type, in other words, integer type, floating point type, physical type, or enumeration type. For instance, BIT'IMAGE('0') yields "'0'" and BIT'IMAGE(A) yields the text representation of the value of the object A of type BIT. The predefined attribute VALUE provides the reverse mapping between a character string and a value of a scalar type; BIT'VALUE("'0'") yields the bit literal '0'.

18.14 VHDL-93: SEQUENTIAL STATEMENT LABELS

As the final topic in this chapter, we will quickly cover statement labels. In VHDL-87, the only sequential statement that may have a label is the **loop** statement. To improve consistency, all sequential statements in VHDL-93 may be given labels. Thus, the sequential statements explained in previous sections can be modified to include labels, as shown below.

```
label: wait ...
label: assertion ...
label: SIG <= ....
label: VAR := ....
label: exit .....
```

A label can be any legal identifier.

18.15 SUMMARY

In this chapter, we introduced algorithmic modeling styles in VHDL, which principally employ the process statement and VHDL's powerful set of sequential statements. It is instructive at this point to offer a few comparative observations concerning data flow and algorithmic behavioral modeling styles. Data flow modeling styles often describe more detailed, or lower-level, abstractions because the nonprocedural semantics can be exploited to describe hardware parallelism. However, the details of concurrent hardware signal transformations can obscure the overall input/output behavior. These trade-offs are reversed for algorithmic modeling. An algorithmic model succinctly describes input/output behavior as a series of state transitions or as a more abstract functional computation, but tends to hide lower-level implementation details. Hence, data flow and algorithmic modeling have somewhat complementary advantages and disadvantages and deciding which modeling style to use depends on the particular situation and individual designer preference.

In algorithmic modeling, input/output behavior is composed of one or more processes. While a process itself is a concurrent statement, a process contains one or more sequential statements. The sequential statements describe signal transformations. Many VHDL sequential statements mimic common software programming constructs: **if** and **case** statements support conditional control flow; **loop**, **next**, and **exit** statements support iteration control flow; and procedures support behavioral decomposition. The **wait** statement controls process execution and the sequential assertion and **report** statements support error reporting.

To improve our information modeling skills, we learned three new user-defined data types: integer, floating point, and file. These new data types are added to the following list that summarizes the data types covered thus far in our study of VHDL and digital systems.

- *Scalar*
  - –Enumeration
    - –*Predefined:* BIT, BOOLEAN, CHARACTER, SEVERITY_LEVEL, FILE_OPEN_KIND*, *and* FILE_OPEN_STATUS*
    - –*User-Defined: Chapter 16*
  - –Numeric
    - –*Integer*
      - • *Predefined:* INTEGER, NATURAL, *and* POSITIVE
      - • *User-Defined: Chapter 18*
    - –*Floating Point*
      - • *Predefined:* REAL
      - • *User-Defined: Chapter 18*
    - –*Physical*
      - • *Predefined:* TIME *and* DELAY_LENGTH*
      - • *User-Defined: Chapter 15*
- *Composite*
  - –Array
    - –*Constrained*
      - • *Predefined: none*
      - • *User-Defined: Chapter 16*
    - –*Unconstrained*
      - • *Predefined:* BIT_VECTOR *and* STRING
      - • *User-Defined: Chapter 16*
- *File*
  - –*Predefined:* TEXT
  - –*User-Defined: Chapter 18*
  - * VHDL-93

Integer and floating point types allow users to define unique sets of integers and floating point numbers, respectively. File types allow a VHDL model to communicate with the host file system.

We also discussed subtypes, which define subsets of values of a given base type. Objects of subtypes of the same base type can be mixed in VHDL expressions and statements. Subtypes also provide the ability to associate a resolution function with a type instead of with individual signals.

Finally, we examined several new VHDL-93 features including new file input/output, text-related predefined attributes, and sequential statement labels. VHDL-93 file declarations and operations declare files as a new object class and provide for explicit open and close operations.

18.16 PROBLEMS

1. Consider the following combinational logic.

$$OUT\_1 = \overline{(IN\_1 \cdot IN\_2) + (IN\_3 \cdot IN\_4)}$$

   a. Describe the logic using data flow modeling and concurrent signal assignment statements.
   b. Describe the logic using algorithmic modeling and a single process statement.

2. Consider the following Moore machine realizing a 2-bit binary counter.

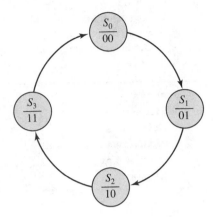

a. Describe the counter using a guarded concurrent signal assignment statement.

b. Describe the counter using a single process from a state transition perspective.

c. Describe the counter using a single process from a functional (in other words, arithmetic) perspective.

d. List all assumptions for parts (a), (b), and (c).

3. Consider the following A and B signal waveforms.

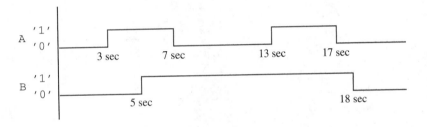

a. Show the waveform for signal C generated by the following process.

```
process
begin
  wait on A, B;
  C <= A xor B;
end process;
```

Assume an initial value of '0' for C. Indicate by arrows when the sequential signal assignment statement executes. Explain your answer.

b. Show the waveform for signal C generated by the following process.

```
process
begin
  wait on A until (A /= B);
  C <= A xor B;
end process;
```

Assume an initial value of '0' for C. Indicate by arrows when the sequential signal assignment statement executes.

c. Show the waveform for signal C generated by the following process.

```
process
begin
  wait on A;
  wait until (A /= B);
  C <= A xor B;
end process;
```

Assume an initial value of '0' for C. Indicate by arrows when the sequential signal assignment statement executes. Explain your answer.

4. Consider the following A and B signal waveforms.

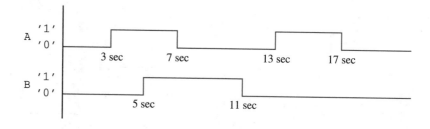

a. Show the waveform for signal C generated by the following process.

```
process (B)
begin
  C <= A xor B;
end process;
```

Assume an initial value of '0' for C. Indicate by arrows when the sequential signal assignment statement executes. Explain your answer.

b. Show the waveform for signal C generated by the following process.

```
process
begin
  wait on A for 5 sec;
  wait until (A /= B);`
  C <= A xor B;
end process;
```

Assume an initial value of '0' for C. Indicate by arrows when the sequential signal assignment statement executes. Explain your answer.

c. Show the waveform for signal C generated by the following process.

```
process
begin
  wait on A;
  wait for 5 sec;
  C <= A xor B;
end process;
```

Assume an initial value of '0' for C. Indicate by arrows when the sequential signal assignment statement executes. Explain your answer.

5. A sequential algorithmic description can be transformed into a concurrent data flow description by decomposing the behavior into several processes. Consider the following Mealy machine state diagram.

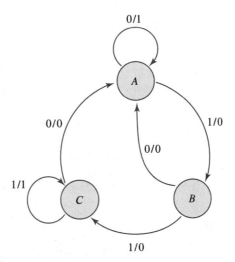

a. Describe the state machine using one process.
b. Describe the state machine using two processes: one process for state transitions and one process for output transitions.
c. Describe the state machine using four processes: three processes for state transitions (one process per state) and one process for output transitions. (*Note*: Since multiple processes are driving state transitions and only one process should control state at anytime, state should be held in a resolved signal. A process can disconnect as a source to a resolved signal by driving the signal with the special value **null**.)

6. For the Mealy machine state diagram in Problem 5, do the following.
a. Describe the state machine using a guarded concurrent signal assignment statement to model state transitions and a concurrent signal assignment statement to model output transitions.

b. Describe the state machine using three guarded concurrent signal assignment statements to model state transitions and a concurrent signal assignment statement to model output transitions. (*Note*: Since multiple guarded concurrent signal assignment statements are driving state transitions and only one guarded concurrent signal assignment statement should control state at any time, state should be held in a guarded signal.)

7. Draw the state diagram for the following Moore machine.

```
entity MOORE_MACHINE is
  port
    (SIG_IN  : in BIT;  CLK  : in BIT;
     SIG_OUT : out BIT);
end MOORE_MACHINE;

architecture MOORE_MACHINE of MOORE_MACHINE is
    type STATE_TYPE is (A, B, C);
    type STATES_TYPE is array (NATURAL range <>) of STATE_TYPE;
    function MUX (DRIVERS : STATES_TYPE) return STATE_TYPE is
    begin
      return DRIVERS(DRIVERS'LEFT);
    end MUX;

    signal STATE : MUX STATE_TYPE register := A;
begin
  CLK_RISE: block (CLK='1' and not CLK'STABLE)
  begin
    STATE_A: block (GUARD and STATE=A)
    begin
      STATE <= guarded STATE when SIG_IN='0' else
                     B;
    end block STATE_A;

    STATE_B: block (GUARD and STATE=B)
    begin
      STATE <= guarded STATE when SIG_IN='0' else
                     C;
    end block STATE_B;

    STATE_C: block (GUARD and STATE=C)
    begin
      STATE <= guarded A;
    end block STATE_C;
  end block CLK_RISE;

  with STATE select
    SIG_OUT <= '1' when A,
               '0' when B,
               '1' when C;
end MOORE_MACHINE;
```

8. Use conditional sequential statements instead of arrays to model the state and output transitions in the sequential adder in Figure 18.5(b).

9. The MUX resolution function in Problem 7 should only receive one driver per invocation. Add a sequential assertion statement to the function to check for its proper use.

10. Add a synchronous clear, CLR, and set, SET, capability to the 3-bit binary counter described in Figure 18.2. If CLR is asserted active-1, the output equals 000_2. If SET is asserted active-1, the output equals 111_2. It is illegal to assert CLR and SET at the same time.
    a. Add a sequential assertion statement to check for the proper use of CLR and SET.
    b. Add a concurrent assertion statement to check for the proper use of CLR and SET.

11. Consider the following resolution function.

```
function WIRED_AND (SOURCES : BIT_VECTOR) return BIT is
 variable RESULT : BIT := '1';
begin
 for I in SOURCES'RANGE loop
  RESULT := RESULT and SOURCES(I);
 end loop;
 return( RESULT );
end WIRED_AND;
```

    a. Recode WIRED_AND using an infinite-loop.
    b. Recode WIRED_AND using a while-loop.
    c. Comment on which form of the iteration statement is preferred in this application.

12. VHDL supports a *concurrent procedure statement* that is a shorthand notation for a process containing a sequential procedure call. The process also contains a signal wait on any signals associated with procedure formal parameters of mode **in** or **inout**. For example, the following concurrent procedure call

```
entity ADDRESS_CALC is
 port
    (ADDRESS, OFFSET : in BIT_VECTOR(7 downto 0);
     VADDRESS : out BIT_VECTOR(7 downto 0);
     OVERFLOW : out BIT);
end ADDRESS_CALC;

architecture DATA_FLOW of ADDRESS_CALC is
  procedure ADDER (LARG, RARG : in  BIT_VECTOR;
                   SUM        : out BIT_VECTOR;
                   CARRY      : out BIT) is

    . . .
  end ADDER;
begin
  -- Concurrent procedure call
  ADDER(LARG=>ADDRESS, RARG=>OFFSET, SUM=>VADDRESS,
        CARRY=>OVERFLOW);
end DATA_FLOW;
```

is equivalent to

```
architecture DATA_FLOW of ADDRESS_CALC is
  procedure ADDER (LARG, RARG : in  BIT_VECTOR;
                   SUM         : out BIT_VECTOR;
                   CARRY       : out BIT) is
    . . .
  end ADDER;
begin
  process
  begin
    ADDER(LARG=>ADDRESS, RARG=>OFFSET, SUM=>VADDRESS,
          CARRY=>OVERFLOW);
    wait on ADDRESS, OFFSET;
  end process;
end DATA_FLOW;
```

    a. Write the procedure `ADDER`. Make the `ADDER` procedure general enough to handle input operands having either ascending or descending array indexing.

    b. Place the `ADDER` procedure into a package and rewrite the entity `ADDRESS_CALC`.

13. Rewrite the concurrent procedure call version of the `ADDRESS_CALC` design entity in Problem 12 to declare `ADDRESS`, `OFFSET`, and `VADDRESS` using a subtype instead of a constrained array type.

14. <u>VHDL-93:</u> Add to the `ADDER` procedure written in Problem 12 "print" statements that record entering the procedure and the values of the formal parameters `LARG` and `RARG`.

15. Generate a VHDL model of a 256x8 ROM using the following interface.

```
entity ROM_256x8 is
  port
    (ADDRESS : in    BIT_VECTOR(7 downto 0);
     CE_BAR  : in    BIT;
     NIB_MD  : in    BIT;
     DATA    : inout BIT_VECTOR(7 downto 0));
end ROM_256x8;
```

`CE_BAR` is an active-0 chip enable signal, and `NIB_MD` is an active-1 nibble mode signal. A read cycle is performed on the falling edge of `CE_BAR`. If `NIB_MD` is deasserted, one byte stored at location `ADDRESS` is read onto `DATA`. If `NIB_MD` is asserted, four contiguous bytes starting at `ADDRESS` are read onto `DATA` on four consecutive read cycles.

    a. Use a binary file to store ROM contents.

    b. Use a text file to store ROM contents.

16. <u>VHDL-93:</u> Make any changes that are required to the ROM model in Problem 15 to make it compliant with VHDL-93.

17. For simulation, input stimuli (in other words, test vectors) are often generated and output responses are monitored via a simulator control command interface *or* a VHDL *testbench*. Like a hardware testbench, a VHDL testbench contains the VHDL entity to be tested and supporting processes that mimic the behavior of signal generators and recorders. Write a testbench for simulating the text file version of `ROM_256x8` in Problem 15(b) by reading bytes at locations 0, 105, and 200 and nibbles starting at locations 4, 64, and 252. Verify the operation.

18. Write a testbench for simulating the VHDL model of the Mealy machine generated in Problem 5(a). Read the input stimuli from a text file having the following contents.

```
1
1
0
0
1
1
```

Test inputs are listed one per line. Record the output response in another text file.

REFERENCES AND READINGS

[1] Benders, L. and M. Stevens. "Task Level Specification and Communication," *18th European Symposium Microprocessing and Microprogramming,* pp. 355–362, September 1992.

[2] Berthet, C., J. Rampon, and L. Sponga. "Synthesis of VHDL Arrays on RAM Cells," *Proceedings of the European Design Automation Conference,* pp. 726–731, September 1992.

[3] Carlson, S. *Introduction To HDL-Based Design Using VHDL.* Mountain View, CA: Synopsys Inc., 1990.

[4] Ecker, W. and S. Marz. "Subtype Concept of VHDL for Synthesis Constraints," *Proceedings of the European Design Automation Conference,* pp. 720–725, September 1992.

[5] Harper, P. and K. Scott. "Towards a Standard VHDL Synthesis Package," *Proceedings of the European Design Automation Conference,* pp. 706–712, September 1992.

[6] Lim, S., D. Hendry, and P. Yeung. "Experiences and Issues in VHDL-Based Synthesis," *Proceedings of the European Design Automation Conference,* pp. 646–651, September 1992.

[7] Nagasamy, V., N. Berry, and C. Dangelo. "Specification, Planning, and Synthesis in a VHDL Design Environment," *IEEE Design and Test of Computers,* vol. 19, no. 2, pp. 58–68, June 1992.

[8] Ramachandra, L., F. Vahid, S. Narayan, and D. Gajski. "Semantics and Synthesis of Signals in Behavioral VHDL," University of California, Irvine, Semiconductor Research Corporation (SRC) Publication C92184, April 1992.

[9] Wolf, H., A. Jakach, C. Huang, P. Manno, and E. Wu. "The Princeton University Behavioral Synthesis System," *Proceedings of the 29th ACM/IEEE Design Automation Conference,* pp. 182–187, June 1992.

19
VHDL: A LAST LOOK

In this chapter, we will finish our discussion of VHDL by examining a variety of advanced behavioral and structural modeling capabilities. New language constructs will be introduced and previously learned language constructs will be reviewed to summarize our understanding of VHDL and to exercise the full power of the hardware description language to model digital systems.

We will complete our study of structural modeling by discussing *mixed structural/behavioral* descriptions and *configurations*. Mixed structural/behavioral descriptions combine structural, data flow, and algorithmic modeling styles within a single architecture. Configurations provide the ability to explicitly bind design entities to components.

We will also complete our study of data types. VHDL data types have been gradually introduced over the course of several chapters. The listing below shows that six of the eight VHDL data types have been discussed, leaving only *record data types* and *access data types* to address in this chapter.

- *Scalar*
  - Enumeration
    - *Predefined:* BIT, BOOLEAN, CHARACTER, SEVERITY_LEVEL, FILE_OPEN_KIND*, *and* FILE_OPEN_STATUS*
    - *User-Defined: Chapter 16*
  - Numeric
    - *Integer*
      - *Predefined:* INTEGER, NATURAL, *and* POSITIVE
      - *User-Defined: Chapter 18*
    - *Floating Point*
      - *Predefined:* REAL
      - *User-Defined: Chapter 18*
    - *Physical*
      - *Predefined:* TIME, *and* DELAY_LENGTH*
      - *User-Defined: Chapter 15*
- *Composite*
  - Array
    - *Constrained*
      - *Predefined: none*
      - *User-Defined: Chapter 16*
    - *Unconstrained*
      - *Predefined:* BIT_VECTOR *and* STRING
      - *User-Defined: Chapter 16*
- File
  - *Predefined:* TEXT
  - *User-Defined: Chapter 18*

* VHDL-93

19.1 MIXED BEHAVIORAL/STRUCTURAL DESCRIPTIONS

In previous chapters, we have presented a variety of VHDL models illustrating several popular hardware modeling styles, including structural, data flow, and algorithmic. It is important to realize that these modeling styles are *not* mutually exclusive. The modeling style of a design entity may be largely structural, data flow, or algorithmic, but this is a restriction imposed by the user, not the language. In general, a VHDL model may contain an arbitrary mix of modeling styles. An architecture may have portions of its behavior described as an algorithm via process statements, a data flow via concurrent signal assignment statements, or a structure of lower-level design entities via component instantiation statements. This flexibility reveals some of the descriptive power of VHDL.

As an example, consider modeling the quad 2-port register discussed in Chapter 10 and shown again in Figure 19.1. A data flow model of the quad 2-port register is given in Figure 19.2. For convenience, the data inputs $A_0/A_1 - D_0/D_1$ are grouped together using the 4×2 array DATA. The first array index (dimension) refers to one of the four D flip-flops; the second array index refers to one of the two multiplexer inputs per flip-flop. For instance, DATA(A, '1') refers to the A_1 input. Notice how user-defined data types can be exploited to align model naming conventions with schematic naming conventions, thus making the

Figure 19.1

A quad 2-port register

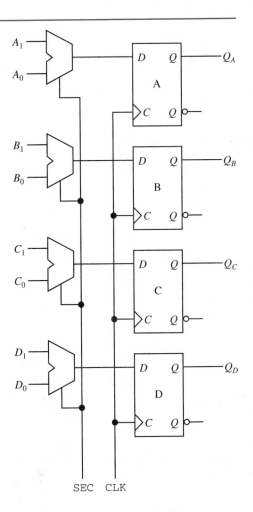

SEC   CLK

Figure 19.2

A data flow VHDL model
of the quad 2-port register

```
package REG_PKG is
   -- Enumeration type
   type FF_INDEX is (A, B, C, D);

   -- Array types
   type INPUT_TYPE is array (FF_INDEX, BIT) of BIT;
   type OUTPUT_TYPE is array (FF_INDEX) of BIT;
end REG_PKG;

use WORK.REG_PKG.all;
entity DUAL_PORT_REG is
 port
    (DATA : in  INPUT_TYPE;
     SEL  : in  BIT;
     CLK  : in  BIT;
     Q    : out OUTPUT_TYPE);
end DUAL_PORT_REG;

architecture DATA_FLOW of DUAL_PORT_REG is
begin
  CLKED: block (CLK='1' and not CLK'STABLE)
  begin
     QUAD: for I in DATA'RANGE(1) generate
        Q(I) <= guarded DATA(I, SEL);
     end generate QUAD;
  end block CLKED;
end DATA_FLOW;
```

model easier to understand. The iteration generate statement generates four guarded concurrent signal assignment statements. On each rising clock edge, the guard expression changes from FALSE to TRUE, and the guarded concurrent signal assignment statements execute to set the four register outputs.

An alternative description of the quad 2-port register is given in Figure 19.3, illustrating a combination of data flow and structural modeling styles. The guarded concurrent signal assignment statements are replaced with concurrent signal assignment statements *and* component instantiation statements.

The DATA_FLOW_STRUCTURE architecture can be viewed as a refinement of the DATA_FLOW architecture because the DATA_FLOW_STRUCTURE architecture reveals more details about how the input/output function is implemented and the inherent parallelism of the associated hardware. The input multiplexing combinational logic is modeled with concurrent conditional signal assignment statements driving the signal DINT. The output edge-sensitive memory is modeled with component instantiation statements and the concurrent signal assignment statements driving the output Q.

The four lower-level D_FF component instances refer to the D flip-flop design entity given in Figure 19.4; this model is based on the D flip-flop model described in Figure 17.9. The component D_FF declared in Figure 19.3 refers to the D_FF design entity given in Fig-

Figure 19.3

The data flow and structural VHDL model of the quad 2-port register	

```vhdl
architecture DATA_FLOW_STRUCTURE of DUAL_PORT_REG is
 component D_FF
  port
    (D, CLK    : in BIT;    Q, Q_BAR : inout BIT);
 end component;

 signal DINT, QINT : OUTPUT_TYPE;
begin
 BIT_LOOP: for I in DATA'RANGE(1) generate
  -- Combinational logic
  DINT(I) <= DATA(I, SEL);

  -- Memory
  D_FFX: D_FF port map (D=>DINT(I), CLK=>CLK, Q=>QINT(I),
                        Q_BAR=>open);

  Q(I) <= QINT(I);
 end generate BIT_LOOP;
end DATA_FLOW_STRUCTURE;
```

Figure 19.4

A positive edge-triggered D flip-flop

```vhdl
entity D_FF is
 port
    (D, CLK   : in BIT;
     Q, Q_BAR : inout BIT);
 end D_FF;

architecture DATA_FLOW of D_FF is
begin
 CLKED: block (CLK = '1' and not CLK'STABLE)
 begin
  Q <= guarded D;          -- Guarded concurrent signal assignment
  Q_BAR <= not Q;
 end block CLKED;
end DATA_FLOW;
```

ure 19.4 by implicitly assuming *default configurations*; components and design entities having matching interface declarations are paired together.

19.2 CONFIGURATIONS

Through alternative architectural bodies and component instantiations, we are able to construct the general model hierarchy shown in Figure 19.5, where a design entity can have multiple architectures and the architectures can, in turn, be composed of lower-level design entities that can also have multiple architectures. This general tree-like model structure

Figure 19.5

General model structure

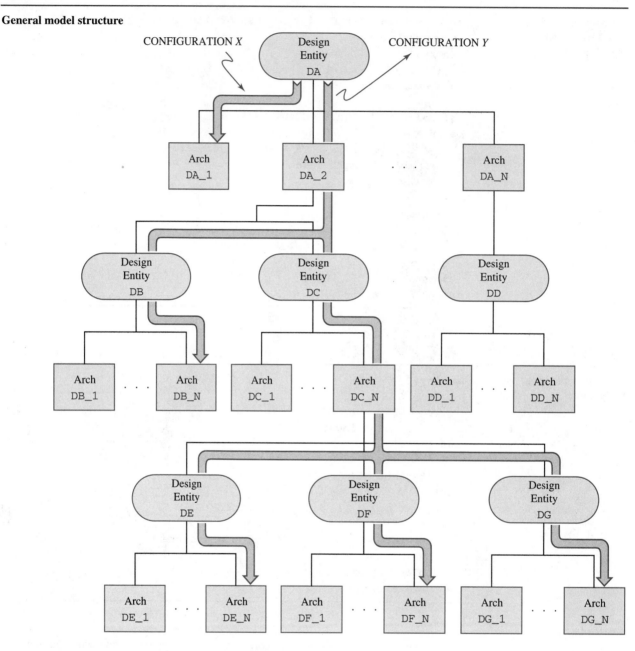

creates a "family" of descriptions for the root design entity, in other words, the design entity DA at the top or apex of the tree. These descriptions typically realize the same input/output function, but may differ in model fidelity or simulation performance.

Although it is useful to define this family of related descriptions, we typically deal with only *one* description at a time. Thus, we need a mechanism to identify a member of a family of descriptions; this mechanism is called a *configuration*. A configuration identifies a set of behaviors that constitute a particular description of the root design entity. For example, the arrows encompassing the interface DA and architecture DA_1 identify a configuration, in other words, a description of the root design entity DA. Alternatively, the arrows encompassing the interface DA, the architecture DA_2, and the five lower-level interface/architecture pairs DB/DB_N, DC/DC_N, DE/DE_N, DF/DF_N, and DG/DG_N identify another configuration.

As we investigate configurations, remember that we are simply trying to define a "profile" or set of branches of a design entity's hierarchical tree structure that identifies one of many possible descriptions. Sections 19.2.1 and 19.2.2 first discuss in more detail the motivation for configurations. Then, Section 19.3 presents several examples showing how to write configurations.

19.2.1 Structural Modeling

A configuration identifies a particular design entity description by selecting architectures and associating *design entities* with *components* instantiated within the architectures. This process delineates a set of behaviors or, equivalently, a path down through the design entity's hierarchical tree structure. To clarify this concept, let us examine the roles of components, design entities, and configurations in structural modeling.

A VHDL structural model starts by defining components via component declaration statements and connecting instances of these components via component instantiation statements. This task does not strictly constitute a complete VHDL structural model because a component is *not* a design entity and a structural model must ultimately describe the interconnection of design entities. A component is really just a "placeholder" for a design entity, an intermediate declaration that will eventually be associated with an actual design entity via a configuration. Recall that Figure 12.3 presents a useful analogy between the interconnection of sockets containing integrated circuits and the interconnection of components associated with design entities.

To summarize, the complete structural modeling process involves the following three tasks.

Structural Modeling Process
1. Define component — component declaration
2. Define component instance — component instantiation
3. Define component/design entity association — configuration

The three tasks generate a structural model in an *indirect* fashion by first describing the interconnection of component instances (tasks 1 and 2) and then describing the association between component instances and design entities (task 3). As previously explained, we thus far have been relying on default configuration rules, which implicitly define required component/entity associations. Component/entity associations can also be explicitly defined via user-declared configurations.

As an example, Figure 19.6 (p. 626) presents a simple user-declared configuration for the DUAL_PORT_REG design entity shown in Figure 19.3. The construction of configurations will be explained in detail in following sections. For the present, it is sufficient to note that the configuration called CONFIG selects the DATA_FLOW_STRUCTURE architecture and associates the design entity D_FF and architecture DATA_FLOW given in Figure 19.4 and located in the WORK library with all (four) instances of the component D_FF, thereby identifying the set of behaviors illustrated in Figure 19.6(b). Figure 19.7 (p. 627) illustrates how component instances, design entities, and configurations work together to define the structure of the quad 2-port register.

Figure 19.6

A configuration for the
DUAL_PORT_REG design
 entity

```
configuration CONFIG of DUAL_PORT_REG is
  for DATA_FLOW_STRUCTURE
    for all : D_FF
      use entity WORK.D_FF(DATA_FLOW);
    end for;
  end for;
end CONFIG;
```

(a)

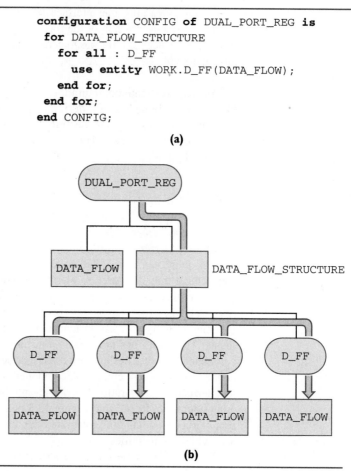

(b)

Figure (a) shows a configuration for the structural description of DUAL_PORT_REG.
Figure (b) illustrates the structural description as a set of paths identifying a set of behaviors in
DUAL_PORT_REG's model hierarchy.

19.2.2 Rationale

Based on our sample configuration, you might justifiably be wondering why VHDL has the
concepts of components and configurations. Would not it be easier to do away with the level
of indirection associated with components and directly instantiate instances of design entities? The answer is yes for simple situations, and VHDL-93 does provide direct design entity
instantiation (see Chapter 12). However, VHDL does not mandate this strategy because there
are several modeling/simulation situations that require the added flexibility of components
and configurations.

Consider the implications of bypassing components and directly instantiating design entities. Checking that a particular design entity instantiation is correct—in other words, that
the right wires are hooked up to the right ports—requires that the interface of the instantiated design entity be defined and under the designer's control. However, for a large design
project, a designer typically does not have control over all design entity interfaces. Lower-
level design entities could be supplied by a different design group or by a different company.
One solution to this dilemma is to provide the designer with the ability to declare "local" design entities, called components. Any differences in the interfaces between local design entities (component declarations) and actual design entities are addressed by configurations.

Figure 19.7

**Definition of the quad
2-port register structure**

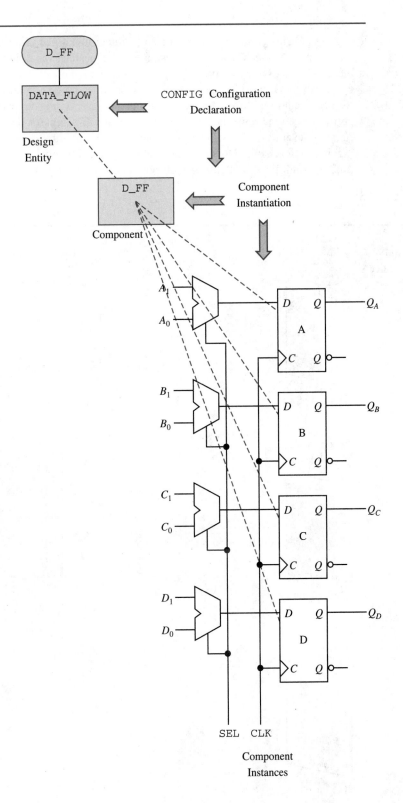

Components and configurations also facilitate investigating design trade-offs associated with using different versions of a design entity. Consider, for example, the VHDL structural model of a digital signal processor containing a component denoting a multiplier given in Figure 19.8. Several integrated circuit vendors offer plug-compatible digital multipliers, so we may want to use a vendor part in our processor design if the part can meet interface and performance requirements. Alternative digital signal processor designs can be evaluated by running several system simulations using models of different vendor multipliers. Changing multipliers simply requires generating new configurations, effectively "plugging in" different multiplier chips into a multiplier chip socket. It is important to note that changing

Figure 19.8

VHDL digital signal processor with a component denoting a mutliplier

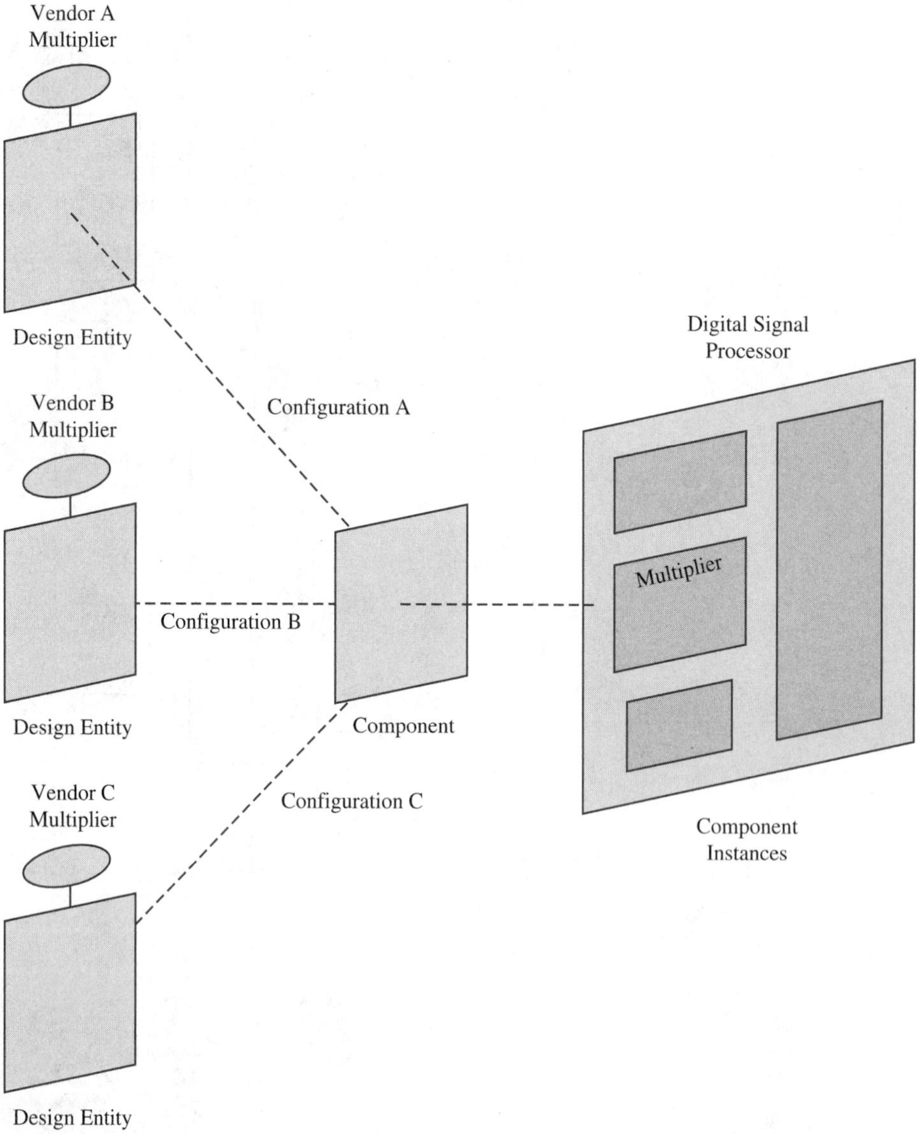

Different configurations bind different design entities to the component.

multiplier models does *not* involve changing the basic structural description of the digital signal processor; the processor always uses a multiplier in the same manner regardless of the particular multiplier model. Thus, components and configurations facilitate generating different descriptions involving different versions of a design entity.

Another benefit of components and configurations is they support *mixed-level simulation*. In mixed-level simulation, different parts of a digital system are simulated at different levels of abstraction or detail. In general, lower levels of abstraction imply more functional parallelism and timing detail, which in turn imply slower simulations. Simulating an entire design at a detailed level just to study a small portion of the design unnecessarily wastes simulation resources. Excessively long system simulations for large designs can be avoided by simulating in detail only the portion of the system being presently studied and/or designed. All other portions of the design are modeled and simulated at a more abstract level. In this manner, simulation is efficiently focused or tailored to match verification needs. Mixed-level simulation can be easily accomplished using the abstract behavioral modeling capabilities of VHDL and configurations. For a particular simulation, a configuration is generated that selects abstract, perhaps algorithmic, architectures for all design entities except the set of design entities comprising the portion of the system of immediate interest.

The final advantage of components and configurations is *delayed binding*. Delayed binding means that decisions concerning which design entities to use or generic parameter values to specify can be postponed. Figure 19.9 shows that there are generally three phases of VHDL modeling during which information may be specified.

The earlier discussion on investigating different multipliers for a digital signal processor demonstrates delayed binding for design entities. The designer does not have to commit at the time the multiplier is used within the digital signal processor model exactly which multiplier design entity is referenced. Rather, the designer can postpone this decision until later in the design cycle. Delayed binding can be similarly applied to generics. Recall that values can be supplied for generic parameters via a **generic map** clause of a component instantiation statement to tailor certain structural/behavioral aspects of component instances. This is a form of delayed binding, postponing model definition until component instantiation. The binding of values to generic parameters can be even further delayed by placing the **generic map** clause in a configuration instead of in a component instantiation, thereby postponing model definition until model operation, in other words, simulation. Several parameter values can be comparatively studied by simply generating new configurations and running several simulations. Later configuration examples will illustrate this practice.

Figure 19.9

Phases of delayed binding

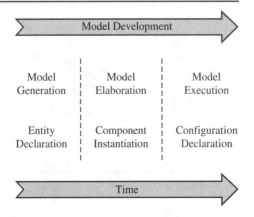

19.3 CONFIGURATION DECLARATIONS

We hope we have presented convincing arguments for components and configurations and their application in using vendor parts, exploring alternative designs, conducting mixed-level simulation, and exploiting delayed binding. In this section we will address the rules for writing various forms of user-declared configurations.

Let us begin by reconsidering the design entity DUAL_PORT_REG. Configurations for the alternative descriptions given in Figure 19.2 and Figure 19.3 are shown in Figure 19.10(a) and Figure 19.10(b), respectively. The configurations in Figure 19.10 are formally called *configuration declarations*.

A configuration declaration has the following form.

```
configuration name of entity_name is
  ....
  -- Statements
  ....
end name;
```

The header of a configuration declaration starts with the keyword **configuration**, followed by the name of the configuration, followed by the name of the design entity being configured. Like an architecture declaration, a configuration declaration is always associated with a design entity interface. Both configurations CONFIG_A and CONFIG_B in Figure 19.10 are associated with or configure the design entity DUAL_PORT_REG.

Statements contained within a configuration declaration define the desired configuration, starting with a *block configuration* statement. As the name implies, a block configuration statement selects a *block*, such as an architecture. For instance, the block configuration statement in configuration CONFIG_A

Figure 19.10

Configurations for the DUAL_PORT_REG design entity

```
configuration CONFIG_A of DUAL_PORT_REG is
  for DATA_FLOW                    -- Block configuration
  end for;
end CONFIG_A;
```

(a)

```
configuration CONFIG_B of DUAL_PORT_REG is
  for DATA_FLOW_STRUCTURE          -- Block configuration
    for all : D_FF                 -- Component configuration
      use entity WORK.D_FF(DATA_FLOW);
    end for;
  end for;
end CONFIG_B;
```

(b)

Figures (a) and (b) present alternative configurations for the design entity DUAL_PORT_REG.

```
for DATA_FLOW
 .....

end for;
```

selects the DATA_FLOW architecture for DUAL_PORT_REG, while the block configuration statement in configuration CONFIG_B

```
for DATA_FLOW_STRUCTURE
 .....

end for;
```

selects the DATA_FLOW_STRUCTURE architecture for DUAL_PORT_REG.

If the selected architecture contains no instantiated components, then a sufficient set of behaviors has been identified and the configuration declaration is finished. This situation is illustrated with configuration CONFIG_A, which contains only a block configuration. However, if the selected architecture contains instantiated components, then these component instances must be further configured with *component configurations*.

19.4 COMPONENT CONFIGURATIONS

A component configuration has three possible forms: *open*, *entity*, and *embedded*.

```
-- Open configuration
for label_identifier : component_name
  use open;
end for;

-- Entity configuration
for label_identifier : component_name
  use entity entity_name(architecture_name)
    generic map (generic_associations)
    port map (port_associations);
end for;

-- Embedded configuration
for label_identifier : component_name
  use configuration configuration_name;
end for;
```

Open and entity component configurations will be explained in this section; embedded component configurations will be explained in the next section.

All component configurations start by identifying component instances. The methods for identifying items in an attribute specification (see Chapter 16) also apply to identifying instances in a component configuration. To illustrate these methods, we will consider a simple structure composed of five instances of the component DSGN_SUBENTITY created by five component instantiation statements.

```
LABEL1: DSGN_SUBENTITY port map (PX => S1, ...);
LABEL2: DSGN_SUBENTITY port map (PX => S2, ...);
LABEL3: DSGN_SUBENTITY port map (PX => S3, ...);
LABEL4: DSGN_SUBENTITY port map (PX => S4, ...);
LABEL5: DSGN_SUBENTITY port map (PX => S5, ...);
```

Component instances can be identified by listing the associated instantiation labels, as shown below.

```
-- Component configuration
for LABEL1, LABEL3, LABEL4 : DSGN_SUBENTITY
  ....
end for;


-- Component configuration
for LABEL2, LABEL5 : DSGN_SUBENTITY
  ....
end for;
```

The first component configuration applies to the DSGN_SUBENTITY component instances labeled LABEL1, LABEL3, and LABEL4. The second component configuration applies to the DSGN_SUBENTITY component instances labeled LABEL2 and LABEL5.

Component instances can also be identified by using the keyword **others**. The keyword **others** picks up all other instances of a component within a block configuration not explicitly identified in previous component configurations. Component configurations for the DSGN_SUBENTITY component instances can be rewritten as follows.

```
-- Component configuration
for LABEL1, LABEL3, LABEL4 : DSGN_SUBENTITY
  ....
end for;


-- Component configuration
for others : DSGN_SUBENTITY
  ....
end for;
```

The second component configuration still applies to the DSGN_SUBENTITY instances labeled LABEL2 and LABEL5. A configuration using the keyword **others** must be the last configuration for a particular component given within a block configuration because a component instance can be configured to only one design entity.

Finally, component instances can be identified using the keyword **all**, which identifies all instances of a particular component within a block configuration.

```
-- Component configuration
for all : DSGN_SUBENTITY
  ....
end for;
```

The single component configuration applies to all component instances, LABEL1 through LABEL5. A configuration using the keyword **all** must be the only configuration for a particular component given within a block configuration because, again, a component instance can be configured to only one design entity.

19.4.1 Open Configuration

Having identified component instances, let us now discuss how to bind component instances to design entities. The simplest binding is *no* binding specified with the *open configuration*.

```
-- Open configuration
for label_identifier : component_name
  use open;
end for;
```

Returning to our DUAL_PORT_REG structural description example, the configuration

```
configuration CONFIG_OPEN of DUAL_PORT_REG is
  for DATA_FLOW_STRUCTURE
    for all : D_FF              -- Open component configuration
    use open;
    end for;
  end for;
end CONFIG_OPEN;
```

specifies that no design entity is associated with the D_FF component instances; in other words, the association or binding is left open. This configuration is obviously not very useful because it leaves the structural description undefined. The open configuration is typically used to define an incomplete or partial configuration that can be analyzed.

19.4.2 Entity Configuration

A more common way to bind component instances to design entities is via an *entity configuration*.

```
-- Entity configuration
for label_identifier : component_name
  use entity entity_name(architecture_name) -- Entity clause
    generic map (generic_associations)      -- Generic map
                                            -- clause
    port map (port_associations);           -- Port map clause
end for;
```

The **entity** clause identifies a design entity and an associated architecture. The **generic** and **port map** clauses associate component/entity generics and ports, respectively. The syntax of the component configuration **generic** and **port map** clauses is the same as the component instantiation **generic** and **port map** clauses. One or more of these three clauses may be omitted by exploiting *implicit* configurations generated by *default entity configuration rules.*

To review, default entity configuration rules implicitly associate a component with a design entity having the *same* name and the *same* interface, in other words, having matching generics and ports. Matching generics and ports means the same number of declarations having identical names and types. If the design entity has multiple architectures, the most recently analyzed (compiled) architecture is selected. For instance, we were able to claim that the D_FF component in Figure 19.3 refers to the D_FF design entity in Figure 19.4 without having to actually write an explicit configuration by making sure the component declaration

```
component D_FF
 port
   (D, CLK   : in BIT;
    Q, Q_BAR : inout BIT);
end component;
```

matched the design entity declaration

```
entity D_FF is
 port
   (D, CLK   : in BIT;
    Q, Q_BAR : inout BIT);
end D_FF;
```

and by taking advantage of default configurations. A default configuration can always be overridden by writing an explicit configuration, as is illustrated with the CONFIG_B configuration shown in Figure 19.6(a) and Figure 19.10(b).

To illustrate the full descriptive capabilities of an entity configuration, let us change the declaration of the D flip-flop design entity so default configuration rules do not apply. For instance, assume that an application-specific integrated circuit (ASIC) vendor will manufacture our quad 2-port register and that the vendor has supplied the following D flip-flop design entity declaration.

```
entity D_FLIPFLOP is
 generic
   (SETUP, HOLD, PROP_DELAY : TIME);
 port
   (DATA, CLOCK : in BIT;
    Q, Q_BAR    : inout BIT);
end D_FLIPFLOP;
```

The vendor-supplied design entity declaration and the user-defined component declaration have different entity names and port names. Also, the vendor-supplied design entity contains generic parameters that are not included in the component declaration. These component/design entity "mismatches" or differences are easily accounted for via the configuration below.

```
configuration CONFIG_VENDOR of DUAL_PORT_REG is
  for DATA_FLOW_STRUCTURE
    for all : D_FF
      use entity WORK.D_FLIPFLOP
      generic map (SETUP=>5 ns, HOLD=>3 ns, PROP_DELAY=>12 ns)
      port map    (DATA=>D, CLOCK=>CLK, Q=>Q, Q_BAR=>Q_BAR);
    end for;
  end for;
end CONFIG_VENDOR;
```

Figure 19.11 shows how the configuration CONFIG_VENDOR binds the D_FF component to the D_FLIPFLOP design entity. The most recently analyzed architecture of D_FLIPFLOP

Figure 19.11

**Binding component and
design entity declarations**

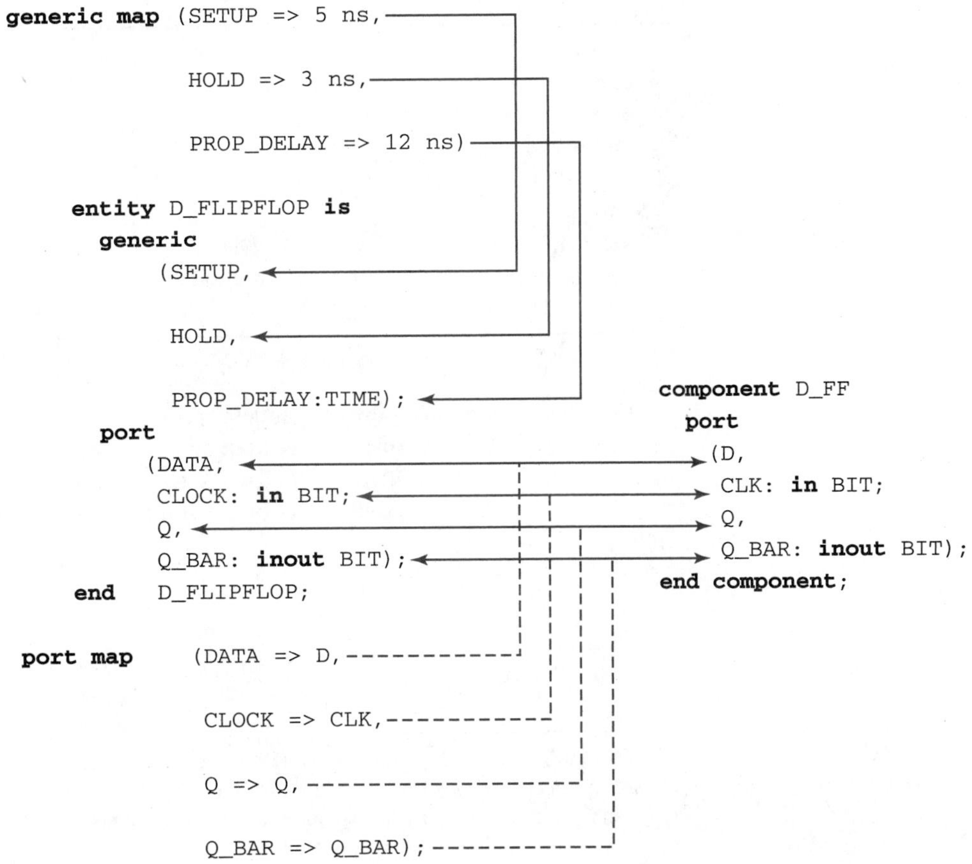

is selected via the default entity configuration rules because the entity clause does not specify an architecture. The **generic map** clause

---

```
generic map (SETUP=>5 ns, HOLD=>3 ns, PROP_DELAY=>12 ns)
```

---

sets the values for the timing generic parameters, demonstrating delayed generic parameter/value binding. The **port map** clause

---

```
port map (DATA=>D, CLOCK=>CLK, Q=>Q, Q_BAR=>Q_BAR)
```

---

associates the *local* or component ports D, CLK, Q, and Q_BAR with the *formal* or design entity ports DATA, CLOCK, Q, and Q_BAR, respectively. Associating an actual signal instead of a local component port to a design entity port in a component configuration **port map** clause provides a convenient way to connect global clocking or test signals.

What if the component declaration and associated design entity declaration have different generic parameter or port *types*? For example, let us again change the vendor-supplied D flip-flop design entity declaration so that the input/output signals use the 9-valued logic type STD_ULOGIC discussed in Chapter 16 instead of the 2-valued logic type BIT.

```
library IEEE;
use IEEE.STD_LOGIC_1164.all;
entity D_FLIPFLOP is
 generic
    (SETUP, HOLD, PROP_DELAY : TIME);
 port
    (DATA, CLOCK : in STD_ULOGIC;
     Q, Q_BAR    : inout STD_ULOGIC);
end D_FLIPFLOP;
```

Type incompatibilities in structural descriptions are addressed by including *type conversion functions* in the **generic** and **port map** clauses.[1] The following new version of the configuration declaration CONFIG_VENDOR uses two type conversion functions TO_BIT and TO_STDULOGIC to associate the ports of the component D_FF with the ports of the design entity D_FLIPFLOP.

---

[1] VHDL-93: In VHDL-93, type incompatibilities between *closely related* types can be addressed by including *type conversion expressions* in the **generic** and **port map** clauses. Integer and floating point types and types differing only in resolution function constraints are defined to be closely related.

```
library IEEE;
use IEEE.STD_LOGIC_1164.all;
configuration CONFIG_VENDOR of DUAL_PORT_REG is
 for DATA_FLOW_STRUCTURE
   for all : D_FF
     use entity WORK.D_FLIPFLOP
     generic map (SETUP=>5 ns, HOLD=>3 ns, PROP_DELAY=>12 ns)
     port map (DATA=>TO_STDULOGIC(D),
               CLOCK=>TO_STDULOGIC(CLK),
               TO_BIT(Q)=>TO_STDULOGIC(Q),
               TO_BIT(Q_BAR)=>TO_STDULOGIC(Q_BAR));
   end for;
 end for;
end CONFIG_VENDOR;
```

The **use** clause gains visibility to the type conversion functions TO_BIT and TO_STDULOGIC provided in the STD_LOGIC_1164 package. The function TO_BIT converts values of type STD_ULOGIC into values of type BIT, and the function TO_STDULOGIC converts values of type BIT into values of type STD_ULOGIC.

The port associations

```
DATA=>TO_STDULOGIC(D)       CLOCK=>TO_STDULOGIC(CLK)
```

specify that information passing into the DATA and CLOCK ports of the D_FLIPFLOP design entity is converted from type BIT to type STD_ULOGIC by the type conversion function TO_STDULOGIC. The port associations

```
TO_BIT(Q)=>TO_STDULOGIC(Q)
TO_BIT(Q_BAR)=>TO_STDULOGIC(Q_BAR)
```

specify that information passing into the Q and Q_BAR ports of the D_FLIPFLOP design entity is converted from type BIT to type STD_ULOGIC by the type conversion function TO_STDULOGIC, and information passing out of the Q and Q_BAR ports of the D_FLIPFLOP design entity is converted from type STD_ULOGIC to type BIT by the type conversion function TO_BIT. Once specified, these type conversion functions are automatically invoked when needed during simulation.

19.5 CONFIGURING MULTIPLE LEVELS OF HIERARCHY

Thus far, we have configured a *single* design entity, selecting an architecture and design entities for instantiated components. This configuration process can be extended to multiple levels of a design hierarchy by nesting block configuration statements. To demonstrate this capability, we will expand the structural description of our quad 2-port register, DUAL_PORT_REG, example by decomposing the D flip-flop, D_FF, into a network of **nand** gates to yield the three-level hierarchy shown in Figure 19.12.

Figure 19.12

The three-level model hierarchy for a quad 2-port register

Dual Port Register   (DUAL_PORT_REG

     D Flip-Flops   (D_FF)

        nand Gates   (NAND_GATE)

As an alternative to the data flow D flip-flop model given in Figure 19.4, Figure 19.13 presents a structural D flip-flop model (see Chapter 8 for a discussion of D flip-flops). A VHDL model for a **nand** gate is given in Figure 19.14.

The CONFIG_MULTI_LEVEL configuration declaration in Figure 19.15 configures the new three-level model hierarchy for DUAL_PORT_REG by specifying (1) the architecture for the quad 2-port register, (2) the design entities for the D flip-flops within the register, and (3)

Figure 19.13

Gate-level structural model of a positive edge-triggered D flip-flop

```
entity D_FF is
 port
    (D, CLK    : in BIT;
     Q, Q_BAR : inout BIT);
end D_FF;

architecture STRUCTURE of D_FF is
 component NAND2_GATE
  port
    (A, B : in BIT;  C : out BIT);
 end component;

 component NAND3_GATE
  port
    (A, B, C : in BIT; D : out BIT);
 end component;

 signal INT1, INT2, INT3, INT4 : BIT;
 begin
   G1: NAND2_GATE port map (INT1, INT2, INT3);
   G2: NAND2_GATE port map (INT3, CLK, INT2);
   G3: NAND3_GATE port map (INT1, INT2, CLK, INT4);
   G4: NAND2_GATE port map (INT4, D, INT1);
   G5: NAND2_GATE port map (INT2, Q_BAR, Q);
   G6: NAND2_GATE port map (INT4, Q, Q_BAR);
 end STRUCTURE;
```

Figure 19.14

A nand gate model

```
entity NAND_GATE is
 port
    (I1, I2, I3 : in BIT := '1';
     O1 : out BIT);
end NAND_GATE;


architecture DATA_FLOW of NAND_GATE is
begin
  O1 <= not (I1 and I2 and I3);
end DATA_FLOW;
```

the design entities for the **nand** gates within the D flip-flops. The first block configuration statement

```
for DATA_FLOW_STRUCTURE
        .
        .
end for;   -- DATA_FLOW_STRUCTURE;
```

Figure 19.15

Multi-level configuration
for the **DUAL_PORT_REG**
design entity

```
configuration CONFIG_MULTI_LEVEL of DUAL_PORT_REG is
  for DATA_FLOW_STRUCTURE
    for all : D_FF
      use entity WORK.D_FF(STRUCTURE);

      for STRUCTURE
        for G3 : NAND3_GATE
         use entity WORK.NAND_GATE(DATA_FLOW);
           port map ( I1=>A, I2=>B, I3=>C, O1=>D );
         end for;
         for all : NAND2_GATE
          use entity WORK.NAND_GATE(DATA_FLOW)
            port map ( I1=>A, I2=>B, I3=>open, O1=>C );
         end for;
       end for; -- STRUCTURE

      end for    -- D_FF;
    end for;  -- DATA_FLOW_STRUCTURE
  end CONFIG_MULTI_LEVEL;
```

Figure 19.16

Embedded configurations

```
configuration CONFIG_DFF of D_FF is
  for STRUCTURE
    for G3 : NAND3_GATE
      use entity WORK.NAND_GATE(DATA_FLOW);
      port map ( I1=>A, I2=>B, I3=>C, O1=>D );
    end for;
    for all : NAND2_GATE
      use entity WORK.NAND_GATE(DATA_FLOW)
      port map ( I1=>A, I2=>B, I3=>open, O1=>C );
    end for;
  end for;
end CONFIG_DFF;
```

(a)

```
configuration CONFIG_MULTI_LEVEL of DUAL_PORT_REG is
  for DATA_FLOW_STRUCTURE
    for all : D_FF
      use configuration WORK.CONFIG_DFF;
    end for;
  end for;
end CONFIG_MULTI_LEVEL;
```

(b)

selects or opens up the DATA_FLOW_STRUCTURE architecture associated with DUAL_PORT_REG to configure instances of component D_FF. The second, nested block configuration statement

```
for DATA_FLOW_STRUCTURE
    .
    .
  for STRUCTURE
      .
      .
  end for; -- STRUCTURE
    .
    .
end for; -- DATA_FLOW_STRUCTURE
```

selects or opens up the STRUCTURE architecture associated with D_FF to configure instances of components NAND3_GATE and NAND2_GATE.

19.5.1 Embedded Configuration

An alternative form of the CONFIG_MULTI_LEVEL configuration declaration is given in Figure 19.16, which divides the hierarchical configuration into two embedded configurations. Figure 19.16(b) illustrates the third form of the component configuration introduced in the previous section, an embedded configuration. An embedded configuration has the general form

```
    for label_identifier : component_name
      use configuration configuration_name;
    end for;
```

and refers to a lower-level configuration to define a unique model description. The configuration declaration CONFIG_MULTI_LEVEL for DUAL_PORT_REG refers to the configuration declaration CONFIG_DFF for D_FF to configure the D flip-flops.

19.6 CONFIGURATION SPECIFICATIONS

VHDL provides configuration *declarations* and *specifications*. Configuration declarations, studied in the previous sections, allow configuration information to be *separate* from structural information. Architectures define the general tree-like model structure hierarchy, while configuration declarations define specific descriptions contained within the model hierarchy. Configuration specifications allow configuration information to be *combined* with structural information by explicitly binding component instances to design entities *within* an architecture.

As an example, Figure 19.17 adds a configuration specification to the quad 2-port register model given in Figure 19.3 to show an alternative way to explicitly define the binding of

Figure 19.17

Configuration specifications

```
architecture DATA_FLOW_STRUCTURE of DUAL_PORT_REG is
  -- Component declaration
  component D_FF
   port
     (D, CLK   : in BIT;   Q, Q_BAR : inout BIT);
  end component;

  -- Configuration specification
  for all : D_FF
    use entity WORK.D_FF(DATA_FLOW);

  -- Signal declarations
  signal DINT, QINT : OUTPUT_TYPE;
begin
  BIT_LOOP: for I in DATA'RANGE(1) generate
   -- Combinational logic
   DINT(I) <= DATA(I, SEL);

   -- Memory
   D_FFX: D_FF port map (D=>DINT(I), CLK=>CLK, Q=>QINT(I),
                            Q_BAR=>open);
   Q(I) <= QINT(I);
  end generate BIT_LOOP;
end DATA_FLOW_STRUCTURE;
```

the D_FF component instances to the D_FF design entity in Figure 19.4. A configuration specification is simply a component configuration contained within an architecture, thereby hardwiring at model development time the pairing of component instances and design entities.

Configuration specifications provide an easy way to define component instance/design entity bindings that the designer knows are not likely to change. However, be aware that configuration specifications negate the benefits of separating configuration and structural information. Once a component is configured via a configuration specification, it is illegal to attempt to configure the component instance via a configuration declaration.

19.7 VHDL-93: CONFIGURATIONS

VHDL-93 offers several enhancements to configurations. First, to make VHDL syntax more uniform, configuration declarations may optionally close with the keyword **configuration**.

```
configuration name of entity_name is
  ....
  -- Statements
  ....
end configuration name;
```

Second, configurations may be directly instantiated. Figure 19.18 shows how the CONFIG_DFF configuration declaration given in Figure 19.16(a) of the gate-level model of the D flip-flop (Figures 19.13 and 19.14) can be used to write yet another version of the quad 2-port register given in Figure 19.1.

Finally, VHDL-93 allows certain cases where *both* a configuration specification and configuration declaration can be given for the same component instance to *incrementally* bind ports and/or rebind generics. Incremental binding for generics supports a popular design practice of successively refining model timing accuracy as more details of the final design emerge.

A VHDL model of a digital system may start out using only delta propagation delays. As synthesis progresses, the delta delays are then refined by replacing them with estimated finite propagation delays. As subsequent implementation progresses, the estimated finite propagation delays are again refined by replacing them with more accurate finite propagation delays extracted from the physical details of the implementation. This latter step of inserting implementation-related information "back" into a design description is called *back annotation*.

The process of successively redefining model timing information is cumbersome in VHDL-87 because a generic parameter may be assigned a value only *once* in a component instantiation, configuration specification, or configuration declaration. This problem is addressed in VHDL-93 by allowing a generic parameter to be assigned a value in either a component instantiation or a configuration specification *and* configuration declaration. In the latter case, the generic value given in the configuration declaration overrides or rebinds the generic value given in the component instantiation or configuration specification.

To illustrate this new capability of incremental binding, consider the following entity declaration for a **nand** gate.

```
entity NAND_GATE is
 generic
    (PROP_DELAY : TIME := 0 ns);
 port
    (I1, I2, I3 : in BIT;
     O1 : out BIT);
end NAND_GATE;
architecture DATA_FLOW of NAND_GATE is
begin
   O1 <= not(I1 and I2 and I3) after PROP_DELAY;
end DATA_FLOW;
```

The default initial value for the PROP_DELAY generic parameter of 0 nanoseconds sets the propagation delay to a delta delay.

Figure 19.18

Structural VHDL model of a quad 2-port register using a configuration

```
package REG_PKG is
   -- Enumeration type
   type FF_INDEX is (A, B, C, D);

   -- Array types
   type INPUT_TYPE is array (FF_INDEX, BIT) of BIT;
   type OUTPUT_TYPE is array (FF_INDEX) of BIT;
end REG_PKG;

use WORK.REG_PKG.all;
entity DUAL_PORT_REG is
 port
    (DATA : in   INPUT_TYPE;
     SEL  : in   BIT;
     CLK  : in   BIT;
     Q    : out OUTPUT TYPE);
end DUAL_PORT_REG;

architecture DATA_FLOW_STRUCTURE of DUAL_PORT_REG is
 signal DINT, QINT : OUTPUT_TYPE;
begin
 BIT_LOOP: for I in DATA'RANGE(1) generate
   -- Combinational logic
   DINT(I) <= DATA(I, SEL);

   -- Memory
   -- Directly instantiate configuration of D_FF design entity
   D_FFX: configuration WORK.CONFIG_DFF
          port map (D=>DINT(I), CLK=>CLK, Q=>QINT(I),
                    Q_BAR=>open);

   Q(I) <= QINT(I);
  end generate BIT_LOOP;
end DATA_FLOW_STRUCTURE;
```

The following INVERTER design entity uses the NAND_GATE design entity and overrides the default propagation delay by setting the PROP_DELAY generic parameter to an estimated, nominal value of 2 nanoseconds via the configuration specification given in the STRUCTURE architecture.

```
entity INVERTER is
 port
    (A      : in BIT;
     A_BAR : out BIT);
end INVERTER;
architecture STRUCTURE of INVERTER is
 -- Component declaration
 component NAND
  port
    (I1, I2, I3 : in BIT; O1 : out BIT);
 end component;

 -- Configuration specification
 for N1 : NAND
   use entity WORK.NAND_GATE
   generic map (PROP_DELAY=>2 ns); --  Nominal delay
begin
   N1: NAND port map (I1=>A, I2=>A, I3=>A, O1=>A_BAR);
 end STRUCTURE;
```

Later in the design process as more implementation details such as loading and interconnect become available, the nominal propagation delay of 2 nanoseconds can be refined using the following configuration declaration.

```
configuration ANNOT_CONFIG of INVERTER is
 for STRUCTURE
  for N1 : NAND
   generic map (PROP_DELAY=>1.4 ns); -- Physical delay
  end for;
 end for;
end ANNOT_CONFIG;
```

Figure 19.19 shows that the ANNOT_CONFIG configuration declaration overrides the configuration specification by redefining the NAND_GATE design entity PROP_DELAY generic parameter to 1.4 nanoseconds.

19.8 RECORD TYPES

Having completed our study of advanced structural modeling, let us now study an aspect of advanced behavioral modeling: *record types*. Recall that an object of a scalar type contains *one* data value, whereas an object of a composite type can contain *many* data values. Like an

Figure 19.19

Combining configuration specifications and declarations

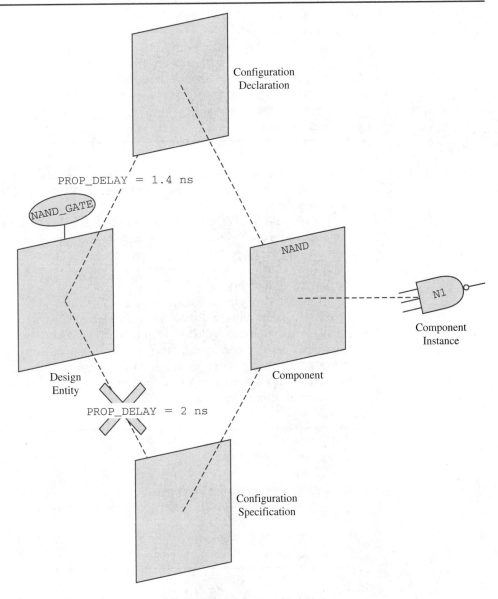

array type, a record type is a composite type and is used to group together related data. A record type is different, however, from an array type in that all the elements of an array type must be of the *same* type, whereas elements of a record type can be of *different* types.

To clarify these comparisons, consider the following example of a record type declaration that might be used as an abstract representation of a communication message.

```
  -- Record type declaration
type MESSAGE_TYPE is
record
  -- Element declaration(s)
  TR_ADDR  : BIT_VECTOR(3 downto 0); -- Sender address
  RCV_ADDR : BIT_VECTOR(3 downto 0); -- Receiver address
  TEXT     : STRING(1 to 20);        -- Data
  PRIORITY : NATURAL;                -- Priority
end record;
```

As with all type declarations, a record type declaration starts out with the reserved keyword **type**, followed by the name of the record type, MESSAGE_TYPE. The type name can be any legal identifier.[2]

The *elements*, also called *fields*, of a record are declared between the keywords **record** and **end record**. MESSAGE_TYPE has the following four elements.

1. TR_ADDR of type BIT_VECTOR
2. RCV_ADDR of type BIT_VECTOR
3. TEXT of type STRING
4. PRIORITY of type NATURAL

All element names must be distinct. Different elements can have different types, allowing for complex data structures including records-of-arrays and records-of-records. However, element types must be constrained types, so unconstrained array types like BIT_VECTOR or STRING must include index constraints to set the bit or character string length, respectively. Elements having the same type may be given in separate element declarations or combined in a single element declaration.

```
  -- Record type declaration
type MESSAGE_TYPE is
record
  -- Element declaration(s)
  TR_ADDR, RCV_ADDR : BIT_VECTOR(3 downto 0);
  TEXT     : STRING(1 to 20);
  PRIORITY : NATURAL;
end record;
```

Having declared a record type, we can then declare objects of this type. Objects of record types are often simply called *records*. Example record declarations are given next.

---

[2] VHDL-93: In VHDL-93, a record type declaration may close by repeating the name of the record type.

```
type MESSAGE_TYPE is
record

    ⋮

end record MESSAGE_TYPE;
```

```
signal SV_REQUEST : MESSAGE_TYPE;

variable TMP_MESSAGE : MESSAGE_TYPE;

constant NULL_MESSAGE : MESSAGE_TYPE :=
  (TR_ADDR=>X"0", RCV_ADDR=>X"0", TEXT=>(others=> ' '),
   PRIORITY=>0);
```

Values for records can be specified using *aggregates*. Recall from Chapter 16 that an aggregate specifies a set of values. Aggregate values are paired up with record elements using positional or named association. The constant declaration for NULL_MESSAGE illustrates named association. For instance, the first association element TR_ADDR=>X"0" pairs the value 0000_2 with the record element TR_ADDR and the third association element TEST=>(others=>' ') pairs the array aggregate value of 20 blank characters with the record element TEXT. With positional notation, aggregate values are paired up with the record elements based on the order in which the values are listed from left to right in the aggregate and the order in which the elements are defined from beginning to end in the record type declaration.

All elements of a record may be referenced by simply using the record name. For example,

```
SV_REQUEST <= NULL_MESSAGE after 10 ms;
```

assigns the entire record NULL_MESSAGE to the signal SV_REQUEST; assignment is on an element-by-element basis. Entire records can be compared using the equality (=) and inequality (/=) operators; the comparison is also on an element-by-element basis. Two records are equal if they are of the same record type and their corresponding elements are equal.

Individual elements within a record may be referenced using a *selected name* notation composed of the record and element names separated by a period. The values, for example, of the TR_ADDR and TEXT elements of the constant NULL_MESSAGE record are respectively referenced by the selected names NULL_MESSAGE.TR_ADDR and NULL_MESSAGE.TEXT. Selected and indexed name notation can be combined. For instance, the leftmost bit of the array element TR_ADDR of the constant record NULL_MESSAGE is referenced by NULL_MESSAGE.TR_ADDR(3).

Let us now look at how to use record types to model hardware. A common application of records is grouping related signals of different types. For example, consider modeling the register file discussed in Chapter 10 and shown again in Figure 19.20 (p. 648). A register file is a random access memory (RAM) configured as a set of registers. The register file in Figure 19.20 contains sixteen 8-bit registers and provides separate read and write ports to support simultaneous register access and update.

A VHDL description of the 16 × 8 register file is given in Figure 19.21 and Figure 19.22. For convenience, the read enable, RE, and read address, RADR, input signals are grouped together as elements of the RD input signal. Likewise, the write enable, WE, and write address, WADR, input signals are grouped together as elements of the WR input signal. RD and WR are signals of the record type CONTROL_TYPE declared in the package UTILITIES_PKG. UTILITIES_PKG also declares the subtype BYTE (see Chapter 18 for a discussion of subtypes). The 16 × 8 register file is modeled as an array having 16 elements; the array's elements are indexed by the integers 0 to 15 and each element is an 8-bit vector. The

Figure 19.20

A 16 × 8 register file

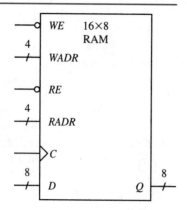

type conversion function TO_INT declared in UTILITIES_PKG converts the input read and write 4-bit addresses into integer array indexes. The definition of the TO_INT function provided in the UTILITIES_PKG body is left as an exercise. Two guarded concurrent signal assignment statements control synchronous, parallel read and write operations.

19.9 ACCESS TYPES

We have made it to the last topic in the last chapter on VHDL: *access types*. Access types, also called *pointer types*, provide for dynamic objects, objects that are created and destroyed during simulation. Access types are more frequently used in modeling large, complex digital systems, such as computers or communication systems, than in modeling the basic combinational and sequential digital systems addressed in this text.[3] However, access types and dynamic data structures can be judiciously exploited in certain sequential applications. Before investigating these applications, we will first discuss how to declare and use access types.

Figure 19.21

The package for the REGISTER_FILE VHDL model

```
package UTILITIES_PKG is
  -- Record type declaration
  type CONTROL_TYPE is
  record
    EN_BAR : BIT;
    ADDR   : BIT_VECTOR(3 downto 0);
  end record;

  -- Subtype declaration
  subtype BYTE is BIT_VECTOR(7 downto 0);

  -- Function declaration
  function TO_INT ( ARG : BIT_VECTOR ) return INTEGER;
end UTILITIES_PKG;
```

---

[3] As a historical note, the original definition of VHDL did not include access types.

Figure 19.22

VHDL model of a register file

```
use WORK.UTILITIES_PKG.all;
entity REGISTER_FILE is
 port
    (RD, WR : in  CONTROL_TYPE;
     CLK    : in  BIT;
     DATA   : in  BYTE;
     Q      : out BYTE);
end REGISTER_FILE;

architecture DATA_FLOW of REGISTER_FILE is
   type MEMORY_TYPE is array (0 to 15) of BYTE;
   signal MEMORY : MEMORY_TYPE;
begin
  -- Write
  WRITE: block (CLK='1' and not CLK'STABLE and WR.EN_BAR='0')
  begin
   MEMORY( TO_INT(WR.ADDR) ) <= guarded DATA;
  end block WRITE;

  -- Read
  READ: block (CLK='1' and not CLK'STABLE and RD.EN_BAR='0')
  begin
   Q <= guarded MEMORY( TO_INT(RD.ADDR) );
  end block READ;
end DATA_FLOW;
```

Thus far, object creation and destruction has been implicitly defined by the language. For example, an object declared within a subprogram is created when the subprogram is invoked and destroyed when the subprogram returns control to the calling parent. As another example, an object declared within an architecture is created at the start of simulation and destroyed at the end of simulation. Access types allow for the explicit definition of object creation and destruction.

Figure 19.23 illustrates the basic idea behind access types. The rectangle on the left-hand side represents an object of an access type, called an *access object*, or a *pointer*. An access object holds the address of an *access**ed** object*, represented by the rectangle on the right-hand side. The accessed object holds a value of some type, for example, INTEGER, REAL, and so on. An access object indirectly references a value by holding the address of an accessed object that in turn holds a value. Keep this two-rectangle picture of an access object and an accessed object in mind as we discuss access types.

Figure 19.23

Access types/objects

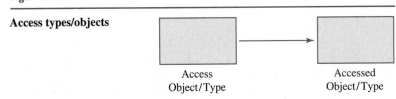

Access
Object/Type

Accessed
Object/Type

Examples of access type declarations are shown below.

```
-- Access type declarations
type INT_PTR_TYPE is access INTEGER;

type BOL_PTR_TYPE is access BOOLEAN;
```

An access type declaration starts with the keyword **type**. Then, the name of the access type is given, followed by the name of the accessed type. Access objects of type INT_PTR_TYPE can point only to accessed objects of type INTEGER; access objects of type BOL_PTR_TYPE can point only to accessed objects of type BOOLEAN.

Having declared an access type, we can now declare objects of the type. INT_PTR and BOL_PTR are examples of access object declarations.

```
-- Access variable ("pointer") declarations
variable INT_PTR : INT_PTR_TYPE;

variable BOL_PTR : BOL_PTR_TYPE;
```

Access objects may only be variables; it is illegal to declare constants or signals of an access type. The variables INT_PTR and BOL_PTR are both automatically initialized to the value **null**. An access variable, or pointer, having the special value **null** points nowhere.

Having declared an access object, we can now create the accessed object via the **new** operator. For example, the variable assignment statement

```
INT_PTR := new INTEGER;
```

creates an accessed object implicitly initialized to INTEGER'LEFT, in other words, the most negative integer supported by the predefined INTEGER type. Alternatively,

```
INT_PTR := new INTEGER'(1_002);
```

creates an accessed object explicitly initialized to 1,002. In other words, INT_PTR points to the integer 1,002. If the accessed object is of a composite type, then we use an aggregate to explicitly specify the initial value. For instance, the following declarations

```
type ARRAY_TYPE is array (1 to 5) of REAL;

type ARRAY_PTR_TYPE is access ARRAY_TYPE;

variable ARRAY_PTR : ARRAY_PTR_TYPE := new
                     ARRAY_TYPE'(6.0, 7.0, 8.0, 9.0, 10.0);
```

create an access variable that points to an array containing the five real numbers 6.0, 7.0, 8.0, 9.0 and 10.0. Note that the **new** operator may be used to create an accessed object in a variable declaration statement, as well as a variable assignment statement.

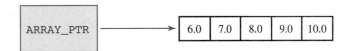

Accessed objects can be destroyed using the predefined DEALLOCATE procedure. DEALLOCATE accepts a variable of any access type, destroys the accessed object, and resets the variable to **null**. For example,

---

```
DEALLOCATE(INT_PTR);
```

---

destroys the accessed object pointed to by INT_PTR and resets INT_PTR to **null**.

Now that we can declare and initialize access objects, how do we reference their accessed values? The rules concerning referencing access values are listed below.

1. If an access variable points to an accessed object that is of a scalar type (integer, real, enumerated, or physical), then append **.all** to the access variable name to reference the accessed object value. For example, if the access variable assignment statement

---

```
INT_PTR := new INTEGER'(-3);
```

---

   is executed, then INT_PTR.**all** yields the value − 3, in other words, the integer value pointed to by INT_PTR.

2. If an access variable points to an accessed object that is of an array type, then use the indexed name notation. For the declaration of the access variable ARRAY_PTR given above, for example, ARRAY_PTR(3) yields 8.0, in other words, the value of the third element of the array pointed to by ARRAY_PTR.

3. Finally, if an access variable points to an accessed object that is of a record type, then use the selected name notation. For example, if the access variable REC_PTR is declared as

---

```
type RECORD_TYPE is
record
  VALUE  : INTEGER;
  DELAY  : TIME;
end record;

type REC_PTR_TYPE is access RECORD_TYPE;

variable REC_PTR : REC_PTR_TYPE := new RECORD_TYPE' (9, 20 ns);
```

---

then `REC_PTR.DELAY` yields 20 ns, in other words, the value of the `DELAY` element of the record pointed to by `REC_PTR`.

There is one more aspect of access types that we need to discuss before looking at a modeling application: *linked lists* and *incomplete type declarations*. A linked list is a dynamic data structure composed of several objects, called entries, "hooked" or linked together to form a list. A linked list entry is typically declared as a record containing at least two elements: (1) a value and (2) a pointer to the next entry in the list. Unfortunately, this record structure poses a problem, as shown below.

```
-- Incorrect example
type ENTRY_PTR is access ENTRY;

type ENTRY is
 record
    VALUE : BIT_VECTOR(7 downto 0);
    NXT   : ENTRY_PTR;
 end record;
```

The declaration of the record type `ENTRY` is required to complete the declaration of the access type `ENTRY_PTR`, but the declaration of the access type `ENTRY_PTR` is required to complete the declaration of the record type `ENTRY`. This "chicken-and-egg" problem is solved by the incomplete type declaration shown below.

```
-- Correct example
type ENTRY;                -- Incomplete type declaration

type ENTRY_PTR is access ENTRY;

type ENTRY is
 record
    VALUE : BIT_VECTOR(7 downto 0);
    NXT   : ENTRY_PTR;
 end record;
```

The first type declaration is an incomplete type declaration, specifying only that `ENTRY` is a type that will be defined later. Establishing `ENTRY` as a legal reference to a type enables the access type `ENTRY_PTR` to be declared in the second declaration. Once the access type `ENTRY_PTR` is declared, the third declaration completes the declaration of the record type `ENTRY`.

Let us now apply what we have learned about access types to construct a model of a first-in-first-out (FIFO) queue of *arbitrary* length. A FIFO queue or stack stores data. A write or *push* operation adds data to the queue and a read or *pop* operation removes data from the queue. The first value written to the queue is the first value read from the queue, hence the name first-in-first-out.

Figure 19.24 illustrates a typical FIFO application: data buffering. System A needs to send data to system B, but system A generates data at a high rate at irregular intervals while system B processes data at a comparatively lower rate at regular intervals. The FIFO mitigates this mismatch in producer/consumer data rates by temporarily storing the high bursts of data from system A and then transferring the data to system B at its lower, regular data rate. The size of the FIFO is important. The FIFO should be big enough to store any data that is produced by system A faster than it is consumed by system B, so data is not lost. However, the larger the FIFO, the higher the hardware costs. These design trade-offs can be investigated by constructing an abstract FIFO model of infinite size, running several system simulations of various production/consumption data rates, and studying statistics of actual FIFO usage.

The interface of such an abstract FIFO VHDL model is given in Figure 19.24(b). The interface declares five ports. W_BAR and R_BAR are active-0 control signals designating FIFO write and read operations, respectively. CLK is a clock signal. EMPTY is an active-1 signal designating the status of the FIFO; EMPTY is asserted when FIFO is empty. Finally, data that is written to or read from the FIFO flows through the DATA port.

The architecture of the abstract FIFO VHDL model is given in Figure 19.25 (p. 654). The architecture INF_LENGTH contains one process statement. Within the process, the first three type declarations define the record type ENTRY to model linked list entries. ENTRY has two elements: VALUE and NXT. VALUE is an 8-bit vector representing the information stored in a FIFO entry. NXT is an access variable that points to another FIFO entry. The access variables FIRST_ENTRY, LAST_ENTRY, and OLD_ENTRY are used to conduct add and delete operations on the linked list in response to write (push) and read (pop) requests. FIRST_ENTRY points to the "head" of the linked list containing the oldest entry. LAST_ENTRY points to the "tail" of the linked list containing the newest entry.

Sample write and read operations are explained using the FIFO linked lists shown in

Figure 19.24

First-in-first-out (FIFO) queue data buffering

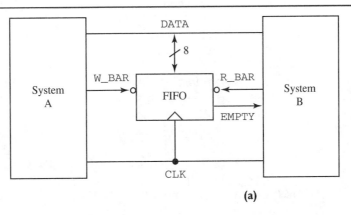

(a)

```
entity FIFO is
 port
    (W_BAR, R_BAR : in BIT;
     CLK    : in BIT;
     EMPTY : out BIT;
     DATA   : inout BIT_VECTOR(7 downto 0));
 end FIFO;
```

(b)

Figure (a) illustrates a FIFO data buffering application. Figure (b) shows a FIFO VHDL model interface.

Figure 19.26. Figure 19.26(a) shows the condition of the pointers FIRST_ENTRY and LAST_ENTRY at the beginning of simulation. Pointers FIRST_ENTRY and LAST_ENTRY are both equal to **null**, denoted by the "ground" symbol, designating an empty FIFO. If W_BAR is asserted on the rising edge of the clock signal CLK, then a write operation occurs. A new linked list entry is created, the contents of the input signal DATA are stored in this entry, and the new entry is added to the end of the linked list. Figure 19.26(b) and Figure 19.26(c) show the condition of the pointers FIRST_ENTRY and LAST_ENTRY after two successive write operations. The FIFO in Figure 19.26(c) can be emptied by two successive read opera-

---

Figure 19.25

A FIFO model using access types

```
architecture INF_LENGTH of FIFO is
begin
 process
   type ENTRY;
   type ENTRY_PTR is access ENTRY;
   type ENTRY is record
     VALUE : BIT_VECTOR(7 downto 0);
     NXT   : ENTRY_PTR;
   end record;

   variable FIRST_ENTRY, LAST_ENTRY, OLD_ENTRY : ENTRY_PTR;
 begin
  wait until (CLK = '1');

   -- If W_BAR is asserted active-0, write FIFO
   if ( W_BAR='0' ) then
     if ( FIRST_ENTRY=null ) then
        FIRST_ENTRY := new ENTRY'(DATA, null);
        LAST_ENTRY := FIRST_ENTRY;
     else
        LAST_ENTRY.NXT := new ENTRY'(DATA, null);
        LAST_ENTRY := LAST_ENTRY.NXT;
     end if;
     EMPTY <= '0' after 20 ns;

   -- If R_BAR is asserted active-0 and FIFO is not empty, read
   -- FIFO
   elsif ( R_BAR='0' and FIRST_ENTRY/=null ) then
     DATA <= FIRST_ENTRY.VALUE after 35 ns;
     OLD_ENTRY := FIRST_ENTRY;
     FIRST_ENTRY := FIRST_ENTRY.NXT;
     DEALLOCATE(OLD_ENTRY);
     if ( FIRST_ENTRY=null ) then
        LAST_ENTRY := FIRST_ENTRY;
        EMPTY <= '1' after 20 ns;
     end if;
   end if;
 end process;
end INF_LENGTH;
```

Figure 19.26

Sample FIFO linked lists

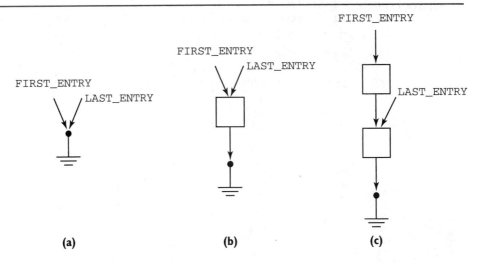

(a)   (b)   (c)

tions. If R_BAR is asserted on the rising edge of the clock signal CLK and FIFO is not empty, then a read operation occurs. The data stored in the oldest linked list entry pointed to by FIRST_ENTRY is assigned to the output signal DATA and then this entry is destroyed. If FIFO is now empty, the FIRST_ENTRY and LAST_ENTRY pointers are reset to **null** and the EMPTY status signal is asserted.

19.10 SUMMARY

In this chapter, we addressed a variety of advanced behavioral and structural modeling practices in our last look at VHDL. With advanced structural modeling practices we addressed mixed behavioral/structural descriptions and configurations. With advanced behavioral modeling practices we addressed defining and using record and access types.

Mixed behavioral/structural descriptions combine data flow, algorithmic, and structural modeling styles within a single architecture. Portions of design entity function can be described as (1) an abstract sequential process, (2) a set of concurrent signal transforms, and (3) the interconnection of lower-level subfunctions. A configuration selects a design entity description from a family of hierarchical design entity descriptions that can be constructed via alternative architectures and component instances. Configurations identify particular design entity descriptions by selecting architectures and binding design entities to component instances. Although configurations can be complex to address arbitrarily complex design hierarchies and component instance/design entity associations, they can often be effectively "skipped" by assuming convenient default configurations.

In Sections 19.8 and 19.9 we discussed record and access types, respectively. Record types are one of two composite types available in VHDL, providing the powerful ability to represent a collection of values of different types that are conceptually related. Access types provide runtime or simulation-time object allocation/deallocation to construct and manage dynamic data structures.

19.11 POSTSCRIPT

As all journeys must eventually come to an end, it is time to bring our exploration of the world of digital systems to an end. Although we have covered a lot of ground, other journeys await you. I wish you well.

In closing, let me leave you with one last "problem-solving skill," an excerpt from a poem that hangs on my son's bedroom wall.

Problem-Solving Skill
Listen to the mustn'ts, child.
Listen to the don'ts.
Listen to the shouldn'ts . . .
the impossible, the won'ts.
Listen to the never-haves,
then listen close to me . . .
Anything can happen, child.
Anything can be.

Shel Silverstein

19.12 PROBLEMS

1. Consider a binary coded decimal (BCD) down counter that counts in the following descending sequence.

$$9_{10}, 8_{10}, 7_{10}, \ldots, 1_{10}, 0_{10}, 9_{10}, 8_{10}$$
$$\mapsto \text{time}$$

   a. Design the BCD counter using positive edge-triggered D flip-flops and the synthesis methods discussed in Chapter 9.
   b. Generate a mixed structural/behavioral VHDL model of the BCD counter by describing (1) the combinational logic using process statements or concurrent signal assignment statements and (2) the memory, using the quad D flip-flop component shown in Figure 10.6.
   c. Write a behavioral model for the quad D flip-flop component and verify that the VHDL model generated in part (b) correctly operates by clocking the BCD counter through 15 cycles.
   d. Draw the VHDL design entity hierarchy.

2. Consider a resetting digital system that detects the prime binary coded decimal (BCD) numbers 1, 3, 5, and 7, realizing the following sample input (X)/output (Z) sequence.

$$X = 0010011100111000 \ldots$$
$$Z = 0000000100010000 \ldots$$
$$\mapsto \text{time}$$

   a. Design the resetting sequence detector using positive edge-triggered D flip-flops and the synthesis methods discussed in Chapter 9.
   b. Generate a mixed structural/behavioral VHDL model of the resetting sequence detector by describing (1) the combinational logic using process statements or concurrent signal assignment statements and (2) the memory, using the quad D flip-flop component shown in Figure 10.6.
   c. Write a behavioral model for the quad D flip-flop component and verify that the VHDL model generated in part (b) correctly accepts the input X and generates the output Z.
   d. Draw the VHDL design entity hierarchy.

3. Change the resetting prime binary coded decimal number detector in Problem 2 to a nonresetting prime binary coded decimal number detector realizing the following sample input (X)/output (Z) sequence.

$$X = 000111010101 \ldots$$
$$Z = 000111000101 \ldots$$
$$\longmapsto \text{time}$$

Repeat the synthesis and VHDL modeling tasks in parts (a), (b), and (c) of Problem 2.

4. Construct a VHDL model of the programmable logic array (PLA) shown below. (*Hint*: See Chapter 5 for a discussion of programmable logic arrays.)

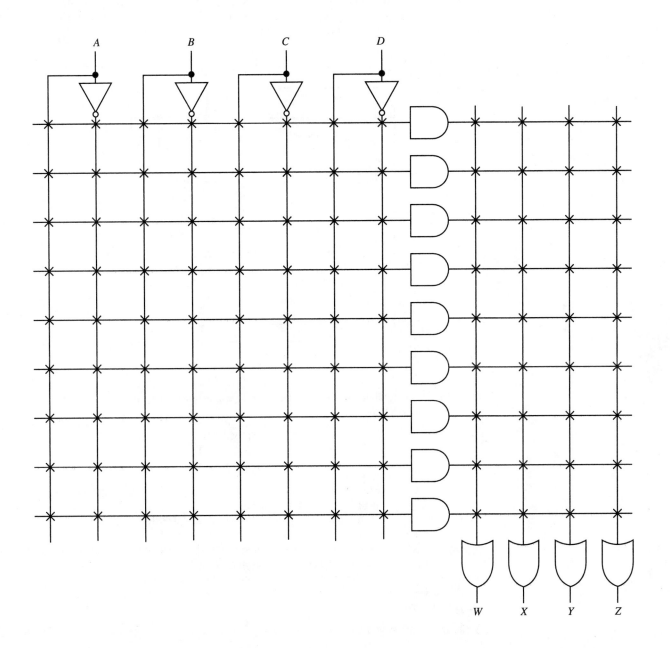

5. Generate a mixed structural/behavioral VHDL model of the BCD down counter designed in Problem 1 by describing (1) the combinational logic using the PLA component generated in Problem 4 and (2) the memory, using process statements or concurrent signal assignment statements. Verify that the VHDL model correctly operates by clocking the BCD down counter through 15 cycles. Draw the VHDL design entity hierarchy.

6. Generate a mixed structural/behavioral VHDL model of the nonresetting prime binary coded decimal number detector designed in Problem 3 by describing (1) the combinational logic using the PLA component generated in Problem 4 and (2) the memory, using process statements or concurrent signal assignment statements. Verify that the VHDL model accepts the input X and generates the output Z given in Problem 3. Draw the VHDL design entity hierarchy.

7. Convert the VHDL model of the combinational programmable logic array (PLA) generated in Problem 4 into a VHDL model of a sequential programmable logic array (PLA) by adding to each output the programmable macrocell shown below. (*Hint*: See Chapter 10 for a discussion of sequential programmable logic arrays and programmable macrocells.)

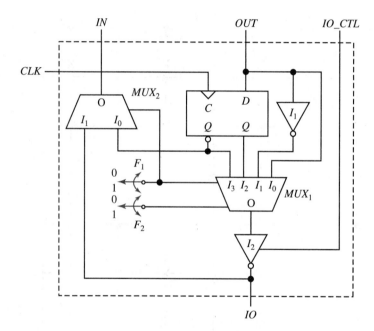

8. Consider a resetting digital system that detects three consecutive rising edges realizing the following sample input (X)/output (Z) sequence.

$$X = 00101011110101010 \ldots$$
$$Z = 00000010000000010 \ldots$$
$$\longmapsto \text{time}$$

   a. Design the rising edge detector as a Moore machine using positive edge-triggered D flip-flops and the synthesis methods discussed in Chapter 9.
   b. Generate a VHDL behavioral description of the rising edge detector design using an algorithmic modeling style.
   c. Generate a VHDL structural description of the rising edge detector implementation using the sequential programmable logic array component generated in Problem 7.
   d. Verify that both the behavioral and structural descriptions accept the X sequence and yield the Z sequence.
   e. Draw the design entity hierarchy.

9. Write two configuration declarations to define the two alternative descriptions developed for the

rising edge detector in Problem 8, parts (b) and (c). Identify the configured descriptions in the design entity hierarchy generated in Problem 8, part (e).

10. a. Generate a VHDL model of the following simple structure using the BUF_INV component declaration shown below.

```
component BUF_INV
  generic (BUF_OP : BOOLEAN);
  port    (A : in BIT;  B : out BIT);
end component;
```

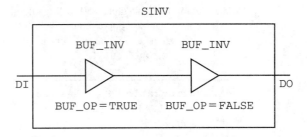

b. Finish the structural VHDL model generated in part (a) by writing a configuration declaration binding all instances of the component BUF_INV to the following design entity.

```
entity EQ_NOT is
  generic (EQ_OP : BOOLEAN;  DELAY : TIME);
  port (A : in BIT;  Z : out BIT);
end EQ_NOT;
architecture BEHAVIOR of EQ_NOT is
  begin
    Z <= A after DELAY when (EQ_OP=TRUE) else
         not A after DELAY;
end BEHAVIOR;
```

c. Simulate the configuration developed in part (b) with EQ_NOT DELAY equal to 10 nanoseconds and inputs '0' and '1'.

11. Repeat Problem 10 by defining the BUF_INV component instance/EQ_NOT design entity bindings via a configuration specification instead of a configuration declaration.

12. VHDL-93: Repeat Problem 10 by defining the BUF_INV component instance/EQ_NOT design entity bindings by writing a configuration declaration for the EQ_NOT design entity and directly instantiating the configuration.

13. a. Generate a VHDL model of the following simple structure using the SINV component declaration shown below.

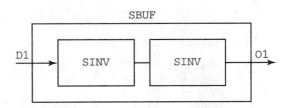

```
component SINV
  port    (DI : in BIT; DO : out BIT);
end component;
```

b. Finish the structural VHDL model generated in part (a) by writing a configuration declaration for SBUF, using the configuration for SINV generated in Problem 10.

c. Simulate the configuration developed in part (b) with inputs '0' and '1'.

14. Write the package body defining the type conversion function TO_INT declared in the UTILITIES_PKG package in Figure 19.21.

15. In addition to architectures, a block configuration can also select **blocks** and **generate** iterations within architectures. For instance, individual instances of the component D_FF created by the **generate** statement in Figure 19.3 can be identified and configured by

```
configuration CONFIG_C of DUAL_PORT_REG is
  for DATA_FLOW_STRUCTURE            -- Block configuration
   for BIT_LOOP(A)                   -- Block configuration
      -- Component configuration
   end for;

   for BIT_LOOP(B to C)              -- Block configuration
      -- Component configuration
   end for;
   .....
 end CONFIG_C;
```

a. Write a configuration for DUAL_PORT_REG given in the Figure 19.3. Bind component instances identified by BIT_LOOP(A) and BIT_LOOP(D) to the D_FF entity in Figure 19.4 and component instances identified by BIT_LOOP(B) and BIT_LOOP(C) to the D_FF entity in Figure 19.13.

16. Modify the FIFO VHDL model given in Figure 19.24 and Figure 19.25 to accept data messages of the following record type.

```
-- Record type declaration
type MESSAGE_TYPE is
record
  -- Element declaration(s)
  TR_ADDR   : BIT_VECTOR(3 downto 0);   -- Sender address
  RCV_ADDR  : BIT_VECTOR(3 downto 0);   -- Receiver address
  TEXT      : STRING(1 to 20);          -- Data
  PRIORITY  : NATURAL;                  -- Priority
end record;
```

Simulate the new FIFO showing two writes followed by two reads.

17. Modify the FIFO VHDL model generated in Problem 16 to be a priority first-in-first-out queue. A read operation yields the message with the highest priority; messages having equal priority are processed on a first-in-first-out basis. Simulate the new FIFO showing six writes followed by six reads. The six writes use the following contents of the signal DATA.

```
(B"0000", B"0000", (others=>'0'), 2)
(B"0001", B"0001", (others=>'1'), 4)
(B"0010", B"0010", (others=>'2'), 2)
(B"0011", B"0011", (others=>'3'), 1)
(B"0100", B"0100", (others=>'4'), 4)
(B"0101", B"0101", (others=>'5'), 3)
```

18. Generate a VHDL model of a last-in-first-out (LIFO) queue.
    a. Use a linked list data structure. Simulate the LIFO showing four writes followed by five reads.
    b. Use an array data structure. When the number of write operations exceeds the size of the array, storage "wraps" and old data are overwritten. Simulate the LIFO showing four writes followed by five reads.

REFERENCES AND READINGS

[1] Auletta, R. and J. Aylor. "A Non-Interpreted Model for Digital System Design," *Proceedings of IEEE Southeastcon*, pp. 296–299, March 1986.

[2] Aylor, J., R. Waxman, B. Johnson, and R. Williams. *Performance and Fault Modeling with VHDL*; chap. "The Integration of Performance and Functional Modeling in VHDL." Englewood Cliffs, NJ: Prentice-Hall, pp. 22–145, 1992.

[3] Benders, L. and M. Stevens. "Petri Net Modeling in Embedded System Design," *CompEuro*, pp. 612–617, 1992.

[4] Chickername, V., J. Lee, and J. Patel. "Design for Testability Using Architectural Descriptions," *Proceedings of International Test Conference*, pp. 752–761, 1992.

[5] Chung, M. and S. Kim. "The Configuration Management for Version Control in an Object Oriented VHDL Design Environment," *IEEE International Conference on Computer-Aided Design (ICCAD)*, pp. 258–261, 1991.

[6] Coppola, A., M. Perkowski, R. Anderson, J. Freedman, and E. Pierzchala. "Tokenized State Machine Model for Synthesis of Sequential Circuits into EPLDs and FPGAs," *IFIP Transactions A (Computer Science & Technology)*, vol. A 22, pp. 34–46, 1993.

[7] Ecker, W. and S. Marz. "System Level Specification and Design Using VHDL: A Case Study," *11th IFIP International Conference on Computer Hardware Description Languages and Their Applications*, pp. 505–522, 1993.

[8] Frank, G., D. Frank, and W. Ingogly. "An Architecture Design and Assessment System," *VLSI Design*, August 1985.

[9] Leung, S. "Organizing Design Alternatives Using VHDL Configurations," *IEEE COMPCON*, pp. 398–402, 1990.

[10] Pierre, L. "VHDL Description and Formal Verification of Systolic Multipliers," *11th IFIP International Conference on Computer Hardware Description Languages and Their Applications*, pp. 225–242, 1993.

[11] Rao, R. *A Building Block Approach to Uninterpreted Modeling Using VHDL*, PhD thesis, Dept. of Electrical Engineering, University of Virginia, 1990. Masters thesis.

[12] Rose, C., M. Buchner, and Y. Trivedi. "Integrating Stochastic Performance Analysis with System Design Tools," *Proceedings of the ACM/IEEE Design Automation Conference*, pp. 482–488, June 1985.

[13] Swaminathan, G., R. Rao, J. Aylor, and B. Johnson. "Colored Petri Net Descriptions for the UVA Primitive Modules," University of Virginia, Semiconductor Research Corporation (SRC) Publication C93052, February 1993.

APPENDIX A
POWERS OF TWO

Table A.1

Powers of 2

n	2^n	2^{-n}
0	1	1
1	2	0.5
2	4	0.25
3	8	0.125
4	16	0.0625
5	32	0.03125
6	64	0.015625
7	128	0.0078125
8	256	0.00390625
9	512	0.001953125
10	1024	0.0009765625
11	2024	
12	4096	
13	8192	
14	16,384	
15	32,768	
16	65,536	
17	131,072	
18	262,144	
19	524,288	
20	1,048,576	

Note: 2^{10} is commonly referred to as "1K," although 2^{10} actually equals 1024. Similarly, 2^{11} is commonly referred as "2K," and so on.

APPENDIX B
VHDL RESERVED KEYWORDS

VHDL reserved keywords are listed below.

- abs
- access
- after
- alias
- all
- and
- architecture
- array
- assert
- attribute
- begin
- block
- body
- buffer
- bus
- case
- component
- configuration
- constant
- disconnect
- downto
- else
- elsif
- end
- entity
- exit
- file

- for
- function
- generate
- generic
- guarded
- if
- in
- inout
- is
- label
- library
- linkage
- loop
- map
- mod
- nand
- new
- next
- nor
- not
- null
- of
- on
- open
- or
- others
- out

- package
- port
- procedure
- process
- range
- record
- register
- rem
- report
- return
- select
- severity
- signal
- subtype
- then
- to
- transport
- type
- units
- until
- use
- variable
- wait
- when
- while
- with
- xor

Additional reserved keywords provided in VHDL-93 are listed below.

- group
- impure
- inertial
- literal
- postponed
- pure

- reject
- rol
- ror
- shared
- sla
- sll

- sra
- srl
- unaffected
- xnor

APPENDIX C
INTRODUCTION TO SEMICONDUCTOR PHYSICS

One way to classify materials is by their *conductivity*, which is a measure of the ability to conduct charge or electricity. Materials are classified into one of the following three general conductivity categories:

1. *metals*
2. *semiconductors*
3. *insulators*

Metals readily conduct electricity, so metals have a relatively high conductivity, on the order of $10^5 \ \Omega^{-1}\text{cm}^{-1}$. Common examples of metals include iron, copper, and aluminum. In contrast, insulators, also called *dielectrics*, do not readily conduct electricity, so insulators have a relatively low conductivity, on the order of $10^{-16} \ \Omega^{-1}\text{cm}^{-1}$.[1] Common examples of insulators include glass, rubber, and plastic. Between the two extremes of metals and insulators lie semiconductors, which exhibit intermediate levels of conductivity.

A material's atomic and crystalline structure determine whether it is a semiconductor. Table C.1 lists several well-known semiconductors, organized according to their position in the periodic table of elements.

The following discussion focuses on *silicon* (Si) because it is presently the most prevalently used semiconductor in integrated circuits. Figure C.1(a) shows the atomic structure of silicon. The atomic number of silicon is 14; hence, a silicon atom has 14 electrons orbiting about its nucleus. According to quantum electronics, the electrons occupy discrete energy levels, or *shells*, in other words, K-, L-, and M-shells. The electrons in the outer M-shell are called *valence* electrons and principally determine silicon's chemical properties.

Figure C.1(b) shows a simplified, two-dimensional view of the crystalline structure of silicon. Silicon crystalizes in a *diamond* structure or lattice, where neighboring atoms are bonded together by sharing their valence electrons. Only the outer M-shell electrons of each silicon atom are shown in Figure C.1(b) to stress the importance of the valence electrons in the bonding mechanism. The inner L- and K-shell electrons are considered part of an "extended" nucleus with an appropriately modified positive charge of $+4$.[2] The bond formed by sharing valence electrons is called a *covalent* bond. Each silicon atom forms four covalent bonds with its nearest four neighbors to borrow four valence electrons to complete its M-shell, which results in a minimum energy configuration.

Figure C.1(b) shows a perfect silicon crystal at a very cold temperature, having no impurities (contamination from other materials) or defects (structural irregularities). Figure C.2(a) illustrates that as the temperature is increased, a valence electron can obtain enough thermal energy to break out of its covalent bond.

Problem-Solving Skill

All physical systems tend toward a minimum energy configuration.

---

[1] It is interesting to note the wide range of conductivities found in nature, varying by more than twenty orders of magnitude.

[2] The unit of charge is the charge of an electron, called the *electronic charge*, equaling $\simeq 1.6 \times 10^{-19}$ coulombs (C). An electron has a negative unit charge; a proton has a positive unit charge.

Table C.1

Semiconductors

Periodic table group(s)	Semiconductor
Group IV	Silicon (Si) Germanium (Ge)
Group III/V	Gallium-Arsenide (GaAs) Gallium-Phosphide (GaP) Indium-Phosphide (InP)
Group II/VI	Cadmium-Telluride (CdTe) Cadmium-Selenide (CdSe)

Note: Semiconductors involving more than one element are called *compound semiconductors*.

Figure C.1

Atomic and crystalline structure of silicon

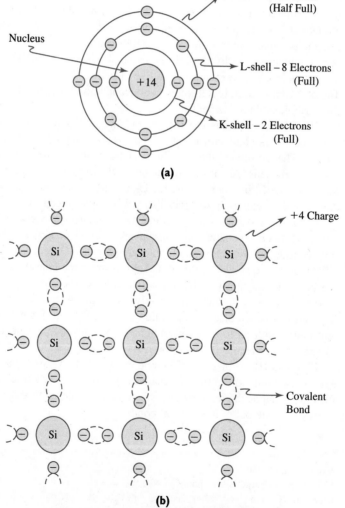

(a)

(b)

Figure C.2

Electrons and holes

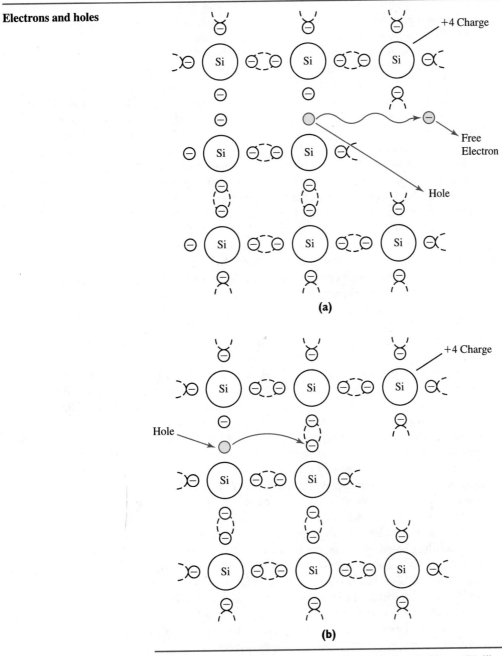

Figure (a) shows the generation of an electron/hole pair. Figure (b) illustrates the hole conduction mechanism.

An electron that escapes the confines of its covalent bond is called a *free* electron because it is "free" to move about the crystal lattice. The vacancy is called a *hole*. The creation of a free electron and an associated hole is called *electron/hole pair generation*. The reverse process of electron/hole pair generation, called *electron/hole pair recombination*, can also occur; a free electron loses energy and returns to the crystal lattice to reform a broken covalent bond. At a given temperature, the dynamic processes of generation and recombination yield a steady-state density of free electrons and holes. For example, at room temperature,[3] approximately one silicon atom in every 10^{12} atoms has a broken covalent bond, yielding $\simeq 10^{10}$ free electrons and holes per cubic centimeter.

A unique property of semiconductors is that both free electrons *and* holes participate in the conduction process. In other words, the current, or flow of charge, in response to an externally supplied electric potential in a semiconductor is due to the movement of free electrons and holes. To understand the hole conduction mechanism, consider Figure C.2(a) and Figure C.2(b). In Figure C.2(b), a valence electron from a neighboring silicon atom has moved over to fill the vacancy shown in Figure C.2(a). The movement of the negatively charged valence electron to the right can be equivalently viewed as the movement of a positively charged hole to the left. Either view yields conventional current flow to the left. Hence, holes can be conceptualized as positively charged particles, providing a conduction mechanism that is separate from the free electron conduction mechanism. In summary, it is useful to visualize semiconductor current as being composed of free electrons sailing along on skateboards and valence electrons traveling at a slightly slower pace on pogo sticks.[4]

A pure piece of silicon, also called *intrinsic* silicon, contains an equal number of free electrons and holes. This balance can be altered by interjecting foreign substances into silicon. The foreign substances are called *impurities* or *dopants*. Adding foreign substances to silicon is referred to as *doping* silicon. There are two kinds of doping: *donor* and *acceptor*. Donor doping increases the number of free electrons relative to the number of holes, while acceptor doping increases the number of holes relative to the number of free electrons. Figure C.3 and Figure C.4 (p. 670) illustrate the basic processes of donor and acceptor doping, respectively.

Donor dopants for silicon are pentavalent elements from Group V of the periodic table having five valence electrons, such as phosphorous (P), arsenic (Ar), and antimony (Sb). Figure C.3(a) illustrates donor doping, showing a phosphorous atom displacing a silicon atom in the diamond crystal lattice.

Similar to the game of musical chairs, four of the five valence electrons of the phosphorous atom form covalent bonds with neighboring silicon atoms, leaving the loosing fifth electron as an "extra" electron. At room temperature, the extra electron's thermal energy easily exceeds its ionization energy; thus, it breaks free of the phosphorous atom and becomes a free electron, and the phosphorous atom becomes a positively charged ion. Note that a free electron has been generated *without* an associated hole, making the density of free electrons greater than holes (usually much greater) in donor-doped silicon. Since there are more free electrons than holes, electrons are called the *majority carriers* and holes are called the *minority carriers* in the conduction process. Figure C.3(b) shows a simplified view of donor-doped silicon, also called *n-type* silicon.[5] The positive charges represent donor ions, not holes. Donor ions have the same positive unit charge as holes, but are *fixed* charges in the crystal lattice. The negative charges represent donated free electrons. Intrinsic free electrons and holes are not shown because it is customary to assume that the doped free electron density is much greater than the intrinsic carrier densities.

Acceptor dopants for silicon are trivalent elements from Group III of the periodic table having three valence electrons, such as boron (B), aluminum (A1), and gallium (Ga). Figure

---

[3] Room temperature is customarily considered to be 300 K (Kelvin).

[4] My apologies to theoretical solid-state physicists, but the skateboard/pogo stick travel – electron/hole conduction analogy is useful.

[5] By convention, "n" is the symbol used for free electron concentration.

Figure C.3

Donor doping of silicon

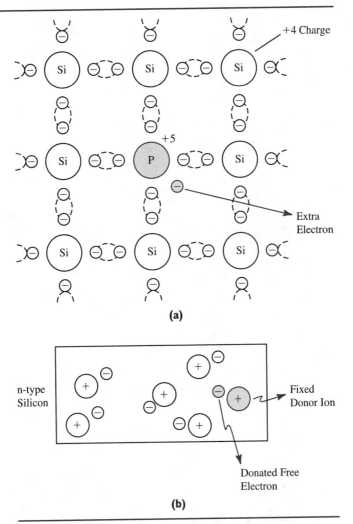

(a)

n-type
Silicon

Fixed
Donor Ion

Donated Free
Electron

(b)

Figure (a) shows the covalent bonding of a phosphorous dopant in the silicon crystal structure. Figure (b) shows a simplified view of donor-doped, also called n-type silicon, highlighting the fixed donor ions and free electron charges.

C.4(a) illustrates acceptor doping, showing a boron atom displacing a silicon atom in the diamond crystal lattice.

   Having only three valence electrons, the boron atom needs an electron to complete its four covalent bonds. The boron atom obtains the needed electron by stealing a valence electron from a neighboring silicon atom. Having accepted an additional electron, the boron dopant becomes a negatively charged ion. The broken covalent bond of the neighboring silicon atom that lost the valence electron becomes a hole. Note that a hole has been generated without an associated free electron, making the density of holes greater than free electrons in acceptor-doped silicon. In this situation, holes are called the *majority carriers* and free electrons are called the *minority carriers*. Figure C.4(b) shows a simplified view of acceptor-

Figure C.4

Acceptor doping of silicon

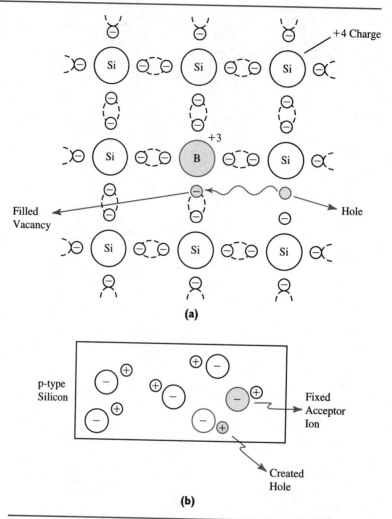

Figure (a) shows the covalent bonding of a boron dopant in the silicon crystal structure. Figure (b) shows a simplified view of acceptor-doped silicon, also called p-type silicon, highlighting the fixed acceptor ions and hole charges.

doped silicon, also called *p-type* silicon.[6] The negative charges represent fixed acceptor ions. The positive charges represent holes. Again, intrinsic free electrons and holes are not shown because it is customary to assume that the doped hole density is much greater than the intrinsic carrier densities.

REFERENCES AND READINGS

[1] Boer, K. *Survey of Semiconductor Physics*. New York: Van Nostrand Reinhold, 1990.
[2] Ferry, D. and R. Grondin. *Physics of Submicron Devices*. New York: Plenum Press, 1991.
[3] Li, S. *Semiconductor Physical Electronics*. New York: Plenum Press, 1993.
[4] Seeger, K. *Semiconductor Physics*. New York: Springer-Verlag, 1991.

---

[6] By convention, "p" is the symbol used for hole concentration.

INDEX